Advances in Signal Processing
for Nondestructive Evaluation
of Materials

NATO ASI Series

Advanced Science Institutes Series

A Series presenting the results of activities sponsored by the NATO Science Committee, which aims at the dissemination of advanced scientific and technological knowledge, with a view to strengthening links between scientific communities.

The Series is published by an international board of publishers in conjunction with the NATO Scientific Affairs Division

A	**Life Sciences**	Plenum Publishing Corporation
B	**Physics**	London and New York
C	**Mathematical and Physical Sciences**	Kluwer Academic Publishers
		Dordrecht, Boston and London
D	**Behavioural and Social Sciences**	
E	**Applied Sciences**	
F	**Computer and Systems Sciences**	Springer-Verlag
G	**Ecological Sciences**	Berlin, Heidelberg, New York, London,
H	**Cell Biology**	Paris and Tokyo
I	**Global Environmental Change**	

NATO-PCO-DATA BASE

The electronic index to the NATO ASI Series provides full bibliographical references (with keywords and/or abstracts) to more than 30000 contributions from international scientists published in all sections of the NATO ASI Series.
Access to the NATO-PCO-DATA BASE is possible in two ways:

– via online FILE 128 (NATO-PCO-DATA BASE) hosted by ESRIN,
Via Galileo Galilei, I-00044 Frascati, Italy.

– via CD-ROM "NATO-PCO-DATA BASE" with user-friendly retrieval software in English, French and German (© WTV GmbH and DATAWARE Technologies Inc. 1989).

The CD-ROM can be ordered through any member of the Board of Publishers or through NATO-PCO, Overijse, Belgium.

Series E: Applied Sciences - Vol. 262

Advances in Signal Processing for Nondestructive Evaluation of Materials

edited by

X. P. V. Maldague

Electrical Engineering Department,
Université Laval,
Québec City, Québec, Canada

Springer Science+Business Media, B.V.

Proceedings of the NATO Advanced Research Workshop on
Advances in Signal Processing for Nondestructive Evaluation of Materials
Québec City, Québec, Canada
August 17–20, 1993

A C.I.P. Catalogue record for this book is available from the Library of Congress

ISBN 978-94-010-4459-2   ISBN 978-94-011-1056-3 (eBook)
DOI 10.1007/978-94-011-1056-3

Printed on acid-free paper

Table of Contents

Direct and inverse problems

Signal processing in thermal NDE

Artificial Intelligence

Applications to emerging NDE techniques

Applications to established NDE techniques

Hardware implementations

Special presentations

Preface

The NATO Advanced Research Workshop on *Advances in Signal Processing for Non Destructive Evaluation of Materials* was held August 17-20, 1993, at Pavilion La Laurentienne, on Université Laval Campus, Québec City (Québec), Canada. This Proceedings volume includes most presentations given at the Workshop.

Scientific content

Non Destructive Evaluation (NDE) is now playing an increasing role in our modern global economy, for instance in security sensitive industries such as in aerospace industries. The complexity of the inspection task and the larger production runs now impose more operator-assisted or fully-automated signal processing. This ARW had this particular goal to bring together people from both fields of expertise: NDE and signal processing with the underlying idea to stimulate mutual cross-fertilization.

An important consideration is that ideal production processes would only produce quality and then, it would be no need for NDE. In fact, NDE is an added cost to the industry and there is an economic trade-off which opposes costs for increased NDE processing to traditional procedures: added processing should pay for itself if to be accepted by the industry. This essential point brings us back to our feet in the sense that NDE solutions must always be realistic: fast, robust, cost-effective.

On the signal processing side, it appears that the technology is now mature to open the plant floor to artificial intelligence. In this respect, sensor fusion, signal knowledge representation, expert systems, neural networks, fuzzy logic, computer vision and integration of numeric and non-numeric informations are expected to be seen more frequently in NDE applications (for implementation of what is now called *soft computing*) although totally intelligent and autonomous NDE systems are still a long way ahead.

Moreover, a trend is now initiated to evaluate noise properties in order to increase signal to noise ratio for better discrimination between false alarms and real flaws. Statistical methods such as evaluation of noise dispersion is an effective tool for this. A new method of decomposition known as PARAFAC (PARAllel FACtor analysis) was also proposed with possible applications in inverse NDE. New filters and transforms (such as wavelet transform analysis) were also discussed to extract relevant features from NDE signals such as non-linear and morphological filters. Preprocessing of NDE data is seen as an area where work is still required. In this respect, adequate calibration of the NDE systems is essential for repetitive and quantitative flaws analysis when coupled with proper finite difference modelling of the direct problem particular to the various NDE techniques.

On the NDE side, new techniques seem now to emerge seriously, in particular niches such as holographic interferometry, microwave resonance or shearography where digital image processing is becoming a routine procedure.

Traditional NDE techniques such as ultrasonics, infrared techniques, X-ray, computed tomography are expending with new fields of applications. For instance, it was reported that computed tomography can now be deployed for the inspection of small diameter pipes (13 mm diam.). Also, new developments in infrared thermography now permit direct competition with ultrasonic: studies report agreement in flaw size estimation in CFRP composites of 10 % between both techniques but in a much quick fashion for thermography. Moreover, in thermographic NDE, new signal processing tools now allow restoration of actual shape of defects taking into account heat flow scattering while correction is also becoming available to correct images for shape curvature of specimens.

Reliability of the NDE procedure is also becoming a key issue which is important to assess, for instance using probability of detection curves (POD). Such curves provide an estimation of the NDE inspection performance.

A session of the workshop was also dedicated to hardware implementations as a physical support to NDE devices. Progresses of the hardware such as parallel processing, DSP chips (Digital Signal Processing) bring unprecedented computing power on the plant floor thus allowing complex algorithms to be implemented in real or close to real time environments. This hardware aspect is fundamental to make NDE devices quick, compact and very cost effective. This trend will be pursued at an even faster pace in the future.

Finally, at the wrap up session, two important resolutions were proposed: first to hold another ARW on the same topic in 3 to 4 years to report on milestone developments. Also, it was proposed to establish an international electronic mailing network among the NDE community, starting with the group of people who were present at this ARW. This internet mailing network is now operating and can be reached at the following electronic address:

`nde@swrinde.nde.swri.edu`

The idea behind this NDE network is to provide an informal forum of exchange and discussion. Moreover, it would bring an easy communication link to exchange experimental data and software tools among the NDE community. On this last point, the European TRAPPIST format discussed at the panel session will be promoted as a new standard to exchange compatible data and programs.

ARW Context

NATO Advanced Research Workshops (ARWs) are particularly interesting since they give the possibility to a restricted group of specialists to interact closely and cross-fertilize their respective research fields of interest in a less formal and constrained set-up than regular international conferences.

The idea to organize this workshop matured at the occasion of the *1992 Quantitative Infrared Thermography* Conference held in Paris, France. However, it had deeper grounds since we have to go back in August 1987 when Professor Chen of University of Massachusetts Darthmouth organized a NATO ARW entitled: *Signal Processing and Pattern Recognition in Non Destructive Evaluation*. This ARW was held at Lac Beauport, a nice resort area located in the hills, at some 20 minutes North of Québec City (*NATO ASI Series F*, vol. **44**). I want to thank Professor Chen for his help in the organization of this present workshop.

This NATO workshop could not have taken place without the decisive participation of the International Organizing Committee whose members were Dr. Douglas Burleigh from General Dynamics in USA, Dr. Denis Gingras from National Optics Institute of Canada, Dr. Ermanno Grinzato from National Research Council of Italy, Prof. Vladimir Vavilov from Tomsk Polytechnic University in Russia and myself.

Obviously this workshop could not have taken place without the financial help of NATO which positively moved in response to our proposal. Special acknowledgments are due to the NATO Science Committee whose continual support helped making this ARW a success.

Thanks are also due to Université Laval's authorities who accepted to officially host the workshop. Moreover, Sciences and Engineering Faculty and Electrical Engineering Department offered the Welcome cocktail. I want also to express my gratitude to the Ministère des Affaires Internationales du Québec which kindly offered the Workshop banquet.

I want also to mention the support of the Canadian Association for Research in Non Destructive Evaluation and of the Canadian Society for Non Destructive Testing.

Forty four people from twelve different countries (including nine NATO countries) participated to the workshop. I want to thank all of them for their kind cooperation and participation which was essential to the success of the meeting.

Finally, on behalf of the International Organizing Committee, I sincerely wish this Proceedings will bring new ideas,   and will help a little the progress of Mankind towards a better world, a flawless world!

Prof. X. Maldague
Workshop Director
Québec, September, 1993.

Group photo of some Workshop participants at the front entrance of Pavilion La Laurentienne, Université Laval, Québec (Canada), on August 17, 1993.

INTELLIGENT SIGNAL PROCESSING

Simon Haykin
McMaster University
Communications Research Laboratory
1280 Main Street West
Hamilton, Ontario, Canada L8S 4K1

1.   Introduction

For the past 40 years or so, interest in the use of (artificial neural networks has been motivated by the recognition that the human brain operates in a manner that is entirely different from a conventional digital computer. A neural network is made up of an interconnection of a large number of nonlinear computation units known as neurons, which operate in a highly parallel fashion. Interest in the use of neural networks was reignited in the 1980s largely due to (1) the popularization of the back-propagation algorithm as a tool for the training of multilayer perceptrons, and (2) the use of attractor neural networks (exemplified by the Hopfield model) as content-addressable memories and optimization networks. For a historical account of neural networks, the reader is referred to Cowan (1990) and Haykin (1994).

Neural networks represent an important constituent of *softcomputers* (Zadeh, 1993). Other constituents include fuzzy logic, genetic algorithms, and probabilistic networks. Indeed, these softcomputing techniques are responsible for a paradigm shift away from the conventional digital computer, which we are witnessing in the design of intelligent machines. Important classes of such machines include intelligent pattern recognition systems, intelligent robots, intelligent controllers, and intelligent signal processors. In this paper, we confine our attention to the latter class of machines.

The paper is organized as follows. In Section 2 we describe the attributes of intelligent signal processors. In Section 3 we outline the important characteristics of neural networks that are basic to the construction of intelligent signal processors. In Section 4 we discuss time-frequency analysis and principal components analysis as important adjuncts to neural networks for preprocessing. In Section 5 we present some results demonstrating the application of these techniques to a difficult radar detection

1

X. P. V. Malague (ed.), Advances in Signal Processing for Nondestructive Evaluation of Materials, 1–12.
© 1994 *Kluwer Academic Publishers.*

problem. Section 6 concludes the paper by presenting some final thoughts on the subject.

2.   Attributes of Intelligent Signal Processing

A signal processing device, operating in a nonstationary environment of unknown statistics, is said to be "intelligent" if it is capable of exploiting the information content of its input in the most efficient manner possible and at all times. For such a capability to be feasible, we may identify the following list of attributes (Haykin, 1994):

1. *Nonlinearity*, so as to capture higher-order statistics of incoming signals. Indeed, nonlinearity is a basic ingredient of a universal computing machine.
2. *Adaptivity*, so as to respond to continuing changes in environmental conditions. This capability usually implies "in-situ learning" which means that adjustments are made to the free parameters of the machine while the input signal is being processed.
3. *Prior information*, the use of which biases the system design. This, in turn, specializes the architectural design of the machine.
4. *Information preservation*, which is maintained until the system is ready for final decision-making.
5. *Multisensor fusion*, which accounts for the combined use of information gathered by a multitude of sensors operating in the environment of interest.
6. *Robustness*, with respect to parameter changes and component failures, which is considered to be essential for practical use.
7. *Attentional mechanism*, whereby through interaction with a user or in a self-organized manner, the machine is enabled to focus its computing power around a particular point in an image or a particular location in space for more detailed analysis.

This is a formidable list of requirements. To provide for them we require to use neural networks with some appropriate tools such as time-frequency analysis, principal components analysis, and fuzzy logic.

3.   Neural Networks as one of the "Principal" Tools

Neural networks are well suited for the design of intelligent signal processors by virtue of a unique combination of properties, as described here (Haykin, 1994):

1. Neural networks are *universal approximators*, in the sense that they can approximate any continuous input-output mapping to any desired degree. Hence, they are ideally suited for model building.
2. Neural networks are inherently *nonlinear*; hence, they have the ability to capture the underlying nonlinearity of physical phenomena responsible for

the generation of incoming data.

3. Neural networks have the capability to learn from their environment, which can be implemented in a supervised or self-organized (unsupervised) manner.

4. Neural networks are ideally suited for pattern classification, detection, and prediction.

5. Neural networks are perfectly made for VLSI implementation by virtue of the highly pervasive use of a neuron as the basic computational unit.

6. Above all, neural networks permit the input data to "speak for itself", and thereby free us from the burden of having to guess suitable input-output functions to represent the data.

For supervised learning, the multilayer perceptron is the most widely used network architecture. It consists of a layered architecture with an input layer of source nodes, followed by one or more hidden layers, and an output layer. The hidden neurons act as feature detectors. For pattern classification, all the neurons (including the output neurons) are nonlinear with a sigmoidal nonlinearity (e.g., logistic function or hyperbolic tangent function) being the most common form of nonlinear activation function. On the other hand, for prediction and modeling, all the neurons (except for the output neurons) are nonlinear; the output neurons are purposely chosen to be linear. The back-propagation algorithm, representing a generalization of the ubiquitous least-mean-square (LMS) algorithm, has emerged as the standard algorithm for the training of multilayer perceptrons (Rumelhart and McClelland, 1986; Haykin, 1994).

Radial basis function (RBF) networks represent an alternative structure for supervised learning (Broomhead and Lowe, 1989; Poggio and Girosi, 1990; Haykin, 1994). RBF networks provide local approximation and therefore faster training. This is to be contrasted with the global approximation provided by multilayer perceptrons, and therefore a better generalization performance. An RBF network uses a hidden layer of radial basis functions (usually Gaussian), which is followed by a linear output layer. The development of RBF networks evolves naturally from regularization theory (Poggio and Girosi, 1990; Haykin, 1994). An optimum design strategy for RBF networks permits the adaptive selection of the Gaussian functions as well as the linear weights. In so doing, it helps overcome the "curse of dimensionality".

How can we combine the nice feature of local approximation exemplified by RBF networks and the nice feature of global approximation exemplified by multilayer perceptrons trained with the back-propagation algorithm? An attractive solution to this important question lies in the use of modular networks (Jacobs and Jordan, 1991; Jordan and Jacobs, 1992; Haykin, 1994). This class of neural networks involves the use of a number of non-communicating expert networks (modules), whose outputs are mediated by a gating network. Their design involves the combined use of supervised learning and competitive (unsupervised) learning, that is rooted in maximum-likelihood estimation theory. The design of modular networks is still in its early stages of development.

4. Preprocessing: Time-Frequency Analysis and Principal Components Analysis

Suppose that we are given a nonstationary time series consisting of a transient component (e.g., target signal) superimposed on a signal-dependent interference (e.g., clutter). How can we process such a signal by means of a neural network? Well, we can proceed by applying the received signal directly to a neural network, thereby forcing the network to discover the inherent features characterizing the signal and then performing the desired detection. A practical drawback of this simplistic approach, particularly in the context of large-scale engineering problems, is that the supervised training of the neural network (e.g., multilayer perceptron) can be painfully slow. A more efficient procedure is to use time-frequency analysis on the nonstationary received signal, and thereby transform it into a two-dimensional image. One dimension of the image is represented by frequency, and the other by time. Recognizing that we ourselves are "visual thinkers", the use of time-frequency analysis should help the neural network "see" the salient features of the received signal more easily than would be the case otherwise. Having computed the time-frequency image, we can go on and apply it directly to the input layer of the neural network. Such an approach may also be inefficient due to the highly redundant nature of the time-frequency image, as is often the case. Now, the redundant components of the time-frequency image may be removed prior to processing, thereby enhancing the efficiency of computation. To do so we may apply principal components analysis to the time-frequency image. Thus, an efficient form of preprocessing consists of time-frequency analysis followed by principal components analysis, as shown in Fig. 1. The use of principal components analysis may also be justified on learning-theoretic grounds by invoking the principle of structural minimization (Vapnik, 1992).

As with neural networks, there is no unique method of time-frequency analysis. We may classify time-frequency analysis as linear and nonlinear. Linear methods of time-frequency analysis include the short-time Fourier transform (Riley, 1989), and the wavelet transform (Daubechies, 1990; Rioul and Vitterli, 1991). Nonlinear (quadratic) methods of time-frequency analysis include the Wigner-Ville transform (Cohen, 1989; Hlawatsch and Boudeaux-Bartels, 1992). In this paper we focus on the use of the Wigner-Ville transform that is an information-preserving mapping procedure. The combined presence of a target signal and clutter gives rise to the generation of cross-product terms. Usually, such a phenomenon is considered to be undesirable. However, in the context of target detection, the presence of cross-product terms is part of the information-preserving property of the Wigner-Ville transformation, and

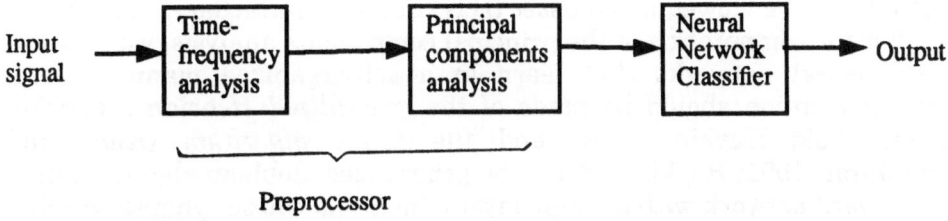

Figure 1: Block diagram of a target detector based on pattern recognition

as such it may be viewed as an asset (Haykin and Bhattacharya, 1992).

For the computation of the principal components analysis, we may use "linear" neural networks that learn in a self-organized manner. Here, particular mention should be made of the *generalized Hebbian algorithm* (Sanger, 1989; Haykin, 1994) and the *APEX algorithm* (Kung and Diamantaras, 1991; Haykin, 1994). The generalized Hebbian algorithm uses a feedforward network with a single layer of neurons whose synaptic weights are adjusted in accordance with certain modifications of Hebb's postulate of learning. The APEX algorithm also uses a single layer of neurons, but this time the network involves the combined use of forward and lateral connections. Despite these differences between the generalized Hebbian algorithm and the APEX algorithm, their long-term, convergence behavior is, in theory, exactly the same.

5. Some Important Radar-related Results

In this section, we illustrate some of the signal processing ideas discussed in the previous two sections by describing some radar-related results pertaining to the detection of a small target embedded in an ocean environment. The particular problem we have in mind concerns the detection of small fragments of ice known as *growlers* in the presence of sea clutter (i.e., radar backscatter from the ocean surface). Typically, the part of a growler visible above the ocean surface is about the size of a grand piano. However, realizing that ninety percent of the volume of a growler lies below the ocean surface, its presence can represent a major hazard to shipping in ice-infested waters, such as those encountered on the East Coast of Canada during late spring or early summer. Presently, there is no radar commercially available for solving this difficult detection problem in a reliable fashion.

Haykin and Bhattacharya (1992) describe the results of some early work involving the use of Wigner-Ville distribution as a preprocessor, and a multilayer perceptron as pattern classifier for solving the growler detection problem. In particular, it is shown that such a receiver performance provides an order of magnitude improvement in overall performance over a conventional Doppler processor. Moreover, its use makes "invisible" targets (due to obscuration by ocean waves) "visible in signal-processing terms".

Figures 2 and 3 present the time-frequency images for two different conditions, one representing sea clutter alone and the other representing sea clutter plus a target signal. Both images were computed using the Wigner-Ville transformation. The difference between these two images is dramatically apparent. Figures 4 and 5 show that the detection performance of the pattern recognition approach using a neural network is superior to that of a conventional constant false alarm (CFAM) processor by order of magnitude (in terms of probability of false alarm for a prescribed probability of detection). In the results presented here, the input applied to the multilayer perceptron

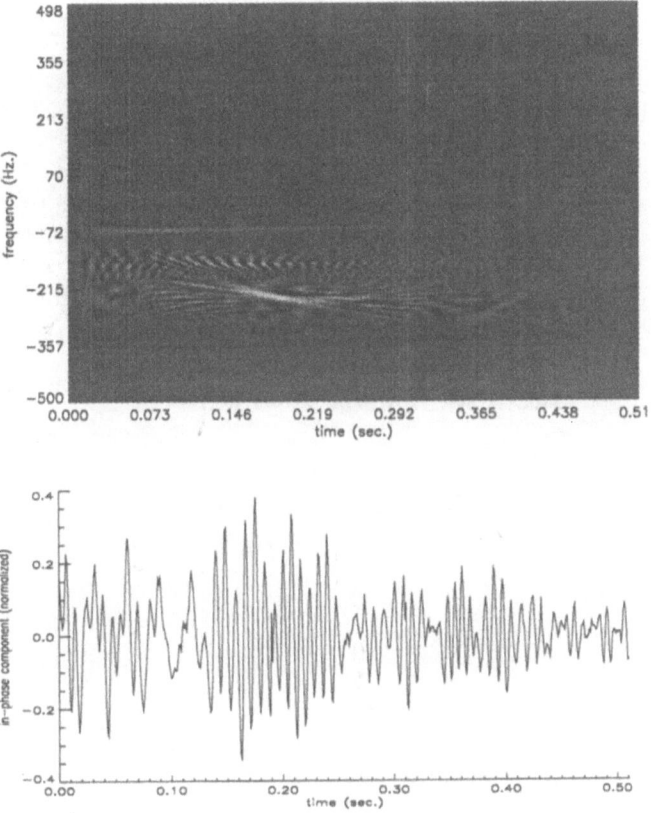

Figure 2: Wigner-Ville distribution for the radar return
from a growler that is partially visible

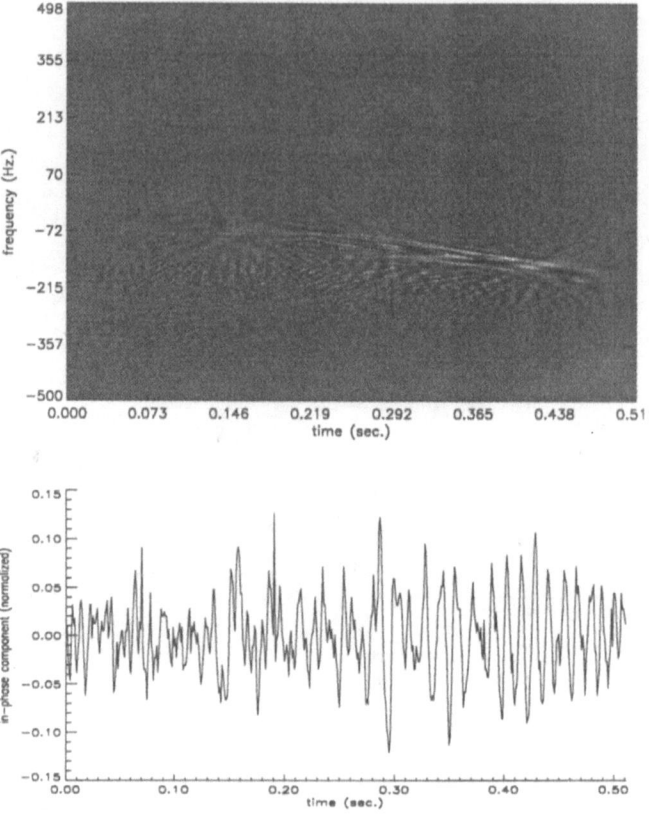

Figure 3: Wigner-Ville distribution for the radar return from the sea

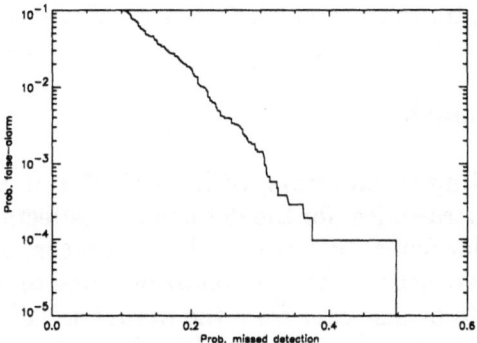

Figure 4: The receiver operating charactreristics (ROC) for the detection of a growler using the classical adaptive detection algorithm

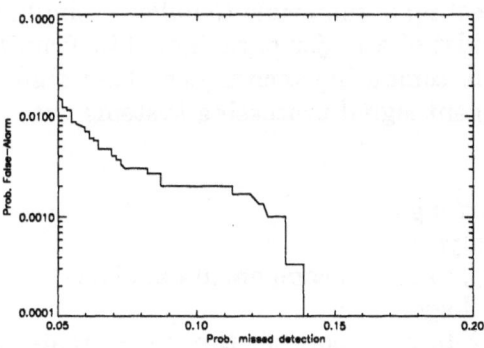

Figure 5: The ROC for the classification of growler using a single hidden layer, fully connected network

consists of an ambiguity function computed from the Wigner-Ville image of the raw radar time series.

6. Concluding Remarks

The nondestructive testing of materials, with particular emphasis on the use of active sensors (e.g., ultrasonics) for the detection of defects, has a great deal in common with the radar detection problem. In both cases, the problem is one of essentially discriminating between two conditions, one corresponding to the presence of clutter due to backscatter from the main body of the medium, and the other condition corresponding to the combined presence of an echo due to a target/defect and clutter.

In this paper we have presented some results demonstrating the computational power of neural networks, working in combination with the Wigner-Ville transformation as a time-frequency analysis preprocessor. The use of such an approach represents a radical departure from conventional procedures used to design radar receivers. Most importantly, they have enabled us to achieve a level of overall system performance far beyond that attainable with conventional methods. As impressive as these results are, they represent merely a first step toward the building of an intelligent signal processing system for ocean surveillance.

It is hoped that the ideas presented in this paper inspire researchers in the nondestructive testing of materials to follow a similar path.

We are in the midst of a major paradigm shift from the conventional digital computer to soft computing techniques. This shift should make it possible to build intelligent signal processing systems with emphasis on the following:

- Nonlinearity
- Adaptation and learning
- Time-frequency analysis
- Data reduction (e.g., principal components analysis)
- Neurobiological analogy

Above all, we are going to rely more on the use of real-life data, bringing signal processing closer to the physical world where it naturally belongs.

Acknowledgements

The author wishes to express his gratitude to NSERC for providing financial support for the work described in this paper. He is also indebted to his research colleagues Tarun Bhattacharya, Graeme Jones, Andrew Ukrainec, Paul Yee, Bob Dony, Bob Li, and Liang Li for many useful discussions.

References

Broomhead, D.S., and D. Lowe, 1988. "Multivariable functional interpolation and adaptive networks", Complex Systems, vol. 2, pp. 321-355.

Cohen, L., 1989. "Time-frequency distributions - A review", *Proc. IEEE*, vol. 77, pp. 941-981.

Cowan, J.D., 1990. "Neural networks: The early days", In *Advances in Neural Information Processing Systems*, 2, (D.S. Touretzky, ed.), pp. 828-842, San Mateo, CA: Morgan Kaufmann.

Daubechies, I., 1990. "The wavelet transform, time-frequency localization and signal analysis", *IEEE Trans. Information Theory*, vol. 36, pp. 961-1065.

Haykin, S., 1994. *Neural Networks: A Comprehensive Foundation*, New York: Macmillan.

Haykin, S. and T.H. Bhattacharya, 1992. "Adaptive radar detection using supervised learning networks", *Computational Neuroscience Symposium*, pp. 35-51.

Hlawatsch, F. and G.F. Bourdeaux-Bartels, 1992. "Linear and quadratic time-frequency signal representations", *IEEE Signal Processing Magazine*, vol. 9(2), pp. 21-68.

Jacobs, R.A. and M.I. Jordan, 1991. "A competitive modular connectionist architecture". In *Advances in Neural Information Processing Systems* 3 (R.P. Lippmann, J.E. Mood, and D.J. Touretzky, eds.), pp. 767-773, San Mateo, CA: Morgan Kaufmann.

Jordan, M.I. and R.A. Jacobs, 1992. "Hierarchies of adaptive experts", In *Advances in Neural Information Processing Systems* 4 (J.E. Moody, S.J. Hanson, and R.P. Hippmann, eds.), pp. 985-992, San Mateo, CA: Morgan Kaufmann.

Kung, S.Y. and K.I. Diamantaras, 1991. "A neural network learning algorithm for adaptive principal component extraction (APEX)", ICASSP, pp. 861-864.

Poggio, T. and F. Girosi, 1990. "Networks for approximation and learning", *Proc. IEEE*, vol. 78, pp. 1481-1497.

Riley, M.D., 1989. *Speech Time-Frequency Representations*, Boston, MA: Kluwer.

Rioul, O. and M. Vetterli, 1991. "Wavelets and signal processing", *IEEE Signal Processing Magazine*, vol. 8(4), pp. 14-38.

Rumelhart, D.E., and J.L. McClelland, 1986. Parallel Distributed Processing, Vol. 1, Cambridge, MA: MIT Press.

Sanger, T.D., 1989. "Optimal unsupervised learning in a single-layer linear feedforward neural network", *Neural Networks*, vol. 12, pp. 459-473.

Vapnik, V.N., 1992. "Principles of risk minimization for learning theory", In *Advances in Neural Information Processing Systems* 4 (J.E. Moody, S.J. Hanson, and R.P. Lippmann, eds.), pp. 831-838, San Mateo, CA: Morgan Kaufmann.

Zadeh, L.A., 1993, Plenary Lecture, IEEE International Conference on Neural Networks, San Francisco, CA.

RECEIVER OPERATING CHARACTERISTIC (ROC) IN NONDESTRUCTIVE INSPECTION

C. NOCKEMANN, G.-R. TILLACK, H. WESSEL,   BAM   Berlin
H. HEIDT   MFPA   Weimar
V. KONCHINA   PIIT   St. Petersburg

BAM
Bundesanstalt für Materialforschung und -prüfung
Laboratory 6.24
Combination and Evaluation of NDT-Methods
Unter den Eichen 87
12200 Berlin
Federal Republic of Germany

MFPA
Materialforschungs und -prüfanstalt
an der Hochschule für
Architektur und Bauwesen (HAB)
Amalienstraße 13
99404 Weimar
Federal Republic of Germany

PIIT
St. Petersburg Institute of Railway Transport Engineers
9 Moskovsky p
St. Petersburg 190131
Russia

ABSTRACT: A suitable and objective method to investigate the performance of inspection systems is the Receiver Operating Characteristic (ROC) which is based on the general theory of signal detection. All diagnostic systems include the task of separating a signal from a background of noise. The idea of the ROC method is to characterise the accuracy of an inspection system by evaluating the correct detection rate versus the false alarm rate for a set of different thresholds between signal and noise applied during this task. This method is called detection ROC because it deals only with the pure detection of a signal (defect). If there are several types of defects one can also measure the accuracy of the defect classification and the correctness of the indicated defect importance with the help of the so-called Joint ROC. For the industries it is particularly important to find the unacceptable defects which has to be repaired. For this case the masked ROC is calculated where only the correct or false indications of critical defects are taken into account.
The power of the method is demonstrated in terms of three examples: ultrasonic testing of fibre reinforced plastic materials used in space- and aircraft manufacturing and of welds typical for railway systems and radiographic weld inspection. Of interest was here the dependence of the testing performance from the ultrasonic frequency, from the experience of the inspectors and from the digitisation procedure, respectively. Special tests - reducing the number of samples - were undertaken to show until which minimum amount of statistical basis the result is stable.

X. P. V. Malague (ed.), Advances in Signal Processing for Nondestructive Evaluation of Materials, 13–30.
© *1994 Kluwer Academic Publishers.*

1.   Introduction

Whereas the main topic of this workshop deals with the processing of signals from NDE methods this paper is concerned with the utilisation of the achievements of signal detection theory for the assessment of the reliability of NDE results.

For a proper description of the efficiency of an NDT system it is important to distinguish between the measurement of the device parameters - as sensitivity and resolution - which guarantee merely the functioning of the NDT equipment and the actual testing performance in defect detection and classification. The latter can only be achieved by a statistics based investigation of the whole testing chain including the physical method, the device and the inspector or the automated system including processing procedure. The trend in the NDE world which takes these facts into account maybe summarised as "Performance Demonstration" and is documented in the PISC reports as well as the ASME, Section XI, Appendix 8

The employment of the signal detection theory for NDE assessment consists in the adoption of the approach of statistical system theory in terms of the ROC which will be shortly described below before the three examples of ROC studies - in ultrasonic testing of fibre reinforced synthetic materials used in space- and aircraft manufacturing and of welds typical for railway systems and radiographic weld inspection - will be described. The questions of interest were here the dependence of the testing performance from the ultrasonic frequency, from the experience of the inspectors and from the digitisation procedure, respectively.

Special tests - reducing the number of samples - will be presented to show until which minimum amount of statistical basis the ROC result are stable.

2.   The ROC method

As an approach arising from system theory the ROC method [1-6] counts the input and output of a system which is in its turn considered as black box. This system consists in our case in the whole testing chain including the physical method, the device and the inspector or the automated system including processing procedure. The input is the "truth": whether the material under consideration contains a defect or not which is indicated in fig. 1. The output - the four possible results of nondestructive diagnosis TP, FN, TN and FP - are also presented in fig. 1. Because the both indication types which belong to the truth "defect present" on the one hand and the both which belong to the truth "defect not present" on the other hand it is enough to choose from each truth type one sort of indication. For ROC evaluations it is usual to count the TP (correct defect indication) and the FP (false alarm) frequencies. The probability of TP indication yields from the TP frequency divided by the total number of defects and the probability of false alarm from the frequency of FP divided by the number of sample sections without defect - for both cases the number of test samples and inspectors must be sufficient.

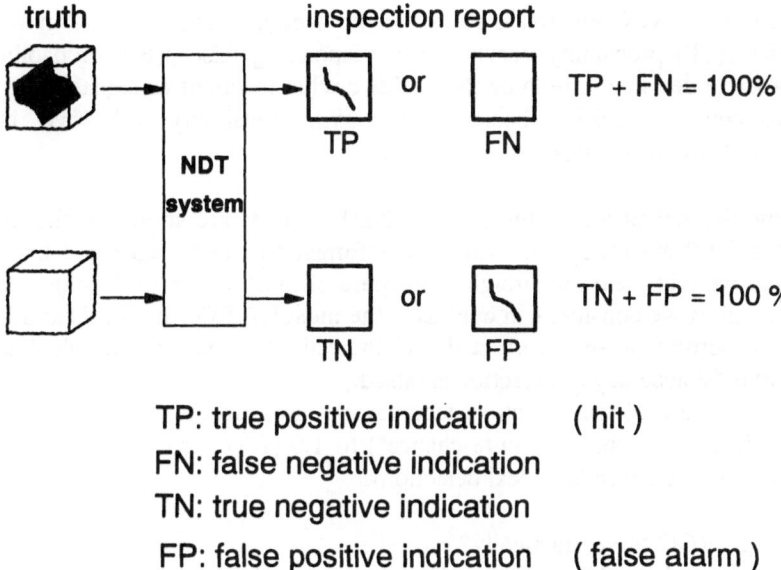

Fig. 1:   The four cases of NDE diagnosis

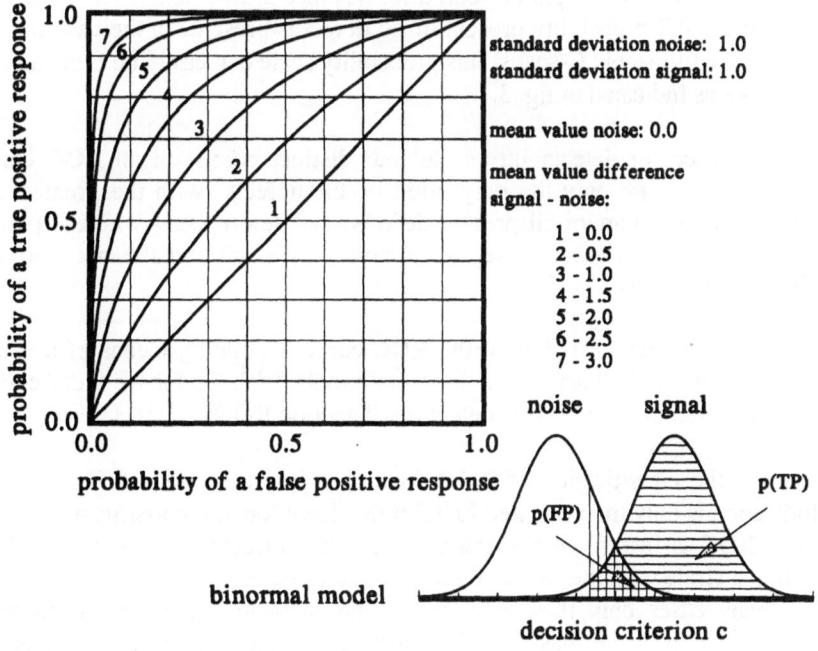

Fig. 2:   A set of theoretical ROCs

Now the idea of the ROC method is to assess the accuracy of an inspection system by evaluating the p(TP) probability versus the corresponding false p(FP) probability for a variety of possible decision criteria or thresholds c what to count as signal (from defect) and what to count as noise - because the varying sensitivity will affect both the detectability of flaws and the false call likelihood.

In fig. 2 some theoretical ROCs for different NDT systems are shown. If the indication rates of TP and FP are always the same ,no information can be extracted by an NDT inspection and we will yield the straight line "pure chance" curve (full line) - the worst ROC curve at all. If we consider - in contrast - the most left ROC it reveals at a false call rate of 10% a portion of 96% correct defect indications. From the straight line to the upper left curve the accuracy of detection is raised.

As an overall-measure of the detection accuracy the area under the ROC-curve is well appropriated: it moves from 0.5 ("pure chance") to 1.0 (edge-curve: 100% true positive calls at a rate of zero false calls = ideal detection).

How to create an ROC curve practically?
As already mentioned the TP and FP indication frequencies must be accumulated (and transformed to relative frequencies) for a variety of the decision criterion c. Or, more efficiently, the inspector can be required to apply several decision criteria simultaneously in associating with a statement (signal present or not) a confidence level which he chooses to adopt in distinguishing nominal "defective" results from nominal "non defective" results. This confidence level might be scored for instance in five ranks: The defect is with 100%, 70%, 50%, 30% probability present or no defect is present. In the case of imaging NDT methods (like ultrasonic C-scans) this probability scale can easily be translated into a detectability scale as indicated in fig. 3.

Using this confidence- or detectability scale four distinct points on an ROC curve are obtained: In point one are only the very good visible defects with the greatest signals included (and so only a part of all present defects) and few noise - whereas point four contains almost all defects (also those which have only small signals) but also a great amount of noise (false calls).

The fit from the experimental points to the ROC curve is done by means of a maximum likelihood regression on the basis of the binormal model [1,3,4]. So far we were concerned only with the assessment of the defect detection (detection ROC).

To validate also the classification of the defects we apply the joint ROC: For the joint ROC an indication is only then counted as TP if the detection and classification is correct. For the local ROC is instead of the classification the correctness of indicated location checked within a circle with the so called radius of acceptance.
Because in many cases only the detection of unacceptable defects is of interest, we developed the concept of the masked ROC [6] where the statistical basis is yielded by the unacceptable defects and the sample sections without unacceptable defects.

While it is a sophisticated task to determine the exact statistical significance of the difference in inspection performance of different experimental modes we use here the paired t-test for simplification - but which turned out to be a good approximation because some correlations cancel each other [6].

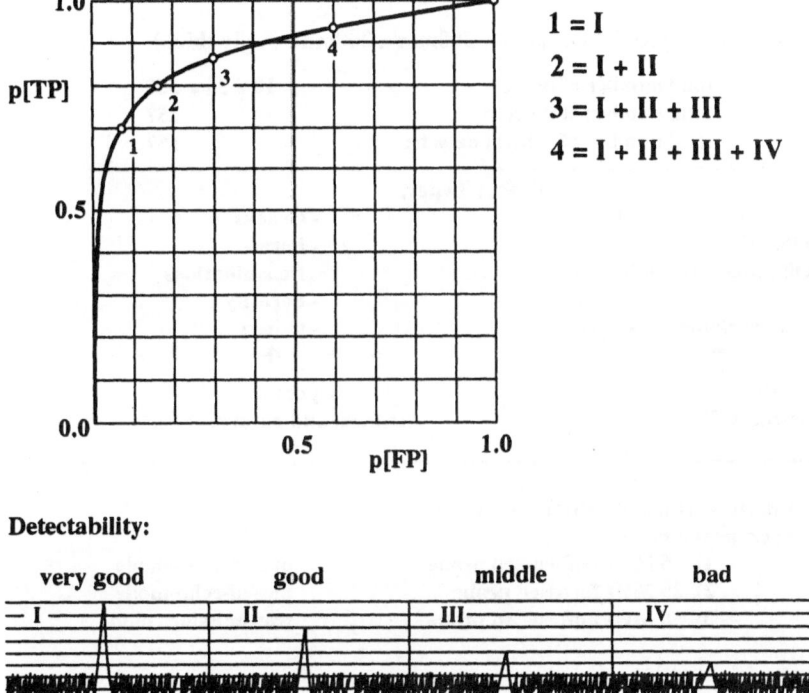

Fig. 3: Practical creation of a ROC curve (confidence rating method)

3.   Examples

3.1   INSPECTION PERFORMANCE IN ULTRASONIC TESTING OF FIBRE-REINFORCED PLASTIC PLATES

Ordered by ESA (ESTEC/Contract No.9289/90/NL/LC(SC)*) the difference in inspection performance due to different experimental modalities in ultrasonic testing of carbon fibre reinforced plastic plates - which are representative for air- and spacecraft industry - was evaluated by an ROC study.

10 test samples CFRP-plates 300mm x 300mm (8 ply thick)

total number of boxes:	10 x 36 = 360
total number of defects:	257
total number of critical defects:	157

Defect Types:

- Backing-paper Faults	- Cracks
- Fibre Cut-outs	- Pores
- Oil Films - Delamination	- Delaminations
	- Porosity
- Impact Damage Defects	- Others
↑	↑
Design =	Report =
Artificial Defects	Artificial and natural Defects

5 Inspectors (Harwell) ASNT level II/III
3 Experimental modes:

1)	5 MHz unfocussed probe	pulsed-echo-mode
2)	25 MHz focussed probe	pulsed-echo-mode
3)	5 MHz unfocussed probe	transmissions mode

Fig. 4:   Overview of the experimental parameters

The special aim was to investigate whether the higher test effort using a 25 MHz focussed probe results also in a higher inspection performance - compared to a normal 5 MHz probe - or not and what is the effect of the transmission mode. The ultrasonic experiments (creation of C-scans and visual evaluation by the inspectors) has been accomplished in the AEA Technology, National NDT Centre Harwell **.
The outline of the experiments is given in fig. 4 (With "box" 5 cm x 5 cm areas on the plates are denoted - into which the total area of the plates has been divided for better overview.) A special C-scan*** of test plate No. 2 is shown in Fig. 5 where the mix of artificial and natural defects is seen.

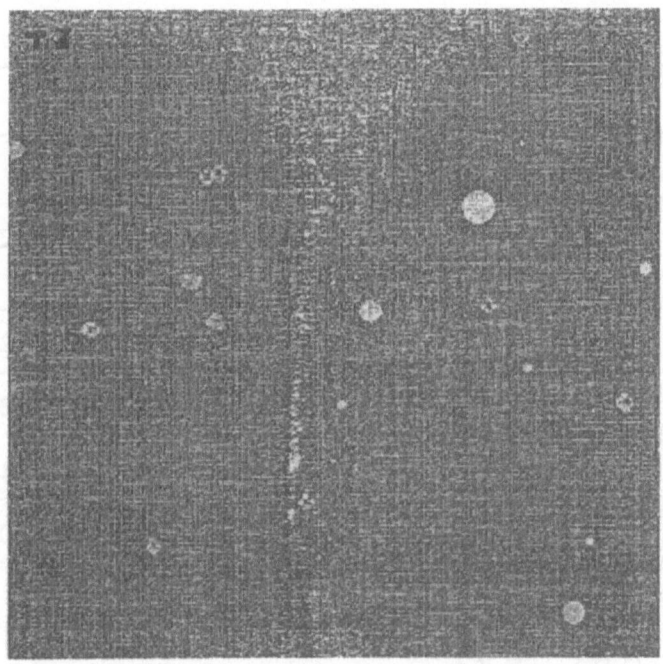

Fig 5: Example of a test plate of CFRP material (special C-scan***)

After scanning the inspectors had to fill in their indications in the report form shown in fig. 6. The confidence level scale is converted to the detectability label d here. In contrast to the case of weld testing where the possible defect types are well defined in ISO-standards the defect types in the case of CFRP plates have to be characterised by a "home made" definition.

Beside the filling of the inspection protocol forms ($10 * 3 * 5 = 150$ forms) there is to fill a form for each test plate with the "true data". Intrinsically these data have to be obtained by destructive testing. Because this mechanical procedure would be doubtful in the case of CFRP material on the one hand and because it would be regrettable to destroy the plates before testing with new methods on the other hand we chose the following way to get the "true data": We merged the information from the design maps, from visual inspection, from additional radiographs and from the "majority decision" of the inspectors from the mode two inspections.

REPORT FORM FOR THE ULTRASONIC EXPERIMENTATION

inspector: date: time:

experiment: sample:

indication table

b	n	x	y	f	s	d	c	i

grid division

x in mm

	0..49	50..99	100..149	150..199	200..249	250..299
0..49	1	2	3	4	5	6
50..99	7	8	9	10	11	12
100..149	13	14	15	16	17	18
150..199	19	20	21	22	23	24
200..249	25	26	27	28	29	30
250..299	31	32	33	34	35	36

y in mm

b box which contains the defect
n indication number within the box
x x-coordinate of the defect centre (0..299)
y y-coordinate of the defect centre (0..299)
f form of the defect
    A - area like
    L - longitudinal
s size of the defect in mm
d detectability of the defect
    1 - very good
    2 - good
    3 - middle
    4 - bad
c classification of the defect
    C - crack
    P - pore
    D - delamination
    R - porosity
    O - other
i importance of the defect
    Y - critical
    N - not critical

Fig 6: Report form for an ROC study of ultrasonic C-scan testing

In fig. 7 are shown the results for the detection ROC for the three experimental modes. As expected the inspection performance of the 25 MHz focussed probe is higher than for the 5 MHz normal probe in pulse echo mode. The maximum reached TP rate is nearly the same for the two 5 MHz modes but the false call rate for the transmission mode is definitely smaller than for the echo mode. The whole ROC curve of the 5 MHz transmission mode lies in the range of the 25 MHz echo mode curve. The comparison of the actual experimental rate of the TP indications for the modes 2 and 3 (triangles for the 5 MHz transmission mode, boxes for the 25 MHz mode) shows that the mode 3 yields lower values than the mode 2. While these experimental points are only the actually reached operating points the theoretical extrapolated ROC curves give the potential capability of the NDT system. That means if the ROC curves of mode 2 and mode 3 are nearly the same and so the area indices a_z as a measure of the capability of the NDT system the two systems do not differ. This area index including their standard error was fitted by the maximum likelihood regression and used to construct confidence intervals and to perform statistical tests of significance. There is no difference in the area index for the modes 2 and 3. The comparison of all area indices for the detection ROC yields that there is no difference between the modes which turned out to be significant on a confidence level of 97.5% (one tailed t-test of the differences of the single inspectors). Considering the maximum reached in the experimental points it is seen that mode two yields a 20% higher TP rate but also a 10% higher FP rate (noise).

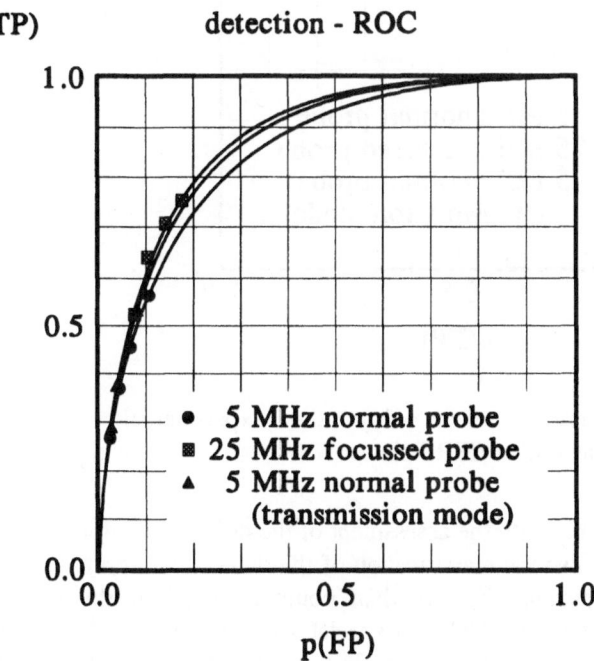

Fig 7: Detection ROC of the 3 experimental modes in ultrasonic testing of CFRP plates

The results for the masked ROC (only critical = unacceptable defects, fig. 8) fall also even more close together. As already seen at the detection ROC the modes 2 and 3 give nearly the same extrapolated ROC curve but the experimental values of the mode 2 represent the upper right part of the curve and the values of mode 3 the lower left part of the curve. The maximum values for the modes 1 and 2 are of about 75% TP rate and 10% FP rate. We have also found that the areas under the masked ROC curves do not differ on the same significance level as for the detection ROC by using the t-test. This gives the hint that the more effective working 5 MHz normal probe is accurate enough for detecting critical defects in its box area. We defined as critical all cracks and all defects greater than 5 mm (recommendation by ESA).

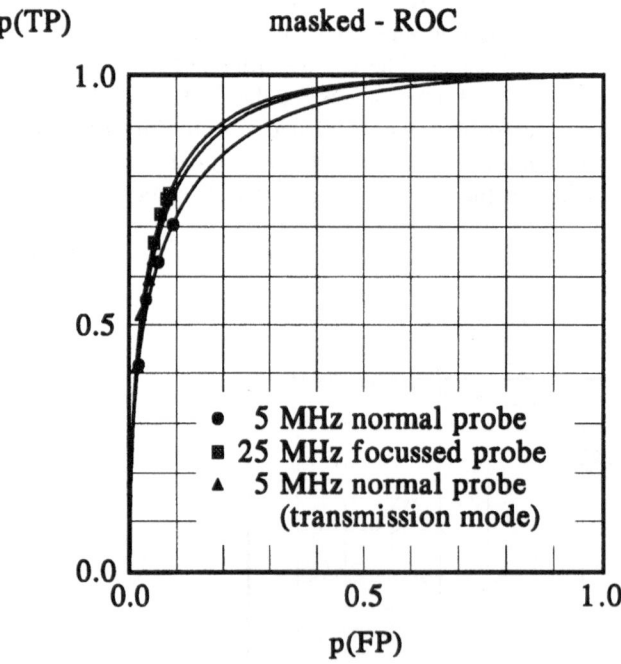

Fig 8: Masked ROC (unacceptable defects only) of the 3 experimental
modes in ultrasonic testing of CFRP plates

So far we were concerned only with the assessment of the defect detection (detection - or masked ROC). To validate also the classification of the defects we apply the joint ROC: For the joint ROC an indication is only than counted as TP if the detection and classification is correct. The local ROC as a special kind of joint ROC is a plot of the probability of a FP detection response on the abscissa, as usual, with the probability of both a TP detection response and correct localisation response on the ordinate. The following results show quantitatively the performance of the used 3 modes in defect classification and localisation.

In the case of the joint ROC (detection & classification, fig. 9) the difference in the TP rate of the three modes becomes higher, because the shape of the defects - which is important for the classification - is better seen in mode two.

The results of the local ROC are shown in fig. 10 in the following way: the maximum p(TP) rate is indicated versus the radius of acceptance (ROA). In this case the local ROC was defined as a joint ROC where a TP occurs only when the detection and location is correct. The FP probability follows from the corresponding detection ROC. All curves start at a very small TP rate for an extreme small radius of acceptance of about 1 mm. The curves increase than for greater ROA's up to a saturation point which is near 10 mm. For ROA's greater than 10 mm the curves increase very slowly. The curves for the two 5 MHz modes are nearly the same because the local resolution doesn't depend from the used mode (pulse echo mode or transmission mode) but only from the used frequency and focussing property. That's why the reached maximum value of the 25 MHz probe is greater than for the 5 MHz modes. So one would prefer a 25 MHz focussed probe for the task to find the defects within an ROA of let's say 5 mm.

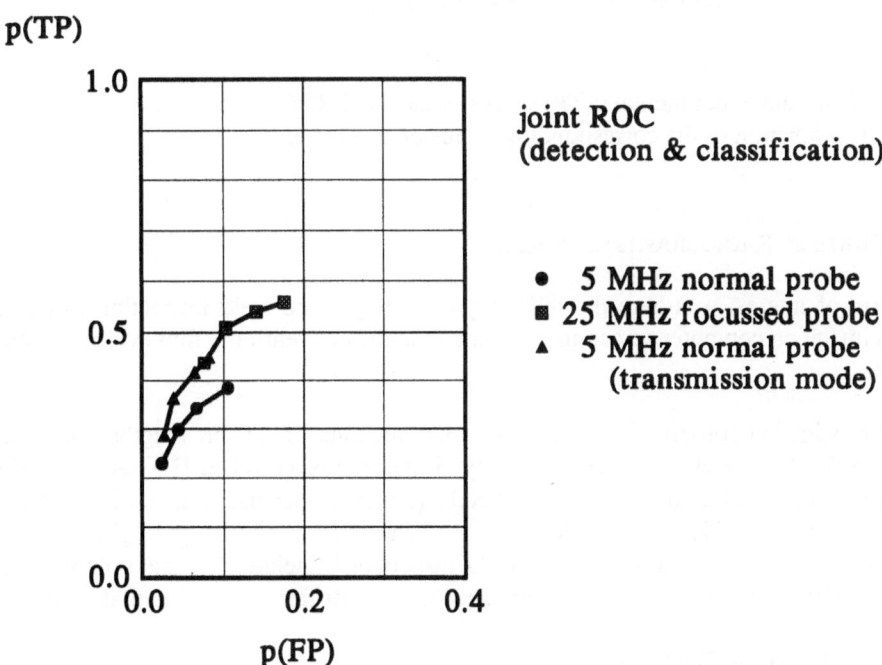

Fig 9: Joint ROC (detection & classification) of the 3 experimental modes in ultrasonic testing of CFRP plates

24

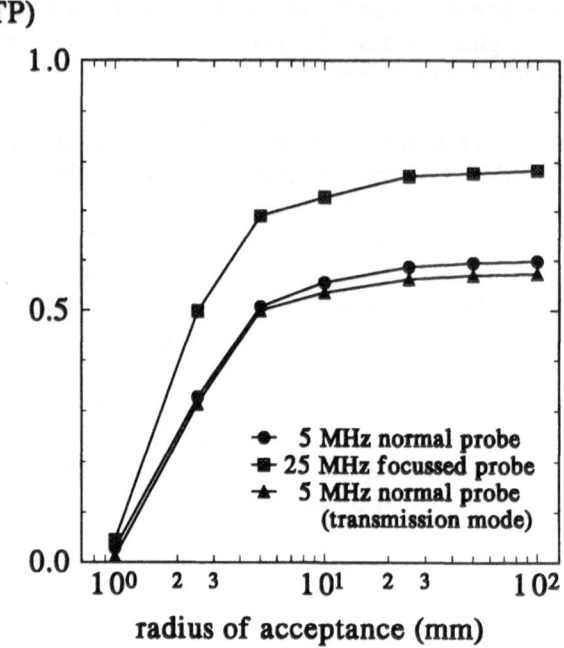

p(TP)

radius of acceptance (mm)

Legend within figure:
- 5 MHz normal probe
- 25 MHz focussed probe
- 5 MHz normal probe (transmission mode)

Fig 10: Maximum experimental p(TP) values of the local ROC as a function of the corresponding radius of acceptance

3.2 DIGITISED RADIOGRAPHS OF WELDS

The centre of interest was here the reliability of radiographic weld inspection using the digitised image on computer screen in comparison to the conventional film evaluation on a light box.

In this investigation (ordered by KWU) 92 weld sections of 24 cm lengths containing more than 1000 defects were examined by 5 inspectors (level II/III). As digitising equipment a high resolution (70 μm) 12 bit NDT (Du Pont) scanner was used (for details see the contribution of Hannelore Wessel et al.. "The influence of image digitisation and calibration of the scanning system upon weld inspection", included in this volume). The evaluation of the radiographs were accomplished in three different experimental modes:

1. Evaluation of the film using a light box
2. Evaluation of the digitised film image on computer screen
3. Evaluation of the digitised film image on computer screen using image improvement utilities

p(TP) **ROC: detection of critical defects**

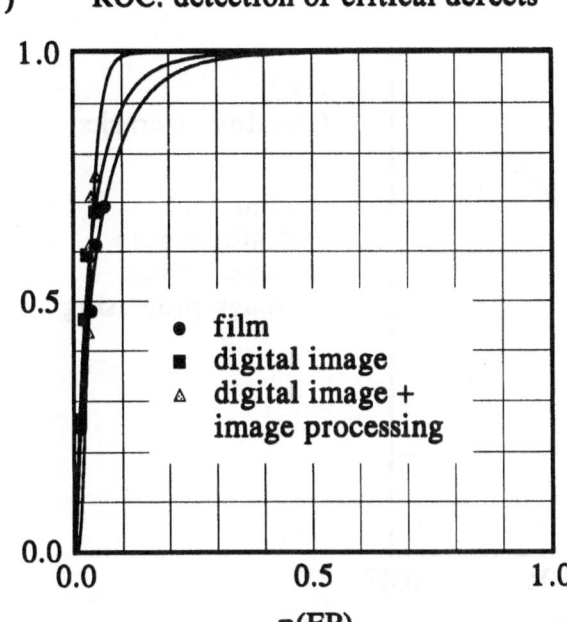

Fig 11: Results of the Masked ROC (detection of unacceptable defects) of X-ray inspections of welds

We concentrate the attention here to the masked ROC (fig. 11) where only the right (TP) and wrong (FP) indications of unacceptable defects are taken into account: The masked ROC curve for the evaluation of digitised images using image improvement raises over the curve of the conventional light box inspection with an amount of about 10% which is not so much but even statistically significant since the huge statistical base-showing the performance of the equipment.

We reduced the statistical basis in further test ROC evaluations from the original set of 92 films and found that the result remains stable until 30 films (10 defects are in the average on each film). In spite of this we recommend to apply the higher basis of about 100 films in order to have the basis rich enough for ROC evaluations for the different defect types and groups of different defect sizes - which are often asked for by industrial consumers. Because cracks are the most important critical defects - where critical can be interpreted in both senses the malfunction of the equipment and the detectability on radiographs - we extracted the indications belonging to cracks and evaluated them in a detection ROC for cracks from which the experimental points are shown in fig. 12.

The fit from the experimental points to the ROC curve is done by means of a maximum likelihood regression on the basis of the binormal model.

p(TP)

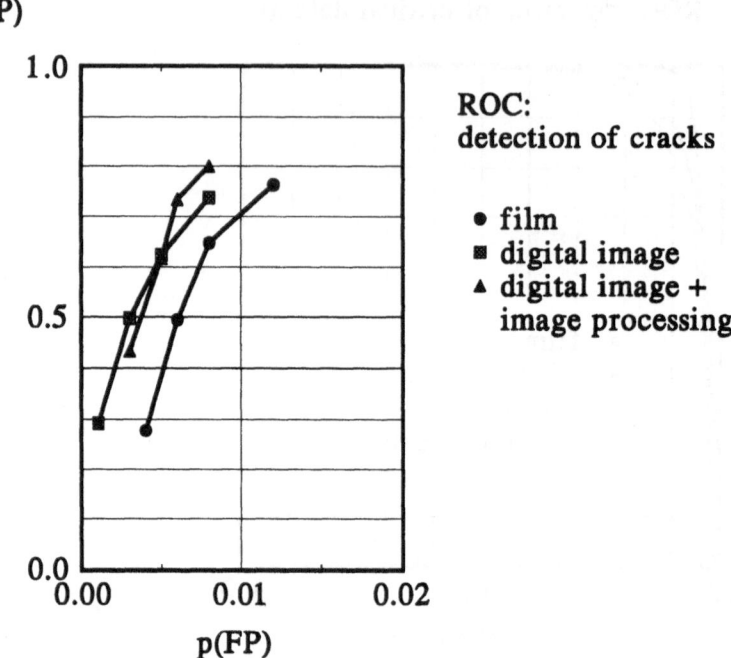

Fig 12: Results for the detection of cracks only in radiographic weld inspections

3.3   THE INFLUENCE OF OPERATOR'S EXPERIENCE ON THE INSPECTION PERFORMANCE
      IN ULTRASONIC TESTING

We had 9 plates with a total number of 62 artificial flaws distributed on them in different depths which are to be detected by means of a 50° and a 65° angel probe, respectively. These probes can be applied from the left and from the right and from the top and from the bottom so that totally 258 defects were to be detected per inspector. The detectability scale was here defined quantitatively in terms of 2dB steps from -6dB to +6dB. The adjustment of the ultrasonic device was done by an administrator before the experiments and then fixed.

We had two groups of inspectors with different experience level:
The first group with "no experience" carried out the experiments immediately after the course in ultrasonic testing. The second group "with experience" had have practical experience for at least 10 years.

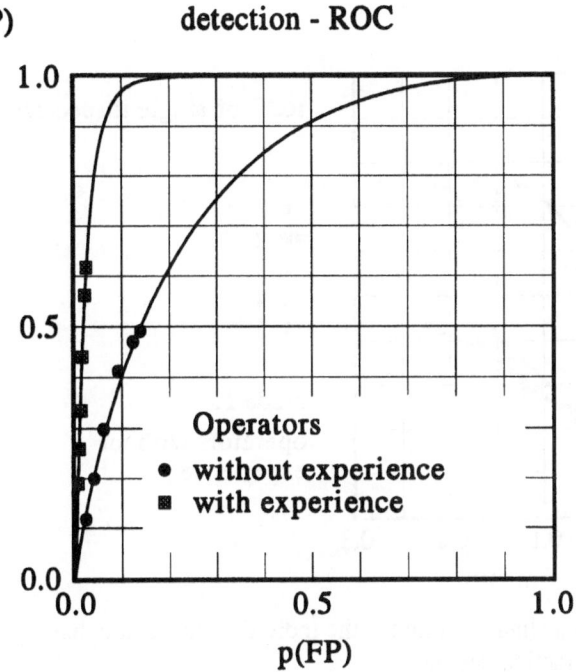

p(TP)   detection - ROC

Fig 13: Results for ultrasonic hand testing by inspections with different testing experience (none, more than 10 years)

In fig. 13 the results for the detection ROC pooled for the two groups are shown. Interpreting the area under the ROC curve as probability in a two alternative test with a defective sample and a sample without defects to say correct which is which we see that this probability is 20% higher for the inspectors with experience. This difference is statistical significant up to a 95% confidence level (paired t-test). In figures 14 and 15 the detection ROC curves for the individual inspectors of each group are presented. The deviations from one inspector to an other is for the unexperienced inspectors considerable greater than for their experienced colleagues. That is also the reason for the confidence level's not being higher than 95%.

Finally, in table 1 the results for the maximum experimental defect detection- and false alarm rates and the area values are listed for different levels of test sample reduction. Especially the lower part shows that the drawing of conclusions from ROC results obtained with a too small statistical basis is dangerous.

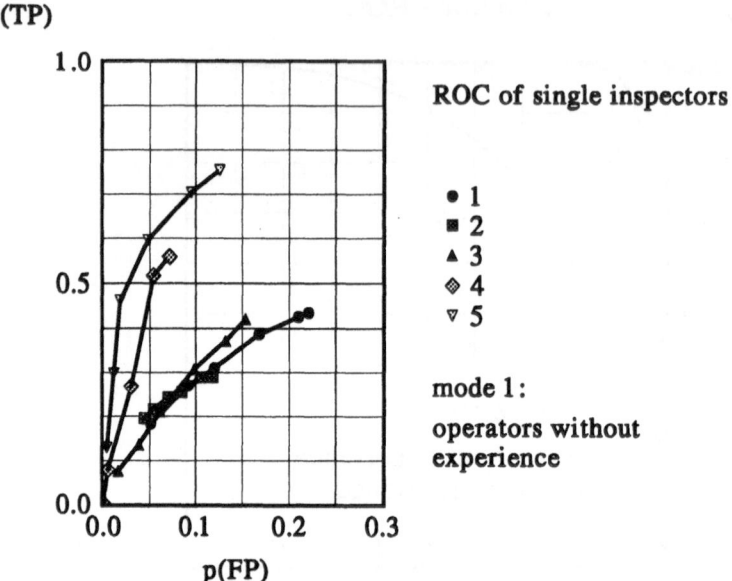

Fig 14: Results for ultrasonic hand testing for the individual inspectors having no experience in practical testing

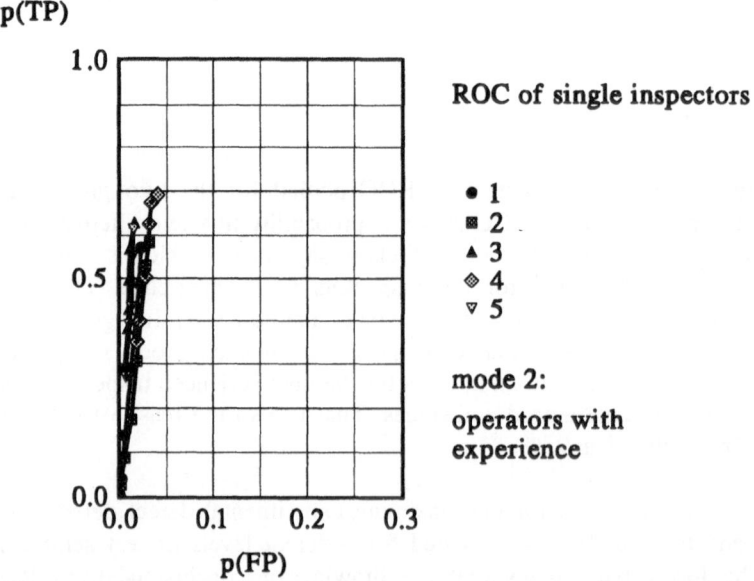

Fig 15: Results for ultrasonic hand testing for the individual inspectors having at least 10 years experience

reduction	n_{TB}	$n_{TB,wd}$	n_{def}	p(TP)		p(FP)		area index	
				mode 1	mode 2	mode 1	mode 2	mode 1	mode 2
1/1	72	24	248	0.491	0.617	0.138	0.024	0.799	0.973
1/2	34	13	106	0.422	0.733	0.071	0.013	0.794	0.978
1/3	24	11	74	0.487	0.653	0.134	0.020	0.795	0.980
1/7	10	4	32	0.486	0.636	0.113	0.014	0.844	0.985
1/36 (I)	2	0	12	0.250	0.388	0.008	0.008	0.991	0.991
				0.373	0.293	0.088	0.016	0.897	0.977
				0.512	0.362	0.067	0.042	0.869	0.951
1/36 (II)	2	1	6	0.378	0.111	0.065	0.052	0.859	0.796
				0.422	0.733	0.071	0.013	0.589	0.989
				0.543	0.629	0.030	0.000	0.949	0.994
				0.514	0.771	0.085	0.000	0.901	0.994

Table 1: Maximum of the experimental reached TP and FP rate for different data reduction factors with the total number of test bodies n_{TB} ($n_{TB,wd}$ test bodies without defects) and n_{def} defects

4. Summary and Conclusion

In three application examples it was shown that the ROC method is capable of distinguishing quantitatively between different testing modalities with respect to their testing performance. The seriousity of conclusions depends sensibly on a sufficient statistical basis. Statistical significance tests should ensure the stability of ROC results. The first two examples show the ROC method's being well suited for device evaluation whereas the third example is a promising demonstration of the ROC's potential for NDE personal certification on the basis of the concept of performance demonstration.

5. Acknowledgement

The work on inspection performance in ultrasonic testing of fibre-reinforced plastic plates was supported by the European Space Agency under contract ESTEC 9289/90/NL/LC(SC). We are especially indebted to *M. D. Judd from ESA Materials Processing Division ESTEC Nordwijk for sponsoring the work making available the results for publication and helpful discussions, to **C. Hobbs from the Harwell AEA Technology, National NDT Centre Harwell (UK), for the delivery of the test plates and the accomplishment of the experiments and help in the design of the report form and to ***S. Gripp from BAM 6.32 for the production of special C-scans.

The sponsoring of the work on the digitised radiographies of welds by Siemens KWU Erlangen (Germany) represented by A. Mattis is gratefully acknowledged.

We give our thank to A. K. Gurvich and G. J. Dymkin from PIIT St. Petersburg (Russia) for the cooperation and helpful hints in evaluating the influence of the operator's experience on the inspection results.

References

1   Metz, C.E. 'Basic Principles of ROC analysis' Seminars in Nuclear Medicine **8** 4 (1978)
2   Metz, C.E. 'Some practical issues of experimental design and data analysis in radiological ROC studies' Invest Radiol **24** (1989) pp 234-245
3   Swets, J.A. 'Assessment of NDT systems-Part I' Mater Eval **41** 11 (1983) pp 1294-1298
4   Swets, J.A. 'Assessment of NDT systems-Part II' Mater Eval **41** 11 (1983) pp 1299-1303
5   Somoza, E. Mossman, D., and McFeeters, L. 'The info-ROC technique: A method for comparing and optimizing inspection systems' Review of Progress in Quantitative Nondestructive Evaluation, vol. **9**, ed. D.O. Thompson and D.E. Chimenti, Plenum, New York (1990) pp 601-608
6   Nockemann, C., Heidt, H., and Thomsen, N. 'Reliability in NDT: ROC study of radiographic weld inspections' NDT&E International **24** 5 (1991) pp 235-245

COMPENSATION OF ULTRASONIC TRANSDUCERS RESPONSE BY ADAPTING EXCITATION SIGNALS, APPLICATION TO DEFECTS EVALUATION.

J.F. DE BELLEVAL, J.M. GHERBEZZA,
UTC, LG2mS, ura CNRS 1505, BP 649, 60206 Compiègne, France
P. JAGNOUX
SNECMA, CTP, BP 81, 91003 Evry, France

ABSTRACT. Despite the precautions taken by mean of calibration procedures, the evaluation of real defects in ultrasonic non destructive testing presents great discrepancies when we change the transducer by an other supposed to be of similar characteristics. These discrepancies are ascribed to the variations in the spectral responses of the transducers utilised. We have achieved a method which permits, by an adapted excitation, to compensate the differences in the transducers responses and we have shown that this permits a 6 dB gain upon the evaluation disparity of real defects.

1. Introduction

Ultrasonic non destructive testing is used by industry for about forty years. The first measures have permitted the detection of defects but very quickly the necessity for quantitative evaluation of these ones has appeared. When the detected defect has a great size, its evaluation is performed by moving the transducer and observing as a function of this one, the variation of the amplitude of the echo. By opposite, if the defect is of small size (smaller than the transversal extension of the beam) such a procedure does not give any information upon the defect size, the seek of links between defect size and amplitude of detected echo has been then looked for. Having noticed, however, the influence of very numerous parameters on this amplitude (characteristics of transducer, measurement system, tested material, depth of the defect inside the piece,...), this has led not only to elaborate norms or test directives specifying the characteristics of equipments to be employed (transducer and electronic apparatus) and the method to employ, but also to define on a very precise way calibration procedures.

These procedures are generally based upon a comparison between the observed echo of the defect with the one of a defect of standard type (flat bottomed hole or cylindrical hole taken aim perpendicularly to its generatrix) into a sample as representative as possible; often in aeronautical industry, the sample is made with same matter than the piece to be tested. This method permits to set free from the biggest discrepancies between measurement system or between transducers, discrepancies which are not always easy to avoid between elements which are supposed to be the same. Despite these precautions, we

31

X. P. V. Malague (ed.), Advances in Signal Processing for Nondestructive Evaluation of Materials, 31–42.
© 1994 *Kluwer Academic Publishers.*

observe still evaluation differences on real defects, they can reach ten dB's when we use several transducers of identical characteristics with the same measurement system.

These differences seem imputable to discrepancies between the signals detected by the transducers. Indeed the shape in time domain of the detected echo upon a given target varies drastically from a transducer to an other one, even when their essential characteristics are similar. This shape, obviously related to the transducer bandwidth, does not seem to be correctly controlled at its making. This fact leads us to propose a new approach which consists to compensate the transducer response at the electronic level, i.e. by excitation signal adaptation. The procedure consists to choose the desired signal shape, compatible with the type of transducer we want use, then to determine for a given transducer the excitation signal which permits to obtain this shape. This excitation signal is obtained by deconvolution type procedures, relatively simple on theoretical point of view. This principle of excitation signal shaping have been already used, generally for an other goal than the one described here, in particular for the improvement of axial resolution of test [1, 2, 3] or for the improvement of long term test reproducibility [3, 4].

We made use of such a process by determining the excitation signal adapted from a calibration on an appropriate target (two kinds of target have been compared). The excitation is created by mean of an arbitrary signal generator. Two kinds of excitation principle have been tested, the first one corresponds to a short signal whose the only function is to compensate the variations between transducers, the second corresponds to a signal of much greater duration obtained by a spectral phase modulation. A demodulation in reception permits to recompress the signal and hence not to reduce depth resolution. This whole process permits to increase the energy emitted for a given excitation level and so to improve the signal to noise ratio.

2. Method set up

2.1. EXCITATION SIGNAL DETERMINATION

The method we propose has for goal to set free from discrepancies in characteristics between transducers supposed to be identical, particularly their spectral discrepancies. In that aim, we define a signal shape, the one we want to obtain as result. This shape has to be compatible with the characteristics of the transducers we use, specially their bandwidth. The transducers then characterised by classical (transient) echography upon a calibration target, which enables us to calculate by means of pseudo-deconvolution methods the excitation signal giving with this target the desired shape. This last one, which is therefore adapted to each transducer due to the calibration, will compensate the discrepancies between transducers.

Two calibration targets have been tested to determine the influence of their shape: a plane target and an hemispherical target (figuring a small target). In the both cases, they are placed at the focal point of the transducers.

The direct result of the computation of the adapted excitation gives a relatively short signal. This signal is used as it is, in the first set of tests. A supplementary improvement can be obtained, like in radar technique [6], by a broadening of excitation duration. This is obtained by a modulation of the spectral phase of the initial signal. It is then possible to have an ultrasonic emission of energy much greater, without increase of the excitation amplitude. Obviously the echoes have a greater duration as well, but a demodulation of the phase according to the same law permits the recompression of the signal.

These two techniques have been set up. The resulting signal (desired signal in reception) has been chosen upon the shape of a sine function weighted by a Blackmann-Harris window, the operator determining from transducers characteristics the central frequency and bandwidth of this resulting signal.

2.2. SAMPLES AND DEFECTS TESTED

The aim of the tests is to achieve a method permitting to compensate the discrepancies found on real defects when we use different transducers given to be identical. In order to validate the method we hence performed tests upon provoked defects as representative as possible of real defects. For that kind of test, SNECMA has made a piece, Titanium alloys (TA6V) disc, obtained by the assembly of two elements welded by diffusion between which have been inserted particles whose dimension before welding vary from 0.2 mm to 3.2 mm. The tests have been performed on 12 defects corresponding to three kinds of different inclusion spotted by the letters X, Y and Z. For each kind of inclusion three defect sizes have been kept, they are spotted by a number. The last kind of inclusion, spotted by the letter Z has been created two times, assuming with the same dimensions.

2.3. MEASUREMENT SYSTEM AND TRANSDUCERS UTILISED

Five transducers have been used, they have a nominal frequency of 5 MHz, a diameter of 19 mm (0.75") and a focusing distance given by the maker varying from 100 mm to 150 mm. The table I gives the main characteristics of these transducers, in particular the estimated values of the dimensions of focal spots for each of them.

Transducer Number	Transducer Type	Focal Length	Dimensions of focal spot
T1	Panametrics V308	150 mm	73 x 2.3 mm
T2	Harisonic 17-0512	100 mm	34 x 1.6 mm
T3	Harisonic 17-0512	150 mm	73 x 2.3 mm
T4	Panametrics V308	150 mm	73 x 2.3 mm
T5	Panametrics V308	120 mm	47 x 1.8 mm

Table I

In order to evaluate the proposed method, we made a comparison between these results and the ones obtained with a classical transient excitation obtained by mean of a Panametrics generator-receiver. This measurement system stands for our reference.

About the adapted excitation, basis of the method, it is set up by mean of an arbitrary signal generator Lecroy 9100 which permits an analogue conversion of any digital signal with a high frequency sampling rate (100 MHz for our case).

2.4. TEST PROCEDURES AND PRECAUTIONS

The procedure consists in first to do a calibration of the measurement system with every transducers. This is performed with the complete set of the system used (in particular the arbitrary signal generator) by choosing as excitation signal a short impulsion. We then obtain the transfer function of the echographic system consisting of the measurement system and the target. We hence determine, for a set of tests and for every transducer, the excitation signal adapted which will be used for all the defects. Three sets of tests have been made, two with rough signals (short signals) but with two different calibration targets, as previously noted, the third by taking a long signal (whose spectral phase has been modulated).

In order to set free from any spurious variation of the detected signals and to obtain a good reproducibility of the results some precautions concerning the positioning and the orientation of the transducers have been taken. In particular, the locations have been spotted in absolute by respect to a reference on the piece, with an accuracy of about tenth of millimetre, and not by searching the maximum of the echo; the transducers orientations have been determined by searching the maximum of the first echo corresponding to the bottom of the piece.

3. Main results

3.1. SHAPE OF DETECTED SIGNALS

The figure 1 presents an example of calibration results for two transducers (T3 and T4). At the top are presented the results with a classical system (Panametrics), the target being quasi punctual (3 mm diameter hemispherical target). We notice, in addition to a sensitivity difference, a discrepancy in the shape of signals relatively important between the two transducers. This discrepancy is the cause of the variations of the echoes levels on real defects, even with taking the usual precautions while calibrating upon standard defect. At the bottom are presented the results obtained upon the calibration target with excitation signals adapted respectively short and long. The short signals do not require any reception processing while long signals need a spectral phase demodulation. We note naturally that the result is quite conform with what was searched for concerning the detected signal shape. The bandwidths slightly different of the transducers are quite compensated on this kind of target.

We note that the detection levels are almost the same (they diminish of a factor of 1 to a maximum of 2) between the classical system which uses an excitation of about 100 volts and the system using the adapted short signal whose peak to peak level is about 10 volts. The use of long duration signals permits a gain of about a factor 5 to 6 (15 dB) on the presented examples. This gain can be moreover increased by a supplementary broadening of the excitation duration. Let us note that the excitation levels used for long signals are two times smaller than for the short signals due to the saturation of the system used which was not optimum, we have taken into account this difference to evaluate the gains quoted.

The figure 2 presents the results obtained with the same transducers on a real defect, with different kinds of excitation. The top of the graph corresponds to the classical test with a Panametrics generator, we can see discrepancies in the shape of the signals

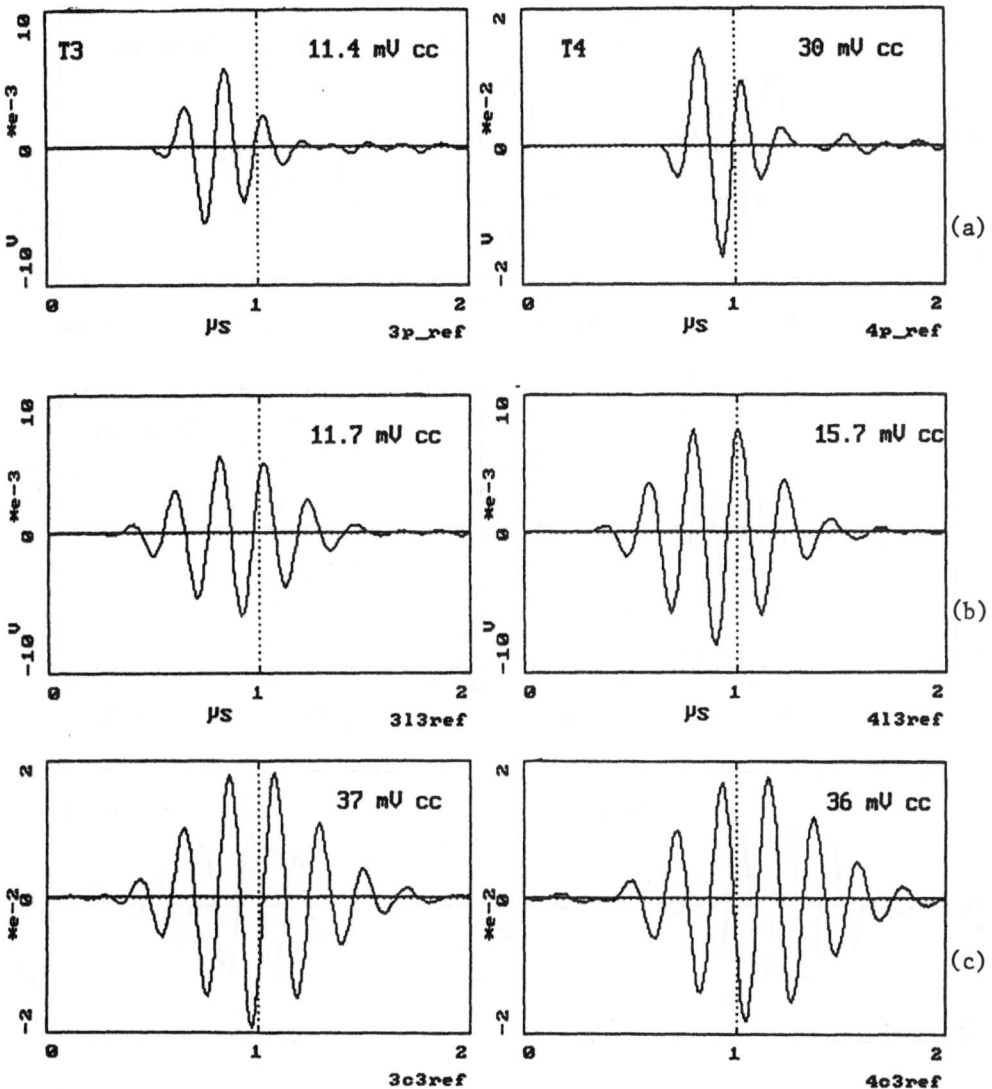

Figure 1   Excitation signal adaptation, results upon calibration target:
                (a) Classical transient excitation, (b) Short adapted excitation, (c) Long
                adapted excitation

36

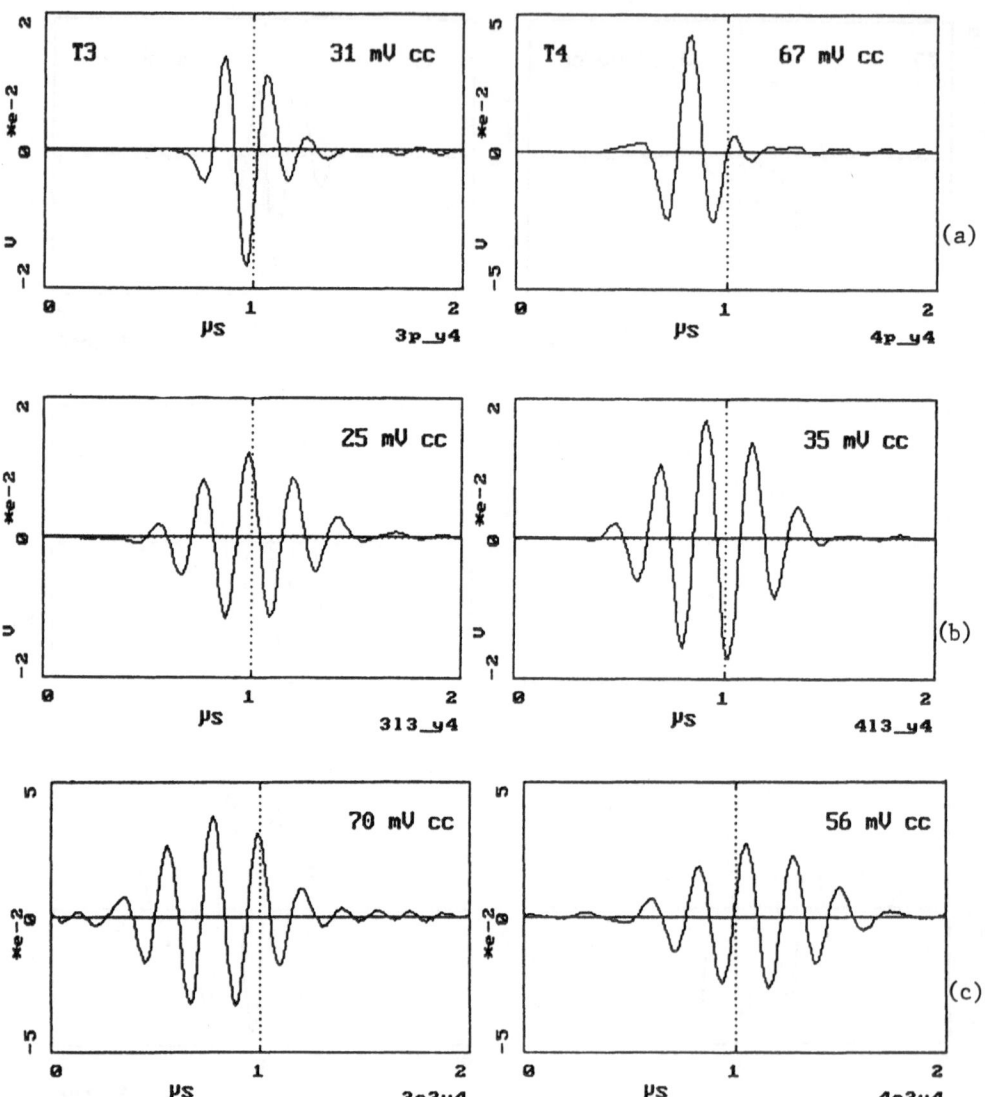

Figure 2   Excitation signal adaptation, results upon real defect:
(a) Classical transient excitation, (b) Short adapted excitation, (c) Long adapted excitation

detected by the two transducers. The centre is related to a short adapted excitation. We notice that even upon a real defect, the compensation of the bandwidth of transducers is quite correct, since the signals of the different transducers have the same shape. We have tested two kinds of excitation, the first (excitation L3) is obtained with a plane as calibration target and the second (excitation L4) with hemispherical target. The differences between the two kinds of excitation are not significant, this seems to indicate that the choice of the shape of target is not critical. At last the bottom of the figure corresponds to a long excitation signal of about 20 μs (excitation C3). Here again we can note a very good compensation of the bandwidth of the transducers by a same shape of the signals of each of them.

As for the hemispherical target, the detection levels are a little weaker for the case of the system using the short adapted signal than for the case of the classical sytem. The use of long signals permits to increase the level of about 5 to 6 times (about 15 dB). We gain still about 40% (3 dB) by doubling the signal duration.

At last the figure 3 presents an example of important improvement that the method can give for defect detection. At the top we see the signal detected in classical echography (Panametrics generator) upon a small defect (X3), by mean of the T1 transducer. The very weak echo corresponding to the defect has a smaller amplitude than a spurious echo located a little further on this echography. This last one, still present for this transducer for such an amplification level, corresponds to an imperfection of the transducer, sometimes called "housing echo". The curve at the bottom points out the improvement brought by signal adaptation. The housing echo has completely vanished and the defect echo is easily detected. The gain in signal to noise ratio in this case is greater than 10 dB. Let us note that a spurious oscillation at 500 kHz is present on this last echogram, it is due to poor adaptation of the electronic used and can be suppressed either by filtering or by an improvement of this electronic. This oscillation will perturb the evaluation of the smallest defects.

3.2. COMPENSATION OF THE DISCREPANCIES BETWEEN TRANSDUCERS

We have tested the method by using the classical procedure of defect evaluation, i.e. the comparison of the echographic signal of a real defect with the one given by a target chosen as reference. As previously explained, it is generally used as reference target a standard defect into a sample made of the same material than the piece to be tested. In our case however, being interested only by the variation of echoes from one transducer to an other, we took for reference the hemispherical target which was used as well, in one of the measure sets, for calibration target to calculate the adapted signal.

An example of compensation between transducers obtained by the method is depicted by the figure 4. On each graph of this figure are quoted the ratios (in decibels) of the signals detected on a defect to the one of the reference target, and this for four real defects of different nature and of similar size. On the same graph, for a given abscissa location corresponding to a defect, is drawn the ratio previously cited for the five transducers, the height of the vertical straight line linking the points is therefore representative of the dispersion of the measure as a function of the transducers. On the left graph which corresponds to a classical transient excitation, we note a great dispersion of the results, which is much smaller on the right graph corresponding to the excitation adapted to each transducer.

38

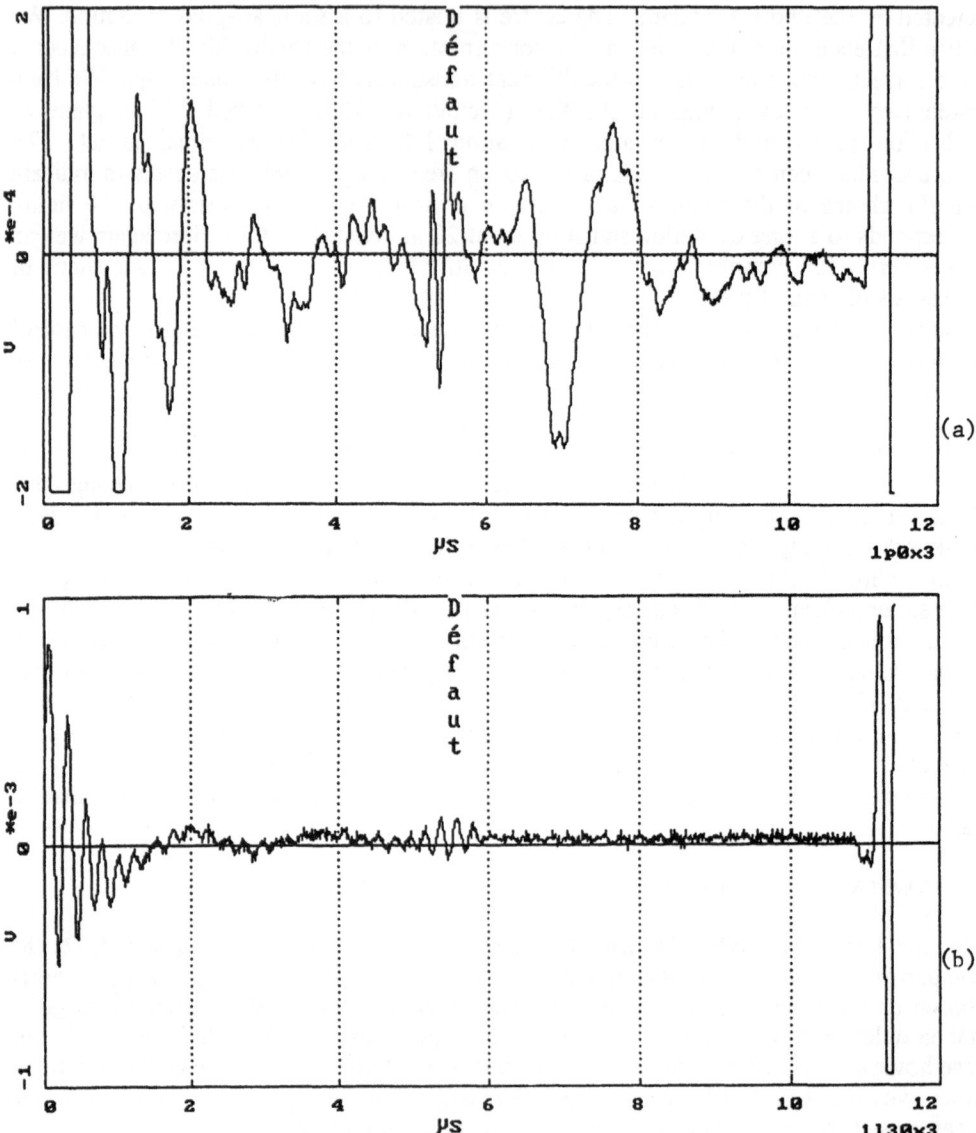

Figure 3   Improvement of real defect detection:
              (a) Classical transient excitation, (b) Short adapted excitation

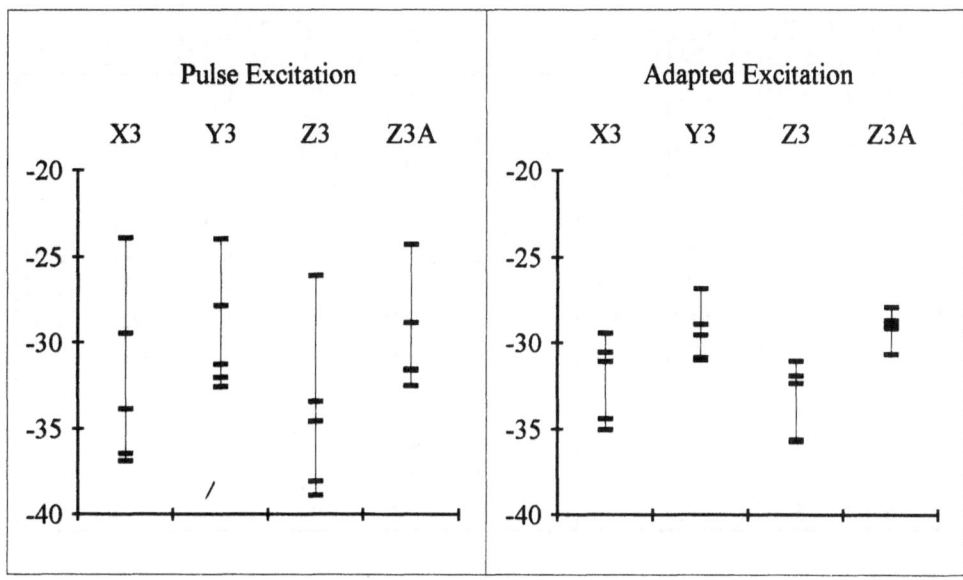

Figure 4   Reduction in dispersion of evaluations of real defects:
(a) Classical transient excitation, (b) Short adapted excitation

In order to better quantize the gain obtained, the table II summarises maximum variations between the different transducers for every defect and every kind of excitation. Two parts arises from this table, the left part lists the variations obtained by considering all the transducers, the right part concerning only the transducers with the same degree of focusing (transducers T1, T3 and T4 which have the same diameter as well as the same focus distance). We note as a general trend a gain which can exceed 6 dB. If we restrict to the three previously quoted transducers, then the residual variation is still below 3 dB, except for the Y3 defect and the L4 excitation for which it is 4.17 dB. In this case the measurement is very strongly perturbed by the 500 kHz spurious oscillation already mentioned. A better electronic set or an adapted filtering would permit to set free from this problem and surely to reduce slightly this variation.

In the studied case, the transducers with different focus distance had the same diameter, so different apertures. The results are less good if they are included into the comparisons. It would be interesting to verify, which is likely, whether the improvement brought is still significant when we include transducers with different focus but with similar aperture.

	Max Variation dB (T1,T2,T3,T4,T5)					Max Variation dB (T1,T3,T4)				
	Impuls.	Exc L3	Exc L4	Exc C3	Exc C30	Impuls.	Exc L3	Exc L4	Exc C3	Exc C30
X3	12.97	6.05	5.62			9.97	1.07	1.67		
Y3	8.64	3.37	4.17			7.32	2.58	4.17		
Z3	12.81	3.88	4.65			8.48	1.40	1.30		
Z3A	8.29	1.54	2.75			7.32	1.24	1.21		
X4	4.87	2.11	1.83	4.60	2.51	3.28	1.16	0.94	2.24	2.51
Y4	6.56	3.65	3.82	5.59	2.82	2.90	0.77	0.94	2.87	2.82
Z4	6.42	4.53	4.52	2.22	3.64	3.74	1.80	1.67	1.72	2.50
Z4A	7.16	5.89	5.52	3.48	3.72	3.24	1.96	1.48	3.32	3.59
X5	6.05	3.26	3.41			3.99	1.63	2.17		
Y5	4.75	2.21	2.50			4.07	2.21	2.50		
Z5	5.97	4.10	3.96			3.22	1.44	1.21		
Z5A	5.42	3.01	2.92			3.50	1.09	1.25		

Table II

An other observation is that the variations in evaluation of defects between transducers by classical method are much more important for small defects tested (X3, Y3, Z3 and Z3a) than for the others; we change, restricted to the transducers with same focus, from 7 to 10 dB for the first ones to a variation less then 4 dB for the others.

3.3. ADVANTAGES AND LIMITATIONS OF LONG DURATION SIGNALS

We already indicated the advantages of long duration excitation signals in term of an improvement of the signal to noise ratio. We observe on table II that this kind of excitation permits as well to decrease on a significant way the discrepancy between the transducers, with however a result slightly lower than the ones obtained with short signals. The increase of the signal to noise ratio should permit to improve the detection of small defects, a limitation nevertheless appeared. This method requires signal processing at reception to separate the input echo (front face of the piece) from the defect echo, these ones are, by the principle of the method itself, partially superimposed. In order to perform this process, it is necessary not to saturate the input echo as it is usually done to increase dynamic. In these conditions, the digitalisation system used (digitalisation on 8 bits) has an insufficient dynamic to detect the smallest defects. For that reason we were not able to obtain valuable results with the smallest defects.

The method apply nevertheless, even with the very limited dynamic of such digitalisers, if the front face echo is not too important (either in the case of material with lower impedance or with oblique incidence). Digitalisation systems with greater dynamic are already available (10 or 12 bits), they would permit to significantly put back this limit. Moreover constant and fast progress in electronics make us to hope in soon future a much greater dynamic with a likely very reasonable price.

4. Conclusion

We have developed a method permitting to set free from the residual discrepancy between ultrasonic transducers supposed to be of identical characteristics with the aim to improve evaluation of real defects. Indeed, despite all precautions taken, it is usual to find up to 10 dB variations in defects evaluation by only a change of transducers, this resulting from an imperfect control of spectral shape of the transducers response. The method we made use of, consists in the adaptation of electrical excitation so that these discrepancies are compensated. For that a calibration of the measurement system upon a "calibration target" and a method relevant to deconvolution, permit to determine for each transducer the excitation signal which gives an imposed shape of the echo upon the target.

We have tested the method upon defects provoked into a Titane alloy supplied by SNECMA and compared, for five transducers, the improvement the method could bring. The five transducers has same nominal frequency, same diameter, three of them have same focus distance but the two others have different focus distance hence a focusing degree (or beam aperture) also different. In order to evaluate the method, a comparison has been done with a classical system using a transient electric excitation.

We observe that the variation in defects evaluations, taking the usual precautions for that kind of test (calibration upon standard defects) reaches in classical method about 10 dB and that the proposed method enables a reduction of this variation up to less than 3 dB if we consider only the transducers of same focus distance. For the other transducers the variation in evaluation is more important with the classical method, the reduction obtained with our method are of about the same order of magnitude, which leads to a final result globally a little less good. It is quite possible, however, that this would be interesting to confirm by means of complementary tests that the global result is so good if we deal with transducers with different focus distances but with same aperture.

The method can be significantly improved concerning the detected echoes level, by a supplementary change of the excitation signal. It consists in a broadening of its duration which can be achieved by modulation of the spectral phase. We have exhibited indeed that this process has permitted in the studied case to obtain a gain of 15 to 18 dB for the same excitation level. This is due to the increase of the energy level of the ultrasonic emission. It is therefore necessary, to get back axial resolution of the testing, to process the reception signal on the inverse way of the emission signal.

We have encountered a limitation to this principle, according to the material utilised. It is no longer possible for that kind of process, like in the classical method, to saturate a part of the detected echoes. The limited dynamic of our digitalisation apparatus hinders us from the detection of very small defects when the input echo is very important. There exists already apparatus which have a greater dynamic than the one we used (digitalisation on 10 or 12 bits instead of 8 bits), moreover the constant and fast progress in electronics makes us hoping even more important dynamic for reasonable price in soon future.

5. Acknowledgments

This was supported by SECMA.

6. Références

[1]   C. Chassaignon, J.F. de Belleval
      Input signal optimisation of ultrasonic transducers for non-destructive testing,
      Ultrasonics International 1985, Londres, july 1985.

[2]   C. Chassaignon, J.F. de Belleval
      Optimisation du signal d'excitation d'un transducteur ultrasonore en contrôle non
      destructif, application à la détection de défauts,
      6th Int. Conf. on NDT Methods, Strasbourg, oct 1986.

[3]   H. Morlo, H.A. Crostack
      Improvements of axial resolution in ultrasonic testing,
      6th Int. Cong. on NDE in Nuclear Ind., Zurich, dec 1983.

[4]   D. Kishoni
      Pulse shaping and extraction of information from ultrasonic reflections in
      composite materials,
      NATO-ASI F44, Signal processing and pattern recognition in NDE of materials,
      C.H. Chen Ed., 1988.

[5]   D. Kishoni
      Application of digital pulse shaping by least squares method to ultrasonic signals in
      composites,
      Prog. Quant. Non Destr. Eval., Vol 5, D.O. Thomson and D.E. Chimenti Eds.,
      Plenum Press, 1986

[6]   M. Carpentier
      Radars, bases modernes,
      Masson

STATISTICAL METHOD TO EVALUATE THE AMPLITUDE OF CYCLICAL NOISY SIGNAL IN IR THERMOGRAPHY

Comparing temporal energy distribution of noise with that of noisy signals allows the evaluation of a periodical noisy signal amplitude

S. OFFERMANN, E. MERIENNE, J L. BEAUDOIN
Laboratoire d'Energétique et d'Optique (G.R.S.M.)
Faculté des Sciences
B.P. 347
F-51100 REIMS CEDEX

ABSTRACT. Noise rejection is of primordial importance in Thermoelasticimetry. The evaluation of stresses by means of the thermoelastic effect requires high accuracy measurement of the dynamic temperature changes of a sample under cyclic mechanical loading. One has chosen a statistical method to reject the noise of sinusoidal signals (quite important for a standard infrared camera (0,05 K)) and to estimate their amplitudes. Digital treatment of a series of thermographic images allows for the calculation of the temporal distribution of every pixel.
Without loading, these distributions are similar to the Gaussian distribution of a white noise. Under cyclical loading one obtains the distribution of the sum of sinusoidal signal and white noise. Therefore this distribution is wider than the Gaussian one. Its width is a function of the signal amplitude. Thus, comparing both distributions, it is possible to estimate the amplitude of the cyclic signal.

1. Introduction

Noise rejection is of great importance in many domains of physics, such also in Thermoelasticimetry.
The relationship between stress change and the induced temperature change is given for a homogenous elastic solid under adiabatic conditions by the following relationship (2 - 6):

$$\Delta T = - K \cdot T \cdot (\sigma_1 + \sigma_2 + \sigma_3)$$

with σ_i : principal stresses; K: thermoelastic constant ($K = \alpha/(\rho.c_p)$).

The expected temperature changes are very small (metals: 0,2-0,3 K for the elastic limit (6)).
The proper noise of a standard infrared camera being rather important (0,05 K), it is consequently primordial to reject the noise to reach a sufficient accuracy.
The usually used method of Lock-In-Detection, correlating the signal of measurement with the reference signal of loading and working with a very good accuracy, has some inconveniences which have lead us to develop a statistical method for noise rejection:

    - the need of a reference signal

    - an IR camera is a scanning system. The level of each pixel of an image is consequently obtained in variable batch mode. This implies a different treatment of each point of the image.

    - our numerical system can only store 125 images with a well known temporal relationship of 40 ms between two images, what is not sufficient for a Lock-In-Detection.

X. P. V. Malague (ed.), Advances in Signal Processing for Nondestructive Evaluation of Materials, 43–48.
© 1994 *Kluwer Academic Publishers.*

2. Statistical method to evaluate the amplitude of cyclical noisy signal

2.1. THEORY OF THE METHOD

Digital treatment of a series of thermographic images allows for the calculation of the temporal distribution of every pixel.Without loading the temporal distribution of a pixel is similar to the gaussian distribution of a white noise (Figure 1).

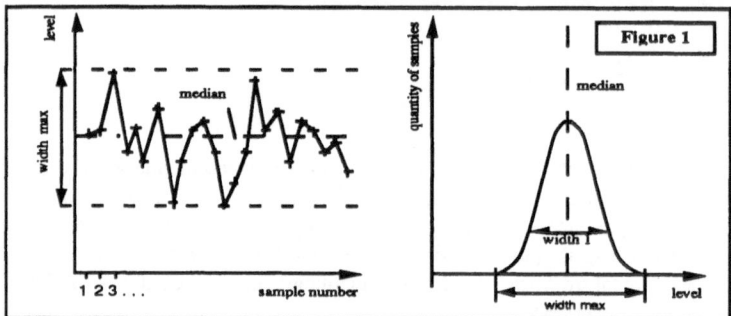

The distribution of a sinusoidal signal without noise has the following form (Figure 2):

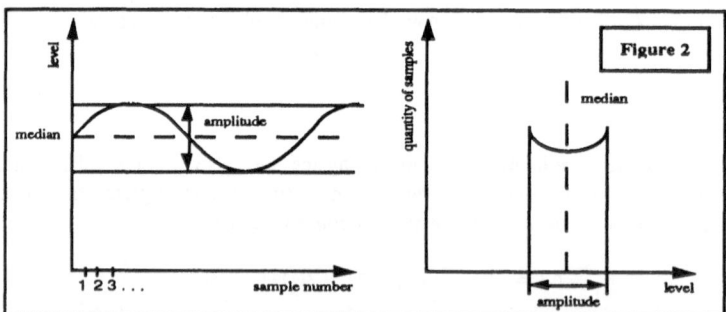

The distribution of noisy signal is wider than the distribution of noise and the distribution of the sinusoidal signal (Figure 3).

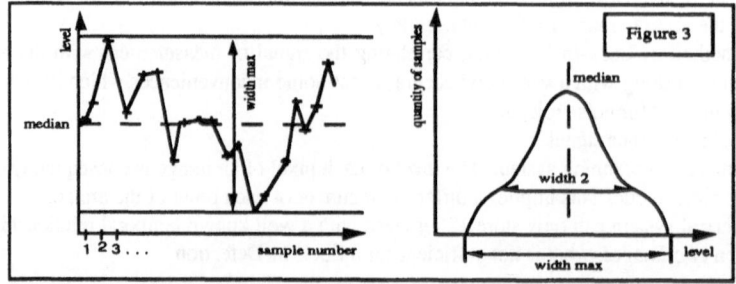

Thus, comparing the distributions of noise and noisy signal, it should be possible to estimate the unknown width of the sine distribution.

Analytical reconstruction of the sine distribution using probability matrix calculation gives only correct results in the limite case of a great number of measures. In the case of small series, one use estimation methods. A good estimation has two important qualities: the median of a series of estimations must be the real value and the standard deviation of the estimation must decrease for an increasing number of measures (1).

Two different methods of estimation have been developped:

2.1.1. *Estimation by means of the width of the distributions of noise and noisy signal (=> Fig. 1-3)*:

$$\text{Signalamplitude} = (\text{width 2}) - C_1 \cdot (\text{width 1})$$

2.1.2. *Estimation comparing the moments of the distributions*

The moment M of a distribution can be obtained by the following equation:

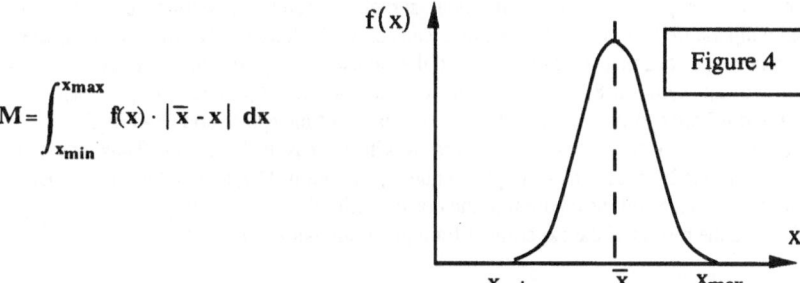

$$M = \int_{x_{min}}^{x_{max}} f(x) \cdot |\overline{x} - x| \, dx$$

For a discrete series of n measures $(x_1, x_2, ..., x_n)$, the relationship between the values of $x_1 - x_n$ and the moment M is given by:

$$M = \sum_{i=1}^{n} |\overline{x} - x_i| \qquad \text{with} \qquad \overline{x} = \frac{1}{n} \sum_{i=1}^{n} x_i$$

The following relationship between the moments of noise and signal,

$$M_{noisy\ signal}^2 = M_{noise}^2 + M_{signal}^2 + C_2$$

allows for the estimation of the moment of the sine distribution and consequenly for the determination of the signal amplitude.

2.2. COMPUTER SIMULATION

A program has been realized, which calculates in a first step a white noise with variable standard deviation, a sinusoidal signal with the amplitude A_{real} and the superposition of both. In a second step the distributions of noise and noisy signal are calculated and the amplitude $A_{estimated}$ of the signal is estimated by means of the two estimations described in 2.1.1 and 2.1.2.

Comparison between A_{real} and $A_{estimated}$ indicates the performance of the estimations and allows for the determination of the parameters C_1 and C_2.

2.3. VALIDATION OF THE METHOD WITH A SINGLE POINT EXPERIMENT

Figure 5 shows the experimental arrangement which has been chosen:

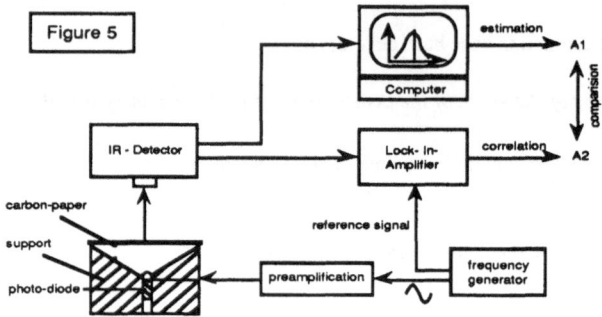

The photo-diode is polarized by a sinusoidal signal. Its radiation warmes up the carbon-paper. The sinusoidal temperature change is measured by means of an IR-detector. The measured signal is processed by two different ways: classical measurement of the sinusoidal signal amplitude by means of a Lock-In-Detection and digitizing on 8 bit (256 levels) to estimate the amplitude using the statistic method. The results A1 and A2 are compared to assert the performance of the estimation (=> Fig. 5).

The signal amplitude of the photo-diode decreases with increasing frequency. Thus, one can easily variate the signal amplitude by means of a simple frequency variation. Using very high frequencies, the periodic amplitude is quasi-zero and the creation of the noise distribution is quite simple.

Figure 6 shows the results of the experiment for a great number of measures (2500):

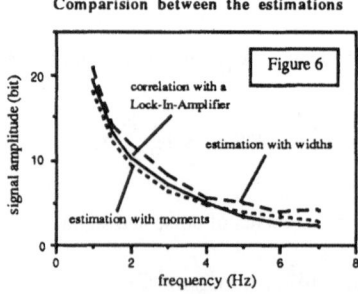

The comparision of estimation results with correlation results shows the validity of the estimations. The estimation by means of the moments of the distributions is a little bit better than the estimation using their widths.

For small series the estimated results have an uncertainty which decreases with an increasing number of measures. For example the standard deviation for a series of 25 measures is 3 bits and for a series of 100 measures it is only about 1,5 bit.

2.4. APPLICATION OF THE METHOD TO A STANDARD IR - CAMERA

After validation of the method for one point (=>2.3.) it has been applied to a full image, replacing the punctual IR-Detector by a standard IR-Camera (AGEMA 880 LW). A series of 100 images is digitized on 8 bit (256 levels) and stored into a computer. A program has been developped estimating the signal amplitude of each pixel. In the aim of testing the performance of the method to recognize forms, a mask of aluminium-foil in the form of a cross has been put between the photo-diode and the carbon-paper (Fig. 7).

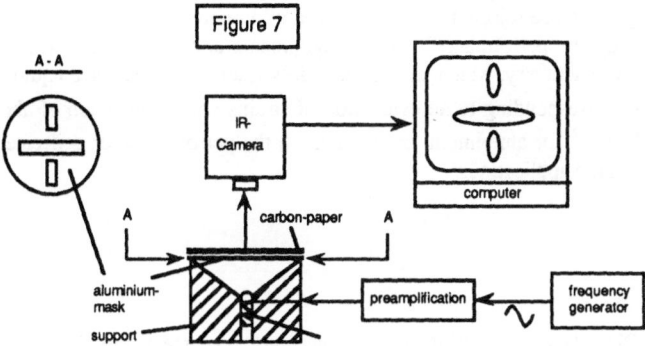

Figure 7

Figure 8 shows a 3D representation of the obtained amplitude image.

Figure 8

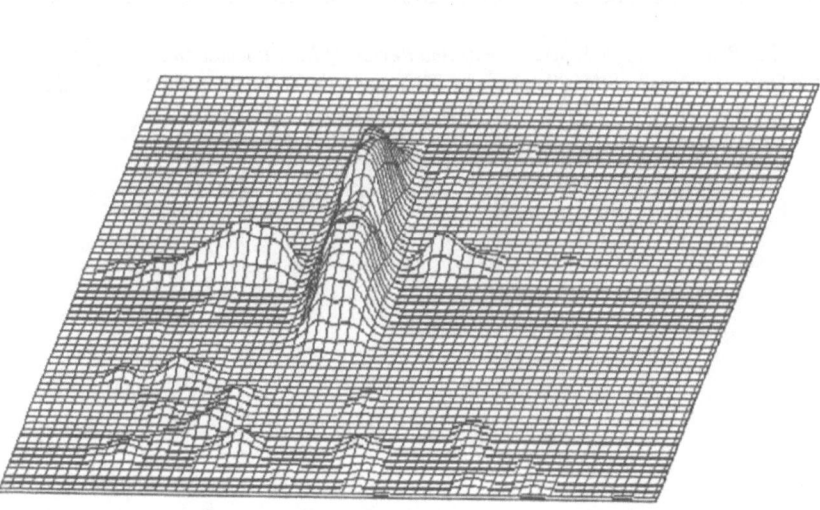

Cyclical temperature changes of 4 bits can still be observed. This is corresponding to a thermal resolution of 0,04 K. Using a higher number of images like 400 or 1000 and realizing eventually a post-treatment of the amplitude images, a thermal resolution better than 0,01 K can certainly be achieved. A 12 bit digitizer will also attribute to put up the performance of the method.

48

3. Conclusions

The increasing performance of computers make possible a fast statistical treatment of full images. Thus the noise of a cyclical signal can be rejected without the need of a reference signal. The application to standard infrared cameras put up the performance of these equipments and offers the possibility for working in domains, like Thermoelasticimetry, which have been mainly reserved to dedicated equipments. The thermal resolution of 0.01 K, corresponding to an evaluation of strains with a precision of about 10 N/mm^2 for steels and about 4 N/mm^2 for aluminium, does not reach the performance of dedecated instruments, but may be sufficient for some applications.

References:

1) G. BAILLAGEON, *"Méthodes statistiques de l'ingénieur"*, Volume 1, Les Editions SMG, TROIS RIVIERES, QC, CANADA, 1990
2) ROCCA & BEVER, *"The Thermoelastic Effect"*, Trans. AIME, 1950, 188, pp. 327 - 33
3) BIOT M.A., *"Thermoelasticity and Irreversible Thermodynamics"*, J. Appl. Phys., 27, n° 3, pp. 240-253, march 1950
4) BREMONT P. *"Développement d'une Instrumentation IR pour l'étude des structures mécaniques"*, Thèse Univ. AIX-MARSEILLE, nov. 1982
5) BLANC R. H. and GIACOMETTI E. *"Infrared Radiometry Study of Thermomechanical Behaviour of materials and structures"*, First.Int.Conf. of Stress Analysis by Thermoelastic Techniques, LONDON, 1984
6) WEBBER J.M.B.,*"Principles of IR Measurements and Review of Instrumentation Techniques for Thermoelastic Stress Analysis"*, Stress Analysis by Thermoelastic Techniques, Proc. SPIE 731, 1987

Quasi Frequency Diversity Processing of Ultrasonic Signals - A Review

T. Stepinski, L. Ericsson, B.J.I. Eriksson and M.G. Gustafsson
Uppsala University, Department of Technology, Circuits and Systems
P.O. Box 534,
751 21 Uppsala,
Sweden

ABSTRACT. Signal processing techniques capable of improving signal to noise ratio in the presence of strong backscattered grain echoes interfering with the echo reflected from material defects are considered in this paper. A number of algorithms belonging to a class of Quasi Frequency Diversity (QFD) techniques are presented and compared. The best known method in this class is the Split Spectrum Processing (SSP), which has been proven successful in suppressing material noise. The SSP is based on the idea of expanding the received ultrasonic signal into a set of split signals corresponding to a number of narrow frequency bands. However, performance of the SSP is highly dependent upon number of split signals, their center frequencies and bandwidths, that may be difficult to estimate properly. Robust and adaptive QFD techniques are proposed as a solution to this problem. Two robust QFD techniques based on removal of narrow frequency bands from the ultrasonic signal spectrum are presented in the paper. The first one, referred to as Cut Spectrum Processing (CSP), does not require any spectral parameters at the price of an increased computational burden. The resultant CSP algorithm includes one single parameter, which can be easily tuned interactively by the operator. The second technique, referred to as Fragment Spectrum Processing (FSP) is based on an idea of fragmentation of the signal spectrum by using a set of irregular comb filters, i.e., filters having a frequency response consisting of a large number of alternating stop- and pass-bands. Efficiency is accomplished by using a set of the comb filters with virtually different (possibly orthogonal) frequency responses. The adaptive QFD technique presented here is based on the fact that the polarity thresholding SSP can be formulated as a multilayer perceptron artificial neural network operating in feedforward mode with binary neurons and bipolar input signals.
The SSP and QFD algorithms described above have been implemented as a software package for PC capable of processing ultrasonic B-scans and evaluated using real ultrasonic signals originating from cracks and notches in welded reference blocks typical for nuclear power plants.

1. Introduction

Ultrasonic pulse-echo inspection of coarse materials, i.e., materials characterized by grains sizes of the order of the ultrasonic waves, is often limited by the presence of backscattered echoes from the material structure, known as material noise [1, p. 451]. Since the frequency spectrum of the received ultrasonic response signal is mainly determined by the relatively narrow-band transducer characteristics, the spectra of the flaw echoes and the material noise overlap to a large extent. Therefore, application of conventional linear filters (such as low- or band-pass filters) is in general not effective in attenuating this type of noise. However, it is well known that the material noise can be suppressed by utilizing its noncoherent nature by means of frequency diversity techniques [2]. The main idea behind frequency diversity, which originates in radar applications, is uncorrelating the multiple backscattering by simultaneously transmitting pulses using two or more channels centered at different frequencies, or by shifting the transmitted frequency between the pulses.

49

X. P. V. Malague (ed.), Advances in Signal Processing for Nondestructive Evaluation of Materials, 49–58.

Similar results can be obtained using a class of signal processing techniques referred to as Quasi Frequency Diversity (QFD) Processing. The QFD algorithms employ one wide-band ultrasonic signal which is then expanded into a set of signals by manipulating the spectrum of original ultrasonic signal. The most renowned QFD algorithm is the Split Spectrum Processing (SSP) [2,3], which has been known for about one decade. The SSP has been demonstrated successful in material noise reduction, provided that certain prerequisites are fulfilled. Unfortunately, our experience has shown that these conditions in many cases cannot be satisfied, which may result in losses of flaw echoes.

This paper is concerned with the presentation and evaluation of various QFD algorithms developed recently for ultrasonic inspection of welds in nuclear power plants. After a short introduction a general structure of the QFD algorithms is presented in the paper, followed by a brief discussion of the SSP as a special case of the QFD concept. Two robust QFD algorithms, Cut Spectrum Processing (CSP) and the Fragment Spectrum Processing (FSP), are presented shortly. Theory of an adaptive QFD algorithm based on neural network is presented in the third part of the paper. Operation of the algorithms discussed in the paper is illustrated using an example of B-scan acquired during weld inspection.

2. Quasi Frequency Diversity Algorithms

A general class of signal processing algorithms which we refer to as Quasi Frequency Diversity techniques share the common structure depicted in Figure 1. The term "quasi" is used because they employ a single transmitted (broad-band) ultrasonic pulse. The methods basically consist of two steps, signal expansion and target echo extraction. The expansion process generates the desired frequency diverse signals by processing the spectrum of the received ultrasonic signal. Generally, each of the signals is generated by manipulating certain parts of the spectrum using, e.g., band-pass or band-stop filters. The idea is to produce a collection of signals where the coherent target echo information is common to all signals, while the material noise exhibits differences due to its noncoherent properties. The extraction algorithm is employed to the signals in order to detect the presence of target echoes embedded in the material noise. The extracted signal may be used directly as an output or it may be multiplied by the original ultrasonic signal, see Figure 1.

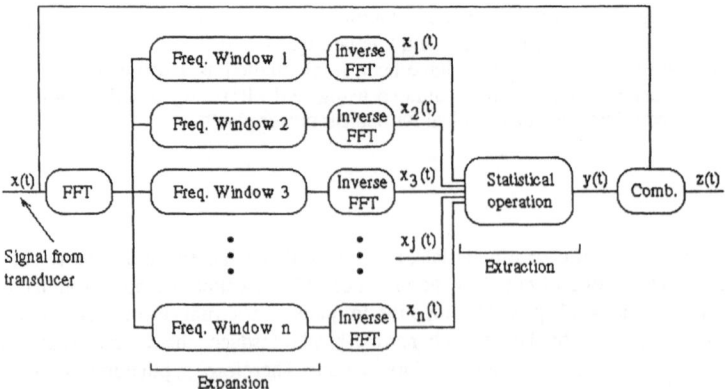

Fig. 1 - General structure of QFD processing algorithm

2.1. SPLIT SPECTRUM PROCESSING

2.1.1. *Expansion Algorithm*. The SSP expansion is typically performed by multiplying the spectrum of the digitized ultrasonic signal by a number of identical and equally distributed Gaussian windows. The split signals are then normalized with respect to amplitude in order to compensate for the frequency response of the ultrasonic transducer. Due to the coherent properties of target echoes, all the split signals should contain similar information at those time instants when the target echo appears in the signal, provided that all the windows have been located within the frequency band of the ultrasonic signal. On the other hand, the material noise should be more or less decorrelated between the split signals. Based on these facts one can expect that a strong correlation between the amplitudes of split signals should be observed at those time instants when the target echo is present in the signals. Thus the material noise can be reduced by introducing a measure of correlation between the split signals.

2.1.2. *Extraction Algorithms.* The target extraction algorithms are used to process the split signals in order to measure the coherence, i.e. the degree of constructive interference, between the split signals at each discrete time instant $t = kT$, where T denotes the sampling period. The use of linear operations, e.g., averaging, for the extraction basically reconstructs the original ultrasonic signal, i.e., only minor noise reduction can be achieved, [2]. Consequently, the extraction algorithms usually used with the SSP are non-linear operations, such as amplitude minimization [2, 3] or polarity thresholding [3, 4, 5].

Minimization is a classical way of extraction based on the observation that for time instants where the destructive interference is present the split signals amplitudes vary much and some of them is very likely to be close to zero. On the other hand for time instants where the constructive interference is observed the amplitudes $|x_i (kT)|$ in Figure 1 are rather constant. The minimization sends the minimum amplitude of the split signals at each time instant to the output.

Polarity Thresholding is another way of taking advantage of the fact that the material noise results in uncorrelated split signals by considering their polarities at each time instant. In the absence of a target reflection at a certain time instant (destructive interference) sign reversals among the split signals will probably be observed. Sign reversal probability will increase with the number of split signals, n. A binary sequence (gating signal), is constructed, which takes the value 1 at a certain time instant if all the split signals have the same sign and takes the value 0 otherwise. The output $z(kT)$ of the polarity thresholding algorithm is combined as a product of the gating signal and the original ultrasonic response signal $x(kT)$, cf Figure 1.

2.1.3. *Robustness of SSP.* If the SSP expansion is performed in such a way that target echo information is not present in at least one of the selected frequency bands (Gaussian windows) the extraction algorithm may cancel a coherent, that is relevant echo. There are two main cases, where a lack of target echo consistency can be observed: if at least one of the Gaussian windows is located outside the frequency band of the ultrasonic signal or if the target echo is more or less completely cancelled by a very intensive material noise level in some of the frequency bands. It is obvious that in the first case the respective window will contain mainly receiver noise and both the minimization and the polarity thresholding will probably fail. This problem can be solved by utilization of automatic spectral estimation prior to the SSP processing [6, 8]. The other case has been observed during our evaluation of the SSP using a large number of real ultrasonic signals, [6]. Because of the frequency dependent nature of the material noise, it is likely that its energy reaches such levels in some local frequency bands, especially in the upper part of the signal spectrum, that the target echo is more or less completely cancelled, cf [7]. SSP extraction based on robust statistics has been proposed as a remedy for such situation which occurs in general for a high level of backscattering in ultrasonic signal, cf [9].

Below we present two QFD algorithms, developed recently in our group, where the expansion process has been modified in order to increase the reliability in the above mentioned cases.

2.2. CUT SPECTRUM PROCESSING

The above discussion leads to the conclusion that although the SSP expansion using Gaussian band-pass filters satisfies the requirement of frequency diversity, there is no guarantee that received coherent information will be present in all, or even in the majority of the split signals. Therefore, one possible way is to design a robust expansion algorithm, i.e., an algorithm producing a set of signals all containing coherent information, provided that the original signal contains that information.

2.2.1. *Expansion Algorithm.* In the algorithm, which we refer to as Cut Spectrum Processing (CSP), the objective mentioned above is met by using a concept which may be thought of as the inverse of SSP, i.e., band-stop filters are used for signal expansion instead of band-pass filters. In order to understand the

CSP concept let us assume that some a priori information concerning frequency range containing spectrum of the received ultrasonic signal is available. In the worst case we can simply exploit the entire frequency range running from zero to the Nyquist frequency. Then the sum composed of all the split signals (obtained using SSP band-pass filters) would definitely contain coherent information, if there is any. Furthermore, any sum excluding one of the split signals would also embrace approximately the same coherent information as any other such aggregate. On the other hand, noncoherent information, such as material noise, would be slightly different in various aggregates, depending on whether the constructive or destructive interference occurs. Qualitatively, excluding one of the split signals from the summation is equivalent to band-stop filtering of the ultrasonic signal.

Although the band-stop filters could be chosen to be Gaussian shaped as in the case of SSP the rectangular windows are preferable due to implementation easiness. Our experiments have verified that they operate very well for real ultrasonic signals. Concerning the width of the band-stop windows, the experimental results indicate that the best choice should be one frequency sample (bin). Thus, the CSP expansion algorithm may be implemented by equating the consecutive bins to zero in the Fourier spectrum of the ultrasonic signal. The cost for the superior robustness obtained by this algorithm is the relatively small signal variation caused by the noise components, compared to the SSP expansion. However, the effects of noncoherency can be increased without affecting the coherent information, by reversing the phase of the complex valued spectrum bins instead of resetting them to zero. Thus, the final CSP expansion algorithm includes two steps: estimation of a relevant frequency band that encompasses the entire frequency spectrum of the ultrasonic signal, and expansion of the ultrasonic signal by changing the signs of the bins included in the frequency range, one at a time. The expansion will result in a number of signals equal to the number of spectrum bins contained in the relevant frequency range. The features of the expanded signal can then be utilized by a suitable extraction algorithm, e.g., the one introduced in the next section.

2.2.2. *Extraction Algorithm.* The CSP expansion technique, which is entirely different from that used in the SSP, requires also the extraction algorithms to be different from the minimization or polarity thresholding commonly used with the SSP. An extraction algorithm using a quotient of the minimum and maximum amplitude found for the n expanded signals for each time instant has been proposed for the CSP, [9]. The quotient is then compared with a threshold value q set by operator, then a binary gating signal similar to that in the polarity thresholding SSP is obtained:

$$y(kT) = 1; \quad \text{if } x_{min}(kT) / x_{max}(kT) > q$$
$$y(kT) = 0; \quad \text{otherwise}$$

where: n - number of split signals
                     q - threshold for coherence determination, $0 < q < 1$.

The output of the processing, $z(kT)$, is constituted by the product of the gating signal $y(kT)$ and the original ultrasonic response signal $x(kT)$, cf Figure 1. Using CSP expansion together with the above extraction algorithm reduces the total number of settings to one parameter q referred to as squelch. Setting $q=0$ results in no action, while $q=1$ exterminates everything. A suitable value may be chosen using operators experience of visual inspection. Practical operation of the Cut Spectrum Processing is illustrated by the example presented below.

2.3. FRAGMENT SPECTRUM PROCESSING

Fragment Spectrum Processing (FSP) is a robust QFD technique which is less computationally intensive than the CSP [10]. The FSP employs more sophisticated filters than the CSP for the expansion, that increase differences between signals in order to reduce number of expanded signals for broad frequency bands.

2.3.1. *Expansion.* The FSP expansion is based on the idea of fragmenting the broad-band ultrasonic signal by using a set of irregular comb filters, i.e., filters having a frequency response consisting of a large number of alternating stop- and pass-bands. Since the pass-bands are distributed throughout the entire spectrum the algorithm is independent of the actual spectral location of the ultrasonic signal. To make the algorithm both robust and efficient the comb filters should meet certain requirements. To achieve robustness, the widest stop-bands should be narrow enough to ensure that target echo information is present in all expanded signals. Effectiveness is accomplished by making the comb-filters as "different" as possible, that is, by ensuring that there is as little spectral overlap between all pairs of frequency windows as possible. This should lead to low correlation between the noise in all pairs of expanded signals. A natural set of comb filters complying with these requirements is obtained by using a collection of sequences known as Walsh sequences as frequency responses. The Walsh sequences are binary valued and orthogonal, with maximum number of sequences equal to the sequence length, see [10] for details.

In order to compensate for variations in transducer frequency response and ultrasonic cross section of the flaws the expanded signals should be normalized, typically with respect to energy or amplitude.

2.3.2. *Extraction*. Since very robust expansion has been used in FSP, the simple and effective minimization procedure should be sufficient for extraction. Minimization algorithm similar to that used in SSP has been successfully adopted for FSP in order to extract the coherent echoes.

The proposed FSP algorithm is expected to achieve the performance close to that of a well-tuned SSP. It should be emphasized however that the FSP is robust against variations in signal spectra, while the SSP must be repeatedly tuned for each processed A-scan. Only a rough estimate of the signal bandwidth is required for using the FSP, in order to set the subsequence length and the maximum allowed length of the stop-bands.

3. Adaptive Split Spectrum Processing

3.1. NEURAL NETWORKS AND POLARITY THRESHOLDING SSP

The neural networks are generally known to be adaptive, massively parallel, nonlinear systems with function and architecture inspired from biology. The common feature of all types of neural networks is that they achieve good performance due to interconnection of a number of simple nonlinear computational elements. The most common are so called supervised neural networks that can be trained using labeled training examples. Here we will consider the most well known neural network - the multilayer perceptron (MLP) which belongs to the class of supervised, feedforward networks. The most accepted method for training the MLP is backpropagation algorithm. Neural networks which have been applied recently to solve many technical problems were presented in a number of overviews and textbooks, e.g. [11] - [13].

The outputs of n band-pass filters in the split spectrum processing (expansion) are to be combined using some statistical algorithm (extraction) in order to suppress the interference noise effects, see Figure 1. The presented above polarity thresholding method is based on the idea that the filter outputs are coherent at the time instants where the echo is present in the signal. When the echo is missing, the output signals should show more or less random values. If, at a certain time instant, all the responses are of the same sign, i.e., there is considerable coherence degree between various frequency bands, the original broadband ultrasonic signal is presented at the output. If one or several signals have different signs from the rest, this indicates that the signals contain significant levels of noise, and the output is set to zero. The algorithm is non-robust in a sense that it requires a high degree of correlation or coherence (same polarity) between the filter signals to let the original signal through. In a situation where some filter outputs only contain random noise, the algorithm yields zero, independently of the rest of the filter outputs. Operation of the polarity thresholding extraction can be presented in a form of scheme shown in Figure 2.

Obviously, this scheme resembles a MLP with binary neurons, and the role of each neuron is very clear. The first neuron in the first layer is a feature detector that fires (outputs one) only if essentially all ($n-m$, $n >> m \geq 0$) the network input signals (filter outputs) are positive. Similarly, the second neuron is a feature detector that fires only if essentially all the input signals are negative. The output neuron implements a logical OR gate, i.e., it fires only if one of the input neurons detects a sign coherence of the input signals. Thus, the polarity thresholding method discussed here, can be treated as a simple two layer perceptron and possibly generalized by introducing variable weight coefficients and trained in order to achieve a desired performance.

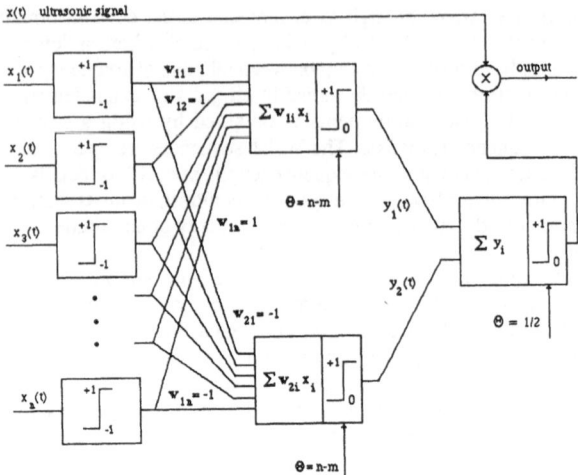

Fig. 2. - The polarity thresholding method in a multilayer perceptron form.

3.2. AN ADAPTIVE POLARITY THRESHOLDING ALGORITHM

The relationship between the MLP and the polarity thresholding method illustrated by Figure 2, opens a way for some straightforward generalizations of the polarity thresholding method. These include: introducing variable weights, replacing the binary neurons with sigmoids (neurons with graded outputs) and using raw data instead of binary data.

The introduction of variable weights (w_{1i}, w_{2i} in Figure 2) amounts to taking into account the relative importance of different filter outputs. It is expected that outputs corresponding to frequencies where the magnitude of the signal spectrum vanishes should be less informative than outputs near the center frequency. Thus, weight values corresponding to uninformative filter outputs should become small during the learning (adaptation) process.

A hidden neuron is nothing but a linear discriminant in the input space which can make a decision unless the weighted input sum equals the threshold value. This will happen for a certain set of input values. Geometrically, this set forms a hyperplane in the input space. A normal vector to this decision plane is given by the weight values of the neuron and the threshold determines the shortest distance from the plane to the origin. Adjusting the weights (and thresholds) of the two hidden neurons corresponds to movements of two hyperplanes in the input space. In this context, it is easy to see that the conventional polarity thresholding method (working on binary data !) corresponds to a MLP with the normals of the decision planes of the hidden neurons fixed and parallel to the $(1,1,1,...,1)^T$ direction in the n-dimensional input space.

Introducing the sigmoid means replacing the hard limiter with a differentiable function, which enables using the backpropagation learning algorithm for adjusting the weights. Further, the network will produce graded outputs on the interval [0,1] and thus will be able to indicate degrees of certainty. It is important to note that the decision plane produced by a sigmoid neuron is exactly the same as for the binary neuron. The only difference is the response across the decision plane which is less abrupt for the sigmoid neuron.

By making decisions based simply on the polarity of the outputs, obviously a lot of information remains unused. Therefore, as a last extension of the conventional polarity thresholding method, the original outputs are fed to the network, not only their signs.

Other possible modifications or generalizations are of course possible. Since the polarity thresholding method only uses the signs of the outputs it can be interpreted, at each time instant, as a sorting/comparison process where the output samples are sorted followed by a comparison of the smallest and largest (minimum and maximum) output samples. If they are of the same sign (polarity), the algorithm outputs a one, otherwise the output is set to zero. This suggests that any more sophisticated operation on the outputs than the simple comparison of the maximum and minimum samples is a generalization of the original polarity thresholding method.

Concluding, we presented here a natural way to extend the polarity thresholding method based on its similarity to the two layer MLP with two hidden and one output neurons. Introducing variable weight coefficients enabled the weight adaptation using the backpropagation learning algorithm. This resulted in an adaptive SSP algorithm which can be trained using a training set of input signals and target output signals, i.e., desired net responses given a certain set of input signals.

Experiments were made in order to compare the performance of this algorithm with the conventional polarity thresholding SSP. Various versions of the adaptive SSP were trained using real ultrasonic signals acquired for reference flaws in austenitic steel, see [14], [15] for details. The results have shown that the adaptive algorithm performs as well as, or significantly better than, the conventional polarity thresholding method. At the same time the algorithm is more robust to the parameters of the split spectrum expansion (such as number of filters and their bandwidth) than the conventional polarity thresholding.

It should be noted that the above presented algorithm is the first step towards adaptive quasi frequency diversity methods based on neural network concepts. Only the extraction has been made adaptive by adding the MLP while the expansion remains virually the same as for any SSP scheme. The weight coefficients obtained during backpropagation can decrease influence of the less informative frequency bands within the frequency range chosen arbitrarily by the operator. The structure of the extraction process in the SSP algorithms is practically limited to a bank of equally distributed identical band-pass filters and only a number of filters and their bandwidth (overlap) can be chosen. As it was shown when discussing robust QFD techniques, various filter types can be successfully used for the extraction. Therefore a fully adaptive QFD should optimize both the extraction and the expansion parameters according to the particular application (i.e., transducer - material combination).

4.   Experiments

Operation of various QFD algorithms presented above is currently verified for materials and flaws encountered in nuclear power plants in colaboration with the Swedish Nuclear Power Inspectorate and the owners of the Swedish Nuclear Power Plants. The QFD program developed for this project can be run on 386 or 486 PCs and enables batch processing of A- and B-scans. The program enables comparing the performance of various QFD algorithms for particular ultrasonic image, transducer, flaw type, etc. The program has been also recently used for processing of ultrasonic B- and D-scans of welments made in alloy steel blocks provided by TWI in Abington, GB.

A reference welded block made of stainless steel part welded to a carbon steel part (see Figure 3) has been inspected using 2MHz, 70°, L-wave ultrasonic transducer. The block was 80mm thick and contained a number of notches in the weld. An example B-scan obtained during inspection of the 8mm deep notch in the weld is presented in Figure 4. The B-scan consists of 140 A-scans, each of them made of 1024 points sampled with the sampling rate 25 MHz.   Classical SSP algorithms, i.e., minimization and polarity thresholding, that are available in the QFD program are provided with the facility for automatic estimation of the spectrum of the ultrasonic signal. During the estimation the amplitude spectrum is approximated by means of a Gaussian function defined by a center frequency and a bandwidth. Based on these parameters the SSP settings, i.e., number of filters and their bandwidth are to be evaluated and used for the SSP expansion in automatic mode, see [6] for details. When processing B-scan images the estimation procedure is operated each time when a new A-scan is processed.

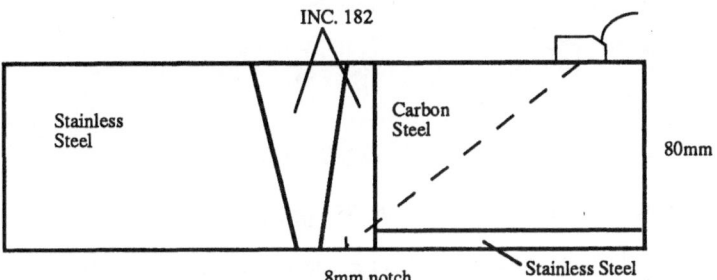

Fig. 3 - One of the reference blocks used for verification of the QFD technique.

Result of processing the B-scan from Figure 4 by means of the minimization and polarity thresholding are shown in Figures 5 and 6 respectively. The mean estimated center frequency for the processed B-scan was 1.9 MHz and the signal 3-dB bandwidth 0.7 MHz. The expansion was made using 29 band-pass filters both for the minimization and the polarity thresholding.

Fig. 4 - Example ultrasonic B-scan of the reference block acquired during inspection by means of 2MHz, 70° transducer.

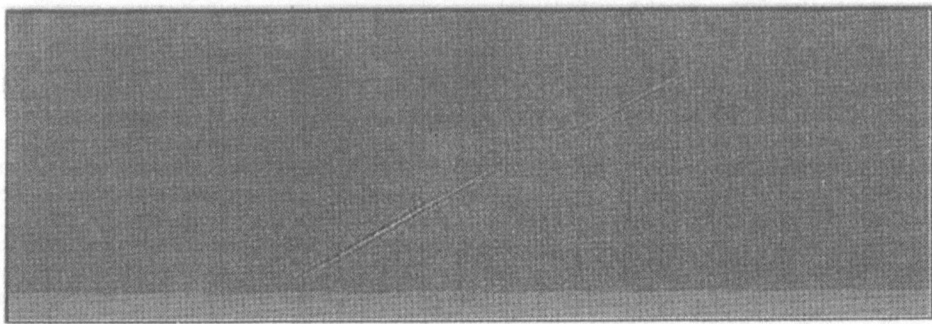

Fig. 5 - B-scan from Figure 4 processed by means of the SSP minimization.

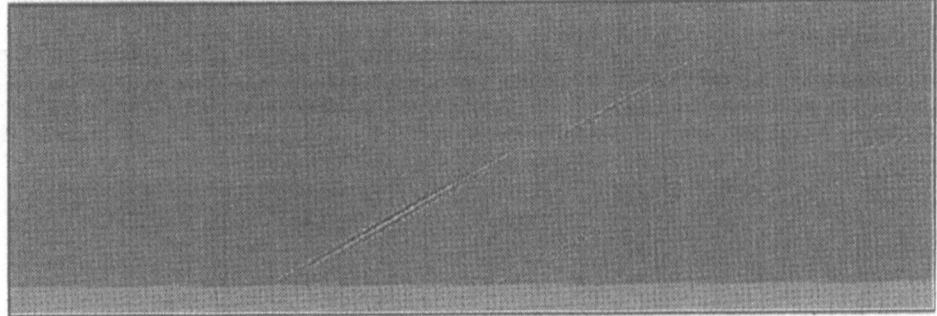

Fig. 6 - B-scan from Figure 4 processed by means of the SSP polarity thresholding.

The CSP technique does not require a precise spectrum estimation and a rough estimation of a frequency range where the ultrasonic signal is present is sufficient. The CSP has only one parameter (i.e., the threshold q for coherence determination) which is set by operator. It is worth noting that change of this setting does not require reprocessing of the CSP expansion, i.e., the result can be presented immediately since only thresholding operation of the extraction is performed again. Operation of the CSP with the expansion made in the frequency range 1 to 3 MHz and the extraction for two different values of the threshold q is illustrated by Figure 7.

Operation of the FSP technique which does not require any settings is illustrated by Figure 8.

Fig. 7 - B-scan from Figure 4 processed by means of the CSP with two values of the threshold: q = 0.3 (upper image) and q = 0.7 (lower image).

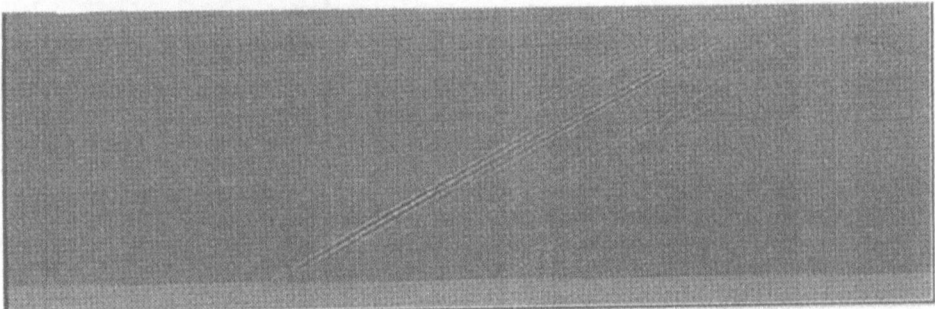

Fig. 8 - B-scan from Figure 4 processed by means of the FSP.

Analysis of Figures 5 to 8 shows that all the algorithms perform well and suppress a great deal of an interference observed in the original B-scan. The CSP works similarly to the polarity thresholding SSP, i.e., it is capable of removing the background noise completely for higher values of the threshold parameter producing "sterilized" images. This effect can be explained by the fact that both algorithms employ similar extraction methods with a hard limiter that generate a binary gating signals at the output. The SSP minimization and FSP, using minimization for the extraction, result in "soft" and natural images containing some amount of spatially uncorrelated background noise.

5.   Conclusion

Well known split spectrum processing technique has been presented in a wider perspective as a member of a general class of Quasi Frequency Diversity algorithms aimed at suppressing of grain noise during ultrasonic inspection of coarse materials. The issue of the SSP robustness, i.e., its sensitivity to settings which in some cases may result in loss of the relevant ultrasonic echoes, has been addressed. One way to solve this problem which has been proposed here is developing the QFD algorithms using robust parameter free expansion methods. Two such algorithms have been presented, the Cut Spectrum and Fragment Spectrum Processing. The second way is making the SSP adaptive by introducing variable parameters that can be estimated during supervised teaching using a set of training signals. One possible adaptive technique based on replacing a polarity thresholding algorithm with a neural network for extraction of coherence information contained in the split signals has been presented. A generalized polarity thresholding algorithm shed a new light on the SSP technique since it enabled using, common in pattern recognition, geometrical models for understanding and illustrating the SSP operation.

Verification of the proposed QFD techniques has been undertaken using specially developed software and a number of test blocks. The preliminary results of the verification illustrated here by processing of an example B-scan are very promising and show that the robust CSP and FSP achieve the performance similar to that of optimally tuned SSP techniques but does require any tuning.

Acknowledgements

This work was supported by both the Swedish Nuclear Power Inspectorate under contract # 13.2-375/91 and by the Swedish National Board for Industrial and Technical Development (NUTEK), under contract # 89-02754P. We would also like to thank ABB TRC in Stockholm for providing the ultrasonic data.

References

1.   Krautkrämer, J., H. Krautkrämer. Ultrasonic Testing of Materials, 4:th Ed., Springer Verlag (1990)
2.   Newhouse, V.L., N.M. Bilgutay, J.Saniie and E.S. Furgason, Flaw-to-Grain Echo Enhancement by Split-Spectrum Processing, Ultrasonics, 20, 59-68 (1982)
3.   Rose, J.L., P. Karpur and V.L. Newhouse, Utility of Split-Spectrum Processing in Ultrasonic Nondestructive Evaluation. Materials Evaluation, 46, 114-122 (1988)
4.   Yeu, L. and Y. Chong-Fu, Two Signal Processing Techniques for the Enhancement of the Flaw-to-Grain Echo Ratio, Ultrasonics, 25, 90-94 (1987)
5.   Bilgutay, N.M., U. Bencharit, R. Murthy and J. Saniie, Analysis of a Non-Linear Frequency Diverse Clutter Suppression Algorithm. Ultrasonics, 28, 90-96 (1990)
6.   Ericsson, L., T. Stepinski, Signal Processing for Ultrasonic Testing of Anisotropic Materials. Swedish Nuclear Power Inspectorate, Report SKI TR 92:1 (1992)
7.   Karpur, P., P.M. Shankar, J.L. Rose and V.L. Newhouse, Split Spectrum Processing: Determination of the Available Bandwidth for Spectral Splitting. Ultrasonics, 26, pp 204-209 (1988)
8.   Ericsson, L. and T. Stepinski, Verification of Split Spectrum Technique for Ultrasonic Inspection of Welded Structures in Nuclear Reactors, European Journal of NDT, 2 (2), 56-64 (1992)
9.   Ericsson, L., T. Stepinski, Cut Spectrum Processing - A Novel Signal Processing Algorithm for Ultrasonic Flaw Detection, NDT&E Intern. 25 (2), 59-64 (1992)
10.   Ericsson, L., B.J.I. Eriksson and T. Stepinski, Efficient Robust Signal Processing for Ultrasonic Detection of Flaws in Coarse Materials, accepted for 7th Asian Pacific Conf. on NDT, Shanghai, Sept. 1993.
11.   R.P. Lippmann, An Introduction to Computing with Neural Nets, IEEE ASSP Mag., April 1987.
12.   R.P. Lippmann, Pattern Classification Using Neural Networks, IEEE Communications Magazine, November 1989.
13.   R. Beale and T. Jackson, "Neural Computing an introduction", Adam Hilger, 1990.
14.   M.G. Gustafsson, Adaptive Split Spectrum Algorithms and Theoretical Aspects of The Split Spectrum Technique, Report UPTEC 92 114R, Sept. 1992
15.   M.G. Gustafsson, T. Stepinski, Adaptive Split Spectrum Signals Processing Using a Neural Network, Research in NDE (Springer-Verlag) vol. 5, 1993, pp.51-70.

SPECIFIC FEATURES OF NDT DATA AND PROCESSING ALGORITHMS : NEW REMEDIES TO OLD ILLS

B. GEORGEL
Électricité de France
Direction des Études et Recherches,
département Surveillance, Diagnostic, Maintenance
6 quai Watier
F-78400 CHATOU, FRANCE

ABSTRACT. Non destructive testing data from in-service inspections have specific features that require the most sophisticated techniques of signal and image processing. Each step in the overall information extraction process must be optimised by using recent approaches such like data decomposition and modelisation, compression, sensor fusion and knowledge based systems. This can be achieved by means of wavelet transform, inverse problems formulation, standard compression algorithms, combined detection and estimation, neural networks and expert systems. These techniques are briefly presented through a number of Électricité de France applications or through recent literature results.

Introduction : Non Destructive Testing at Électricité de France

Électricité de France (EDF) is a national company in charge of production and distribution of electricity in France. In 1992 EDF had thirty-four 900 MW-reactors and twenty 1300 MW-reactors of the same nuclear technology (PWR's). To be exhaustive one must add several 1400 MW-reactors of a new generation ("palier N4") and numerous coal or fuel-fired power stations. This large electricity program makes EDF the biggest utility in the world and raises France's energy independence up to 50 % [1]. Electricity production is mainly nuclear (75 %).
 Non destructive testing plays a vital role in the overall maintenance program of EDF. Many types of components are concerned : steam generators (SG), casted elbows, vessel, pressurizer, turbine blades. In service inspection (ISI) is performed mainly by ultrasonics (US), eddy currents (EC) and gammagraphy (γ) [2]. Roughly speaking US is aimed at vessel and nozzles testing, EC at tubes and γ at nozzles, bent parts of the primary circuit and welds.

X. P. V. Malague (ed.), Advances in Signal Processing for Nondestructive Evaluation of Materials, 59–70.
© 1994 *Kluwer Academic Publishers.*

1. Stating the problem

This paper deals with the difficulty inherent in non destructive testing (NDT) data processing and the dramatic increase in performance that can be (and will be) achieved thanks to the new signal and image processing (SIP) techniques.

The idea that only sophisticated and complementary methods of information processing will make it possible to face the specific challenges of NDT has already been expressed in [11]. But space considerations in that paper precluded discussion of very interesting though more recent topics such as data modelisation and decomposition, compression, data fusion, cooperation betweeen numeric and symbolic representations.

So the main point of this paper is the following : there is no hope to solve the difficult problems of NDT by means of brute force techniques. As a first objective, **routine tasks** can be automatized in order to avoid operators' fatigue and errors. This is not easy to achieve, either because of the amount of data or because of noise and artifacts. As pointed out in [11], NDT data often are corrupted by noise. Moreover noise doesn't obey models that are easy to treat mathematically. It is amusing that such an application field whose name is defined by a negation ('non' destructive testing) generates data that also are 'non something' : non stationary, non linear, non gaussian, etc !

Even more challenging is **diagnosis** (feature extraction, classification, interpretation). To demonstrate the need of NDT for new methods whose performances are related to complexity of data the following sections will discuss some issues that can help enhancing processing and interpretation of real world NDT situations. These sophisticated methods will be new remedies to old ills.

Some results presented below come from EDF's experience in NDT, some others don't. In this case a comprehensive survey of recent literature is given. References [12] and [13] will also be useful to the reader. Our presentation starts from the requirements and then details some relevant algorithms to fulfil them. We will successively talk about data decomposition and modelisation, compression and computer aided diagnosis.

2. Data Decomposition

NDT field operators seldom are familiar with tools that mathematicians call 'transforms' : they better use A-scans, Lissajous diagrams and photographs than Fourier, Hilbert or Wigner-Ville decompositions. However no-one can contest that Fourier mapping between time (or space) domain and frequency ('spectral') domain is now a basic tool to analyse signals and images, providing a deep insight into periodic or random phenomena (Fourier is ideal for stationary phenomena). But more recently a revolution has occurred : it is called *wavelets*. Wavelet decomposition is a transform that replaces Fourier decomposition when analysing non stationary phenomena such as transients. Indeed wavelet is just a collective name for a number of parent operators, ranging from continous wavelets discovered by Morlet (1984) up to discrete orthogonal wavelets of Meyer, Daubechies, Mallat. Wavelet analysis is well suited to phenomena that can be examined at different *scales*, more or less like *fractals* are convenient to represent phenomena that are *self similar*.

Briefly speaking the idea of wavelet transform is to replace the famous Fourier relation :

$$X(f) = \int_{-\infty}^{\infty} x(t) e^{-2i\pi t f} \, dt$$

by $X(a,b) = \int_{-\infty}^{\infty} x(t) \Psi_{ab}^{*}(t) dt$, where $\Psi_{ab}(t) = \frac{1}{\sqrt{a}} \Psi \left[\frac{t-b}{a} \right]$.

Wavelet decomposition now has numerous potential applications like analysis, denoising, compression of very different signals coming from turbulence or astronomy. Wavelets are fascinating because of the following reasons :

i) the theory unifies numerous discoveries in a coherent framework, both in 'time' (back to the Haar basis -1909- !) and in 'space' (function theory, quantum theory, quadrature mirror filters, pyramid decomposition, multiresolution analysis, windowed Fourier analysis, time-frequency analysis),

ii) it has tight connections with fractals, noise filtering, compression,

iii) Meyer and his coworkers have proposed *wavelet packets* and a general scheme for choosing optimally *the best basis* to decompose a given signal for a given purpose [21].

NDT is not in such an advanced position. Few attemps have been made to apply wavelets to NDT signals, although they contain transients. In [22] the Meyer-Jaffard wavelet is used to detect under-cladding cracks by means of echography. Discrimination is based on the fact that a crack response is the second derivative of the incident signal, which results in a double peak in the representation. This work is a good illustration of the versatility of the method because discrete wavelet decomposition is computed not only for integer indices but also for fractional ones, exhibiting distinct peaks in the region of interest.

3. Data Modelisation

In this section we will present an approach where one considers that information is not entirely contained in the available data and hence that ' a priori ' information could be gathered from external knowledge and thus incorporated into the problem solving procedure. Sources of a priori information can be various : constraints on data (e.g. image values are positive), results from a finite elements software, hints about noise nature (e.g. noise is Gaussian or Poisson-like), knowledge about materials or types of degradation (e.g. cracks are likely to be aligned).

The second step is to convert prior information into an appropriate function in order to deal the same manner with both raw data and prior information. This is called 'modelisation' (in a somewhat restricted sense). It can use deterministic (e.g. smooth functions) or stochastic (e.g. Gaussian distributions) tools.

Data modelisation is a very general approach : it finds application in restoration, reconstruction, inversion, etc. As a general rule the more relevant the model is, the more performant the results are.

We will discuss two approaches recently used for **reconstruction** purposes. Both tend to compensate for a lack of data. Their common objective is to reconstruct in 3 dimensions the volume surrounding a potential flaw in a thick wall steel tubing, given only 3 noisy projections (radiographs).

3.1. A STOCHASTIC MODEL

This is based on inverse problems theory. The unknown f (representing values of object voxels) can be found by resolving

$$p = A.f + n \quad (3.1.1)$$

where p stands for projections (2D-radiographs) and n for noise.

Among the various methods that can be investigated (analytic reconstruction —Convolution BackProjection—, discrete deterministic Algebraic Reconstruction Technique —ART—, ART combined with a Markov modelisation, etc) the most powerful in the case of very few projections available are the Bayesian statistical methods :

i) object and noise properties are taken into account by means of probability distributions (each part of a radiograph is considered as a particular realization of a 2D-random process),

ii) one chooses the solution of lowest expected cost using the Bayes theorem :

$$a \text{ posteriori probability } P(F = f | \Lambda = \lambda) = \frac{P(\Lambda = \lambda | F = f).P(F = f)}{P(\Lambda = \lambda)} \quad (3.1.2)$$

where F is the unknown and Λ the input (both are random processes). One assumes that values read on radiographs (p) equal photon counts (λ), ie

$$p = \log(\frac{\lambda_q}{\lambda})$$

A classical choice is to search for f that maximises the *a posteriori* probability (MAP) :

$$\hat{f}_{MAP} = \arg \max_f P(F = f | \Lambda = \lambda) \quad (3.1.3)$$

Our goal is to introduce models into (3.1.3). Using (3.1.2), the following equation is derived from (3.1.3) :

$$\hat{f}_{MAP} = \arg\max_f \{\log P(\Lambda = \lambda | F = f) + \log P(F = f)\}$$

$$= \arg\max_f \{L(\lambda | f) + M(f)\} \quad (3.1.4)$$

This equation allows for a balancing effect between two terms : the first one encourages solutions which match the observed data (this is equivalent to a maximum likelihood estimation, but the solution may have properties which are incompatible with the material to be imaged), as the second one encourages solutions which have higher probabilities under the prior model assumed [31].

The mathematical choice for $L(\lambda | f)$ is dictated by physical setting (due to 3.1.1. it is simply $L(n)$, ie a model for noise). A reasonable approximation is

$$L(\lambda | f) \approx -\tfrac{1}{2}(\hat{p} - A.f)'.D.(\hat{p} - A.f) \quad (3.1.5)$$

which is a quadratic in the entries of f.

The prior density of the object $P(F=f)$ is a much more subjective matter. We will choose Markov random fields (MRF) because pixels in our radiographs depend only on a small neighbourhood. This leads to :

$$P(F = f) = \frac{1}{Z}\exp(-\sum_{c \in C} V_c(f))$$

where V is a potential over the sets c (pairs of neighbouring pixels). This is an application of the Hammersley-Clifford theorem (equivalence between MRF and Gibbs distributions).

Gaussian MRF can be used but they are not a good model for many realistic objects, although they are well suited to easy calculations. We will prefer generalized Gaussian MRF for which :

$$P(F = f) = \frac{1}{Z}\exp\left\{-\sum_i \sum_j b_{ij}(v.|f_i - f_j|)^q\right\}$$

To sum up the global procedure becomes :
find the minimum of

$$\tfrac{1}{2}(\hat{p} - A.f)'.D.(\hat{p} - A.f) + \beta\sum_{i<j} b_{ij}(v.|f_i - f_j|)^q$$

with $1 \le q \le 2$ and β is a weighting between the two terms of the criterion.

This can be solved by a Gauss-Seidel procedure and provides satisfactory results for both simulated and real world situations with few projections, thus proving the interest of adequate models when few data are available. Yet some work is still needed to take into account such phenomena as non-point source or scattering [32].

3.2. A GEOMETRIC MODEL

Assume the flaw is unique and threadlike. Its ideal skeleton will be approximated by a continous curve ω of R^3. ω will be found by minimizing an 'energy' U. Starting from an initial curve ω^0, a sequence of ω^1, ω^2, ... will converge to the desired ω. The method is known as 'snakes' because the initial guess about ω is iteratively distorted or bent to give an admissible approximation of ω. At each step minimising U will tend to decrease distance between actual projections of the flaw on radiographs and projections calculated according to a simple γ-ray propagation law.

In the original algorithm developed in [33], ω is decomposed on a cubic splines basis B as:

$$\omega = B\theta \quad \text{(in matrix form)}.$$

The Expectation-Maximisation method of Dempster is used to get a maximum likelihood estimate of ω from incomplete data. At step (k+1), the solution in θ is :

$$\theta^{k+1} = \arg\min_{\theta} U_\theta(\omega^k, g)$$

where g represents initial information (raw data or edges detected by an operator).

First experiments show that this method is efficient in reconstructing the skeleton of a flaw (and hence its orientation), as long as prior information remains realistic : thin isolated flaw, no scattering. A generalisation to multiple flaws is in progress.

4. Compression

Compression will be in the next future a major challenge in NDT as it has been in remote sensing, in digital telecommunications and more recently in medical imaging.

The need for compression comes from two trends in NDT : first, long term storage of digitized information, secondly, efficient transmission between on-site workstations and central diagnosis facilities. We will mainly comment on the storage objective.

Compression is a question of how much we accept to lose in order to decrease the amount of values needed to correctly represent a signal or an image. It is implicit in the process of compression that measured data include some kind of redundancy so that a tricky coding could be efficient. Most users (in NDT but not only) cannot figure out that high rate compression implies choices (what feature of my data collection could be discarded ? what are the very meaningful parameters in my problem ? ...). Otherwise one is lead to non lossy schemes and these have limited performance (compression rate typically $\tau < 3$). Not to mention is the fact that if data are to be used for human diagnosis tasks they must be compressed without loss. Another important point is to adapt theoretical results to actual data to compress and therefore to test them on real data. In particular each class of algorithms is dedicated to a specific type of data and degree of freedom for the implementation (images with or without textures, real time specifications, required resolution for reconstruction, etc).

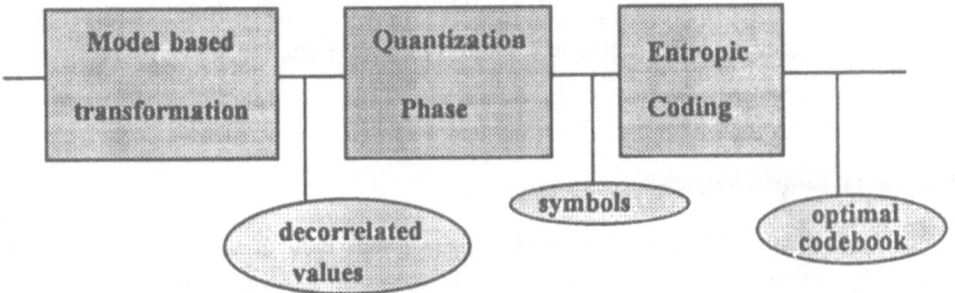

fig 1. general block-diagram of compression algorithms

An overal compression scheme comprises different functionnal blocks (see fig.1). We will only comment on the first one (model based transformations).

Today image compression benefits from a great job done by an ad'hoc working group named JPEG (for Joint Photographic Experts Group). This group has adopted a number of SIP algorithms for coding fixed images and proposed them as an ISO international standard. Among them the DCT one (for Discrete Cosine Transform) has become famous. As such it is far from new ! DCT is just the even part of a discrete Fourier transform calculated on 8 by 8 blocks of the image. In fact the JPEG standard is a set of four very different procedures including DCT, hierarchical DCT, progressive DCT and non lossy differential pulse code modulation (DPCM). The DCT transforms a 8x8 block of image $ima\ (x,y)$ into $IMA(u,v)$ according to :

$$IMA(u,v) = C(u)C(v)\sum_{x=0}^{7}\sum_{y=0}^{7} ima(x,y)\cos(\frac{(2x+1)u\pi}{16})\cos(\frac{(2y+1)v\pi}{16})$$

with $C(u), C(v) = \dfrac{1}{\sqrt{2}}$ if $u,v = 0$ and 0 otherwise

Another method is known as LZW (for Lempel, Ziev and Welch). It is implemented in the routine 'compress' on every UNIX system.

Other types of data or other requirements in terms of quality (compression rate, block effect, etc) may ask for other transforms. The wavelet decomposition (see section 2) is one of the most promising one. It often performs better than the DCT [41],[42] : the results in [42] although focused on landscape images show a dramatic increase in quality when using wavelet packets instead of JPEG (for $\tau = 30$), while [41] compares the different versions of JPEG, LZW and two implementations of wavelets (laplacian pyramid decomposition, QMF dyadic wavelet). The conclusion is the following : wavelet coding overcomes JPEG drawbacks (block effect), reaches compression rates as high as JPEG and allows more versatility, but it is not standard as JPEG is, it is not implemented in low cost boards and it is also more time consuming. For NDT (like for medical applications) it is vital that not only some subjective visual quality could be achieved, but also given figures of merit in order to allow for further processing and interpretation. Non lossy JPEG is therefore a good choice. If losses are allowed, the new algorithms (wavelets, fractals) are worth being considered.

5. Computer Aided Diagnosis

The final aim of any NDT procedure is **diagnosis** (what type of flaw, where, of what dimension ?), although fracture mechanics and return of experience have also to be considered in an overall maintenance program.

It is now clear that only an aid to diagnosis is possible, not an automated diagnosis. Not only individual powerful techniques such as sensor fusion, classification (section 5.1. to 5.3), are necessary, but also cooperation between them.

Computer aided diagnosis (CADIA) can be addressed through different approaches : classification or pattern recognition (classical such like K-nearest neighbours with distance reject or novel such like neural networks), data inversion (classical such like numerical optimisation or novel such like symbolic optimisation [56]), artificial intelligence.

The versatility needed when dealing with these issues is well illustrated by [53] and [55] among others.

In [53] the authors describe two very different methods to analyse eddy current SG data : the first one is an attempt to solve an inverse problem (the probe response to a small defect is modeled as a Gaussian curve) and to recover physical (electrical) parameters of the material from EC measurements (impedances). The second one considers a Lissajous pattern as an image and applies a neural network technique ('retina') to classify artifact types. Unfortunately these results concern only testbench data. An interesting idea of combining both methods is proposed.

Another attempt with the same kind of data is described in [55] : the authors propose a three-level automatic diagnosis system whose implementation on a commercial Unix-workstation is in progress. Lissajous figures are coded according to a Freeman-like technique and fed to a syntactic logic (a grammar is used to select potential flaw-events). Then a multilayer backpropagation neural network is able to classify events as flaw or not. A final level based on deep heuristic knowledge of human experts is dedicated to flaw characteristics estimation. Although this ambitious program seems to be only at the beginning, its aim and orientation remarkably fit in with the spirit of this paper and of [58].

5.1. SENSOR FUSION

When every single method of NDT has been optimized a natural idea is to combine them in order to take advantage of complementary characteristics : for example a given component can be inspected both by US and by X-ray. Each NDT technique has its own range of performance, its precision, its limitations. Data fusion is an attempt at doing better information extraction from heterogeneous data properly mixed together.

Data fusion is not a new mathematical theory. It takes its roots in probability, possibility, evidence, fuzzy logic, artificial intelligence theories and revisits the classical items of statistical detection, estimation, 3D-reconstruction, interpretation.

This point of view is new in NDT where US, EC and X-radiography are traditionally separate specialities, but has been promoted long ago by the battle field community (C^3I for Command, Control, Communication and Intelligence). The objective is to take the best decision, given a number of partially redundant sources of information (radar, infra-red sensors, video camera, etc), without overloading the operator nor system capacities : a reliable information of one sensor can compensate for the lack of precision or resolution of some other. More recently data fusion has also been used in robotics and medical imaging.

The question of how to fuse data is obviously open, as well as the level at which combination has to be done : raw data level, detection results level or decision-making level (conflict solving) ? Fusion can also take into account other sources of knowledge (eg computer aided design data or finite elements software results).

NDT could benefit of these new ideas, for example US data could be used as initialisation for 3D reconstruction. The European project TRAPPIST (for Transfer, Processing and Interpretation of 3D NDT data in a standard environment) is a first step towards a versatile environment allowing multiple sources fusion, although this fusion will be only a graphical combination in the first version of the prototype [51], [57].

5.2. CLASSIFICATION

Classification has been tried long ago in all cases where computers were supposed to be able to mimic human : speech, hand written characters, scenes recognition. Three classes of algorithms were used : statistical, syntactic pattern recognition.

A sudden revival of these topics has been motivated in the early eighties by the work of Hopfield (1982) and discovery of the generalized delta rule (1985). Numerous papers have been published about **neural networks** (ANN), trying to assess their superiority over classical methods, in terms of either performance or speed, to understand their internal way of learning and generalizing, to establish whether or not input data have to be pre-processed, etc.

Ten years after many people agree that, as far as classification of NDT data is concerned, ANN are equivalent to known methods, but faster if dedicated processors can be used. In fact in industrial situations difficulty doesn't come from algorithms but from the lack of representative data. Several experiments at EDF (with EC signals, γ images and also vibration measurements on reactor components and gears) illustrate this conclusion. It is also the conclusion in [54]. Windsor et al. have found that no pre-processing is needed and that performance is of no use to choose a method for a real world application (they propose instead speed, ease of use or implementation).

5.3. KNOWLEDGE-BASED SYSTEMS

In the field of infra-red thermography, SEQUOIA [52] is a very attractive example of what could be aimed in terms of CADIA when combining different kinds of data, methods and knowledge. First a geometrical model of the sample under inspection is built from computer-aided design. Then a theoretical model of thermal effects is used to discriminate between different regions on a thermogram. Several data processing algorithms are applied: filtering, compensation for thermal variations, subtraction of a reference image, elimination of background, segmentation. Segmented images are registered to CAD images and finally relevant parameters are extracted and fed to an expert system. Validation experiments have proved good behaviour. Processing times seem pretty short (a few minutes) without special hardware. Complementary use of more precise sensors and of video camera is planned. The authors conclude that such a system can be developed for radioscopy, shearography and US as well. Obviously this is a belief shared by the authors of [51], [57] and [58].

Conclusion

We have presented the most recent techniques of information processing applied to non destructive testing problems. Their performances obviously come from better mathematical taking into account of information that is not provided by the data themselves. Reasonable hints about data structure may lead to wavelet decomposition ; hypotheses about noise and flaws characteristics allow for powerful geometric or stochastic models ; choices concerning what is worth storing in signals or images, provide efficient compression schemes ; combination of complementary NDT methods can enhance dramatically individual processing techniques ; adequate embedding of both signal processing and NDT expertises into smart architectures dedicated to aided diagnosis will permit to face new challenges due to ever increasing amount of data and required reliability of inspections.
Specialists of NDT and signal processing must join their efforts to solve difficult industrial problems with more and more powerful algorithms, and particularly to implement full scale, realistic demonstrators.

Aknowledgements

The author wishes to thank Mrs. B. Lavayssière, Mr. E. Fleuet and Mr. R. Zorgati (from EDF) for carefully reading and commenting this paper.

References

[1] Revue générale nucléaire, **"Aspects of the French nuclear program"**, n°4, July-August 92 —in English and French—, SEE ed., Paris.

[2] EDF Research and Development Division, **" Les contrôles non destructifs "** (Non Destructive Testings), — in French — , revue ÉPURE, n°16, octobre 87.

[11] B. Georgel, **"Digital Signal Processing for NDT"**, 13th World Conference on NDT, São Paulo, Brazil, October 92, C.Hallai and P. Kulcsar eds., Elsevier.

[12] NATO ASI series F, vol.44, **" Signal Processing and Pattern Recognition in Nondestructive Evaluation of Materials "**, Québec, Canada, August 87, C.H. Chen ed., Springer Verlag.

[13] NDT International, " Special issue on **Digital Signal Processing** ", vol.19 n°3, June 86, Butterworths.

[21] Y. Meyer, **" Les ondelettes, algorithmes et applications "**, — in French —, A. Colin, Paris, 1992 (English version to be published by SIAM).

[22] J.P. Lefebvre, P. Lasaygues, **" Application de l'analyse en ondelettes à la détection de fissures par échographie ultrasonore "**, Journal de Physique IV (colloque C1), — in French —, avril 92.

[31] K. Sauer, C. Klifa, **" Bayesian Tomographic Reconstruction of 3D Objects from Limited Numbers of Radiographs "**, EDF Technical Report HP-21/92-61, Nov. 92.

[32] C. Klifa, K. Sauer, **" Bayesian Tomographic Reconstruction of 3D Objects from Limited Numbers of Radiographs "**, Proc. IEEE 1992 Nuclear Science Symposium & Medical Imaging Conf., Orlando FL, October 92 (EDF Grant # EP 542).

[33] B. Chalmond, **" Approche déformable en reconstruction 3D "** (Snakes approach in 3D reconstruction), — in French —, EDF Technical Report HP-21/93-28, June 93.

[41] E. Fleuet, **" Compression d'images en radiographie industrielle "** (Image Compression in industrial radiography), — in French —, submitted to the 6th European Conference on Non Destructive Testing, Nice France, October 24-28, 94 (also EDF Tech. Report HP-21/93-22).

[42] S. Mensah, " **Compression d'images par ondelettes** " (Image Compression with wavelets), — in French —, Tech. Report Digilog Company, Aix-en-Provence France, Mar.93.

[51] A. Mac Nab, I. Dunlop, " **Advanced visualisation and interpretation techniques for the evaluation of ultrasonic data : the NDT workbench** ", British Journal of NDT, vol.35 n°5, May 93.

[52] H. Tretout et alii, " **Application des techniques d'intelligence artificielle au contrôle non destructif par thermographie infrarouge stimulée thermiquement** " (Application of artificial intelligence techniques to NDT with thermally stimulated infrared thermography) — in French —, Matériaux : science et industrie ; colloque national du Ministère de la Recherche, Paris, juin 92.

[53] C.V. Dodd, J.D. Allen, " **Automated analysis of eddy current steam generator data** ", 11th international conference on NDE in the nuclear and pressure vessel industries, Albuquerque, NM, USA, May 92.

[54] C. G. Windsor et alii, " **The classification of weld defects from ultrasonic images : a neural network approach** ", British Journal of NDT, vol 35 n°1, January 93.

[55] K. Kim, Y. Huh, H. Woo, S. Choi, " **A hybrid system for eddy current signal evaluation of steam generator tubes in nuclear power plants** ", 11th international conference on NDE in the nuclear and pressure vessel industries, Albuquerque, NM, USA, May 92.

[56] J.C. Benas, " **Système à base de connaissances et traitements numériques pour l'analyse automatique des signaux de tubes de générateur de vapeur contrôlés par courants de Foucault** ", thèse de l'université de technologie de Compiègne —in French— juin 92 (see also a paper submitted to the Revue européenne Diagnostic et sûreté de fonctionnement, HERMES, Paris).

[57] B. Georgel, C. Nockemann, "**the European project TRAPPIST**", this issue, August 93.

[58] B. Georgel, R. Zorgati, " **EXTRACSION : a system for automatic Eddy Current diagnosis of steam generator tubes in nuclear power plants** ", 13th World Conference on NDT, São Paulo, Brazil, October 92, C.Hallai and P. Kulcsar eds., Elsevier (see also Zorgati, Georgel, Duvaut, Doligez, Garreau in Actes du colloque GRETSI, Juan-les-Pins, septembre 93, —in French—).

EDGE DETECTORS USING GRECO-LATIN SQUARES

YAGNESH C. TRIVEDI
Department of Natural Science,
Florida Community College at Jacksonville, Kent Campus,
3939 Roosevelt Blvd.,
Jacksonville, Florida 32205

LUDWIK KURZ
Department of Electrical Engineering,
Polytechnic University,
333 Jay St.,
Brooklyn, New York, 11201

ABSTRACT. In this paper a 5x5x5x5 Greco-Latin Squares experimental design model has been applied to detect horizontal, vertical and both diagonal edges in a 5x5 pixels mask. After detecting a type of edge (i.e. horizontal, vertical or diagonal), an edge is localized by using the linear one-way analysis of variance model. Comparisons between row or column or diagonal means called contrasts are carried out by F-test on contrasts. These, in turn are used to detect the location of an edge within the mask. Decision is made about presence or absence of an edge for a pixel located in the center of a 5x5 mask and then the mask is moved throughout the image to trace the edge map. The procedure is simple, efficient and robust as supported by computer generated edge detection examples.

1. Introduction

The field of digital image processing has grown considerably during the last two decades with the increased utilization of images in a wide variety of applications.

Edge detection, though a sub-field of image processing, has become a field in its own right. Local (3x3) window edge detectors in the form of difference operators were proposed by Roberts [1], Sobel [2], Prewitt [3], Robinson [4], Frei and Chen [5], MacLeod [6], Argyle [7], Rosenfeld [8], Marr [9], and Kirsch [10]

To overcome effects of noise in edge detection, smoothing in the form of averaging preceding difference operation was suggested by Rosenfeld [8], Rosenfeld and Kak [11].

Statistical analysis of variance (ANOVA) tests were applied by Suk and Hong [12] for edge detection in noisy images. Analysis of variance for

X. P. V. Malague (ed.), Advances in Signal Processing for Nondestructive Evaluation of Materials, 71–83.
© 1994 Kluwer Academic Publishers. Printed in the Netherlands.

experimental designs originally developed by Fisher [13] and further expanded by Scheffe [14], Cochran and Cox [15], Anderson and McLean [16], Johnson and Leone [17] and Montgomery [18] is particularly suitable to edge detection of noisy images because it treats the image pixels as statistical data corrupted by experimental errors (noise) and while performing hypothesis tests averaging over all data (smoothing) is achieved which is the desired goal. Kadar and Kurz [19] applied Latin squares masks for edge detection.

A new method of edge detection using a (5x5) mask based on Greco-Latin squares experimental designs is proposed in this paper. It requires only one mask to determine edge orientation in horizontal, vertical, 45 diagonal and 135 diagonal directions.

After detecting the edge orientation using the Greco-Latin squares mask, edge is localized and thinned to one pixel level by ANOVA F-test applied to contrasts between edge direction means.

2.   Digital Image Characterization

A two-dimensional image is represented by grey levels of pixels $\{Y_{i\,j}, i = 1$ to I, j = 1 to J$\}$ where I, J= 256, 512 or even 1024 for high resolution images. For I, J = 256 and $Y_{i\,j}$ taking on one of the 256 levels, an image in digital form is represented by $2^8 \times 2^8 \times 2^8 = 2^{24}$ bits (or 65536 pixels of 256 grey levels represented by 8 bit bytes). Such high dimensionality of a data vector for real-time processing of images requires the development of localized operators, processing small parts of images in parallel and in spatial domain.

When statistics of noise are not known completely (which generally is the case), robust localized operators should be used to process images. The emphasis in this paper is on robust techniques for the mixture noise model having the noise density, $f(x) = (1-c)f_1(x) + c f_2(x)$ where c, $(0<=c<=1)$ is the % of contamination by high variance $\sigma^2>>1$ (heavy tailed) gaussian noise $f_2(x)$ in gaussian noise density $f_1(x)$.

This description of noise density indicates slight departure from gaussian assumption. Also, the salt and paper noise is treated here as a heavy tailed gaussian noise.

3.   Experimental Design Models for Image Processing

A two-dimensional image was described earlier as a tabular array $Y = \{Y_{i\,j}$, i = 1 to I, j = 1 to J$\}$. $Y_{i\,j}$ may take on q discrete values (quantization levels). $Y_{i\,j}$ represents an outcome of an experiment in the language of experimental designs. The experimental design partitions variations observed in set $\{Y_{i\,j}\}$ into different parts, each is assigned a source, cause, factor or treatment. The terminology used in experimental designs comes from agriculture. The word 'treatment' refers to treatment of an agricultural plot with a kind of fertilizer or pesticide. Early developments in the experimental designs were done by the late Sir Ronald Fisher when he was in charge of statistics at the Rothamsted Agricultural Experiment Station near London.

The earliest mention of Latin squares seems to have been made in 1624 in connection with puzzles. Fisher pointed out in 1926 that Latin squares and families of mutually orthogonal Latin squares could be used systematically as experimental designs which allow for more than one source of variation in an experiment. Euler, who wrote the two arrays in terms of Greek and Roman letters respectively, called the pair of superimposed arrays a Greco-Latin square. A pair of Latin squares of the same size which when superimposed contain every possible ordered pair of symbols are said to be orthogonal.

Grey levels of the image data represent "treatments" whose distribution creates the various features in the pattern.

3.1 Greco-Latin Squares Designs

The Greco-Latin squares model is represented by

$$y_{ijkl} = u + R_i + C_j + L_k + G_l + e_{ijkl} \qquad \text{....(1)}$$

$i,j,k,1 = 1,2,....,p$ N= pxp where y_{ijkl} is (ijkl)th observation (pixel) p is the number of rows or columns of Greek letter treatments or Latin letter treatments u is the overall mean, R_i is the ith row effect, C_j is the jth column effect, L_k is the kth Latin letter treatment effect, G_l is the 1th Greek letter treatment effect, e_{ijkl}
is the (ijkl)th error or noise term.   N is the total number of pixels in the mask.

Fig. 1 shows a 5 x 5 x 5 x 5 Greco-Latin squares design.

Aα	Bε	Cδ	Dγ	Eβ
Bβ	Cα	Dε	Eδ	Aγ
Cγ	Dβ	Eα	Aε	Bδ
Dδ	Eγ	Aβ	Bα	Cε
Eε	Aδ	Bγ	Cβ	Dα

5 x 5 x 5 x 5 Greco-Latin squares design

Fig. 1

$$\sum_{i,j,k,l} (y_{ijkl} - y_{...})^2 = \sum_{i,j,k,l} [(\bar{y}_{i...} - \bar{y}_{...})$$
$$+(\bar{y}_{.j..} - \bar{y}_{...}) + (\bar{y}_{..k.} - \bar{y}_{...}) + (\bar{y}_{...l} - \bar{y}_{...})$$
$$+(y_{ijkl} - \bar{y}_{i...} - \bar{y}_{.j..} - \bar{y}_{..k.} - \bar{y}_{...l} + 3\bar{y}_{....})]^2 \qquad ....(2)$$

$$SS_T = SS_R + SS_C + SS_L + SS_G + SS_E \qquad ....(3)$$

where $SS_T = \sum_{i,j,k,l} [y_{ijkl}]^2 - [y_{...}]^2/N$

$$SS_R = \sum_i [\bar{y}_{i...}]^2/p - [\bar{y}_{...}]^2/N$$

$$SS_C = \sum_j [\bar{y}_{.j..}]^2/p - [\bar{y}_{...}]^2/N$$

$$SS_L = \sum_k [\bar{y}_{..k.}]^2/p - [\bar{y}_{...}]^2/N$$

$$SS_G = \sum_l [\bar{y}_{...l}]^2/p - [\bar{y}_{...}]^2/N$$

$$SS_E = \sum_{i,j,k,l} [y_{ijkl} - \bar{y}_{i...} - \bar{y}_{.j..} - \bar{y}_{..k.} - \bar{y}_{...l} + 3\bar{y}_{...}]^2 \qquad ....(4)$$

The 'dot' subscript indicates summation over the subscript it replaces and the 'bar' represents the average.

Table 1 is ANOVA table for Greco-Latin squares.

SOURCE OF VARIATION	SUM OF SQUARES	DEGREES OF FREEDOM	MEAN SQUARES	F_o
Row effect	SS_R	$p-1$	$MS_R = SS_R/(p-1)$	$\dfrac{MS_R}{MS_E}$
Column effect	SS_C	$p-1$	$MS_C = \dfrac{SS_C}{p-1}$	$\dfrac{MS_C}{MS_E}$
Latin letter treatments	SS_L	$p-1$	$MS_L = \dfrac{SS_L}{p-1}$	$\dfrac{MS_L}{MS_E}$
Greek letter treatments	SS_G	$p-1$	$MS_G = \dfrac{SS_G}{p-1}$	$\dfrac{MS_G}{MS_E}$
Error	SS_E	$(p-3)(p-1)$	$MS_E = \dfrac{SS_E}{(p-3)(p-1)}$	
Total	SS_T	$p^2 - 1$		

Notice that Latin letters are on 45° diagonals and Greek letters are on 135° diagonals.

Random effects model

$H_{OR} : \sigma_R^2 = 0$
$H_{1R} : \sigma_R^2 > 0$
$H_{0C} : \sigma_C^2 = 0$
$H_{1C} : \sigma_C^2 > 0$
$H_{0L} : \sigma_L^2 = 0$
$H_{1L} : \sigma_L^2 > 0$
$H_{0G} : \sigma_G^2 = 0$
$H_{1G} : \sigma_G^2 > 0$

e_{ijkl} is Normal Independently Distributed $N(0, \sigma^2)$.

Hypothesis testing:

F-statistic for rows, columns, Latin letters and Greek are compared with

$F_{\alpha, p-1, (p-3)(p-1)}$
$F_0 < F_{\alpha, p-1, (p-3)(p-1)}$ accept H_0
$F_0 > F_{\alpha, p-1, (p-3)(p-1)}$ reject H_0

We have $F_{OR}, F_{OC}, F_{OL}, F_{OG}$ to compare with $F_{\alpha, p-1, (p-3)(p-1)}$ to find if any of row effect, column effect, Latin letter treatment effect or Greek letter treatment effect is different from all others.

4. Contrasts

Suppose in conducting an ANOVA test for a random effects model, the null hypothesis is rejected. Thus there are differences between treatments but exactly which treatments differ is not specified. In this situation, further comparisons between treatment means may be useful. The ith treatment is defined as $u_i = u + t_i$ and u_i is estimated by $\bar{y}_i$. Using one way classification analysis of variance model,

$$y_{ij} = u + t_i + e_{ij} \qquad \qquad ....(5)$$

$i = 1$ to p, $j = 1$ to n, $N = np$

Comparisons between treatment means are usually made in terms of treatment average $\{\bar{y}_i\}$.

We would like to test hypotheses:

$H_0 : u_i = u_j$
$H_1 : u_i \neq u_j$ for a pair of means.

For the sake of convenience neighbors will be used as pairs. These tests on pair of means are also called contrasts.

In general, the comparison of treatment means of interest will imply a linear combination of treatment totals such as

$$C = \Sigma_i \, c_i \bar{y}_i . \qquad\qquad ....(6)$$

with the restriction that $\Sigma_i c_i = 0$. Such linear combinations are called contrasts and ci are called contrast coefficients.

The sum of squares for any contrast is

$$SS_C = (\Sigma_i \, c_i \bar{y}_i \,)^2 / (n\Sigma_i \, c^2_i) \qquad\qquad ....(7)$$

and has a degree of freedom. A contrast is tested by comparing its sum of squares to the mean square error. The resulting statistic would be F-distributed with 1 and N -a degrees of freedom.

Contrasts can be created between consecutive pairs of means using

$$\bar{y}_1. - \bar{y}_2. = 0, \bar{y}_2. - \bar{y}_3. = 0$$
$$\bar{y}_3. - \bar{y}_4. = 0, \bar{y}_4. - \bar{y}_5. = 0$$

	Contrasts
$H_0 : \bar{y}_1. = \bar{y}_2.$	$C_1 = \bar{y}_1. - \bar{y}_2.$
$H_0 : \bar{y}_2. = \bar{y}_3.$	$C_2 = \bar{y}_2. - \bar{y}_3.$
$H_0 : \bar{y}_3. = \bar{y}_4.$	$C_3 = \bar{y}_3. - \bar{y}_4.$
$H_0 : \bar{y}_4. = \bar{y}_5.$	$C_4 = \bar{y}_4. - \bar{y}_5.$

$$SS_{C1} = (C_1)^2 / (5 \times 2)$$
$$SS_{C2} = (C_2)^2 / (10)$$
$$SS_{C3} = (C_3)^2 / (10)$$
$$SS_{C4} = (C_4)^2 / (10)$$

The analysis of variance table is given below
Table - 2

Source of Variation	Sum of Squares	D. F.	Mean Square	F_0
C_1	SS_{C1}	1	SS_{C1}	SS_{C1}/MS_E
C_2	SS_{C2}	1	SS_{C2}	SS_{C2}/MS_E
C_3	SS_{C3}	1	SS_{C3}	SS_{C3}/MS_E
C_4	SS_{C4}	1	SS_{C4}	SS_{C4}/MS_E
Error	SS_E	20	$MS_E = SS_E/20$	
Total	SS_T	24		

F_{0C1}, F_{0C2}, F_{0C3} or F_{0C4} is compared with $F_{0.05, 1, 20}$. If any of them is greater than $F_{0.05, 1, 20}$ contrast exists between groups of treatment means. This will be shown to be useful for edge location within the 5x5 mask.

5. Robustness of F-statistics

The assumption underlying the analysis of variance are that treatment effects are additive and that experimental errors are normally and independently distributed with the same error variance. Generally, we can never be sure that all of these assumptions are satisfied, and often we may have reason to suspect that some are false. Extensive studies have been made concerning the consequences of departures from assumptions in the analysis of variance. For example see Cochran and Cox [14], Scheffe [15], Anderson and McLean [16], Johnson and Leone [17] and Montgomery [18] note that the effect is minor when the assumptions and underlying the analysis of variance are not exactly satisfied, i.e., slight departures form the assumptions are of little concern.

The assumption of independence is generally critical. Proper randomization of the experiment introduces independence in the assignment of treatments to experimental units, and the resulting experimental errors may be regarded as independent. Usually proper randomization is achieved by randomly assigning an order in which the observations are to be taken.

Box and Anderson [20] have shown that if the kurtosis, $K_u = E\{(x-u)^4\}/\sigma^4 - 3$ of the population is between +1 or -1 which indicates a deviation from normality with either a highly peaked p.d.f. with long thin tails (Ku positive) and a relatively flat p.d.f. which drops off rapidly to zero (Ku negative); or if skewness, S, $S = \{E[x-u)^3]/\sigma^3\}$ is between zero and plus one, which indicates p.d.f's with long tails to the right, the F-test remains robust and the level of significance is essentially unchanged. The F-test also tends to be robust with respect to the assumption of equality of variances as long as each group contains the same number of observations although it introduces a small bias. The above statements also apply to the power of the F-test, under hypothesis H_1, in which case the distribution is given by the non-central F.

The robust properties stated above for the F-test clearly indicate that feature extraction based on the linear experimental design models of ANOVA are insensitive to the changes in the underlying C D F'S in a manner similar to the robustness properties of nonparametric tests. The robustness here is not meant in a strictly mathematical sense but in a practical sense.

6. Edge Detection

Changes in discontinuities in an image attributes such as luminance, intensity, tristimulus values (for color images) or texture are fundamentally important primitive features of an image since they often provide an indication of the physical extent of objects within the image. Local discontinuities in image luminance or grey level are called luminance or intensity edges or simply edges. Edge detection in noisy images creates a problem of false edges. To avoid false edges smoothing of images before the edge detection is desired. The methods of experimental designs are well suited to this problem.

6.1   Greco-Latin Squares Mask

A Greco-Latin squares is arranged as shown in fig. 2 to detect edges in horizontal(rows), vertical(columns), the Latin letters (45° diagonal) and the Greek letters(135^{c} diagonal) directions.

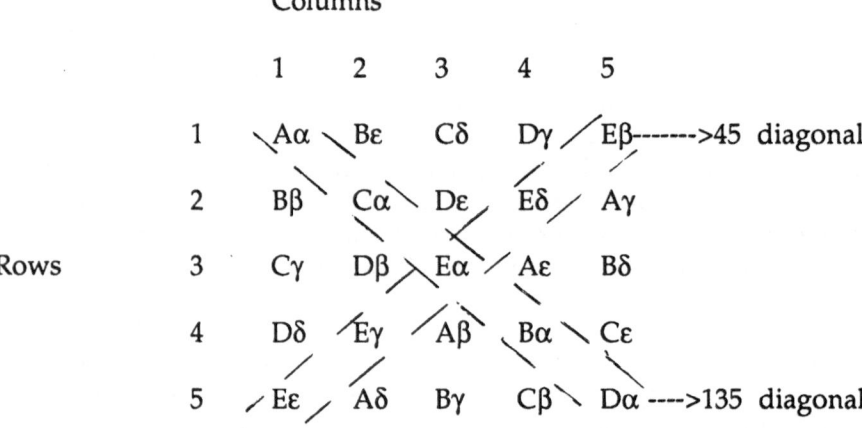

5x5x5x5 Greco-Latin square mask

Fig. 2

Notice that all the Latin letters treatments are parallel to the 45° diagonal and the Greek letter treatments are parallel to the 135° diagonal. Combinations of the Greek and the Latin letters represent location of a pixel, i.e., in row i, column j, parallel to 45^{c} diagonal at k, and parallel to 135° diagonal at 1. Pixels take on any one value of the q grey levels.

The Greco-Latin squares Random Effects model given in section 3 is

$$y_{ijkl} = u + R_i + C_j + L_k + G_l + e_{ijkl} \qquad\qquad ....(1)$$

i,j,k,l = 1 to 5, N = 25

Any two subscripts from i,j,k,l give the pixel location in a 5x5 mask. Variance edge orientation is detected if F statistic F_{OR}, F_{OC}, F_{OL} and/or F_{OG} for rows, columns, the Latin letter treatments, the Greek letter treatments respectively is greater that $F_{0.01,4,8} = 7.01$. Table 3 gives the edge detection scheme.

Table 3

STATISTIC	EDGE ORIENTATION
$F_{OL} > F_\alpha$ and $F_{OG} < F_\alpha$	Parallel to 45 diagonal
$F_{OG} > F_\alpha$ and $F_{OL} < F_\alpha$	Parallel to 135 diagonal
$F_{OR} > F_\alpha$ and $F_{OL}, F_{OG}, F_{OC} < F_\alpha$	Horizontal
$F_{OC} > F_\alpha$ and $F_{OL}, F_{OG}, F_{OR} < F_\alpha$	Vertical
$F_{OR}, F_{OC} > F_\alpha$ and $F_{OL}, F_{OG} < F_\alpha$	Horizontal and Vertical
$F_{OL}, F_{OG}, F_{OR}, F_{OC} < F_\alpha$	No Edge

In the above scheme the edges are not thinned yet. Edges in this case will be 5 pixels wide. To thin edges, one can develop many approaches. One such approach is presented in the next section.

6.2   F-tests on Contrasts for Thinning Edges

After detecting an edge in a horizontal(row), vertical(column), 45° diagonal and/or 135° diagonal direction using the Greco-Latin squares, one-way classification ANOVA (see eq. (5) is used for localizing edges in a given direction. Using the theory discussed in section 4, contrasts are created for the edges detected. For example, if a horizontal(row) edge is detected by the Greco-Latin squares mask, then contrast to be tested are

pair of Means

$C_2 = y_2. - y_3., C_3 = y_3. - y_4.$
$SS_{C2} = (C_2)^2/10, SS_{C3} = (C_3)^2/10$
$y_{ij} = u + t_i + e_{ij}$ ....(5)
$F_{TH} = F_{0.05,1,20} = 4.35$

If $[F_{0C2} > F_{TH})$ & $\bar{y}_3. > \bar{y}_2.)]$ OR $[F_{0C3} > F_{TH})$ & $(\bar{y}_3. > \bar{y}_4.)]$ then the center pixel is on high to low row edge, otherwise it is not. Similar F-tests are generated using a one-way classification model for column, 45° diagonal and 135° diagonal edge directions. Two-way classification model is used when presence of horizontal and vertical edges are determined by the Greco-Latin squares design.

If the center pixel is on any edge than 1 is stored in the edge map, otherwise 0 is stored.

Fig. 3 and Fig. 4 show the results of the above edge detection method. The Fig. 3 and Fig. 4 are the U.S.C. images of the girl and the house. The proposed edge detector performance in no noise and noisy environment is tested to see its capabilities and limitations. Robustness of the Greco-Latin squares edge detector is demonstrated by comparison of its performance in only gaussian noise(variance equal to 100) environment and mixture noise environment where small percentage(1%) of salt and pepper(impulsive noise, i.e., gaussian noise with variance equal to 10000) is added to gaussian noise(variance equal to 100). The edge detector performance is not degraded significantly. Robustness of the edge detector is not in a strictly mathematical sense but in a practical sense supported by the simulation results.

6.3 Mask Movement

After processing the top left corner of the 5x5 pixels, the mask is moved to the right by one column. The process is repeated till all columns are exhausted and then the mask is moved down by one row. The horizontal movement of the mask is continued. The whole process is repeated till the whole image is processed. Thus, for a N x N pixel image, we obtain a (N - 4)x(N -4) pixels edge map.

7.   Conclusions

A new robust edge detection technique based on the Greco-Latin square experimental design for operation in a noisy environment is introduced. When the noise departs form the gaussian assumption, the robust edge detector still performs the task of edge detection successfully. Thinning of edges to only one pixel level is carried out by F-tests on treatment means or contrasts.

Simulation results demonstrate the usefulness of the edge detection scheme developed here. Only one 5x5 mask does the work of many masks described in the literature to perform edge detection for a horizontal, vertical, 45 diagonal and 135 diagonal edge orientation. The proposed edge detector is computionally simple and efficient.

Acknowledgements

We are grateful to Dr. Yap Chua of Computer and Information Science Department, University of North Florida and Mr. Allan G. Weber of Signal and Image Processing Institute, University of Southern California for their help in software development for displaying U.S.C. images. Many thanks to Mr. Eric Reinhardt, Ms. Ann Freudenthal and Ms. Kathy Hughes of Computer Center, University of North Florida for their generous help in software development and image recordings. Special thanks to Pragna Trivedi for typing the manuscript.

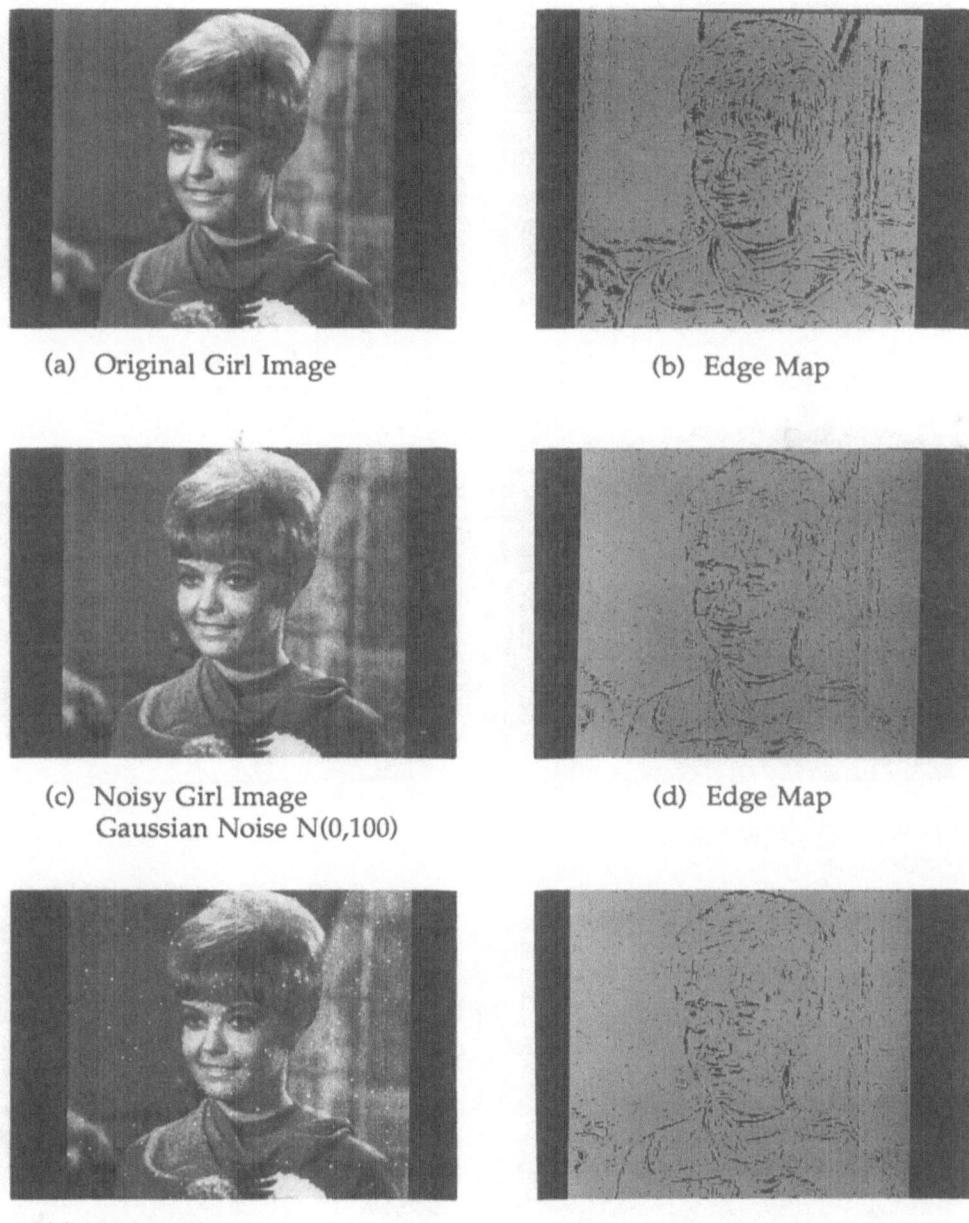

(a) Original Girl Image

(b) Edge Map

(c) Noisy Girl Image
Gaussian Noise N(0,100)

(d) Edge Map

(e) Noisy Girl Image
Mixture Noise
0.99 N(0,100) + 0.01 N(0,10000)

(f) Edge Map

Fig. 3

(a) Original House Image

(b) Edge Map

(c) Noisy House Image
Gaussian Noise N(0,100)

(d) Edge Map

(e) Noisy House Image
Mixture Noise
0.99 N(0,100) + 0.01 N(0,10000)

(f) Edge Map

Fig. 4

REFERENCES

1. Roberts, L. G., Machine Perception of Three Dimensional Solids, Optical and Electro-optical Information Processing, J. T. Tippet etal., Eds. MIT Press, Cambridge, Mass., pp. 159-197, 1965

2. Sobel, Eds. Duda and Hart, Pattern Classification and Scene Analysis, Wiley, N.Y., 1973, pp. 271

3. Prewitt, J.M.S., Object enhancement and extraction, in Picture Processing and Psychopictorics, eds. Lipkin and Rosenfeld, Academic Press, N.Y., 1970

4. Robinson, G. S., Color edge detection, Proceeding SPIE Symposium on advances in image transmission techniques, Vol. 87, San Diego, Ca., August, 1976.

5. Frei, W., and Chen, C. C., Fast boundard detection: a generalization and a new algorithm. IEEE Tran. Computers, Vol. 26, pp. 988-998, 1977

6. MacLeod, I.D.G., On finding structure in pictures, in Picture Language Machines, Ed. Kaneff, S., Academic Press, N.Y., 1970 pp. 231

7. Argyle, E., Techniques for edge detection, Proc. IEEE, Vol. 59, 2, Feb. 1971, pp. 285-287

8. Rosenfeld, A., A nonlinear edge detection technique, Proc. IEEE Letters, 58, 5, May, 1970, pp. 814-816.

9. Marr, D., Early processing of visual information, Philos. Tran. Roy. Soc. London, ser. B, 275, pp. 483-524, 1976.

10. Kirsh, R. A., Computer determination of the consistent structure of biological images, Comput. Biomed. Res. 4, pp. 315-328, 1971,

11. Rosenfeld, A. and Kak, A.C., Digital Picture Processing Vol. 1 and 2, Academic Press, N.Y., 1982.

12. Suk, M. and Hong, S., An edge extraction technique for noisy images, CG & IP, Vol. 25, N.Y., 1966.

13. Fisher, R. A., The Design of Experiments, 8th ed. Hafner Publishing Co., N.Y., 1966.

14. Scheffe, H., The Analysis of Variance, iley, N.Y., 1959.

15. Cochran, W. G. and Cox, G. M., Experimental Designs, 2nd Ed. Wiley, N.Y., 1957.

16. Anderson, V. L. and McLean, Design of Experiments, Marcel Dekker Inc., N.Y., 1974.

17. Johnson, N. L. and Leone, F. C, Statistical and Experimental Designs in Engineering and the Physical Sciences, 2nd Ed., Vol. 1 and 2, Wiley, N.Y., 1977.

18. Montgomery, D. C., Design and Analysis of Experiments, John Wiley & Sons, N.Y., 1976.

19. Kadar, I. and Kurz, L., A class of roust edge detectors based on Latin squares, Pattern Recognition, Vol. 11, pp. 329-339, 1979.

20. Box, G.E.P., and Anderson, S.L., Permutation theory in the deviation of robust criteria and study of departures from assumptions, J. Royal Soc., Seriees B., Vol. 17, pp. 1-34, 1955.

DIGITAL IMAGE PROCESSING IN HOLOGRAPHIC NONDE-STRUCTIVE TESTING (HNDT)

W. JÜPTNER and W. OSTEN
Bremen Institute of Applied Beam Technology
Klagenfurter Straße 2
D-28359 Bremen
Germany

ABSTRACT. All interferometric methods as for instance holographic interferometry, speckle metrology, shearography and topometry are based on the transformation of phase differences of interfering wavefields into observable intensity variations, that is, interference fringes. The analysis of fringe patterns by eye, however, is highly subjective and time consuming. To use interferometric methods for optical shop testing in an industrial environment a drastic decrease of the evaluation time combined with increased reliability of data is necessary. In recent years, various methods for computer aided quantitative and qualitative evaluation of fringe patterns have been developed. Looking back, one may state that the crucial step from the laboratory to the factory floor became possible by applying efficient computational equipment and algorithms combined with new data reduction techniques for the digital evaluation of interferograms. On two examples - the unwrapping of noisy mod 2π images and the knowledge assisted recognition of material faults in interferograms - it will be demonstrated how digital image processing and analysis can improve the reduction of relevant data in holographic nondestructive testing.

1. Introduction

In holographic and speckle interferometry two sets of data are to be measured for the quantitative evaluation of fringe patterns - the phase values $\delta(x,y)$ and the sensitivity vectors $S(x,y,z)$ for each point to be measured of the object under test. During the last ten years a lot of research has been done to reduce the interferometric measurement process to the digital reconstruction of interference phase values. Especially the increased availability of image processing equipment has stimulated the development of a whole class of digital evaluation techniques for high resolution analysis of fringe patterns.

Different approaches can be used to automatize the phase reconstruction process [1]. The most accepted techniques involve calculating the phase $\delta(x,y)$ at each image point, either by shifting the fringes through known phase increments ϕ_n [2] (phase sampling technique) or by adding a substantial tilt to the wavefront causing carrier fringes and Fourier-transformation of the resulting pattern [3] (spatial heterodyne). In either case, the phase is calculated modulo 2π as the principal value. The result is a so called saw-tooth phase image and phase unwrapping has to be carried out to remove any 2π phase discontinuities. This step contains the actual difficulties of the processing scheme. In conventional phase unwrapping algorithms the procedure starts at a single pixel and works along lines or rows to detect pixels where the phase returns abrupt to a value of zero. Then 2π must be added to or subtracted from the present value. This is performed by constructing a stair-case like overlay - the so called BIAS-image, considering the integer multiples of 2π for the entire phase map. For data without noise a threshold of π is adequate to detect relevant discontinuities in the

X. P. V. Malague (ed.), *Advances in Signal Processing for Nondestructive Evaluation of Materials*, 85–101.
© 1994 *Kluwer Academic Publishers*.

phase difference between adjacent pixels. But in general the acquired data set is influenced by noise (e.g. speckle and electronic noise) and the object structure as well as noncontinuous deformations (e.g. cracks) contribute also to distortions of the saw-tooth profile and inconsistencies in the phase field. In such cases conventional phase unwrapping algorithms fail and the reconstructed phase profile will be incorrect. Consequently unwrapping noisy data is one of the key challenges facing digital interferometry. Some more robust and noise insensitive algorithms were described in literature [4-6]. Other authors [7-9] have analyzed the limitation in the measuring range of phase sampling resulting from the fact that sampled imaging systems are used. To avoid aliasing of the fringe pattern it is necessary to keep the fringe frequency lower than the Nyquist frequency of the detector. The purpose of the first part of this paper is to give a systematization of possible sources for an incorrect unwrapping, to describe some procedures to by-pass them or to reduce their influence and to present a new approach considering the phase discontinuities field in a global way.

In the second part a knowledge assisted approach for the qualitative evaluation of interferograms with respect to the automatic recognition and classification of material faults is described. It is well known that holographic and speckle interferometry can be applied not only for high precision quantitative analysis of strain fields but also for nondestructive inspection of single objects and complex structures with regard to surface and internal flaws. These flaws are recognized by the evaluation of the resulting fringe patterns regarding characteristic irregularities as for instance "bull eye"-fringes, distorted fringes, local compressed fringes, cutted and displaced fringes. Based on practical experience and knowledge about the material behaviour the flaw can be classified as e. g. void, debond, delamination, thin area or crack. The methods have been applied successfully for industrial relevant testing problems in different fields: quality control of circuit boards and electronic modules /10,11/, inspection of satellite fuel tanks and pressure vessels /12/, tire testing /13,14/, glass and carbon fibre reinforced material testing /15/, investigation of turbine blades /16/, investigation of large technical components /17,18/ and motor car inspection /19/. But in contrast to the big step forward which was done in quantitative evaluation during the last few years the current situation in qualitative evaluation of fringe patterns is unsatisfactory. The problems to be solved here are more difficult. The task consists not only in the automatic recognition of complex patterns within noisy interferograms but also in the objective evaluation of these patterns with regard to the flaws caused them. First results were reported by Glünder /20/ in 1982. In this work an opto-electronic hybrid processor was used for a fast and very effective data reduction. A complete digital system for the analysis of misbrazing in brazed cooling panels was proposed by Robinson /21/ in 1983. When the plate is pressurized misbrazing is observed as a closed-ring fringe pattern, but the evaluation procedure is very time consuming. Another approach /14/ was proposed which applies parallel hardware. In less than one minute the skeleton of the fringe pattern is derived and two line features are determined to recognize flaw induced patterns.The most obvious deficit in all published procedures is the insufficient flaw classification. Two modern approaches are very prospective to overcome this unsatisfactory situation. The first one applies knowledge based systems or neural networks to "learn" different kinds of flaws from simulated or practical examples /22/. The second approach combines theoretical simulation methods and practical measurements /23/. Our knowledge assisted approach is based on a three-step procedure which is oriented on the pattern information contained in the fringe skeleton (pre-processing, model based pattern recognition and knowledge based flaw classification)and which includes the experience of a model based simulation of fringe patterns taking into account the calculation of the influence of subsurface flaws on the surface deformations using FEM.

2.   Unwrapping of Noisy Interferograms

2.1   PHASE SAMPLING AND 2-D PHASE UNWRAPPING

For our description a simplified model of the intensity distribution in an interferogram is sufficient:

$$I(x,y,t) = a(x,y,t) + b(x,y) \cos [\delta(x,y) + \varphi(x,y,t)]. \qquad (1)$$

The additive disturbances (e.g. varying background intensity and time dependent electronic noise) are summarized by the variable $a(x,y,t)$. Multiplicative noise as for instance the speckle noise and local variations of the fringe modulation contains the term $b(x,y)$. Besides, the time varying reference phase $\varphi(x,y,t)$ is introduced which is significant for the distinction between different kinds of phase reconstruction methods. In double exposure technique the reference shift φ between the wavefronts representing the two states of the object is introduced by stepping or ramping the phase in either of the two separate reference beams during the reconstruction of the interferogram. If e.g. φ is shifted in n steps of a constant value φ_0, n intensity values can be measured for each pixel in the interferogram:

$$I_n(x,y,t) = a(x,y,t) + b(x,y) \cos [\delta(x,y) + \varphi_n] \quad \text{with} \qquad (2)$$

$$\varphi_n = (n-1)\varphi_0, \quad \varphi_0 = 2\pi/m, \quad n = 1,...,m \ (m \geq 3) .$$

The data set needed for the phase sampling detection scheme is a set of three or more interferograms. In the case of four interferograms ($\varphi_0 = \pi/2$) it follows:

$$\delta'(x,y) = \arctan [(I_2-I_4)/(I_3-I_1)] , \qquad (3)$$

$$a(x,y) = 1/4 \ (I_1+I_2+I_3+I_4) , \qquad (4)$$

$$b(x,y) = 1/2 \ [(I_1-I_3)^2 + (I_2-I_4)^2]^{1/2} . \qquad (5)$$

If the signs of the numerator and denominator in Equ. (3) are considered the principal value $\delta'(x,y)$ can be calculated over a range of 2π. The simplicity of Eq. (3) offers a computer version which was

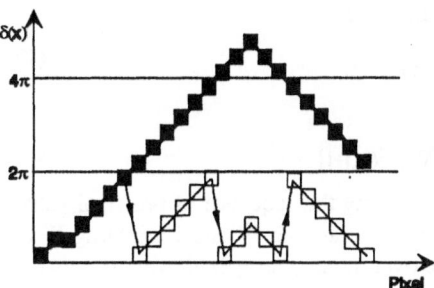

realized in different ways. Only two subtractions, one division and a look-up-table calculation are necessary to calculate the phase δ', lying in the range from $-\pi$ to π. This can be performed in video rates using special designed pipeline processors [24] taking into acount the necessity of a high accuracy of data. The next step in the phase sampling procedure leads to its major difficulties and limitations. The 2π phase discontinuities must be removed to obtain the unwrapped result:

$$\delta(x,y) = \delta'(x,y) + N(x,y) \cdot 2\pi. \qquad (6)$$

This process is illustrated in Fig. 1. An analysis of the 2-D unwrapping algorithm was provided by Ghiglia et al. [25]. The generation of principal values can be regarded as the result of applying a wrapping operator W_L on the true phase samples $\delta(i,j)$ with (i,j) as pixel location:

$$W_L[\delta(i,j)] = \delta'(i,j)$$
$$= \delta(i,j) - 2\pi \cdot N_L(i,j) , \qquad (7)$$

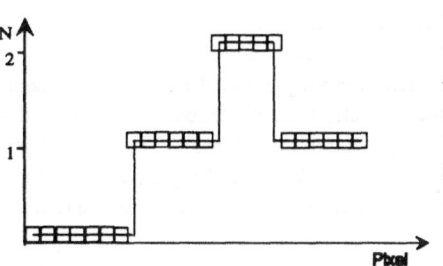

Fig. 1: 1-D phase unwrapping scheme

where L is a label to indicate different wrapping operators and N_L is a proper chosen sequence of integers that

$$-\pi \leq W_L[\delta(i,j)] < \pi. \tag{8}$$

Moreover differencing operators Δ_i and Δ_j in line and column direction, respectively, are defined

$$\Delta_i\delta(i,j)=\delta(i,j)-\delta(i-1,j) \tag{9}$$

$$\Delta_j\delta(i,j)=\delta(i,j)-\delta(i,j-1). \tag{10}$$

Computing the differences of principal values Δ_iW_1 in line and Δ_jW_1 in column direction and applying the wrapping operator again $W_2(\Delta_iW_1)$ and $W_2(\Delta_jW_1)$ yield principal values of the differences. Because W_2 produces values lying in the range from $-\pi$ to π it follows

$$\Delta_i\delta(i,j) = W_2\{\Delta_iW_1[\delta(i,j)]\} \tag{11}$$

$$\Delta_j\delta(i,j) = W_2\{\Delta_jW_1[\delta(i,j)]\} \tag{12}$$

if neighbouring phase samples satisfy the conditions

$$-\pi \leq \Delta_i\delta(i,j) < \pi \tag{13}$$

$$-\pi \leq \Delta_j\delta(i,j) < \pi. \tag{14}$$

Both equations (11) and (12) open several possibilities of 2-D unwrapping of the phase field $\delta(i,j)$. The choice of the starting condition is arbitrary. For instance, the phase of the pixel $(1,1)$ is assumed to be zero and the first column $j=1$ can be unwrapped to obtain

$$\delta(i,1) = \sum_{k=2}^{i} W_2\{\Delta_iW_1[\delta(k,1)]\}. \tag{15}$$

With this initial information all lines can be unwrapped using the scheme

$$\delta(i,j)= \delta(i,1)+ \sum_{p=2}^{i} W_2\{\Delta_jW_1[\delta(i,p)]\}. \tag{16}$$

Line and column operations may be exchanged, so that unwrapping the column direction starts with the preparation of the initial information along the first line:

$$\delta(1,j) = \sum_{p=2}^{i} W_2\{\Delta_jW_1[\delta(1,p)]\}. \tag{17}$$

$$\delta(i,j) = \delta(1,j) + \sum_{k=2}^{i} W_2\{\Delta_iW_1[\delta(k,j)]\}. \tag{18}$$

Eqs. (16) and (18) state the processing scheme for unwrapping the sequence of principal values consisting of a cascade of three operations: differencing Δ, wrapping W and integrating Σ. On condition that the two expressions (13) and (14) are true over the 2-D array both ways of unwrapping described by Eqs.(16) and (18), respectively, and any other interpretation of Eqs. (11) and (12) as for instance the unwrapping along a spiral starting from a certain point within the pixel matrix yield identical results. That means, the process of unwrapping is path independent. Otherwise inconsistent values exist in the principal value phase field. The checking of saw-tooth image for such possible inconsistencies is an approved mean for the identification of erroneous areas and the selection of suitable unwrapping pathes. Unfortunately, in practice the conditions (13) and (14) are not always satisfied due to the presence of noise, the violation of the sampling theorem and the influence of the object shape as well as the effect of its deformation. These phenomena are discussed in the next section.

2.2 REASONS FOR INCORRECT UNWRAPPING

In this section the main error sources are classified into four classes: noise, undersampling, object discontinuities and noncontinuous phase fields. For each of these classes a description of the related phenomena is given. Some simulations of the influence of these types of error are given in [25].

2.2.1 *Noise.*

Noise - mainly caused by the electronic acquisition and processing of the interferogram and the interference of coherent wavefronts scattered from rough surfaces (speckle) is the most familiar error source in automatic evaluation of interference patterns. In distinction to other evaluation techniques which operate with only one interferogram (e.g. fringe tracking) the stability of the intensity distribution during the reconstruction is very essential for phase sampling excepted changes in the intensity due to the phase shift φ. That means, the noise configuration must be identical for all considered frames. Consequently, a real-time environment with interferometric stability in contrast to photographic stability for conventional double-exposure technique has to be guaranteed during the reconstruction of the m (m≥3) interferograms. In most practical situations, this condition is not fulfilled because of the time dependent character of electronic noise, speckle displacements and so on. Hence, computed principal value phases are corrupted by noise to such an extent that local inconsistent regions are caused. A simple method to detect inconsistent regions in the saw-tooth image is given by computing the sum of the wrapped phase differences $\Sigma W\Delta_i$ for all 2x2-pixel windows along the path depicted in Fig. 2 [6]. All points within the window are consistent if the sum is equal zero. Otherwise the points are labeled as inconsistent. Data in regions well away from these points are influenced by the errors if they are not considered. Beginning from the inconsistent regions obvious streaks are drawn through the image. The magnitude of the related phase error depends on the location of the noise generated discontinuity within the saw-tooth. In difference to inconsistent regions errors from isolated pixels don't propagate along the direction of unwrapping.

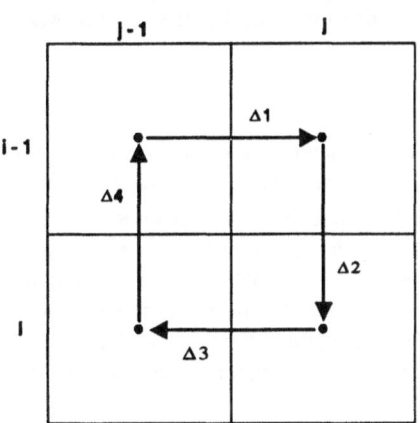

Fig. 2: Path for inconsistency check in a 2x2-window

2.2.2 *Violation of the Sampling Theorem.*

The presupposition for the validity of the phase unwrapping scheme described above is that the phase between any two adjacent pixels does not change by more than π [7]. This limitation in the measurement range results from the fact that sampled imaging systems with a limited resolving power are used. There must be at least two pixels per fringe - a condition that limits the maximum spatial frequency of the fringes to half of the sampling frequency (Nyquist frequency) of the sensor recording the interferogram. Fringe frequencies above the Nyquist frequency are aliased to a lower spatial frequency. In such cases the unwrapping algorithm is unable to reconstruct so modified data. In the case where the fringe frequency is higher than the Nyquist frequency the unwrapping algorithm fails. The simplest solution is to improve the sampling using sensors with higher resolution or to process magnified regions of the interferogram.

90

2.2.3 *Object Discontinuities.* Discontinuities in the object as for instance gaps and shaded regions as well as irregular boundaries give also reasons for an incorrect unwrapping because the intensity does not change in such regions according the phase sampling principle. Consequently, abrupt phase changes in the range of π can appear in the principal value. To avoid such errors corresponding regions should be excluded from the unwrapping process.

2.2.4 *Noncontinuous Phase Fields.* Noncontinous deformations as for instance cracks due to the material behaviour under load give rise to dislocations in the phase distribution. In the surrounding of the crack the structure of the fringe pattern is influenced in such a way that the fringes are cutted and displaced to each other. The noncontinuous behaviour of the phase surrounding the discontinuity results in inconsistent areas and consequently in an incorrect unwrapping if straightforward procedures are used.

2.3   UNWRAPPING OF NOISY INTERFEROGRAMS

In the previous section the main error sources for incorrect unwrapping of principal value phase maps (mod 2π images) were discussed in separate manner. In practice, however, these factors appear in complex. Fig. 3 shows an interferogram of a vibrating industrial component stored with double pulse laser technique. The fringe pattern is corrupted with speckle noise, electronic noise and light reflexes which contribute to a distortion and fusion of the fringes. In some sections the signal -noise ratio and contrast is very low. Other influences are object structure and irregular boundaries. A conventional unwrapping procedure starting with the column at the right border and following the lines will meet a lot of inconsistent areas and consequently generate streaks with incorrect phase offset over the whole phase map.

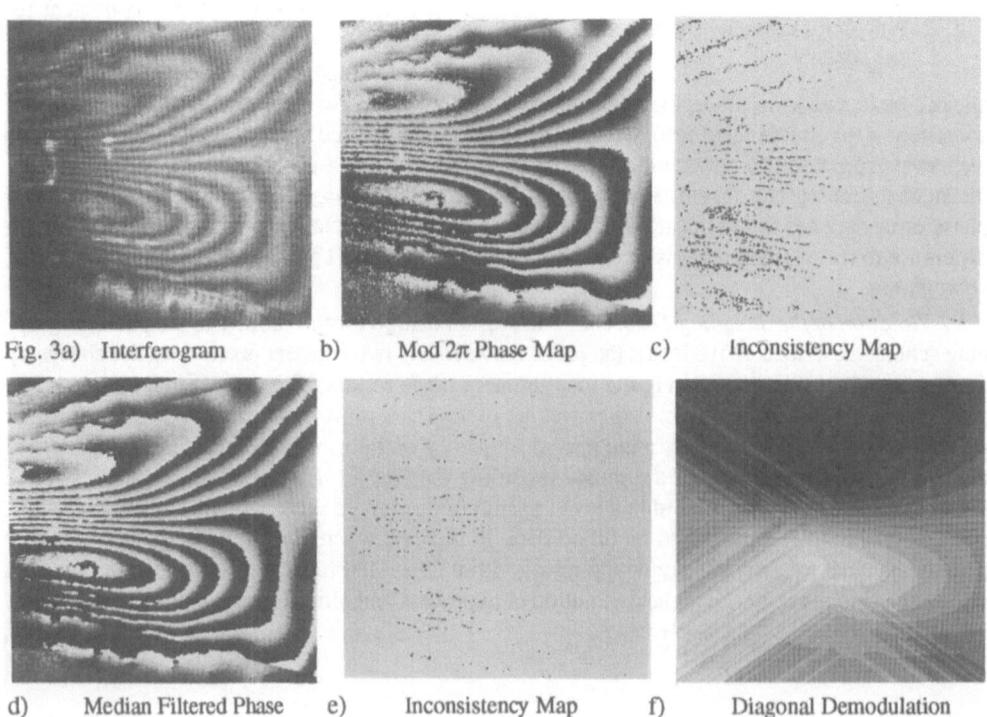

Fig. 3a)   Interferogram   b)   Mod 2π Phase Map   c)   Inconsistency Map

d)   Median Filtered Phase   e)   Inconsistency Map   f)   Diagonal Demodulation
         Map

It is obvious that alternative strategies for a correct unwrapping including preprocessing of the fringe patterns and smoothing the mod 2π phase map are necessary. In this section we want to describe some already known procedures and a new approach. We classify them into three groups according to the dependence of the processing from the content of the phase image: image content noncontrolled procedures, local and global methods of image content controlled procedures.

2.3.1 *Preprocessing.* To avoid faulty demodulations caused by noisy data preprocessing of the phase shifted interferograms is necessary. To reduce electronic noise frame averaging should be performed. For the reduction of speckle noise spatial low-pass filtering is sufficient in most cases.

We prefer an anisotropic filtering using different oriented filter masks which are controlled by a previously derived direction map of the fringe pattern [26]. An other approved mean for the noise reduction is the smoothing of both images resulting from the subtractions in Equ. (3):

$$I_4 - I_2 = 2 \cdot b(x,y) \cdot \sin[\delta(x,y)] \tag{19}$$

$$I_3 - I_1 = 2 \cdot b(x,y) \cdot \cos[\delta(x,y)] \ . \tag{20}$$

Noisy mod 2π images should be smoothed by median filtering capable to preserve true phase discontinuities and to decrease the number of inconsistencies considerably. Anisotropic filtering is also more convenient in this case.

2.3.2 *Image Content Noncontrolled Procedures.* These methods are characterized by walking along a continuous route in a constant direction independent on the phase vallues. The above described conventional procedure (see Eqs. 15-18) belongs to this class. Better results can be obtained by more than one unwrapping direction. Nakadate et al [4] compute four phase values for each pixel using four different unwrapping directions, two in line and two in column direction.. From these four phase values the pair with the smmallest difference is chosen to commpute its mean value as the resulting phase. Another strategy is to start from a proper chosen point in the centre of the interferogram and to follow continuous paths in diagonal direction (see Fig. 3f)

2.3.3 *Image Content Controlled Procedures (Local Methods).* These procedures are characterized by tracing along consistent paths with changes of the direction dependent on the processed phase value. That means the detected inconsistencies are considered directly or differences of neighboured phase values are used. One way is to look for the longest path (e.g.. in a line) without inconsistencies and to start the further unwrappping procedure. form these pixels. (e..g.. in column direction). The unwrapping is stoped if the process reaches an inconsistent area. This area is excluded and a better path is chosen in the neighbourhood (see Fig. 4).

 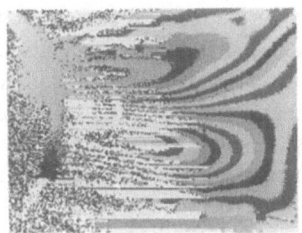 

Fig. 4: Unwrappping along Consistent Paths

This method can be improved by combining detected inconsistencies before unwrapping. Huntley [6] combines inconsistencies with opposite sign to build cutting lines bordering inconsistent areas in the phase map so that the condition $\Sigma W\Delta=0$ is fulfilled for the unwrapping path. Isolated inconsistend points are connected if their distance is minimal. Another algorithm was proposed by Ettemeyer et al. [27]. Here at first a net of neighbourhood relations is generated based on the comparison of all pixels with their upper and right neighbours. These results are stored in a list taking into account that the results with the smallest difference are placed at the top (following the principle that the phase difference between neighbouring pixels is usually small) and with increasing differences further behind. Using this list an unwrapping path is generated with a rank order of connecting pixels controlled by their place in the list. Pixels with differences about π are probably inconsistent, consequently they are considered at the end of the process. Other strategies include more regional information to devide the interferograms in regions and to prevent phase unwrapping along inconsistent paths. Ghiglia et al. [5] have proposed an interesting method based on cellular automata that can accomodate arbitrary boundary conditions due to the region partitioning. Gierloff's partitioning procedure [28] minimizes the number of inconsistent crossings between the regions by shifting the phase between them against each other. Most of these algorithms use interaction - especially in the presence of natural or aliasing-induced dislocations. Both cannot be removed by addition or subtraction of 2π what is the sense of phase unwrapping. Without a-priori knowledge aliasing-induced dislocations cannot be distinguished from true dislocations and the removal of the latter would actually change the phase. Taking into account such non-triviall cases a commpletely automatic unwrapping procedure must include a lot of a-priori knowledge about the phase distribution. Greivenkamp [7] proposed a technique based on the concept of sub-Nyquist sampling. A-priori knowledge about the wavefront (i.e. its shape) yields more than two orders of magnitude improvement in measurement range over the conventional phase-sampling technique. Stahl [8] reports about completely new noise-insensitive lgorithms implemented with elements of artificial intelligence capable to avoid the problems of unwrapping. But the bandwidth of problems in holographic interferometry is so large that it must e an extremly intelligent tool working correct in all imaginable constellations. The main question is - how to decide between physical and noise induced dislocations.

2.3.4 Image Content Controlled Procedures (Global Method). Our new approach consists in the consideration of the topology of the discontinuity. Similar the 2D fringe tracking a skeleton is derived from the mod 2π phase map using two different segmentation operators to detect all relevant edge-like discontinuities. Then a vector approximation of the line structures is performed and this vector description of the discontinuity is stored in a list file. The conversion of raster images into lists makes it possible to separate the metrical and topological properties of the image with respect to its subsequent structure analysis [29]. Based on this new description a better restoration of the noisy discontinuity field is possible. Whereas the consistency check described above is a local test this analysis gives a global approach to the structure of the mod 2π-image. Regional phenomena as for instance phase dislocations can be detected and considered in a better way. Furtheron the image can be cleaned from noisy and isolated discontinuities and true discontinuities can be improved. Taking into account the properties of true discontinuity lines as contour lines of a potential field all these operations can be implemented with respect to the automatization of unwrapping noisy interferograms. This global approach is explained more detailed now.

a) *Region Partioning.* A useful segmentation step lies in a global partitioning of the interferograms nto areas with and without relevant data. According to the phase sammpling principle described bove more information than the phase distribution can be reconstructed from the four phase shifted interferograms. The result of unwrapping in Fig. 3 suggests the generation of masks to exclude areas with inconsistencies autommatically. The generation of masks for labeling irrelevant areas can be done with help of the computed background intensity distribution a(x,y) and the modulation V(x,y)=b(x,y)/a(x,y). The derived background intensity and modulation distribution (see Fig. 5a and 5b) are compared with proper thresholds to receive binary masks. Both masks are combined (see Fig. 5c) to frame the relevant area for further processing (see Fig. 5d).

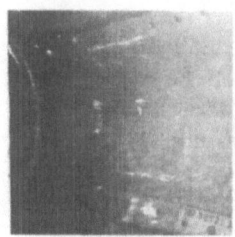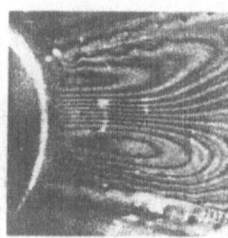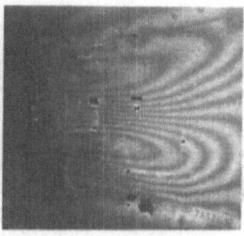

Fig. 5a) Background Intensity   b) Modulation   c) Binary Mask   e) Masked Interferogram

b) *Derivation of Line Structures.* For an effective fringe pattern analysis the conversion of the raster-like mod 2π-image into a list-like data strukture is very often performed by line images. Therefore we consider the discontinuities in the field of principal phase values as edges. For this purpose automatic derivation of line structures from the phase discontinuities by means of edge or region based segmentation operators is necessary. Edge detection is a complicated task because the images are corrupted by noise and it is difficulties to distinguish the different kinds of discontinuities. In image analysis there are a number of so called optimal detectors based on different optimality criteria.The operator for the unwrapping problem must meet following requirements:

- pixel precise localisation of the 2π discontinuities,
- evaluation of discontinuity direction (monotony,fringe order),
- distiinction of the different kinds of discontinuities.

After discontinuity detection the line image includespphase discontinuities as skeleton lines and masked regions that are excluded from further processing (see Fig. 6). The gray values of the skeleton lines represent the local monotony of the phase discontinuity. They can be used to consider the discontinuity direction between the regions.

c) *Vector Appproximation And Data Conversion.* The following context-information of the derived line image can be used as a-priori knowledge for further processing:

- skeleton lines are contourlines of the phase function $\delta(x,y)$,
- fringe order differences of neighbouring skeleton lines can be 0 or +/−1,
- skeleton lines with different order can neither cross nor touch each other,
- skeleton lines with equal order can neither cross nor touch each other, unless closed contour lines; they can only divide,
- skeleton lines can nowhere end, unless at the border of the considered area or of the object.

94

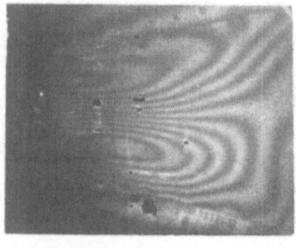 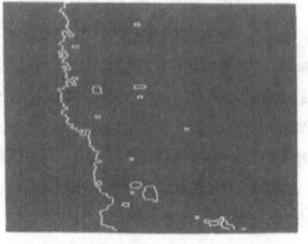 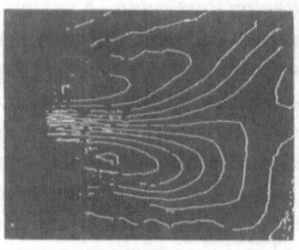

Fig. 6a) Interferogram   b)   Boundary Line Mask   c)   Relevant Contour Lines

In the next step all contour lines are traced and stored in a coded list formm (e.g. chain code). The pixel accurate contour code ensures a precise overlay of the bias-image and the mod 2π-image.Further topological and metric information of the structure of the phase discontinuity map is derived from the approximated contour lines. The aim of this step is the automatical improvement of the true discontinuity lines.The above mentioned error sources cause false lines and disturb true lines (gaps, isolated segments, staggered lines, ...) as shown in Fig. 6. They can be recognized in the list and should be improved with respect to the global structure analysis of the contour map. That procedure is based on the introduction of topological separated regions using the already derived binary mask and the list stored coordinatesof the end points of contour segments. Fig. 7 shows an example with a recognized object discontinuity (gap) and crack (cutted and displaced lines). These recognized structures are considered during the unwrapping as physical caused discontinuities and are not corrected.

 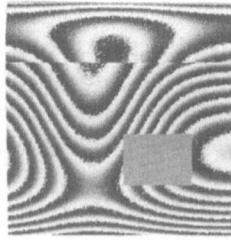  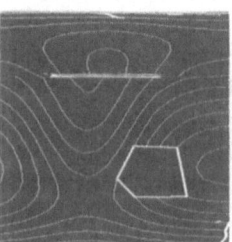

Fig. 7a) Interferogram   b) mod 2π-Image   c) Contour-Line Image   d) Recognized Structures

3.   Knowledge Assisted Fault Detection in Interference Patterns

Our approach is based on a three step procedure which is oriented on the evaluation of the pattern information of the reduced fringe skeleton. The detected irregular patterns are pre-classified for flaw presence in the second step. Final classification with the objective to derive the kind of flaw is made on the base of additional knowledge about the pattern formation and the experimental conditions.

3.1 PREPROCESSING

For the recognition of fault indicating irregularities in holographic interferograms it is necessary to process the rough fringe pattern with regard to image improvement and data reduction. Only the pattern information must be preserved which can be done by reducing the 8-bit digitized interferogram to a binary fringe skeleton.Using adapted algorithms for temporal and local filtering the signal-noise ratio in the interferogram can be improved considerably. The chosen skeleton method can be compared to manual fringe evaluation: by automatic identification of fringe loci with local extreme intensity the complex information of the rough interferogram is reduced to contour lines with equal phase difference. For the segmentation into regions of minima (dark fringes), maxima (bright fringes) and transitions between them two-dimensional algorithms based on local grey value differences and resulting neighbourhood relations are used [30]. Additionally, for each pixel the fringe direction is estimated and stored in a so called direction map. This map is used in a fringe flow controlled improvement of the derived binary image (anisotropic filtering) and in the intended structure analysis of the pattern. By region growing the binary line pattern is improved further, and by line thinning it is reduced to a skeleton. Fig. 8 shows several intermediate results of the processing chain.

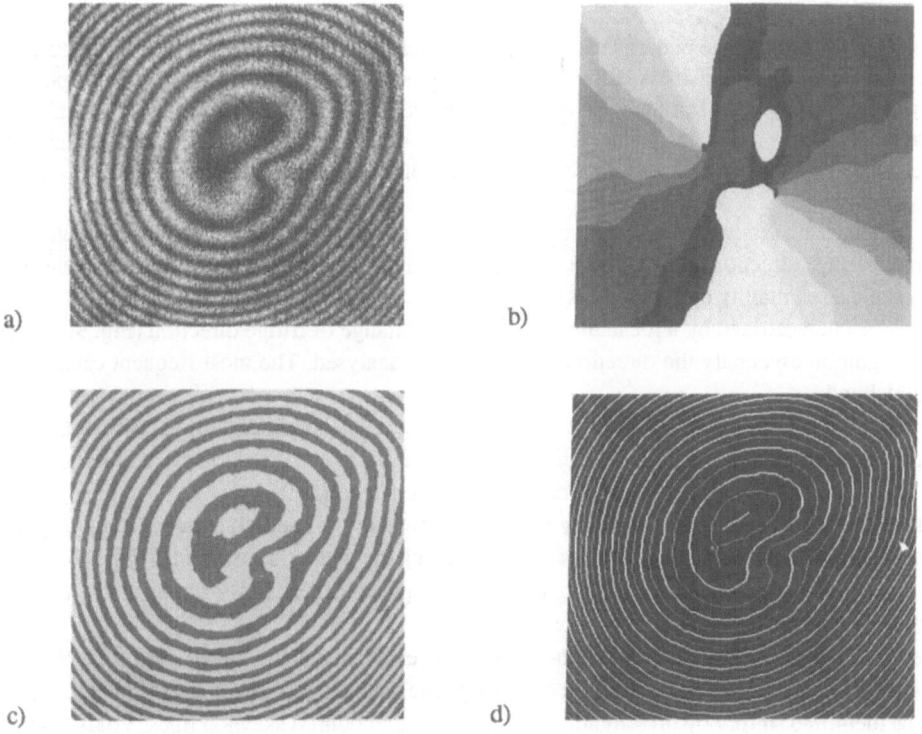

a) b) c) d)

Fig. 8: Processing of Fringe Patterns

      a) Interferogram,   b) Direction Map,   c) Segmented Pattern,   e) Skeleton

Using edge detection algorithms skeletons can be derived also from mod 2π-images in phase shifting interferometry. Qualitative evaluation based on skeleton lines requires a global description of the skeleton. After approximating skeleton lines by vectors a hierarchical data list is generated which contains coordinates of all nodes and information about the shape of each skeleton line. Closed lines or branchings are recognized when tracing the skeleton lines. Line attributes distinguish ring-shaped lines from non-closed ones and register the number of branches. In this way metrical and topological properties of the pattern are stored in a data file which can be accessed easily. Consequently, the resulting list is the basis for the determination of pattern properties necessary for fault classification.

3.2 MODEL BASED PATTERN RECOGNITION

Qualitative analysis of interferograms is based on the identification of characteristic partial patterns within the fringe system (*first classification step*) and on their evaluation with resppect to the kind of flaw (*second classification step*). However, interferometric measurement techniques are sensitive only for deformations on the surface. Therefore flaws under the surface can be detected just by their affects on the surface. Because the conclusion from the pattern to the subsurface flaw is ambiguous additional knowledge has to be taken into account in the second classification step. The aim of the first classification step is to identify and to localize fault indicating patterns in the skeleton. Starting from practical experience our approach is based on the hypothesis that the morphological appearance of faults in fringe patterns can be reduced to a set of typical pattern types or irregularities [31]. So far we distinguish six different pattern types: *compression, bend, groove, displacement, eye* and *chain of eyes*. For the description and detection of these patterns morphological and geometrical properties like density, curvature, direction and shape of the lines are used. Because of the great variety of pattern realizations the application of template matching techniques used in OCR is not possible.

- The pattern class *compression* is defined by a marked change of spatial frequency with almost constant fringe direction and can be detected by evaluating the line density within the skeleton. It can be caused mainly by separations or weak points in the tested material. (Fig. 9a).

- A *bend* is characterized by a local and non continuos change of fringe direction (Fig. 9b). For its recognition especially the direction map should be analysed. The most frequent cause is a local debond.

- The third basic pattern is called *groove* and is defined by a systematic and directed distortion of neighboured fringes (fig. 9c). Local changes of line curvature and direction point to its presence. It is caused mainly by subsurface cracks or extended debonds.

- Cracks on the surface can be proved by local fringe breaking off and systematic lateral *displacements* of line ends belonging together. Near the crack the fringes seem to end (Fig. 9d). Usually the derived skeleton shows many branches at this place. Linear edges can be distinguished from irregular cracks by Hough transform.

- Circular or elliptical structures well known as "bull-eye" fringes are called *eyes* (Fig. 9e). If several of them occur as a sequence on a seeming line we have a *chain of eyes* (Fig. 9f). They can be identified simply by investigating the shape of neighboured skeleton lines. Voids, inclusions and local debonds can cause them.

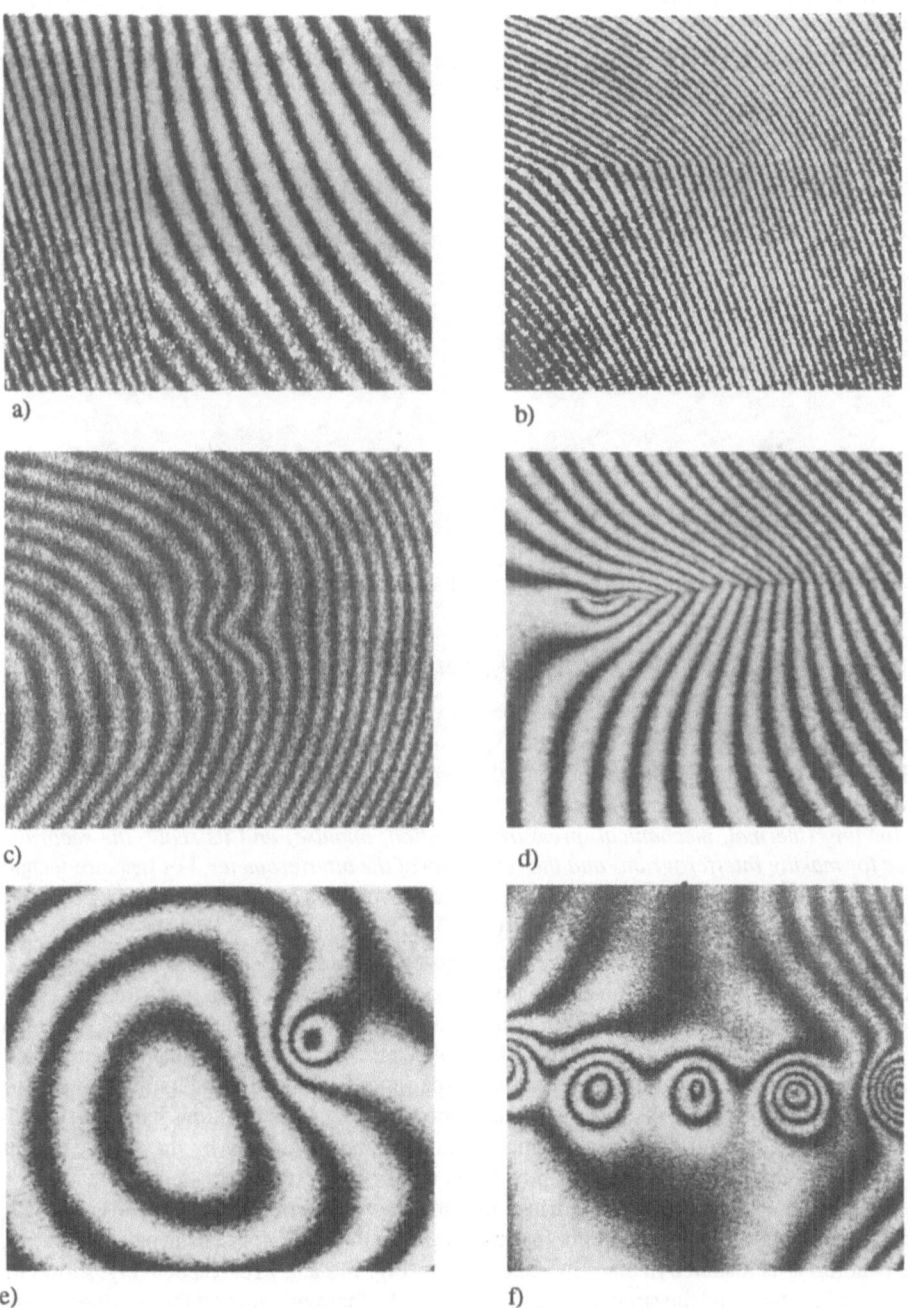

Fig. 9: Pattern Types

a) Compression,   b) Bend,   c) Groove,   d) Displacement,   e) Eye   f) Chain of Eyes

The first classification step results in the detection of regions with irregular patterns pre-classified for flaw presence. Some typical examples are shown in Fig. 10. In the first one the pattern type groove is localized. The second example shows an identified displacement. The identification results are added to the data list describing the skeleton and they are available for the next processing step.

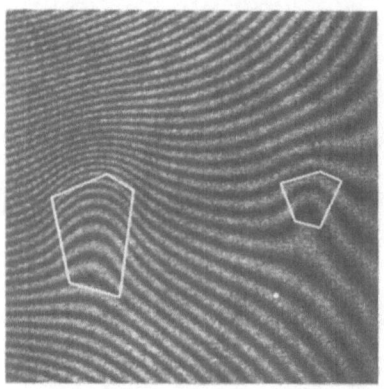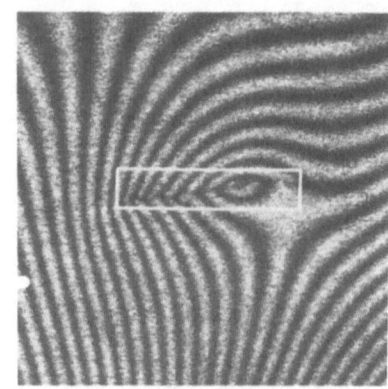

Fig. 10 a) Identified Groove   b) Identified Displacement

3.3   KNOWLEDGE BASED FLAW CLASSIFICATION

In general the conclusion from the identified partial pattern to the type of flaw causing it can be drawn only if additional knowledge about the experimental conditions and the object properties is available. Following factors influence mainly the appearance of the fringes: *the material of the object under test, the construction of the object, the fixing of the object in the loading device, the kind of loading (thermal, mechanical, pressure, vibration, impulse) and its scale, the recording technique for making interferograms and the sensitivity of the interferometer.* For instance a change in loading conditions causes quite different patterns. Also flaws of the same type having different scales may result in very different patterns (see [31]).

Aimed at the development of a knowledge-based system for the detection and classification of faults it is necessary to investigate the connection of pattern realizations with measurement conditions. Knowledge about possible fringe patterns for all considered fault classes (void, inclusion, debond, separation, delamination, crack) should be collected. For this purpose the computer simulation of holographic interferograms and practical experiments are used. For simulated interferograms on the basis of Equ. 1 the surface displacements resulting from different kinds of flaws in different materials and at different measurement conditions are computed by the finite-element-method (FEM). In this way the relationship between an assumed subsurface defect and resulting surface displacements and corresponding fringe irregularities can be studied extensively. The used simulation model is confirmed by making interferograms in the laboratory ensuring similar experimental conditions as assumed in the FEM-calculations. Fig. 11a and 11b, respectively, shows the calculated displacement and the simulated interferogram of a thermally loaded faulty adhesive layer. At the same time a large number of experimentally generated interferograms of prepared objects is investigated to study the dependence between different kinds of flaws and detectable effects in the fringe pattern. Fig. 11c and 11d show a acrylic glass plate with a circular void in different phases of thermmal loading. The fringe pattern corresponds to the simulated one because of the same experimental conditions. Different materials ranging from plastics to steel are considered. Regarded faults are prepared cracks, voids (holes inside), debonds (welding seams) or delaminations in layered structures. In our experiments especially thermal and mechanical loading or the combination of both are used. Expert knowledge concerning the relation cause-pattern and concerning the influence of experimental conditions and object properties will be expressed in form of decision rules. These rules are embedded in a suitable shell for a knowledge-based system.

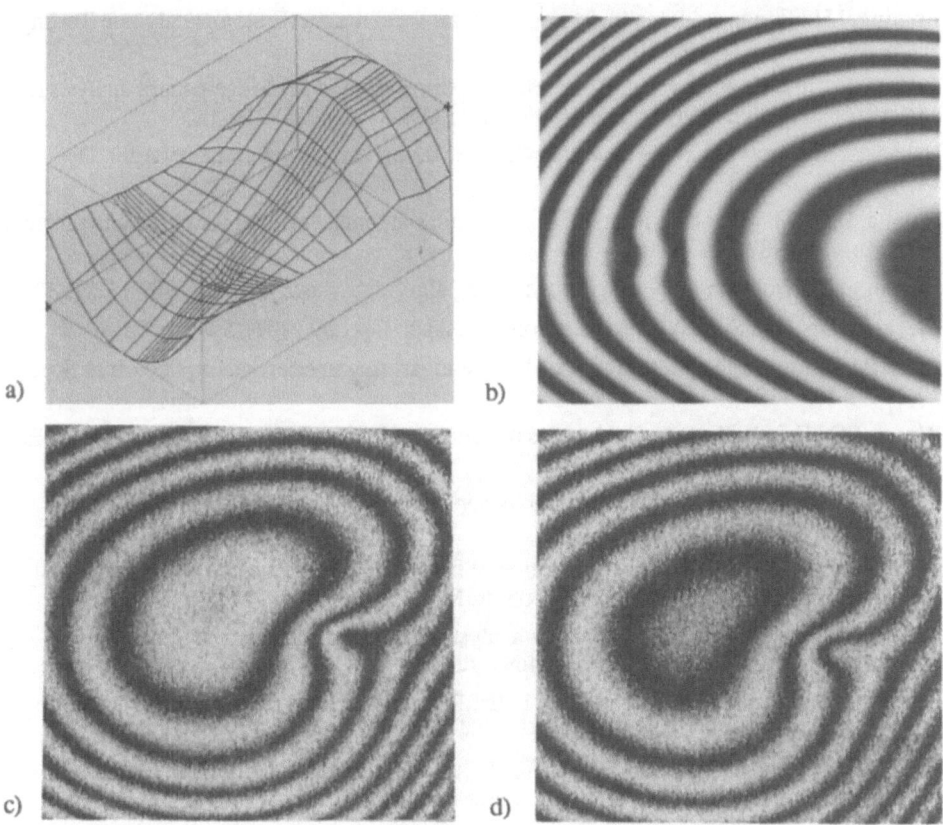

Fig. 11:a) FEM-Mesh and Calculated Deformation   b) Simulated Interferogram

c) and d) Thermal Loaded Object Prepared with a Void in Different Loading States

4. Summary

In this paper the application of digital image processing for the quantitative and qualitative evaluation of interferograms in holographic nondestructive testing was demonstrated. Both examples - the unwrapping of noisy interferograms and the detection of material faults in fringe patterns are in the focus of international investigations. Our approach is based on the derivation of line structures from such noisy fringes, their vectorization and conversion into list structures storaging metrical and topological features. Using these hierarchical lists the improvement of discontinuity fields in mod 2π-images and the recognition as well as the classification of fault indicating patterns is simplified.

5. Literature

[1] Osten, W.: Digitale Verarbeitung und Auswertung von Interferenzbildern. Akademie Verlag Berlin 1991

[2] Bruning, J.H.: Fringe scanning interferometers. In: D. Malacara (Ed.): Optical Shop Testing. Wiley, New York 1978, pp. 409-437

[3] Takeda, M.; Ina, H.; Kobayashi, S.: Fourier-transform method for fringe-pattern analysis for computer based topography and interferometry. J.O.S.A. 72(1982),156-160

[4] Nakadate, S.; Hiroyoshi, S.: Fringe scanning speckle-pattern intereferometry. Appl. Opt. 24(1985), 2172-2180

[5] Ghiglia, D.C.; Mastin, G.A.; Romero, L.A.: Cellular-automata method for phase unwrapping. J.O.S.A.(A) 4(1987), 267-280

[6] Huntley, J.M.: Noise-immune phase unwrapping algorithm. Appl. Opt. 28(1989), 3268-3270

[7] Greivenkamp, J.E.: Sub-Nyquist interferometry. Appl. Opt. 26(1987), 5245-5258

[8] Stahl, P.: Testing large optics: high-speed phase-measuring interferometry. Photonics Spectra 12(1989), 105-112

[9] Wyant, J.C.; Creath, K.: Recent advances in interferometric optical testing. Laser Focus / Electro Optics 11(1985), 118

[10] Pryputniewicz, R.J.: Heterodyne holography applications in studies of small components. Opt.Eng.,24(1985),849

[11] Dudderar, T.D.; Doerries, E.M.: Application of holographic interferometry to real-time studies of heat effects in multilayer circuit boards. Mater.Eval.,37(1979),41-47

[12] Tichenor, D.A.; Madsen, V.P.: Computer analysis of holographic interferograms for non-destructive testing. Opt.Eng.,18(1979)5,469-472

[13] Brown, G.M.: Pneumatic tire inspection. in Erf,R.K. (Ed.): Holographic nondestructive Testing. Academic Press, New York,1974,355-364

[14] Osten, W.; Saedler, J.; Wilhelmi, W.: Schnelle Auswertung von Interferogrammen mit Verfahren der digitalen Bildverarbeitung. Laser Magazin,2(1987),58-66

[15] Roesener, K.; Wagner, S.; Jüptner, W.: Holografische Untersuchungen an großflächigen CFK-Bauteilen. Proc.Laser'89,327-331

[16] Parker, R.J.; Reeves, M.: Holographic flow visualization in rotating turbomachinery. Proc. 1990 SEM Fall Conference, Baltimore,500-507

[17] Steinbichler, H.; Franz, T.; Engelsberger, J.; Sixt, W.; Sun, J.: Industrieller Einsatz der digitalen Bildverarbeitung in der Holographie und ähnlichen Meßtechniken. Proc. Laser'89, 282-291

[18] Trolinger, J.D.; Weber, D.C.; Pardoen,G.C.; Gunnarson,G.T.; Fagan,W.F.: Application of longe-range holography in earthquake engineering. Opt.Eng.,30(1991)9,1315-1319

[19] Füzessy, Z.: Application of double-pulse holography for the investigation of machines and systems. in Frankowski,G et al (Eds).: Application of metrological laser methods in machines and systems.Akademie Verlag, Berlin,1991,75-107

[20] Glünder, H.; Lenz, R.: Fault detection in nondestructive testing by an opto-electronic hybrid processor. Proc. SPIE 370(1982),157-162

[21] Robinson, D.W.: Automatic fringe analysis with a computer image-processing system. Appl.Opt.,22(1983)14,2169-2176

[22] Kreis, Th; Osten, W.; Mieth, U.: Untersuchungen zum automatischne Nachweis von Materialfehlern in Interferenzmustern mit wissensbasierten Systemen und neuronalen Netzen. DFG - Report Kr 953/5-1, Bremen 1991

[23] Bischof,Th.; Jüptner,W.: Determination of the adhesive load by holographic interferometry using the result of FEM-calculations. SPIE Proc. 1508(1991),90-95

[24] Osten, W.; Höfling, R.; Saedler, J.: Two computer-aided methods for data reduction from interferograms. Proc. SPIE Vol. 863(1987), 105-113

[25] Andrä, P.; Mieth, U.; Osten, W.: Strategies for unwrapping noisy interferograms in phase-sampling interferometry. Proc. SPIE Vol. 1508(1991), 50-60

[26] Winter, H.; Unger, S.; Osten, W.: The application of anisotropic and adaptive filtering for the extraction of fringe pattern skeletons. Proc. FRINGE'89, Akademie Verlag Berlin 1989, pp. 158-166

[27] Ettemeyer, A.; Neupert, U.; Rottenkolber, H.; Winter, C.: Schnelle und robuste Bildanalyse von Streifenmustern - ein wichtiger Schritt der Automation von holografischen Prüfprozessen. Proc. FRINGE'89, Akademie Verlag Berlin 1989, pp. 23-31

[28] Gierloff, J.J.: Phase unwrapping by regions. Proc. SPIE Vol. 818(1987), 2-9

[29] Kiesewetter, H.; Osten, W.; Saedler, J.: Automatische Ableitung und Verarbeitung von Linienbildern. Informationstechnik 32(1990), 376-386

[30] Eichhorn, N.; Osten, W.: An algorithm for the fast derivation of line structures from interferograms. J. Mod. Opt. 35(1988)10, 1717-1725

[31] Osten, W.; Jüptner, W.; Mieth, U.: Knowledge assisted evaluation of fringe patterns for automatic fault detection. Proc. SPIE Vol.2004(1993) (to be published)

Design of Morphological Processors for Ultrasonic Nondestructive Evaluation of Materials - A Review

J. SANIIE, M. A. MOHAMED and K.K. CHIN
Department of Electrical and Computer Engineering
Illinois Institute of Technology
Chicago, Illinois 60616

ABSTRACT.   Sequential morphological operations are capable of extracting signal features while suppressing random noise and undesired signal patterns (e.g., speckles in ultrasonic imaging). They utilize a structuring element which interacts with the signal in order to suppress noise and enhance certain desirable information. In this paper we identify a group of sequential morphological processors that exhibit a performance similar to lowpass, bandpass and highpass filters. Furthermore, a class of morphological processors are presented which offer peak detection and edge detection. Deterministic and stochastic properties of combinational (parallel and/or serial) morphological processors have been studied. In particular, the information content of ultrasonic signals has been used to design a suitable structuring element with optimal performance.   The results obtained by applying morphological processors to experimental ultrasonic signals show that combinational morphological processors can improve flaw detection when the signal is contaminated by impulsive thermal noise and/or microstructure scattering echoes.

1. Introduction

Ultrasonic flaw detection is an important technique to assure the integrity of materials nondestructively. The goal in detection is to isolate the flaw echo from its background noise (e.g., microstructure scattering echoes and/or instrumentation noise) and to estimate its exact location. The interfering noise often becomes significant to the point that it masks the presence of flaw echoes. This paper presents several designs of morphological filters suitable for the ultrasonic nondestructive evaluation of materials.   Morphological filters are a class of nonlinear filters which have recently become popular in signal and image processing [1-9].   Morphological filters are attractive for their relatively simple computational demands consisting predominantly of simple addition, subtraction, and comparison operations. They are robust and offer an excellent noise suppression capability.

The word morphology refers to the study of forms and structures. The theoretical foundations of mathematical morphology and a wide range of its applications have been introduced systematically by Matheron [1] and Serra [2].   The primitives of morphological operations are erosion and dilation. All other morphological operations

X. P. V. Malague (ed.), Advances in Signal Processing for Nondestructive Evaluation of Materials, 103–116.
© 1994 *Kluwer Academic Publishers.*

opening) when applied to signals with uniform distributions. Also, the mean, variance and skewness are estimated in order to determine the noise suppression capability of morphological filters and their biasing effects.

The dilation, closing, clos-erosion, and clos-opening density functions when the samples of the inputs are independent and identically distributed (iid) with uniform density functions (distributed between zero and one) and when a flat structuring element with a width M=7 and height equal to zero is used are shown in Figure 1. The mean, normalized variance (ratio of output variance over input variance) and skewness as functions of widths of the structuring element are shown in Figure 2. These figures indicate that dilation shifts the input signal toward the maximum values and the signal variance is reduced, resulting in a smooth operation. The closing operation (i.e., dilation followed by erosion) is a processing step toward recovering the original signal. Overall, the closing operation tunes down the bias caused by the dilation operation, and slightly increases the signal variance with respect to the dilation operation. The clos-opening operation results in further smoothing of the signal, where the signal mean and variance are less than those of the closing signal. The mean of dilation, closing, clos-erosion and clos-opening is shifted to the right (see Figure 2a) due to the effect of dilation as a first operation in the sequence of the above operations. Note that the height of the flat structuring element has no effect on the closing and clos-opening density function, but shifts the dilation density function to the right and the clos-erosion density function to the left by a predictable value depending on the height. In general, the above figures suggest that by increasing the width of the flat structuring element the mean (bias) is increased and the variances of dilation, closing, clos-erosion and clos-opening output density functions is decreased. Furthermore, dilation has been found to be the most effective step in the smoothing operation. The effect of the other operations following dilation is rather small, specially for larger widths.

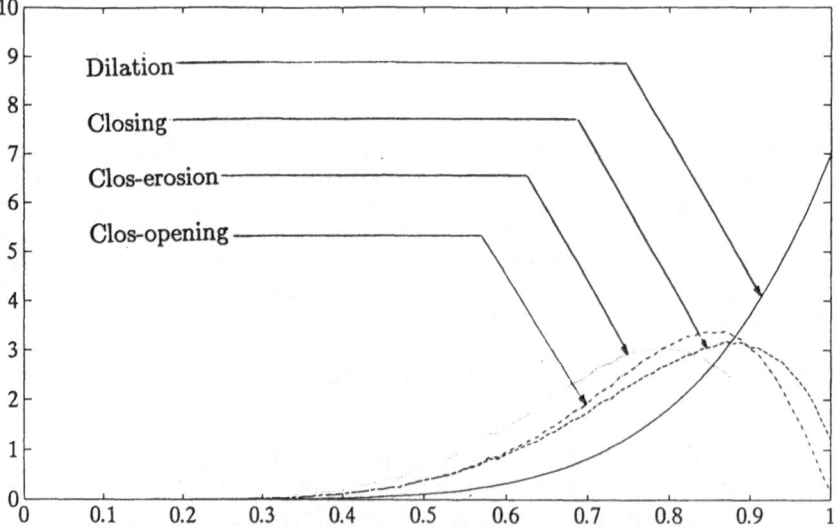

Figure 1.   Probability density functions of dilation, closing, clos-erosion and clos-opening when the input signal has a uniform density function and a flat structuring element with a width of 7.

are derived by combining erosions and dilations in parallel and/or serial order. These operations use a structuring element to interact with the signal and extract information. A structuring element is another signal of a somewhat simpler nature than the signal under processing. The structuring element interacts with the signal under study and transforms it into a new signal which is in some way more expressive than the original. By varying the structuring element, different types of information can be extracted from the signal.

In this paper we present a group of morphological processors with performance similar to lowpass, bandpass and highpass filters. Furthermore, a class of morphological processors is presented that offers peak and edge detection. Deterministic and statistical properties of morphological filters are examined in order to design morphological filters with optimal performance for flaw detection.

2. Statistical Properties of Morphological Operations

There are two fundamental morphological operators: erosion and dilation. Other operations such opening and closing are derived operations defined in terms of erosion and dilation. The multilevel dilation of function f by a multilevel structuring element s can be computed in terms of a maximum operation and a set of addition operations.

$$(f \oplus s)(x) = MAX\,[f(x - z) + s(z)] \quad \textit{for all } z \, \varepsilon \, S \textit{ and } x - z \, \varepsilon \, F \qquad (1)$$

where F and S are the domain of functions f and s respectively. The multilevel erosion of a function f by a multilevel structuring element s can be computed in terms of a minimum operation and a set of subtraction operations.

$$(f \ominus s)(x) = MIN\,[f(x + z) - s(z)] \quad \textit{for all } z \, \varepsilon \, S \textit{ and } x + z \, \varepsilon \, F \qquad (2)$$

In general, both erosion and dilation operations are not invertible. A derived operation using erosion and dilation is known as an opening which is defined as:

$$f \circ s = (f \ominus s) \oplus s \qquad (3)$$

The dual operation to opening is known as closing. Closing is defined as:

$$f \bullet s = (f \oplus s) \ominus s \qquad (4)$$

The goal of any morphological operation is to manage the loss of information through successive transformations. The opening operation is used to suppress positive pulses while a closing is used to suppress negative pulses. A general morphological filter consists of several morphological operations performed in tandem with the same or different structuring elements. It is important to point out that the order of the operations is critical because different results are obtained depending on whether an opening or closing is done first.

In order to properly apply morphological filters to ultrasonic signals, their statistical properties have been examined. In particular, we present the output density functions of sequential morphological operations (i.e., dilation, closing, clos-erosion and clos-

106

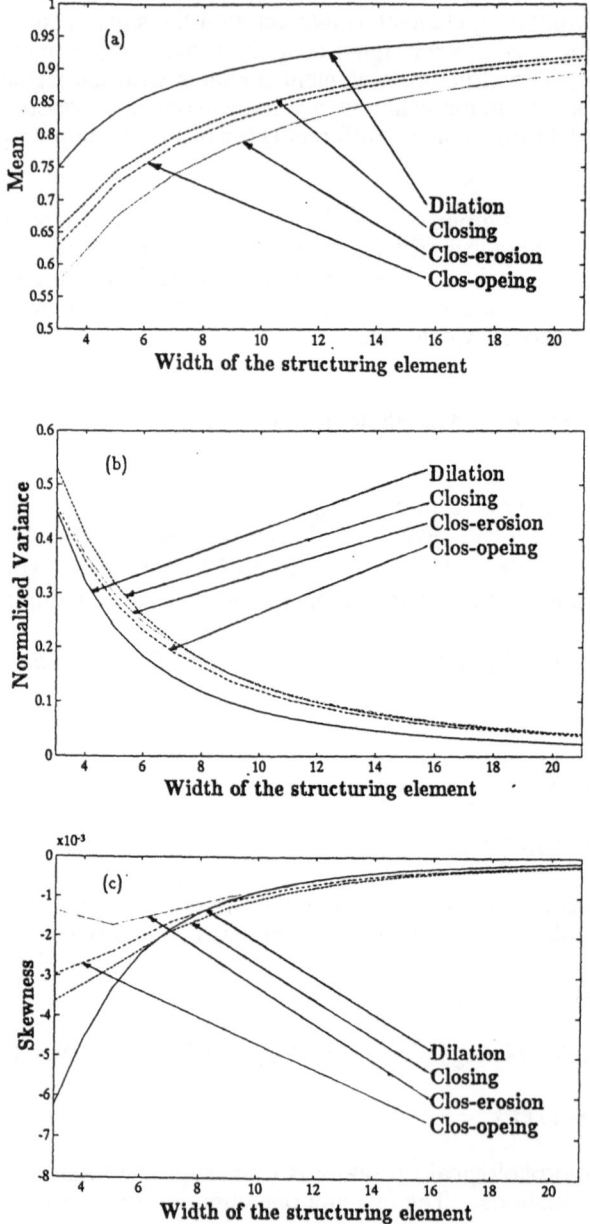

Figure 2.   Mean, normalized variance and skewness of dilation, closing, clos-erosion and clos-opening when the input signal has a uniform density function for a flat structuring element.

3. Morphological Filters

A common noise in ultrasonic measurements is the thermal noise generated in the receiver amplifier. An ultrasonic echo signal,and that echo signal contaminated by impulsive noise (representing the thermal noise) with an amplitude of 0.5 and a duration of 7 samples are shown in Figures 3a and 3b respectively. Figure 3b is processed using an open-closing operation with a flat structuring element (width of 8 samples). Note that the width of the structuring element is larger than the width of the impulsive noise and significantly less than the number of samples per cycle (80 samples per cycle) of the ultrasonic echo. The opening operation is used to suppress the positive impulses while the closing operation suppresses the negative impulses. Figures 3c and 3d show the recovered signal and the error (difference) between the actual echo and the recovered signal respectively. These figures show the ability of an open-closing operation for suppressing positive and negative impulsive noise and recovering the original echo signal.

A model of a lowpass morphological filter for noise suppression is shown in Figure 4 which minimizes the bias due to the extensiveness property of opening and closing. In this model, a positively biased estimate of the output signal is obtained by processing the input signal using an open-closing operation. A second negatively biased estimate of the output signal is obtained by processing the input signal using a clos-opening operation. The average of these two estimates results in a minimum biased output signal. As a practical application of morphological filters for thermal noise suppression, a typical measured ultrasonic signal with poor SNR (Figure 5a) is processed using the lowpass morphological filter (see Figure 4) and the result is shown in Figure 5b. Figure 5c shows the same signal processed using an ensemble averaging of 500 measurements. These results indicate that morphological filters can be used as a replacement for ensemble averaging which requires numerous measurements. In conclusion, the ability of each method to smooth the signal is comparable, although morphological filtering is more efficient.

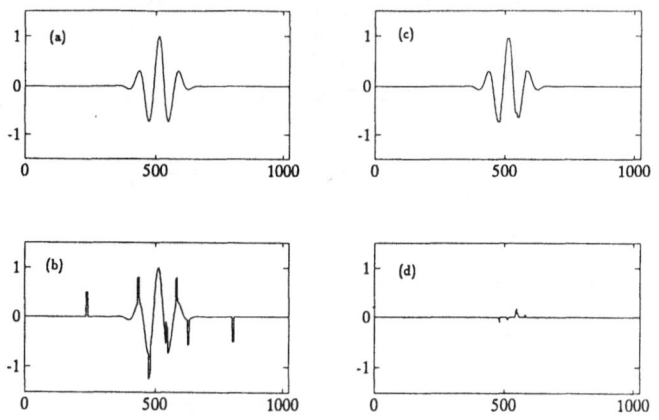

Figure 3.   Impulsive noise suppression using an open-closing operation.
(a) Simulated ultrasonic echo
(b) Echo contaminated by random impulsive noise
(c) Recovered signal using open-closing operation
(d) Error signal

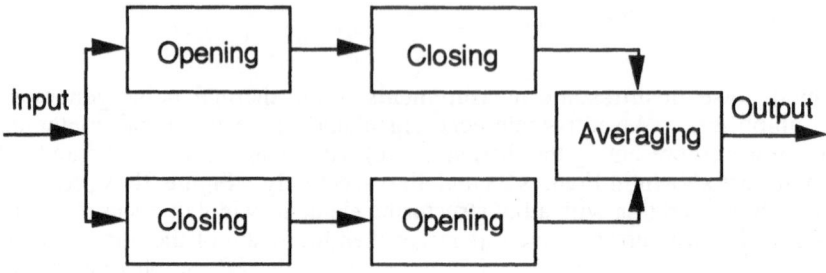

Figure 4.   Block diagram of a lowpass morphological filter for noise suppression.

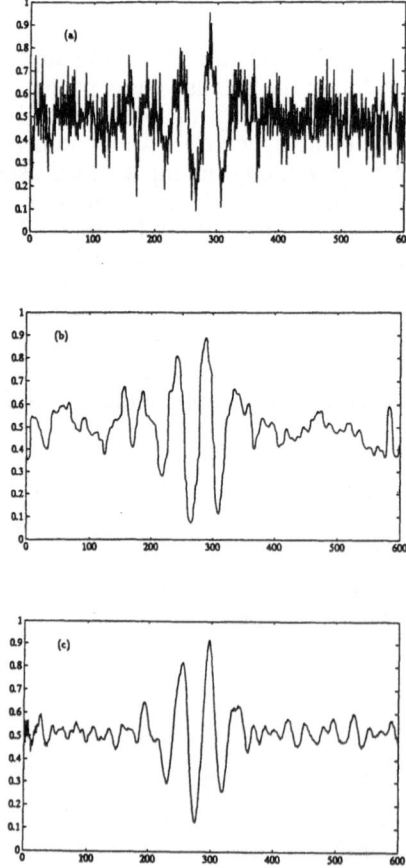

Figure 5.   (a) A typical measured ultrasonic echo with poor flaw-to-clutter ratio.
            (b) The processed echo using a lowpass morphological filter.
            (c) Ensemble averaging of the echo using 500 measurements.

To demonstrate the effect of the parameters of the structuring elements, an echo is processed using the same sequence of morphological operations shown in Figure 4. Figure 6a shows the processed signal using a flat structuring element with a width M and a constant height equal to zero. The output signal is processed for different values of M. When M ≥ 7, the output signal begins to attenuate with small distortion until M = 21. When M ≥ 21 the output echo becomes very small in value and the echo shape becomes highly distorted. Therefore, using a structuring element with a width less than half the period of the sine function is desirable to preserve the original shape without severe distortion. Figure 6b shows the result of applying the echo to the same morphological operations with a half-cycle sine structuring element. In this case, the output echo is less distorted than the output echo with a flat structuring element of the same width. The sinusoidal structuring element has a shape similar to the original signal which allows it to penetrate inside the peaks or valleys of the input signal. Therefore, the surface of the output takes the shape of the structuring element.

Sequentially alternating the application of opening and closing with the same structuring element removes details of the signal that are small relative to this structuring element. Using classical terminology, one can denote these alternating sequential operations as morphological lowpass filters (for example processing shown in Figure 4). Morphological highpass and bandpass filters can also be designed using lowpass filters with different cutoff frequencies. In summary, recommended designs of lowpass, highpass and bandpass filter are:

- The lowpass filter is defined as the average of open-closing and clos-opening.

- The highpass filter is defined as the original signal minus the average of open-closing and clos-opening.

- The bandpass filter is defined as the difference of two averages of open-closing and clos-opening with two different structuring elements.

It is important to mention that the filtering characteristics of a morphological processor cannot be generalized in terms of frequency response due to nonlinearity. Furthermore, the superposition principle does not apply to these nonlinear filters and defining the filter bandwidth is a meaningless quantity. This study only indicates that the filtering process depends on the input signal as well as the parameters of structuring elements.

4. Morphological Filters for Ultrasonic Flaw Detection

An important application of the morphological filter is in ultrasonic flaw detection when the flaw echo has been contaminated by the microstructure scattering echoes. The goal in detection is to isolate the flaw echo from its background clutter (e.g., microstructure scattering echoes) and to estimate its exact location. To illustrate the performance of morphological filters in ultrasonic flaw detection, a broadband transducer has been used to detect a simulated flaw embedded within a steel block with strong grain scattering. The measurement was achieved using the contact technique and data was acquired with a 100 MHz sampling frequency. The measured ultrasonic A-Scan is shown in Figure 7a. This signal was processed by the same sequence of opening and closing operations discussed in the previous section (see Figure 4). Processed results are shown in Figures 7b-7d for a flat structuring element with different widths (M=4, 6, and 8 samples).

110

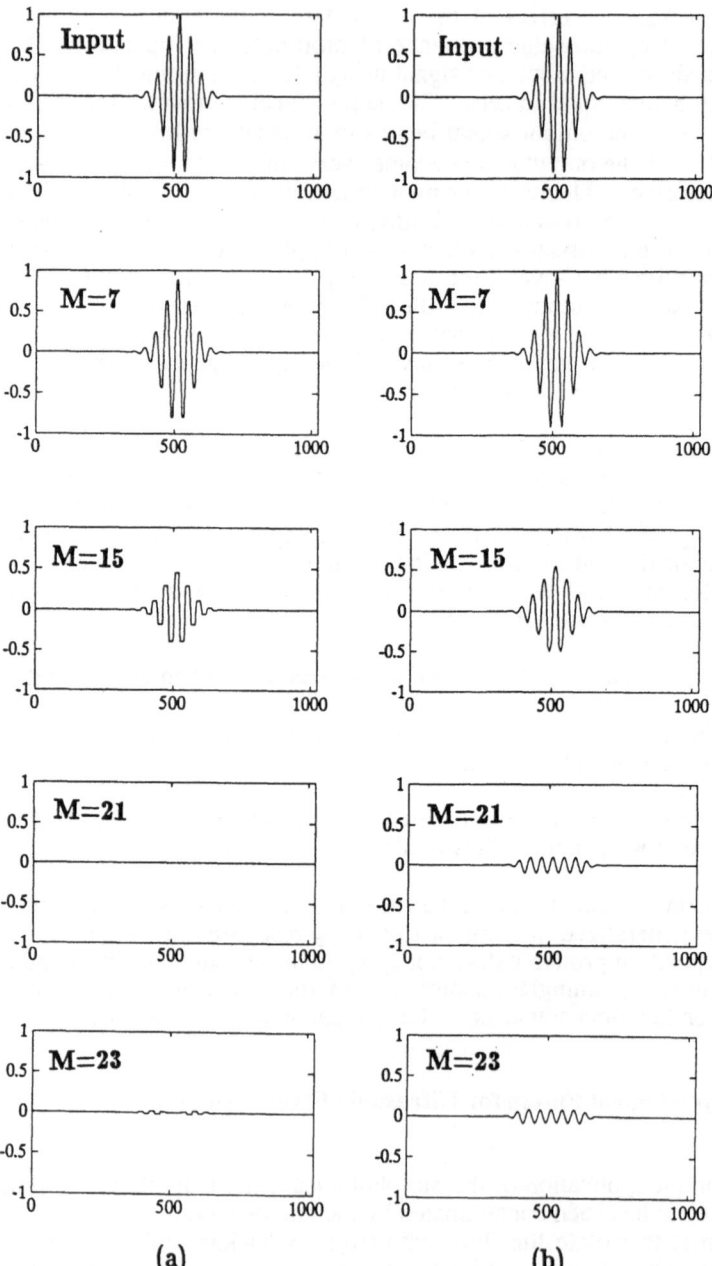

Figure 6.   The effect of the width of a structuring element when applied to an ultrasonic echo for
(a) Flat structuring elements
(b) Half-cycle sine structuring elements

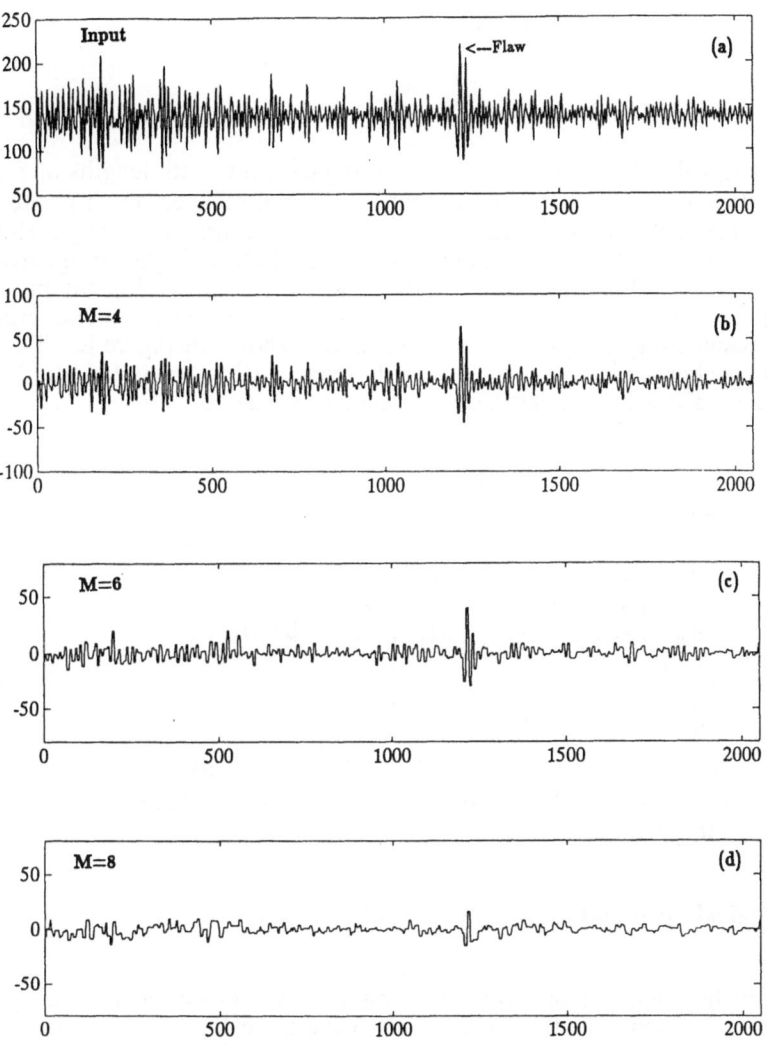

Figure 7.   Flaw echo detection and clutter suppression using lowpass morphological
            filters with flat structuring elements of different widths (M). Figure (a) is
            the original measured signal.

These results indicate that morphological filters are capable of detecting targets while
suppressing clutter.

To further improve the resolution and FCR for flaw detection, it is effective to estimate
the background echoes (clutter) and then subtract this estimate from the preprocessed
signal. This technique can be represented as

$$\hat{y} = y - [(y \circ s_2) \bullet s_2 + (y \bullet s_2) \circ s_2] / 2 \tag{5}$$

where

$$y = [(x \circ s_1) \bullet s_1 + x \bullet s_1) \circ s_1] / 2 \tag{6}$$

x is the input signal, s_1 and s_2 are flat structuring elements with lengths of 8 and 10 respectively. It is important to point out that Equation 6 reduces the high frequency information associated with microstructure scattering (i.e., lowpass filtering). However, the highpass filtering of this result (Equation 5) reemphasizes higher frequencies and broadens the bandwidth of the flaw echo. The above morphological processing (Equations 5 and 6) shows a pseudo bandpass filter response. The processed experimental result using the pseudo bandpass filter is shown in Figure 8. This result shows that the background noise is tremendously reduced (flaw-to-clutter ratio (FCR) is improved by 5.66 dB) while the detection resolution is enhanced.

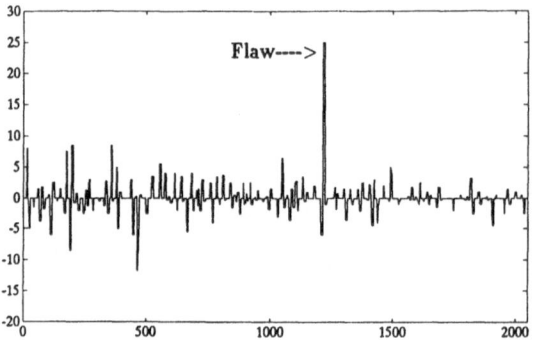

Figure 8.   Processed backscattered flaw signal using a pseudo bandpass morphological filter.

5. Morphological Peak and Edge Detection Algorithms

Peak detection has many applications dealing with the detection of timing events particularly in radar, sonar, and ultrasound. By locating peak positions certain targeted features can be identified. Combinations of rectification, reflection, and morphological erosion lead to the peak detection algorithm as shown in Figure 9. Larger structuring elements yield a narrower detection pulse due to the shrinking nature of the morphological erosion operation. Therefore, choosing a suitable structuring element with the proper width (W) is highly signal dependent.

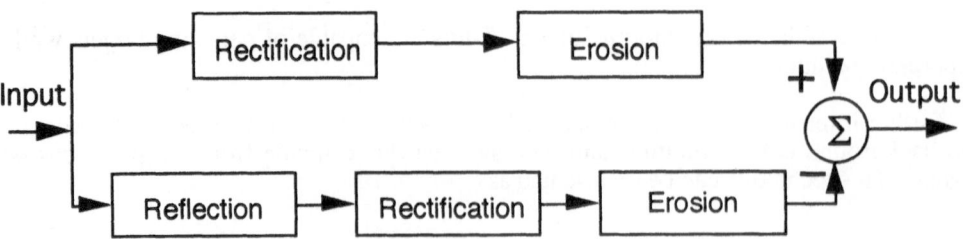

Figure 9.   Block Diagram of the morphological peak detection algorithm.

113

In the presence of noise, the peak detection algorithm is significantly effective as can be demonstrated in the following example. The input, created by adding a sinusoidal signal with an uniformly distributed random noise between -0.5 and 0.5, is displayed in Figure 10a. Then, in Figures 10b-10d, outputs are plotted against the original sinusoidal signal to show the effectiveness of this detection scheme for structuring elements of various widths. Note that as the width of the structuring element increases, the output shrinks to make the peak positions easily identifiable.

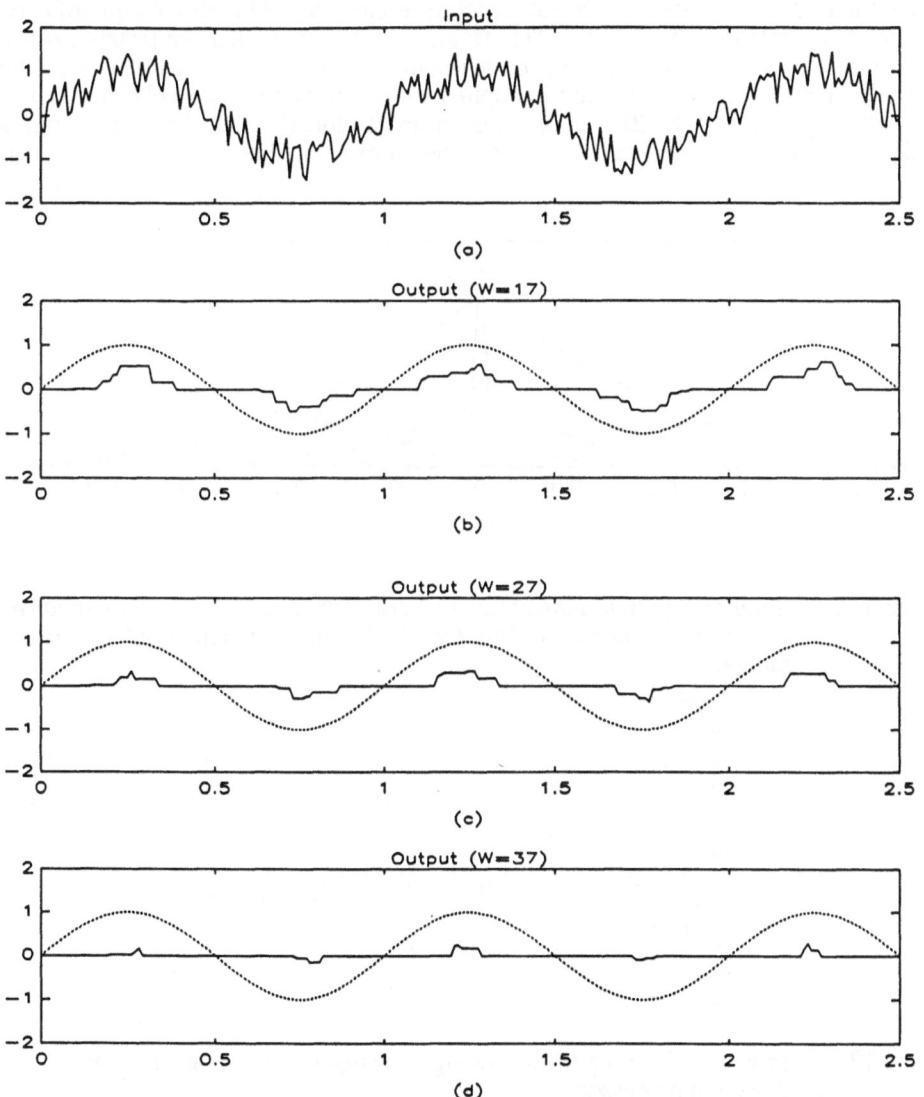

Figure 10.   Peak detection of a noisy signal using flat structuring elements of various widths. (a) Input, (b) Original sinusoidal signal (dotted line) and output (w=17, solid line), (c) output (w=27), and (d) output (w=37).

114

The reduction in noise variance can be demonstrated by analyzing the output density function as a function of the width of the structuring element. When the input noise is independent and uniformly distributed in the interval (-0.5 and 0.5), the output density function of peak detection becomes:

$$f(x) = M(\tfrac{1}{2} - x)^{M-1}U(x) + M(\tfrac{1}{2} + x)^{M-1}U(-x) + [1 - (\tfrac{1}{2})^{M-1}]\delta(x) \qquad (7)$$

where $U(x)$ and $\delta(x)$ are unit step and unit impulse response function respectively. The parameter M is the width of a flat structuring element. The above equation for different M is plotted in Figure 11. This figure clearly shows that the peak detection reduces the noise variance, and this reduction is a monatomic function of the width of the structuring element. Peak detection is applied to the ultrasonic flaw echo masked by grain scattering (see Figure 12). The processed result indicates that the morphological peak detector is capable of reducing grain noise and improving the visibility of flaw echo.

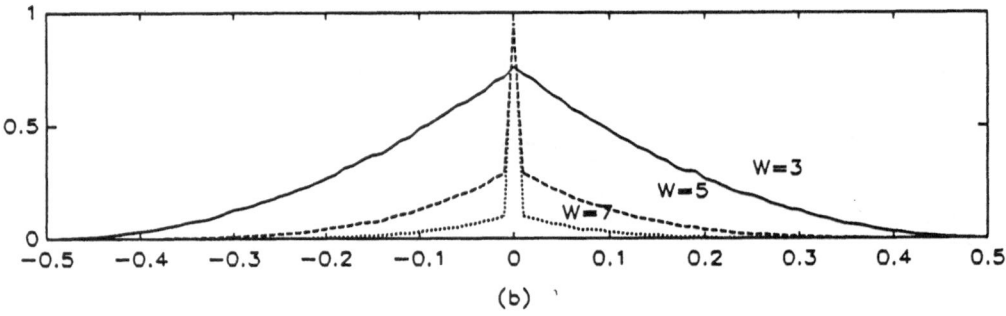

Figure 11. Simulated output density functions of peak detection for flat structuring elements of various widths (w= 3, 5, and 7) when input is uniform random noise.

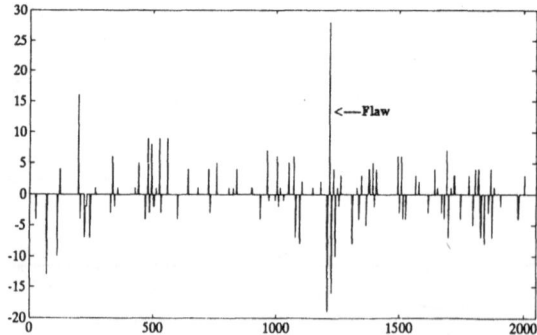

Figure 12. Processed backscattered flaw signal using a morphological peak detection processor.

The edge detection method consists of a sequence of operations which lead to the determination of specific geometric features displaying unique signatures (such as step edge, ramp, ridge). A block diagram of the edge detection process is shown in Figure 13. The lowpass morphological filtering process is used not only to reduce grain noise but also to maintain flaw echoes that are less impulsive and often exhibit lower frequency content when compared to grain scattering echoes. The second stage of the processing enhances the visibility and resolution of the detected edges associated with flaw echoes. This algorithm is applied to the ultrasonic signal for flaw detection (see Figure 14). The edge detection algorithm reduces the grain scattering echoes significantly while it improves the detection of flaw echoes.

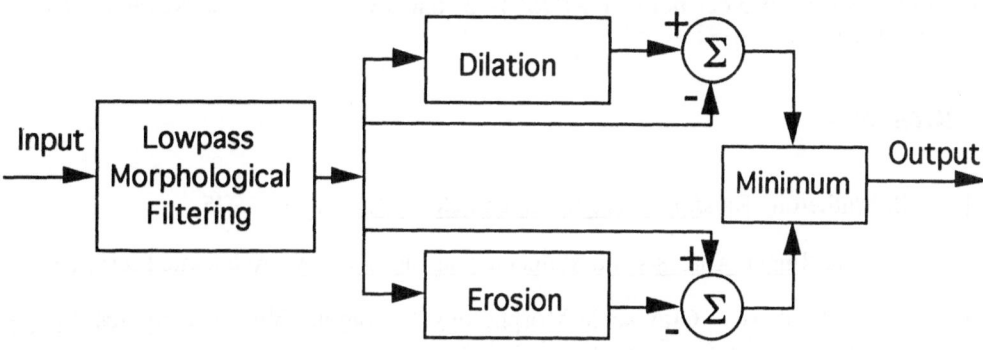

Figure 13.   Block diagram of the morphological edge detection algorithm.

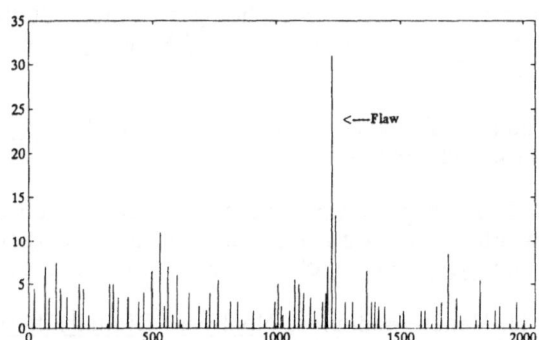

Figure 14.   Processed backscattered flaw signal using a morphological edge detection processor.

6. Conclusion

This paper has dealt with the properties of morphological processors (lowpass, highpass and bandpass filters, peak and edge detection) and their applications for improving the flaw-to-clutter ratio of ultrasonic signals. Our study has focused on the deterministic and stochastic properties of morphological operations using different structuring elements. From both a deterministic and a stochastic point of view, it has been shown that the effectiveness of the filtering process depends on the frequency content of input signals as well as the parameters of the structuring elements. Morphological filters have been applied to detect flaw echoes in ultrasonic signals contaminated by grain scattering noise. The processed experimental results show that morphological filters can detect flaw echoes while suppressing microstructure noise. Moreover, it has been shown that morphological filters can replace ensemble averaging which requires numerous measurements.

ACKNOWLEDGMENTS. This work has been supported in part by the Electric Power Research Institute project number RP2614-75, and by the Office of Naval Research project number N00014-92-J1735.

7. References

[1]   G. Matheron, <u>Random Sets and Integral Geometry</u>, Wiley, 1975.

[2]   J. Serra, <u>Image Analysis and Mathematical Morphology</u>, Academic Press, 1982.

[4]   S. R. Sternberg, ``Gray scale Morphology," <u>Comput. Vision, Graphics, Image Process.</u>, vol. 35, No. 3, pp. 333-355, 1986.

[5]   R. M. Haralick, S. R. Sternberg, and X. Zhuang, ``Image Analysis Using Mathematical Morphology," <u>IEEE Trans. Pattern Anal. Mach. Intell.</u>, vol. PAMI-9, pp. 532-550, Jul. 1987.

[6]   R. L. Stevenson and G. R. Arce, ``Morphological Filters: Statistics and Further Syntactic Properties," <u>IEEE Trans. Circuits and Systems</u>, vol. CAS-34, No. 11, pp. 1292-1305, Nov. 1987.

[7]   C. R. Giardina and E. R. Dougherty, <u>Morphological Methods in Image and Signal Processing</u>, Prentice Hall, 1988.

[8]   C. H. Chu and E. J. Delp, ``Impulsive Noise Suppression and Background Normalization of Electrocardiogram Signals Using Morphological Operators," <u>IEEE Trans .on Biomedical Eng.</u>, vol. BME-36, pp. 262-273, Feb. 1989.

[9]   I. Pitas and A. N. Venetsanopoulos, <u>Nonlinear Digital Filters: Principles and Applications</u>, Kluwer Academic Publishers, 1990.

MATERIALS NDE BY NON LINEAR FILTERING APPLYING HEAT TRANSFER MODELS

E. GRINZATO, S. MARINETTI
CNR-ITEF
Corso Stati Uniti, 4
35100- Padova
Italy

ABSTRACT: In active thermal Non Destructive Evaluation, properties of interest are obtained analysing time evolution of an informative parameter (e.g. excess temperature, absolute or relative temperature contrast, etc.). The fundamental problem to get rid of different kinds of noise is not deeply studied up to now.
Popular filtering techniques, both in space and time domain, are discussed.
The analysis of the thermal signal through a suitable heat transfer model has been suggested by many authors. This technique has been applied to NDT of layered materials if a simple analytical expression describing the real problem is found. New more powerful digital tools allow to extend this approach to other applications. For instance the extraction of useful information from raw data, in case of non selective maximum of the contrast signal, is taken into account.
In this paper the use of non linear filter in time-domain is described to improve the signal to noise ratio of thermal-infrared NDE. The filter is based on the heat transfer model of the test procedure, taking into account actual boundary conditions. The description of the heat transfer allows to separate the useful signal from noise. Therefore the analytical models of classic thermal/infrared NDE are reported.
The suggested method allows to filter data and, at the same time, estimate useful quantities. The basic procedure is an iterative non-linear best fit of experimental data.
Solving the direct problem it is possible to simplify the testing procedure, tuning the filter to the actual experimental conditions. The NDE accuracy is increased taking into account heat losses. Finally it is possible to choose the parameters which have to be

117

X. P. V. Malague (ed.), Advances in Signal Processing for Nondestructive Evaluation of Materials, 117–132.
© 1994 *Kluwer Academic Publishers.*

estimated. In such a way the same approach can be used for different tests as thickness or absorbed energy or surface heat exchange coefficients measuring.

Thermal diffusivity measurements are carried out according with the flash method and processed by the suggested procedure. A filter suitable for thermal/infrared NDE on homogeneous porous materials is designed and tested for the moisture mapping. Experimental results are reported and discussed.

1. Introduction

Thermal/Infrared Non Destructive Evaluation (T/IR NDE) has been becoming more competitive compared with other testing procedures. New fields, apart of aerospace industry, as restoration activities and ship production are attractive applications of NDE. The transfer of methodology and techniques from different subjects as signal processing to NDE may speed up this evolution. There are various reasons for this convenience, e.g. taking advantage in thermal image treatments of the great deal of research developed about video signal processing. In general it is interesting to consider the thermal signal as the electric signal when the leading law is the same. The use of the Laplace and Fourier transforms as mathematical tools in both subjects is an example of a close relationship. All affinities are probably not yet fully explored and the filter design is a challenging field.

The heat transfer modelling adoption as a guide to interpret the thermal signal in NDE is an old idea, limited in its application by mathematical complexity. Nowadays the availability of more powerful computers and dedicated hardware supplies the needed tools to extend this approach. Parallel processing, very large and fast digital memories accomplish to solve involved computations in an acceptable amount of time. Thermal problem inversion has became realistic also in actual cases of increased complexity. Numerical simulations of heat and mass transfer can be performed in 3D, taking into account different boundary conditions. Furthermore facilities dedicated to process image sequences by means of advanced techniques are now available /1/, working in 3D data structure.

2. NDT, NDE in situ and lab: Classic problems

A well known problem in T/ IR NDT is the difficulty to identify a deep defect into a known material. The established procedure, looks at the absolute or relative thermal contrast (C^r) curve versus time (t) for each point of the surface (x,y) /2/. The absolute thermal contrast represents the difference between the temperature ($T_{x,y}$) measured on the considered position and the temperature on a sound zone, chosen as reference (T^{ref}_{sound}). The relative thermal contrast corresponds to the absolute contrast, normalised on a reference temperature as in eq.(1)

$$ C^r_t = \frac{T_{x,y,t} - T^{ref}_{sound,t}}{T^{ref}_{sound,t}} \tag{1} $$

or eq. (2), where $T^{ref}_{\infty,t}$ is the asymptotic temperature /3/ to get rid of the absorbed power density (Q_a) knowledge.

$$ C^{r*}_t = \frac{T_{x,y,t} - T^{ref}_{sound,t}}{T^{ref}_{\infty,t}} \tag{2} $$

Adopting an empirical approach /4/ it is possible to detect a defect and estimate its depth. This approach is limited to one dimension (1D) applications, that is when the defect has a small thickness/ size ratio, the heat flux impinges uniformly on the whole investigated surface and all involved parameters are constants. The occurrence of a maximum in time evolution of relative contrast shows the presence of a defect, while position and amplitude of the maximum give its depth.
As flaws becomes deeper, maxima are less selective. In fig.1 we can see two absolute contrast plots versus time. The upper curve is relative to an artificial void of 2 *mm* thickness, at the bottom of a 10 *mm* plaster layer, the presence of many relative maxima affects the measure accuracy. In the case of the lower plot, relative to a 0.2 *mm* void thickness, it is extremely difficult to obtain any useful information because it is quite impossible to extract

120

the "thermal signal" from the noise. This is still an open problem in phenomenological T/ IR NDT /5/.

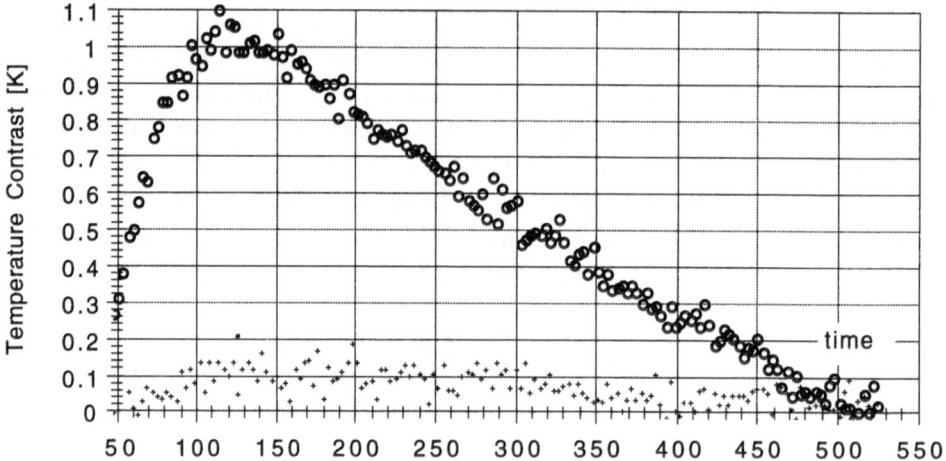

Fig. 1: Absolute thermal contrast corresponding to two artificial voids of 2 and 0.25 mm thickness under a 10 mm plaster layer.

A possible solution of this problem consists in trying to fit data by means of least square method. Unfortunately results obtained with basic functions as polynomial or exponential one is a frustrating exercise, probably in the former experience of any researcher in this field.

Improving the analysis by a more quantitative approach allows not only to identify defective zones but also to find parameters and characteristics as the thermal resistance ($R=s/\lambda$), thickness (s), size of defects, etc. /6,7/. This increased complexity is compensated with the promise to solve open problems as to select different kinds of defect using a set of possible models or to recognise multiple defects /3/.

A relevant aspect of T/ IR NDE is the possibility to recover actual thermophysical properties of materials in situ or in lab. Thermal diffusivity (α) and effusivity (e) are very useful index of the state of the investigated material /8/. In the final stage T/ IR NDE-NDT is applied in situ to perform non-contact inspection. In this case the thermal properties varying in consequence of certain condition have to be determined. Before that, thermal and optical parameters must be measured on bulk samples at laboratory. For instance the influence of the moisture content on thermal conductivity (λ) is on going at our lab for different porous materials. A robust T/IR NDE is needed also if complex parameters as the diffusivity tensor /9/ or the effusivity profile in depth (z) /7/ are sought.

In this paper we point out this aspect of NDE, presenting some results as an example of application of the suggested methodology to process experimental data.

The main problem we are involved in is the NDE of the moisture content in works of art like frescoes or simply the wall of an heritage building. For this purpose the excess temperature induced on the heated surface by T/IR NDE have to be maintained at the minimum level (a few degree). Therefore the noise reduction of a typical NDE becomes important, requiring the adoption of suitable filters.

2.1. INFORMATIVE PARAMETERS MEASURING

The quantitative approach is normally based on the modelling of heat transfer within the material. The use of numerical solution is normally the first step. The solution of direct and inverse problems is very useful but too time consuming and affected from stability problems to be applied to on-line NDE.

Analytical methods has been giving an important contribution for the definition of thermal parameters and the solution of inverse problems. All inverse diffusion problems, as heat conduction, are very sensitive to small fluctuations of the analysed signal. Therefore it is extremely important to process raw data in order to reduce at minimum the noise effect. It was demonstrated /10,11/ that an iterative analytical approach gives a robust estimation of diffusivity and defect thermal resistance.

The thermal diffusivity or effusivity are the main informative parameters for active thermal test, due to the nature of transient heat transfer. We realize immediately the importance to recover its value in situ. Anyway the increase of accuracy is important also if the test is performed in the workshop, according with the classical "flash method" /12/ or any modified version. Considering the Flash method as a reference for the diffusivity measurement, we will show in par.5 how to improve the measure, simply taking into account actual experimental conditions, by means of a suitable thermal model.

3. Noise filtering, state of the art

Up today many techniques has been attempted in order to filter the noise content of thermal images processed in active thermal tests /13/. Basically any analysis algorithm is mainly devoted to increase the signal to noise ratio. Therefore we widely extend the concept of filter including any algorithm capable to extract the searched parameter from raw data.

Most of referred techniques are performed in space domain. The simplest are devoted to increase the visibility of the defect, operating on the whole image grey levels, through tresholding or modifying the histogram.

More effective are filters based on local operators /14/ acting as low-pass filter, to reduce the noise or as high-pass filter to identify the edge of the defect. A combination of the two was also tried /15/ using a Laplacian of Gaussian filter. Generally the use of high-pass filter in not recommended due to the noise increasing. The more popular procedure adopted to filter noise in T/ IR NDT consists in smoothing filters. Linear average or Gaussian filter or non-linear Median filter are spreadly used but their effectiveness is normally poor because high-frequencies content both in the noise and in the signal is reduced.

Others techniques based on local operators as morphological filtering /16/ give results only in specific case, mainly in removing artefacts.

Point operators /14/ have been also used and among them the most popular is the difference. A possible use is the subtraction of the temperature history in a certain point from the analogous signal recorded on a "normal" point. The contrast image is another application of the same idea. The use of the relative thermal contrast instead of the temperature curve implements an effective high-pass filter which highlight the defect presence.

Filters in time domain has been also proposed. The simplest is the averaging of multiple images, it improves the signal to noise because the noise sources are not correlated. Unfortunately this useful procedure is limited to thermal signal whose dynamic in time is slower than the measuring device dynamic. Another heuristic approach based on a point operator is the time domain subtraction /17/. It was suggested to get rid of biased noise by means of time-domain referencing.

The temporal moment method /18/ gives an additional example of processing in time. The temporal moment of zero order (M^0) is defined as the integral of the absolute contrast according with the following expression, for a slab absorbing a pulsed energy density (Q):

$$M^0_{x,y} = \int_0^\infty \left[T_{x,y,t} - T_t^{ref} \right] dt = Q \cdot \frac{s}{\lambda} \left(1 - \frac{z}{L} \right)^2 \qquad (3)$$

A moments image can be approximated by a simple normalised addition of images, giving an improved thermal resolution.

The thermal modelling approach has been adopted as a base for other algorithms. The crucial point is how to describe the heat transfer in a simple analytical form. An example is the logarithmic expression for the temperature increase after heat step pulse, versus the logarithm of time /17/. The correlation between a large part of this curve and experimental data was used to detect delaminations. Also the linear relationship between the temperature increasing due to heat step, versus the square root of time, has been used for the effusivity measure in a semi-infinite body /19/.

At last, filters in frequency domain may be mentioned, mainly related to pulsed heating tests /20/. However FFT filtering is time consuming and critical for the window effect.

4. Algorithm description

In this paper we make a preliminary description of the use of data fit to improve the signal to noise ratio of thermal-infrared NDE. The basic idea is to follow the behaviour of the thermal signal, adopting the heat transfer model as filter, also if this statement does not strictly agree with the classic definition.

A non-linear regression of the experimental data to analytical expression of the thermal model is used within an iterative routine to estimate the searched parameters. The Levenberg-Marquardt method is used to minimise the least square approximation in the space of parameters /21/. This procedure (NLBF) has been implemented on a thermal image workstation /22/, and applied to raw data, obtaining automatically an estimation of thermal diffusivity or effusivity.

Therefore the analytical models of classic thermal infrared NDE procedure is needed /23/. Basically this approach is applicable whenever the mathematical expression of the heat transfer is known and interesting parameters are isolated. Different mathematical tools are available to describe the heat transfer as quadrupole formalism in Laplace space or Green functions.

The suggested method allows to filter data and, at the same time, estimate useful quantities. So the description of the real heat transfer, taking into account boundary conditions, allows to estimate also the linear heat exchange coefficient (h) on the irradiated surface.

Another important advantage is to use the direct solution to predict temperature and time evolution avoiding any possible damage to the tested object. This is of primary importance in NDT of works of art. In such a way it is simple to simulate real test and therefore be aware of the applicability limits. The experimental conditions can be so tuned to different tested objects, which are extremely varying, optimising the sampling period and test duration.

Some simple applications of the method to plane layer of homogeneous material are presented.

5. Application and results

The proposed procedure has been tested measuring the thermal diffusivity and effusivity of dry bricks. These measures allow to evaluate experimentally the influence of the water content in historical monuments. The main application of this T/ IR NDE is the possibility to control the decay of painting and frescoes.

Applying a multilayer thermal model it would be possible also to detect defects under the plaster, and to identify its depth and nature, as different authors report.

5.1. DIFFUSIVITY

In fig.2a it is shown the temperature increase measured on the rear face of a flat plate of 10 *mm* thickness (*L*), insulated on the back, tested according with the "Flash method" /12/. Adopting the eq.(4) and heating the sample for a time (τ_h) of 3 *s*, a diffusivity 4.326×10^{-7} *m² s⁻¹* is obtained.

$$\alpha = \frac{1.38 \cdot L^2}{\left(\tau_{0.5} - \frac{\tau_h}{2}\right)\pi^2} \tag{4}$$

Processing the same experimental data by means the NLBF algorithm a diffusivity equal to 4.004×10^{-7} is measured. We consider the second as the correct value for two reasons: the accordance with figures reported in literature and secondary because the test was not performed in adiabatic conditions as the Flash method states. This hypotheses was confirmed substituting the eq.5 in NLBF with an adiabatic thermal model and obtaining 4.305×10^{-7}. We can conclude that also strongly reducing heat losses occurring during the test, an error of 10 % is induced if they are not considered.

The adopted analytical model of the heat transfer is:

$$\frac{\Delta T \cdot h}{Q_a} = \sum_{n=1}^{\infty} \frac{2 \cdot Bi \cdot cos[\mu_n(1-\frac{z}{L})]sec\mu_n}{Bi(Bi+1)+\mu_n^2} \cdot e^{-\mu_n^2 Fo} \cdot \left(e^{\mu_n^2 Fo_h} - 1\right) \tag{5}$$

This active NDE can be executed measuring temperature on the heated face too, as shown in fig. 2b, using an infrared detector. The diffusivity value recovered in the front-side test and applying NLBF is: 4.015×10^{-7}. The accordance between results of rear and front-face tests is very good, also considering that energy used for the second test was reduced to one/third ($Q = 12 \ kJ \ m^{-2}$). On the other hand, computing time for the diffusivity measurement is longer in this case than for the rear-face, due to slower convergence of series.

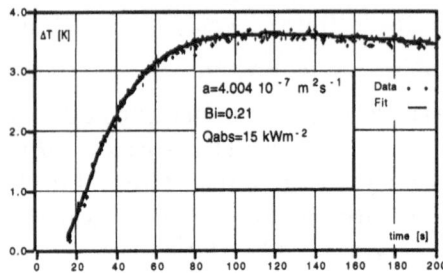 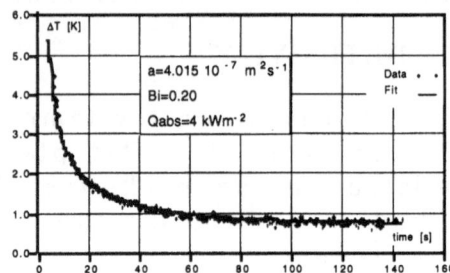

Fig. 2a: Temperature increasing on the rear face and thermal model plot, for diffusivity measurement.

Fig. 2b: Temperature increasing on the heated face and thermal model plot, for diffusivity measurement.

We can imagine this kind of curve repeated several times for any thermal image element. A significant saving of time can be achieved applying the NLBF on a few homogeneous zones, segmented from the whole image /24/.

5.2. EFFUSIVITY

Effusivity measurement for the moisture mapping of wall is here presented. The experimental set up was described in previous papers /19, 25/. The fig.3 gives the temperature increasing of a point belonging to the step-heated surface of a brick wall. The thick layer can be assumed as a semi-infinite body for the test duration (400 s). The related analytical thermal model has the following expression:

$$\frac{\Delta T \cdot Q_a}{h} = 1 - exp^{\left(\frac{h}{e}\right)^2 t} \cdot erfc\left(\frac{h}{e} \cdot \sqrt{t}\right) \tag{6}$$

It is important to realize that the absorbed power density of step irradiation has been reduced to 300 $W \ m^{-2}$ and the temperature increase is less than 7 K. The improvement in the accuracy using NLBF instead of linear regression on adiabatic model is of the order of 10%, in real conditions.

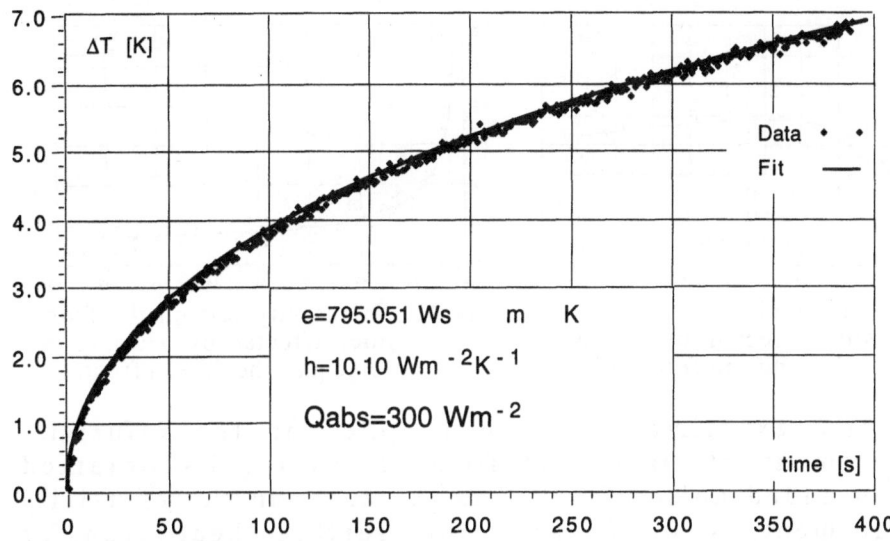

Fig.3: Excess temperature history for a dry brick wall and NLBF curve giving effusivity measurement.

Multispectral optical measurement as described in /19/ allows to map absorbed power density, this implies the linear heat exchange coefficient ($h_{x,y}$) of any point can be inferred. Alternatively an expression for the unknown effusivity, relative to the dry material, can be derived /25/.

5.3. PERFORMANCE ANALYSIS

In order to verify robustness of the proposed procedure, a white Gaussian noise with $\sigma = 1K$ has been added to the experimental data of fig.2a. The fig. 4a shows how NLBF may act as a filter, also in this extreme case where noise and signal have the same amplitude. The estimated liffusivity is 3.934×10^{-7} giving an error less than 2%.
t is more difficult to filter noise affecting the front-face est, because more high-frequencies are contented in the analytical function. In fig.4b the heat transfer model and data are plotted to visualise this point. The error induced in the diffusivity estimation by a $\pm 0.75\ K$ noise addition in

excess temperature ($\sigma=0.25$) is 8.5 %. On the contrary the Biot (*Bi*) number is quite insensitive to noise.

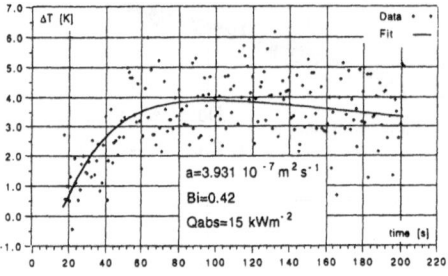

Fig.4a: Diffusivity test on the rear side affected by a ±3 K. noise and the filter effect.

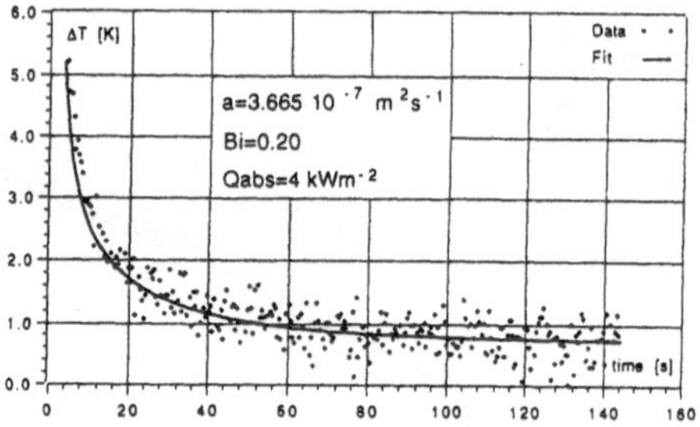

Fig.4b: Diffusivity test on the front side, affected by a ±0.75 K noise and the filter effect.

The same experiment has been done for the effusivity measurement of fig.3. The fig.5 shows results obtained with an added noise of ± 1.5 *K*. The error in the effusivity measurement is 0.7%. Also the surface heat transfer coefficient given by NLBF method practically not vary with noise increasing.

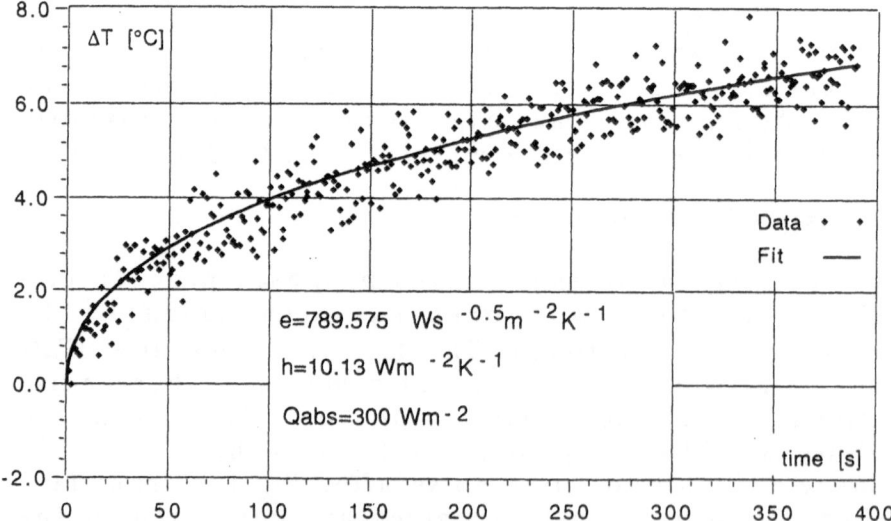

Fig.5: Excess temperature history for a dry brick wall and NLBF curve giving effusivity measurement.

Lastly, conventional low-pass filters are compared to the proposed procedure. The fig.6 shows the result of a

polynomial interpolation of order 7 and the NLBF, applied
to noisy data of fig.4a for diffusivity measurement.

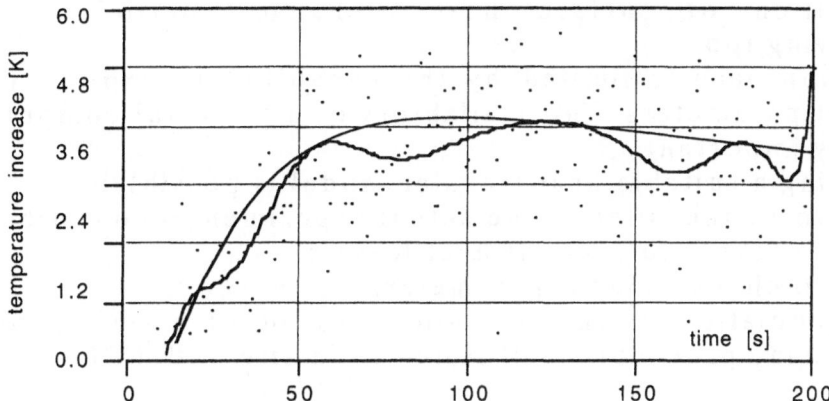

Fig.6: Polynomial interpolation of noisy data of fig.4a for the diffusivity
measurement of a brick and NLBF plot.

In fig.7 curves obtained by means of a smoothing filter and NLBF
are plotted. The data are obtained from the effusivity measurement
of fig.5, superimposing a noise of $\sigma=0.5$.

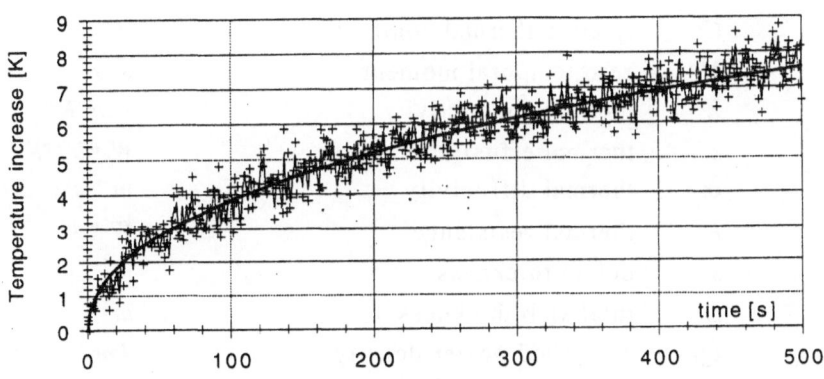

Fig.6: Comparison between smoothing filter and NLBF effect on noisy
data for a brick effusivity measurement.

6. Conclusions

The use of the suggested processing of data in time-domain
allows to extract useful signal from very noisy signal. The

diffusivity and effusivity measurement are very robust also if a great deal of noise is contained in the data. The estimation of surface heat exchange coefficients is satisfying too.

This approach is limited by the capability to explicit the sought parameters. The availability of a powerful computer is very important.

Applying a suitable heat transfer model is possible:

- to fit the filter to the actual experimental conditions;
- to take into account heat losses;
- to choose sought parameters.

The simulation of the test allows to adopt a testing and processing procedure well suited to in situ T/ IR NDE.

Symbols:

z	depth coordinate	m
x, y	surface coordinate	m
t	time	s
$\tau_{0.5}$	half rise time	s
τ_h	irradiation time	s
T	temperature	K
ΔT	excess temperature (above initial)	K
C^r	relative thermal contrast	$\%$
M^0	zero temporal moment	$K\ s$
λ	thermal conductivity	$Wm^{-1}K^{-1}$
e	thermal effusivity	$Ws^{0.5}m^{-2}K^{-1}$
α	thermal diffusivity	$m^2\ s^{-1}$
R	thermal resistance	$W K^{-1}$
s	defect thickness	m
L	total slab thickness	m
Q_a	Absorbed power density	$J\ m^{-2}$
Q	Absorbed energy density	$W\ m^{-2}$
h	linear heat exchange coefficient of the surface	$Wm^{-2}K^{-1}$
B_i	Biot number ($B_i = h\ L/\lambda$)	
μ_n	roots of transcendent equations $\mu \cdot tg\mu = Bi$	
Fo	Fourier number ($Fo = \alpha t/L^2$)	
Fo_h	Fourier number for $t = \tau_h$	
σ	standard deviation	

References

1/   U. Netzelmann, G. Walle, G. Dobmann: "Real-time 3D representation of time-resolved infrared thermographic data". Eurotherm Seminar n°27, pp.239-242; Ed. Europeennes thermique et industrie, Paris 1992.

2/   D. Balageas: "Le controle non destructif par methodes thermiques". Colloque S.F.T. 90, Nantes, 1990.

3/   A. Bendada, D. Maillet, A. Degiovanni: Non destructive transient thermal evaluation of laminated composites: Discrimination between delaminations, thickness variations and multidelaminations." Eurotherm Seminar n°27, pp.218-223; Ed. Europeennes thermique et industrie, Paris 1992.

4/   V. Vavilov, R. Taylor: "Theoretical and practicalaspects of the thermal NDT of bonded structures. Res. Techn in NDT vol.5, pp.239-279, Academic press, London, 1982.

5/   V. Vavilov, P.G. Bison, C. Bressan, E. Grinzato, S. Marinetti: "Informative parameters and noise in transient thermal NDT." Advances in signal processing for non destructive evaluation of materials, Quebec City, 1993.

6/   P. Cielo: "Pulsed photothermal evaluation of layered materials." J. App. Physics, vol.56 n°1, pp.230-234, 1984.

7/   D. Balageas, A. Déom, D. Boscher: "Controle non destructif des composites carbone-époxy par méthode photothermique impulsionnelle." revue Géneral de thermique n°301, 1987.

8/   P. Cielo, X. Maldague, A. Déom, R. Lewak: "Thermographic Nondestructive Evaluation of industrial materials and structures." Material Evaluation, Vol.45 n°6, pp.452-460, 1987.

9/   C. Welch, W. P. Winfree, D. M. Heath, E. Cramer, P. Howell: "Material property measurements with post-processed Thermal image data." SPIE Vol. 313 Thermosense XII, pp124-132, 1990.

10/   J.P. Krapez, P. Cielo: "Thermographic Nondestructive Evaluation: Data inversion procedure. Part I: 1D Analysis." Nondestructive Evaluation, n°3, pp.81-100, springer-Verlag, New York, 1991.

11/   J.P. Krapez, X. Maldague, P. Cielo: "Thermographic Nondestructive Evaluation: Data inversion procedure. Part II: 2D Analysis and experimental results." Nondestructive Evaluation, n°3, pp.101-124, Springer-Verlag, New York, 1991.

12/   W.J. Parker, R.J. Jenkins, C.P. Butler, G.L. Abbot: Flash Method of Determining Thermal Diffusivity, Journal of Applied Physics, vol.32 n° 9, p.1679-1684, 1961.

13/   P.M. Patel, S.K. Lau, D.P. Almond: "A review of image analysis techniques applied in transient thermographic nondestructive testing. Nondestructive testing". Nondestructive testing Evaluation, vol.6, pp.343-364; Gordon and Breach S.A., 1992.

14/   E. Grinzato: "Dedicated image processing for thermographic non-destructive testing." Infrared techniques for NdT, Ed. X. Maldague; cap.V, Gordon and Breach Science Publisher, 1993.

15/   F. Lotti, A. Casini, S. Baronti, T.S. Durani, A. Rauf, K. Boyle: Flaw recognition in composites by processing sequences of infrared images." Proc. 3rd Int. Workshop on time varying processes and moving object recognition. Florence, 1989.

132

16/   V. Vavilov, P.G. Bison, C. Bressan, E. Grinzato, S. Marinetti: "New aspects of dynamic thermal tomography." Sov. J. NDT (Defectoscopia), n°7, 1992.

17/   Cielo, P.Lewac, D.L. Balageas: "Thermal sensing for industrial quality control." SPIE 581, Thermosense VIII,pp. 47-54, 1986.

18/   D. Balageas, D. Boscher, A. Déom, J. Fournier, R. Henry: "La thermographie infrarouge: un outil quantitatif à la disposition du thermicien." revue Géneral de thermique n°322, pp.501-510, 1988.

19/   E. Grinzato, C. Bressan, P. Bison, A. Mazzoldi, P. Baggio, C. Bonacina: "Evaluation of moisture content in porous material by dynamic energy balance, Proc.SPIE, vol.1467, Thermosense XIV, 1992.

20/   L.D. Favro, D.J. Crowther, P.K. Kuo, R.L. Thomas: "Inversion of pulsed thermal-wave images for defect sizing and shape recovery". SPIE 1682 Thermosense XIV, pp. 178-181, 1992.

21/   W.H. Press, S.A. Teukolsky, W.T. Vetterling, B.P. Flannery: "Numerical Recipes in C". p.680, Cambridge University Press, 1992.

22/   P.G. Bison, A. Braggiotti, G.M. Cortelazzo, E. Grinzato, A. Mazzoldi, G.A. Mian: "Open architecture for Multispectral Computer Vision applied to both Visual and Infrared bands." SPIE 1349, pp. 323-329, 1990.

23/   V. Vavilov: "Transient Thermal NDT: Conception in formulae". Eurotherm Seminar n°27, pp.222-234; Ed. Europeennes thermique et industrie, Paris 1992.

24/   E.Grinzato, A.Mazzoldi. "Infrared detection of moist areas in monumental buildings based on thermal inertia analysis". SPIE 1467, Thermosense XIII, pp.75-82, 1991.

25/   P.G. Bison, C. Bressan, E. Grinzato, S. Marinetti, V. Vavilov "Active thermal testing of moisture in bricks"; SPIE 1933 Thermosense XV, pp. 207-214, 1993.

THE INFLUENCE OF IMAGE DIGITISATION AND CALIBRATION OF A SCANNING SYSTEM UPON WELD INSPECTION

HANNELORE WESSEL, CHRISTINA NOCKEMANN, GERD-RÜDIGER TILLACK
Federal Institute of Material Research and Testing (BAM)
Unter den Eichen 87
12200 Berlin
Germany

ABSTRACT. The capability of a high-resolution scanning system for industrial radiographic films was tested dependent upon the specification and calibration of the system using ROC method.

1. Digitisation of industrial radiographs

1.1. REQUIREMENTS TO INDUSTRIAL DIGITISATION SYSTEMS

In the area of non-destructive testing of materials radiography is used frequently. Usually, the radiographic film of, for instance, the weld of a pipe of a nuclear power plant is tested using a lightbox. The inspector tests the film using a lightbox, a magnification glass, his eyes and above all his experience. During the inspection of critical parts of a nuclear power plant 600 and more radiographic films have to be tested - a tiring work with the possibility of misinterpretations. Nowadays the advances of the development of film digitisers allow the operation of these systems in the field of non-destructive material testing.

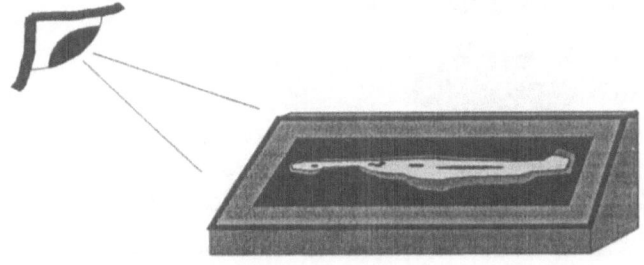

Fig. 1: Conventional inspection of radiographic films

133

X. P. V. Malague (ed.), Advances in Signal Processing for Nondestructive Evaluation of Materials, 133–144.
© 1994 *Kluwer Academic Publishers.*

Therefore digitisation of industrial radiographs becomes more and more important. It allows space-saving and high speed archieving and retrieving of digitised films on optical disks and comfortable interactive image interpretation in combination with image processing hardware. Another aim can be a computer-aided or automatic defect detection of radiographic films of welds or castings.

In contrast with the requirements for medical radiographs scanning systems of high resolution are needed for industrial radiographs. The transformation of the optical densities of the film into digital data (grey values) by the scanning system must be performed without any loss of information. Smallest density differences have to be resolved to enable the detection of very thin flaws in welds or castings. Although in areas of large density differences - a defect detection problem especially of castings - small defects have to be resolved. One of these available high-resolution scanning systems will be discussed in detail below.

1.2. PRINCIPLE OF A HIGH-RESOLUTION SCANNING SYSTEM

The radiographic film is placed onto a glass slide of the digitiser (Fig. 2) and covered by an opalescent glass plate, which provides diffuse light. During the scan operation the glas slide is moved under the light line source, which is fed by a simple projector lamp. The CCD-camera is situated below the light line. It receives the light intensities with a local resolution of 70 by 70 micrometres, which were not absorbed according to the densities of the radiographic film. These intensities are converted into the corresponding digital data of 12 bits depth and stored in the system computer.

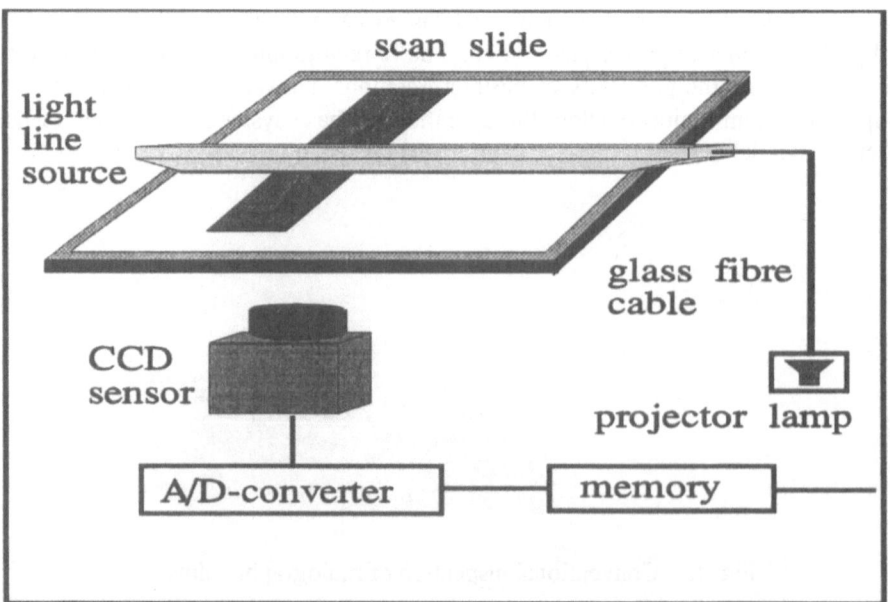

Fig. 2 : Principle of a scanning system

2.   Influence of calibration

In the first instance the quality of the digital image depends upon the quality of the radiographic film. Furthermore it is dependen upon the geometrical and densitometrical resolution of the digitiser, the accuracy of the scanning operation and the calibration of the system. The specification of a system is only as well as its operation in the context of the concerning application.

2.1.   NORDTEST PROJECT

In the field of non-destructive material testing these high-resolution scanning systems gain importance in the area of computer-aided defect detection and image archieving. Within a NORDTEST project, among others the institute BAM studies the influence of calibration of a scanning system upon defect detection of welds using ROC (receiver operating characteristic) method /1,2/. NORDTEST is an internordic (joint Northern European) agency which has the task of promoting development in the field of technical testing. The study even not yet finished, first results should be discussed in this context.

First a short NORDTEST project overview: the influence of calibration of a high-resolution scanning system is to be determined by scanning a catalogue of weld radiographs under different calibration conditions. To determine the influence of calibration upon the whole test operation (scanning, defect detection, recording) the ROC method will be used. After scanning and computer-aided defect detection the results of each calibration mode has to be compared with the *true data* of the catalogue. The true data include the list of the description of all defects within the films of the weld catalogue. Using ROC method, the results of the calibration study will be compared with these data. If there is an influence of calibration the results will differ for the used scanning conditions. The contrast of the digitised radiographic film correlates with the calibration of the scan system. Changes will lead to poorer contrast which means worse defect detection and deviation from the true data. The deviation mainly depends upon the density distribution of the film. The lower the contrast of the film the higher the amount of misindications or missing indications will be. The curves of the ROC method will directly show the influence of calibration.

2.2.   CALIBRATION FUNCTION OF THE SCANNING SYSTEM

The capability of the tested high-resolution scanning system mainly depends upon the light level calibration of the scanning system under the condition that the system is operating properly. The digitiser permits seven different modes, which provide increasing light intensities for the first five light levels and highest scan light intensity together with slower scan time for light level *extended* and *maximum*. The base of the calibration is the adjustment of the voltage supply of the projector lamp to a value that guarantees the highest possible grey value for a scan without film according to the selected light level. Normally, the scanner is calibrated after starting the system, after some operation hours or before use of other lightlevels like *extended* or *maximum*.

2.3.   DETERMINATION OF INFLUENCES OF CALIBRATION

To check the influence of calibration a catalogue of radiographs of welds was scanned by the system in different modes:

calibration mode 1:   Calibration of the scanner each time before a radiograph is scanned

calibration mode 2:   Calibration of the scanner at the beginning of the study and scanning of the catalogue distributed over a period of 100 hours without new calibration of the system

To exclude different test results by different inspectors an automatic image processing and defect detection algorithm /3/ was used. The interest of the studies is focussed to the influence of calibration and not to the performance of the system together with the inspectors. It is notably very complicated (and still not realised) to find an universal algorithm for defect detection of any kind of welds. Using a catalogue of 104 artificially produced welds of the same plate thickness and radiation parameters simplifies the problem.

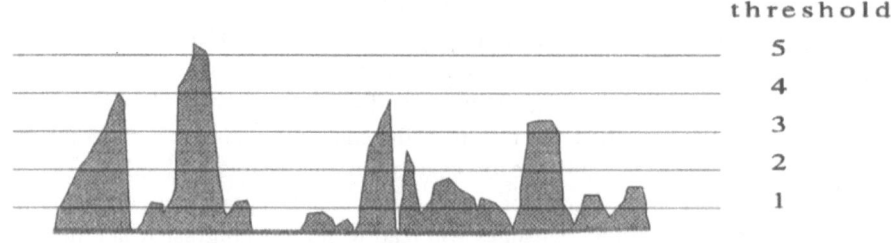

threshold
5
4
3
2
1

Fig. 3:   Different thresholds for defect detection

The catalogue was scanned by a high-resolution digitiser with pixel information depths of 12 bits and a resolution of 70 by 70 micrometres per pixel. Due to the 8 bits image processing this data have to be converted. Therefore the grey value range of the images was established by a histogram function. The relevant grey value range was determined - for instance a range of 52 to 2086 - and compressed to 8 bit which means a range of 0 to 255. Afterwards the digitised films were tested by an automatic algorithm with five different thresholds (Fig.3) for the defect segmentation to control the recognisability over a small range of grey values. This means, that the first threshold will include true indications but also a great amount of noise (false indications) while the last one includes true indications in the ideal case.

The five different thresholds replace the assessment of the recognisability by the inspector. A defect indicated by each of the five thresholds can be assessed *very well* visible. Indication by four thresholds leads to the assessment *well* visible etc. The thresholds are the base for the sensitivity of the test system. Beside the amount of defect detections per films these assessments are used by ROC method.

2.4.   ROC STUDY

The results of the first study (see Fig. 4) lead to grave doubts about the method. The calculated ROC curves did not look as expected. Two events coincided: during the scan operation for the NORDTEST study the projector lamp failed. This entails that there was no defined calibration for mode two. The expected lower ROC curve of mode 2 degraded by the deterioration of the projector lamp of the digitiser. The consequence will be the repetition of the study this fall and to set up a guideline for calibration and calibration tests of digitisers.

On the other hand the ROC curves are degenerated. Two explanations exist: The transformation of optical densities of the films to grey values was not linear and the films of the catalogue differ in their quality. The radiographs with lower dynamic lead to less TP-indications together with a high rate of FP-indications especially for the first segmentation threshold.

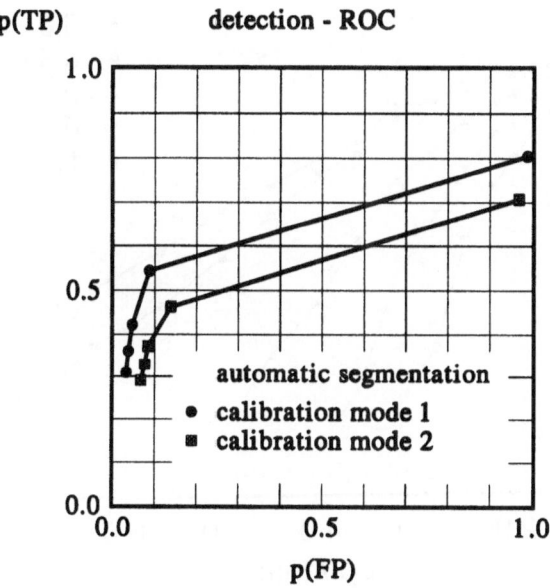

Fig. 4:   ROC curves for the influence of calibration

3.   Reliability of a high-resolution scanning system

Besides calibration the reliability of scanning systems is an important aspect for the operation in the field of non-destructive material testing. Therefore the high-resolution scanning system was tested within a project of the German company SIEMENS/KWU and BAM. The project consisted of two parts: the determination of the scanning system parameters (modulation transfer function MTF, signal-to-noise ratio and characteristic curve) and the determination of the reliability of the scanning system by an ROC study.

3.1. DETERMINATION OF THE SYSTEM PARAMETERS

3.1.1. *Modulating transfer function MTF.* The modulating transfer function MTF describes the reduction of the signal amplitude with growing three-dimensional frequency. For receiving the MTF a radiographic film with the picture of line groups was used. For every line group the difference of the grey values between dark line and light space was measured. Figure 5shows the obtained curves for different light levels of the scanning systems.

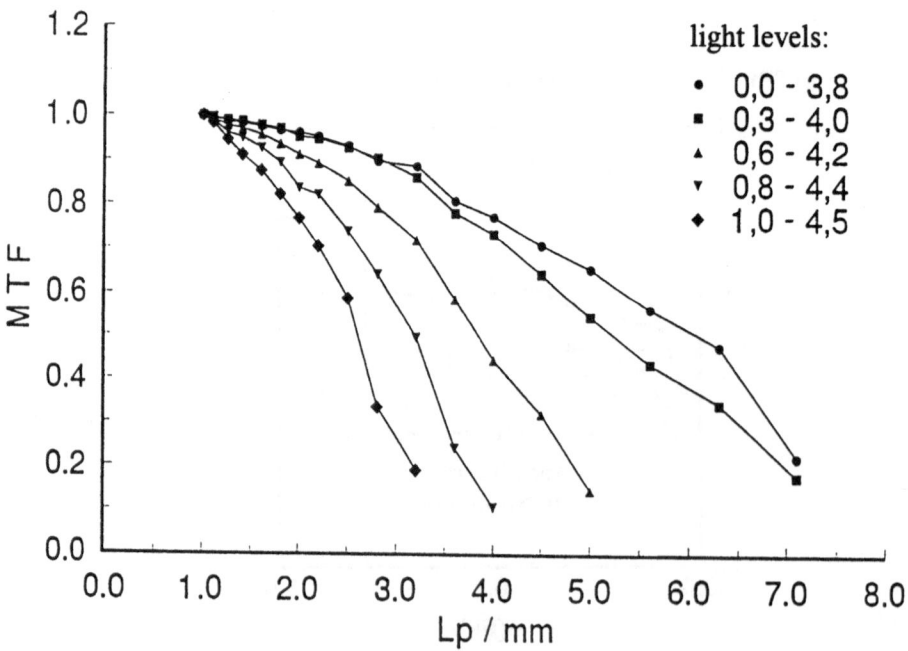

Fig. 5   MTF of different light levels

At the location of small local frequencies (around 1.0 lp/mm) the modulating transfer function MTF values are located near 1.0. The lower the light level of the scanner the steeper the decrease of the curves will be. The scanner is able to resolve 7 to 8 lp/mm at a pixel-resolution of 70 x 70 micrometres per pixel using the 5th light level.

The scanner examination showed that there is a difference concerning the direction of the line pairs. Is the scan direction perpendicular to the location of the line pairs the local resolution is a little smaller. The reason is the influence of the neighbouring elements of the CCD-camera in dependency upon the incoming light intensity.

3.1.2. *Dynamic of the scanning system.* The characteristical curves show the grey value resolution of the system. To obtain the dynamic of the scanning system a radiographic film with a step wedge was scanned. For every density step the mean grey value within the area of 50 by 100 pixels and a resolution of 70 by 70 micrometers per pixel was measured.

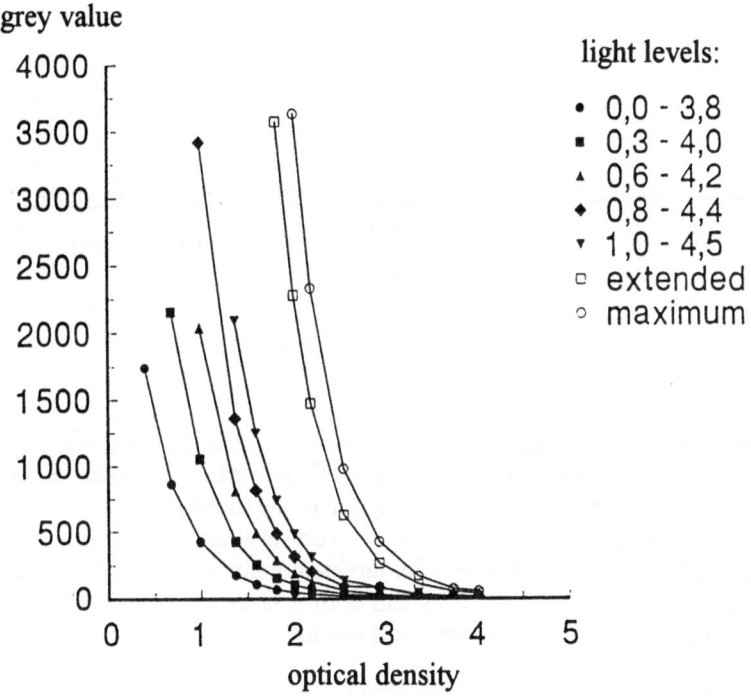

Fig. 6:   Characteristic curve for each light level

Figure 6 shows the dynamic of the seven light levels of the scanning systems. The scanner shows a steep decrease of the curve with increasing densities which means large geometrical resolution. This means, small density differences involve large grey value differences. The curves truncate with increasing densities and small density differences involve smaller grey value differences. Even in areas of high density (greater than 4.0 o.d.) grey values can be resolved.

140

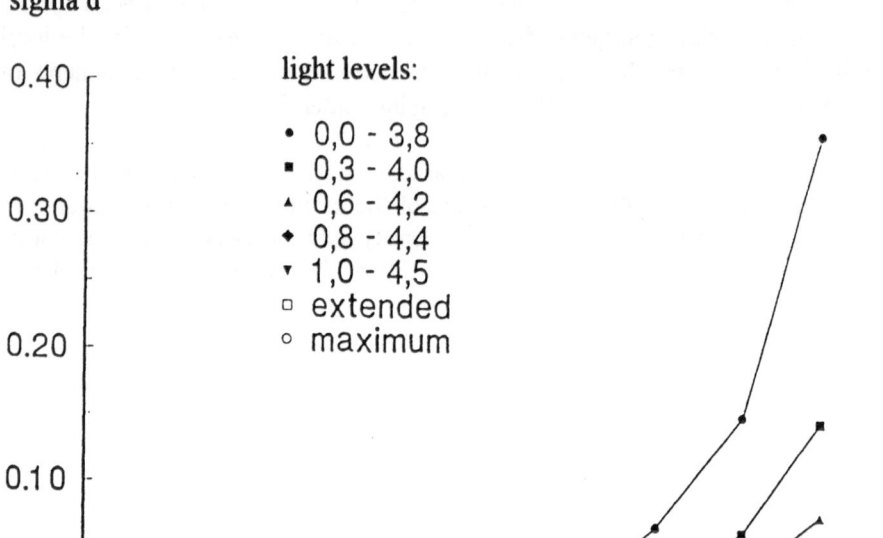

Fig. 7:   Signal-to-noise ratio

3.1.3. *Signal-to-noise ratio*. The third examination was the measurement of the signal-to-noise ratio. Together with the measurement of the mean grey value of each density step the standard deviation was calculated relative to the density.

Figure 7 shows the results. The noise level is very low and nearly equal all over the density range of the scanner. These are favourable conditions for scanning radiographic films with very small density differences and with a large density range. Low contrast radiographic films need a high densitometrical resolution and a large dynamic.

3.2.   DETERMINATION OF THE SYSTEM RELIABILITY USING ROC METHOD

The reliability of test systems can be testet by ROC method. In the above mentioned project ROC method was used to check whether or not the defect detection of digitised radiographic films is influenced by different calibration modes of the system. The aim of this project was to find out wether or not the defect detection of digitised radiographic films of welds leads to the same detection results as conventional defect detection with the light box.

3.2.1.   *Contents of the ROC study.* The ROC study involved the defect detection of a catalogue of 90 radiographic films. These radiographic films included over 1000 weld defects. Two thirds of the catalogue consisted of artificially produced welds, which represent weld defects out of the area of nuclear power plants. The last third consisted of elliptical shootings of power plant pipes. This catalogue was tested by five experienced inspectors (level III), two of them familiar with testing of digitised radiographic films of welds and image processing applications.

The ROC study included three steps (modes):

1.   Defect detection of the film catalogue using a light box (mode 1)

2.   Defect detection of the film catalogue using the digitised images of the films with standard adjustments (mode 2)

3.   Defect detection of the film catalogue using the digitised images and digital image processing for image enhancement (mode 3)

For each step of the study each film (film length 24 cm) was divided in parts of one centimetre length. Each part was testet by each inspector and for each mode. The results (defect classification, assessment of defect recognisability and assessment of defect gravity) were written into ROC forms. Furthermore, the true data of this catalogue were determined in another study. Comparing these true data with the results of the inspectors, following ROC curves were calculated by a computer program (for further information about ROC see /1,2/).

3.2.2.   *Detection ROC.* The detection ROC describes the reliability of the test system related to defect detection. The more a curve approximates the ordinate the better the reliability of the system will be.
Figure 8 shows the curve for each test mode. The difference between these ROC curves is minimal. This means, principally there is no significant difference , in terms of reliability, between conventional defect detection using a light box and defect detection using the digital image of the film with or without image processing.

3.2.3.   *Detection ROC of critical weld flaws.* The result of the detection of critical flaws in welds is of particular interest. A flaw assessed critical means, that the weld of this part of the object has to be repaired. That implicates a longer shutdown of the plant and further inspection of the repaired weld. Figure 9 shows the ROC curves of the reliability of critical flaw detection of the tested scanning system. The ROC curve of mode three (digital image with image processing) shows the best parameter of the test capability which is given by the area below the ROC curve /4/. The reason for this good result is the possibility to detect very small flaws of low contrast using image enhancement, especially contrast enhancement, filtering and zooming.

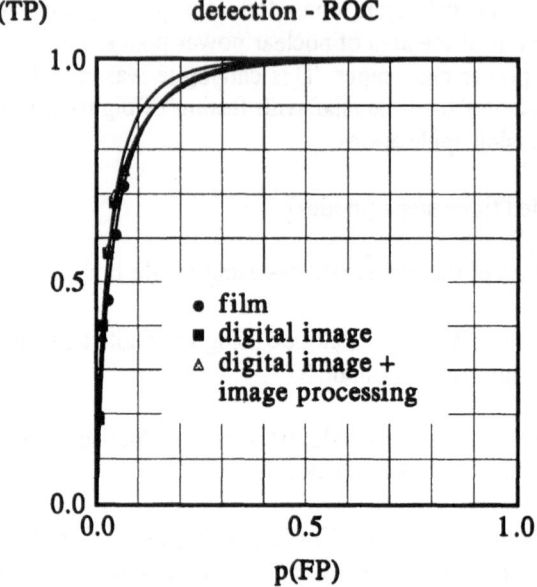

Fig. 8: Detection ROC

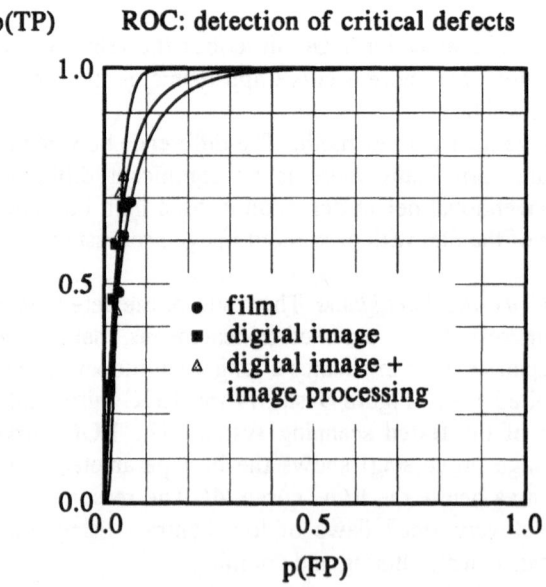

Fig. 9: Detection ROC of critical defects

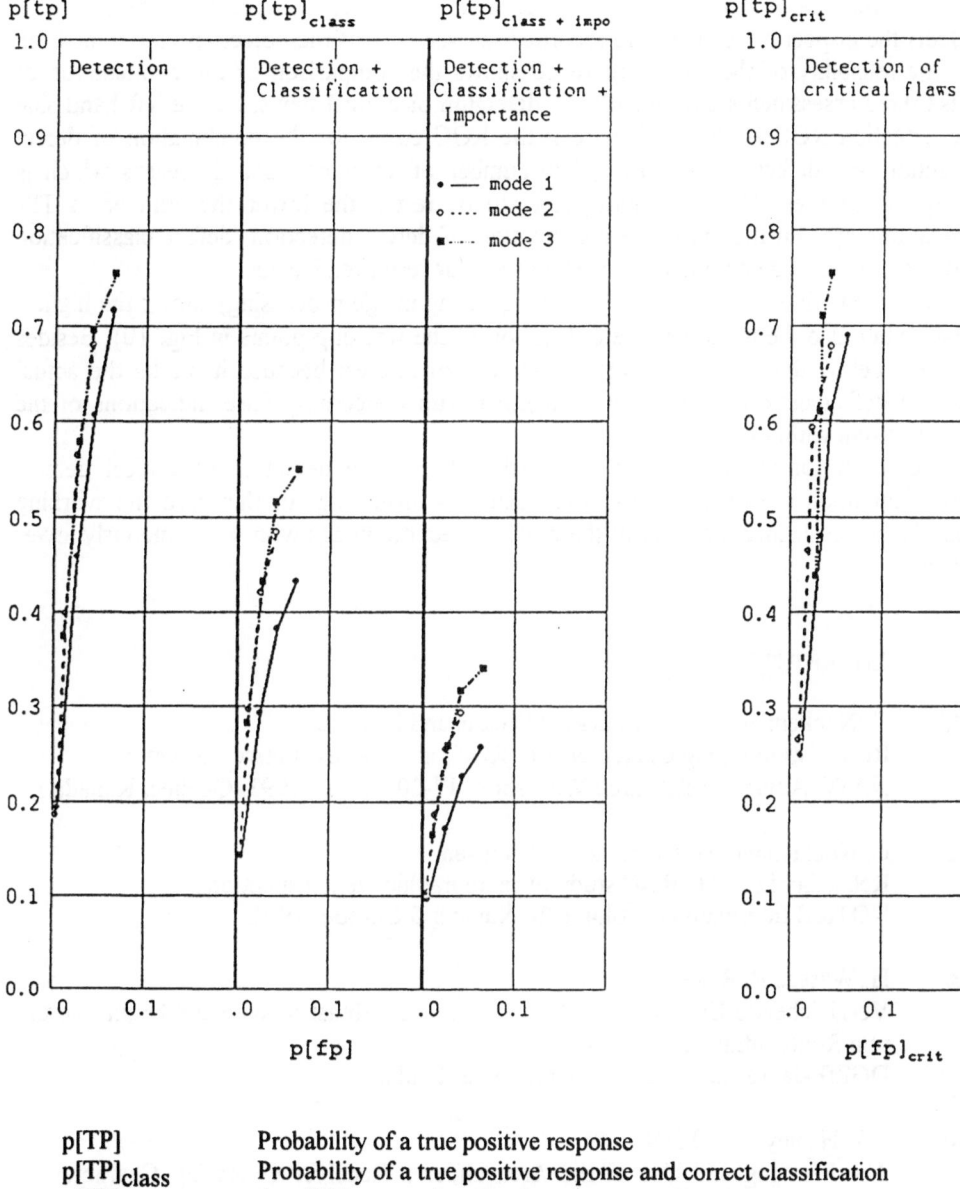

p[TP]	Probability of a true positive response
p[TP]$_{class}$	Probability of a true positive response and correct classification
p[TP]$_{class+impo}$	As p[TP]$_{class}$ and correct importance assessment
p[TP]$_{crit}$	Probability of true positive response for detection of critical defects

Fig. 10: Experimental results

3.2.4. *Assessment of different defect descriptions.* Besides the detection of the weld defects the inspectors furthermore recorded the visibility of the defect and the dimension for the reliability of the weld. Figure 10 shows the comparison of the combination of these three assessments and sets off the detection of critical flaws. At the left hand side the detection ROC is shown. Beside it the ROC curve for the combination of defect detection and defect classification. The number of TP-indications decreases which is easily to interpret. The more parameters to be tested the lower the number of TP-indications are. Furthermore the combination of defect detection, defect classification and assessment of defect importance shows similar result tendencies.

ROC mode three (testing the digital image using image processing) shows the highest TP-rate but also the highest FP-rate (location of the working points in Fig. 10). Besides the area below the curve the working point is of interest because it marks the actual reached defect indication probability together with concerning false indications of the tested system within this study.

The result for mode three of the masked ROC, the detection of critical defects dependent upon the tested radiographic films, is promising. In this case the working point is located higher than that of the other inspection modes with simultaneously lower FP-rate.

4.   LITERATURE

/1/   C. Nockemann, G.-R. Tillack, H.Wessel and H. Heidt.
      Receiver Operating Characteristic (ROC) in Non-destructive Inspection.
      NATO Advanced Research Workshop, 17-20 August 1993, Quebec, Kanada.

/2/   C. Nockemann, H. Heidt and N.Thomsen.
      Reliability in NDT: ROC study of radiographic weld inspections.
      NDT&E International Volume 24 Number 5 October 1991.

/3/   H. Wessel, P. Rose.
      Vergleichende Untersuchung hochauflösender Abtastsysteme zur Digitalisierung von Röntgenfilmen.
      DGZfP-Jahrestagung 27.-29.April 1992, Fulda.

/4/   J.A. Hanley, B.J. McNeil.
      The Meaning and Use of the Area under a Receiver Operating Characteristic (ROC) Curve.
      Radiology 143 (1982) 29.

NEW DEVELOPMENTS IN THE DETECTION OF DEFECTS BY PHOTOTHERMAL THERMOGRAPHY

R. DANJOUX   Groupe LG-MECICA
D. COLLET   Groupe LG-ETNA
5 rue de l'Industrie BP 23
ZA Les Nuisements
51350 Cormontreuil - France
Phone   (33) 26 82 04 00
Fax   (33) 26 85 28 48

J. L. BEAUDOIN

LEO
BP 347
51062 Reims Cedex - France
Phone   (33) 26 05 32 47
Fax   (33) 26 05 32 50

1. Introduction

The general principle of the control by thermography consists in submitting an object to an external excitation, forcing it to react, and then observing the thermal reaction by an infrared camera. A large variety of systems can be used as an excitation : vibrations, micro-waves, magnetic induction, air, water, and light. The results we present here exclusively concern this last point; thus, the technique is called Photothermal Thermography (Photo = luminous excitation and thermal = induced effect). Others also call it Video Pulse Thermography. When used in non destructive testing, the technique permits the detection of subsurface singularities. We present, hereafter, some visualization techniques and programs which help NDT operators to give a better diagnosis : faster, more reliable, and more precise.

2. Recallings

Photothermal radiometry is a recent technique in non destructive testing and analysis of materials. Its principles have been known for quite a long time. However, its developments only date back a couple of years. The first published works on the subject date back to the end of the 70's and the middle of the 80's ([1], [2], [3]). Actually, approximately a tenth of international teams (university or private companies) are active. ([4], [5])

The equipment we currently use is issued from a close collaboration between the University of Reims, Mecica and Dassault Aviation [5]. The parts under control are submitted to a impulsional light flux delivered by a four-Xenon flash-illuminator (we call a flash a shot). The main characteristics of the system TIRADE are the following :

- field of work   380 X 380 mm^2
- homogeneity of the energy deposit   greater than 95%
- flash time duration   8,5 ms
- maximum energy density   ca. 22000 Joules/m^2

This illuminator has been designed to work with any kind of infrared camera. In Reims we use non modified Agema cameras, short wave as well as long wave models.

145

X. P. V. Malague (ed.), Advances in Signal Processing for Nondestructive Evaluation of Materials, 145–160.

The illuminator induces front face temperature increases of several tenths of degrees, for carbon epoxy or highly absorbing materials. The measurement time is extremely short : a few seconds. The signals are stored under the form of a matrix, each representing the emittance map measured by the camera at a given instant

Eventually, one obtains a substantial amount of data, typically several megabytes. We present, hereafter, various data treatment procedures.

3. Automatic data treatment procedures

At first, our team tried to set up automatic defect characterization and detection procedures. We present the most performing one, which was named DETECT.

3.1. DETECTION

The detection is based on the calculation of a spatial gradient, by a matrix convolution of each thermogram with a bidimensional derivative operator (for example, one can use PREWITT or SOBEL operators). The convolution is done for each thermogram of the original emittance sequence, actually leading to a sequence of new images. These ones are summationned to give a unique image, representative of the singularity contours.

The next step is a thresholding, in order to obtain a binary image. The threshold is determined statistically, from the mean value and the standard deviation.

The ultimate contour is obtained after erosion/dilatation. Such an operation is well known for those who work with image treatment systems.

So, the software gives a map, where all the singularity contours are numbered by order of importance.

The time for a complete treatment depends on the number of thermograms, but usually it is smaller than 10 seconds with a PC486/33 compatible.

3.2 CHARACTERIZATION

The automatic characterization requires a thermal transfer modelling, using a monodimensional geometry.

The temperature vs. time curve of each point on the field is rapidly analyzed to give the following information :
- depths of the defects
- importance of the defects, in terms of thermal contact resistance

3.3. EXAMPLES

We have chosen to illustrate the procedures with two examples.

3.3.1. *Academic part.* The first results concern an academic part in carbon epoxy (dimensions : 200 X 200 X 1.2 mm). Four teflon round sheets were inserted at the fabrication, in order to simulate delaminations. They are arranged at the corners of a 10 cm square; their depths of implantation vary from 150 micrometers to 1.5 mm. Moreover, a superficial defect was also induced by a YAG laser pulse. This part is in rather good accordance with the theoretical model : monolithic material, monodimensional thermal transfer.

Figures 1 to 4 depicts the four stages of the procedure.
　　　　　fig 1　　: Gradients image
　　　　　fig 2　　: Figure 1, with a statistic threshold
　　　　　fig 3　　: Contoured image
　　　　　fig 4　　: Characterization

One can note that only three of the four defects are detected. The last one is implanted so close to the rear face, that the relative contrast it induces is insufficient to be detected by the method.

3.3.2. *Non academic part.* This time, the specimen is a section of a helicopter cabin. It is a complex part, with various materials, reinforcements, rivets, etc... Taking into account its size - the complete part is a couple of m^2 wide- several tests were necessary, and we only present partial results. Figures 5 to 7 show the first three stages of the procedures. The characterization is impossible due to the non-compatibility of the part with the model.

3.4 INTERMEDIATE CONCLUSION

The technique we have presented is of high performance. When the parts under control are in good accordance with the theoretical criteria, the procedure leads to a complete and automatic characterization of the defects. However, this must be balanced by two factors :
　　a) when a part is complex, the technique not only detects defects, but also all kinds of
　　　　singularities (for example structures). The characterization is thus impossible.
　　b) small size or deeply buried defects are hardly detected.

Different solutions can be chosen to improve the performances of such a NDT equipment. At first, one can be concerned by automation. A project in collaboration with Dassault and Eurocopter consisted in integrating more complex modellings in an expert system. Unfortunately, artificial intelligence codes can not be implemented on a simple PC, and although results exist it is a heavy operation

The second course consists in giving the operator more liberty in his decisions, but facilitating his task. In fact, it is illusory to think that an operatour could be easily replaced by a expert software, even working on a powerful machine. That is why we have developed new "decision aiding softwares".

4. Decision aiding softwares

Since the development of NDT by thermography, the majority of teams have focused their researches on characterizations of physical parameters :
　　- depth and size of defects
　　- measurement of the thermal diffusivity
　　- measurement of the thermal effusivity, etc...

People were mostly interested by the metrological aspects of thermography. One of the reasons was probably a will to understand phenomena. However, one must not forget that infrared thermography is an imagery technology in itself. Considering this fact, we have tried

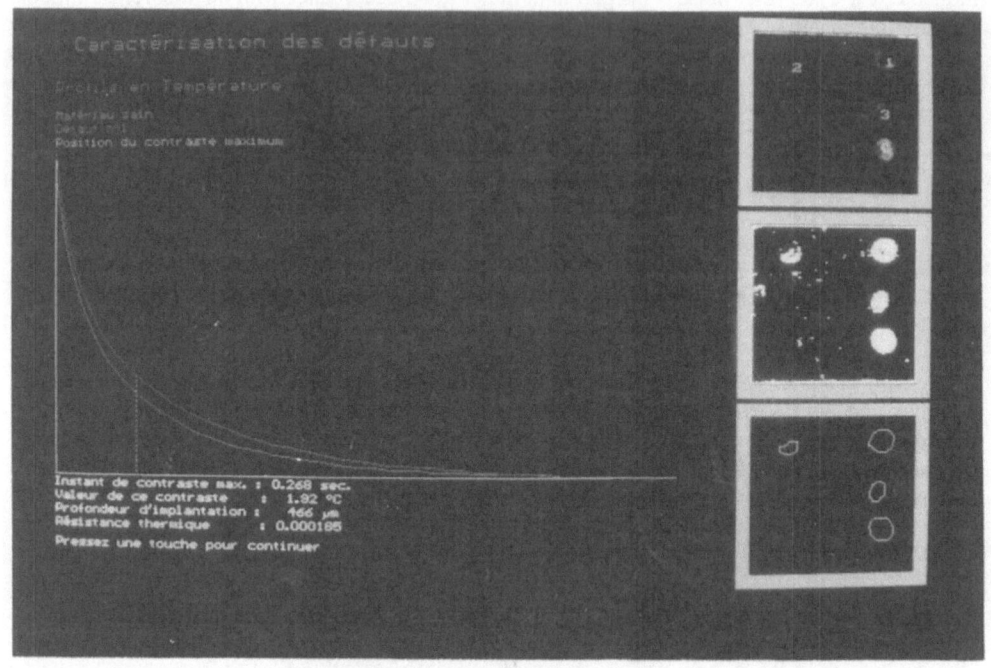

Fig 1 Academic part in carbon epoxy. Image of gradients (top right).
Fig 2 Result of a satatictic threshold applied on fig.1 (middle right).
Fig 3 Result of contour applied on fig. 2 (bottom right).
Fig 4 Automatic characterization of defects. Academic part (left).

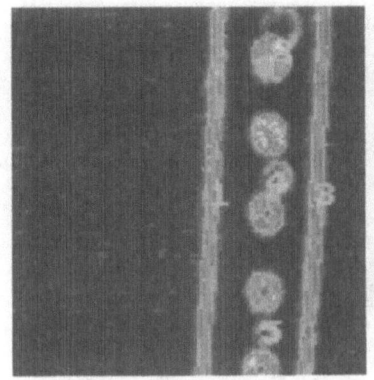

Fig 5   Production part (helicopter cabin floor)
           Image of Gradients.

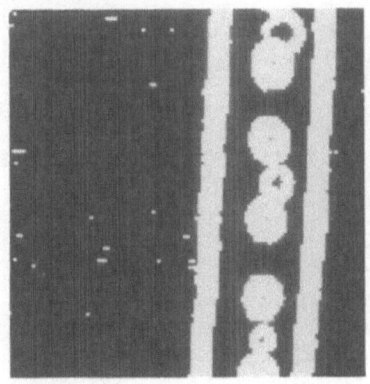

Fig 6   Result of a statistic threshold applied on fig 5.

150

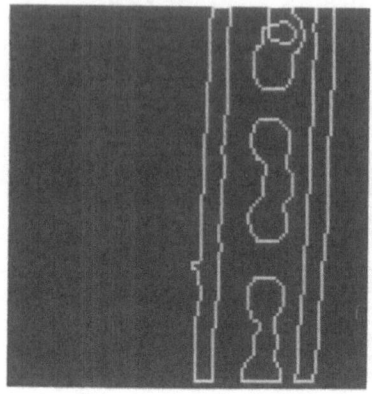

Fig 7   Result of a contour applied on fig 6.

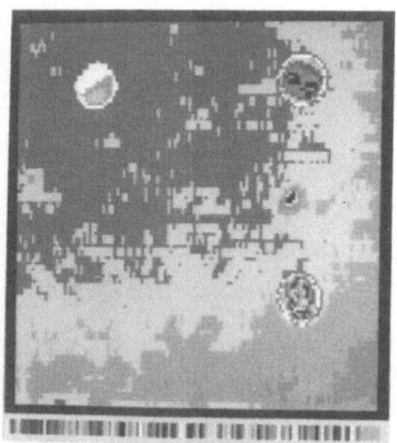

Fig 8 Thermogram in 256 linear gray levels (left).
Fig 9 The same thermogram as fig. 8, but with a random color map (right).

to make better use of the imagery. Let us precise that here, the term imagery is associated with the subjectivity in a perception of an image, rather than the Slit Response Function (SRF) at 50% and the Limit Spatial Resolution Power.

The role of "decision aiding softwares" is to amplify certain factors, proper to the perception of an external phenomenon. In the case of non destructive testing by imagery, a defect appears in contrast to normal zones. The detectivity improvement includes a greater discrimination of local contrasts.

4.1. DIGITAL RECORDER

It is an old technique, as the first version applied to thermography was developed at our company in 1988.

The principle consists in re-loading in RAM all the grabbed thermograms and making them run on the screen, as if it were a never ending film. Thus the name 'recorder' was decided.

Some functions are proposed to the operator :
- fast wind and rewind   >> and <<
- front and rear run   > and <
- pause on image   []
- speed   + and -
- color map

The image files are displayed on the screen in 256 colors, without any particular treatment. This means that the information represented is the analogic signal the sensor delivers, then digitalized. In the case of infrared thermography, it is the emittance, coded on 8 bits.

Two distincts phenomena have to be considered :
     a) the local contrast on an image
     b) the effect of temporal dynamic

Figures 8 and 9 are two representations of the same thermogram. They come from a shot on the part defined on paragraph 3.3.1. The only difference between the two figures is the color map :
     - 256 linear gray levels for figure 8
     - 256 random colors for figure 9

The information in these two images is absolutely the same, however, it is obvious that the naked eye easily detects the presence of singular zones in figure 9, but not in figure 8. The explanation is simple : a human eye is not able to distinguish simultaneously more than a fortieth gray levels. On the contrary, it has no difficulty to make the difference between highly contrasted shades.

This makes the role of the color map evident; its choice must not be taken too lightly. Figure 10 shows several maps, calculated with a specific editor we designed. Everything can be imagined; for example, why not try to use the chromatic sensitivity of the eye ?

The retina persistance explains the second phenomenon. If the refreshing frequency of the contrasted images is such as the sensitive cells of the eyes are saturated, the operator will have the feeling of constantly seeing the defects.

152

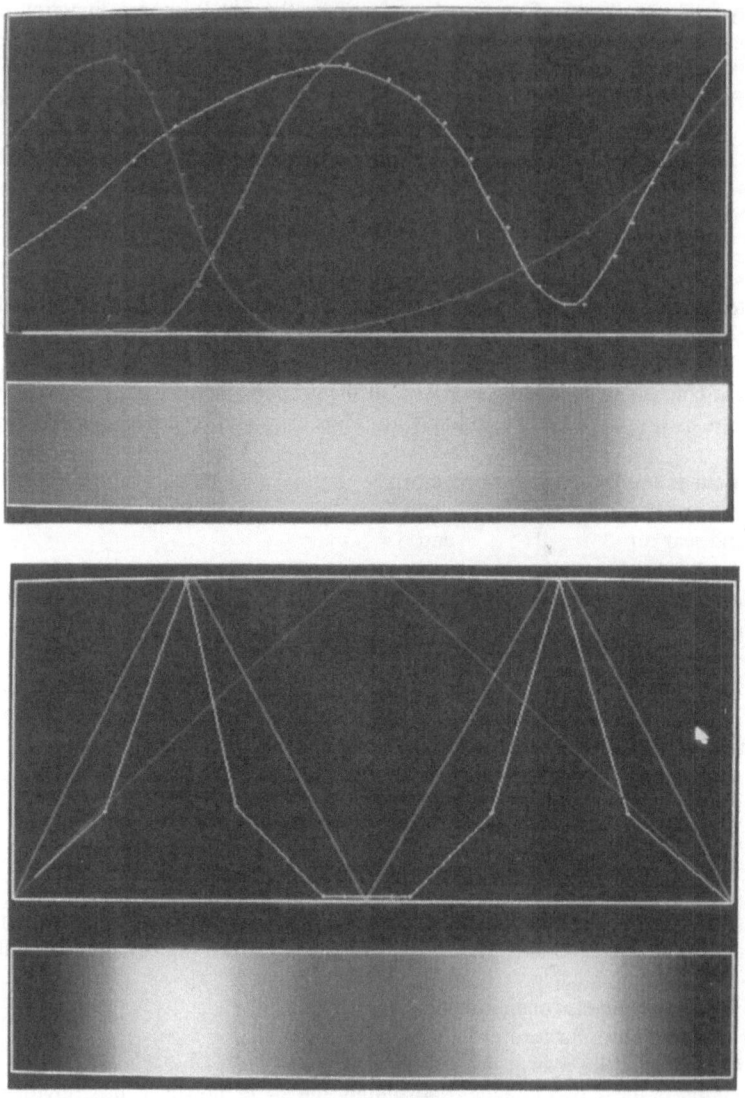

Fig 10 Examples of color map.

The basic version of the digital recorder permits a display at a maximum frequency of ca.200 Hz, for 128 X 128 files (performances measured on a PC 486/33, in 256 colors VGA mode). This is a very high limit, which is greater than the refreshment frequency of most of the usual screens. One can thus say that "several thermograms are displayed simultaneously".

4.2. HI-SIZE RECORDER

The previous program was used as a basis to design a hi-size recorder. It adopts the same great lines, but permits the simultaneous display of several thermogram sequences, obtained when controlling large size parts. In this case, one passes on several shots, according to a pre-defined program.

The composite offers the operator a global vision of the part (see figure 11). Of course, due to the large number of data to be treated, the display frequency goes down : 16 Hz with a total format 512 X 384, and 4 Hz with a format 1024 X 768.

Figure 11 is a reconstitution of 12 shots, seized on the part described in paragraph 3.3.2. One distinguishes many singular points. One also notes that geometry shifts appear. They are due to variations in the relative positioning of the camera to the part. This was absolutely embarassing, and led us to introduce geometric re-clamping functions (figure 12) :
- homothety
- rotation / center
- horizontal translation
- vertical translation

Each sequence of thermograms is repositioned to its neighbors. Such an operation is time consuming, and therefore only takes place when the files are loaded. The data manipulation in RAM is thus facilitated;

4.3. FLOATING POINT RECORDER

The functioning of the recorder has been explained in paragraph 4.1. The same principle is used to represent, this time, image sequences calculated in floating point. This means that the information represented on the screen is directly a numerical value : temperature, emittance, or any data issued from a treatment. This gives the operator the possibility to visualize in optimum conditions a sequence, in order to give a decision in accordance with a criterion.

In parallel to the development of this tool, certain treatments have been introduced, the most interesting ones are :
- calculation of a sequence in temperature, from a digitalized
  sequence in emittance, using experimental parameters
  (atmospheric transmission, lens transmission, ambient and
  environment temperatures, distance, ..
- substraction of a file to a sequence
- calculation of a sequence of temporal derivative files
- calculation of a sequence of moment quotient files (both
  directions)
- calculation of a sequence of effusivity files
- calculation of a sequence of adimensioned files
- transformation of any linear file in a logarithmic format

154

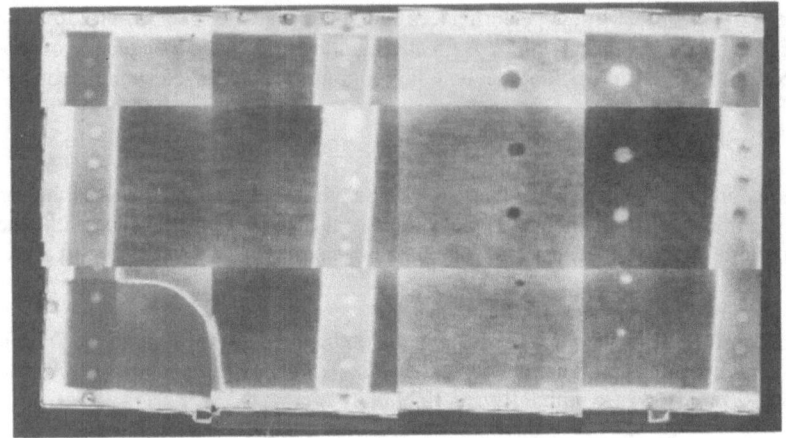

Fig 11   Composite recorder, for large parts.

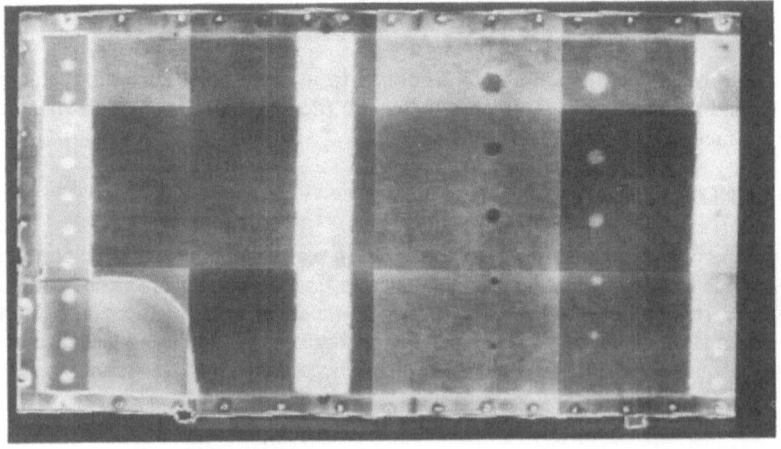

Fig 12   Figure 11, after geometric treatments.

These functions are commanded by the operator, thanks to lines of command. They are all chainable.

The time evolution of any data (on one point of coordinates x,y) can be immediately displayed on the screen, and saved under a numerical format directly readable by a spreadsheet (non exhaustive examples : Quattro from Borland and Excel from Microsoft).

This tool is illustrated with the sequence of thermograms, issued from the control of the part described on 3.3.1. On Figure 13 (issued from a sequence in temperature), it is easy to distinguish with the naked eye the signature of three internal defetcs, as well as the superficial one.

The temperature vs time curves obtained on the four internal defective zones and a sound area are shown on figure 14. It is clear that the contrast induced by the deepest defect is not exploitable. In fact, we give the curve, as we know that the defect is present.

Figure 15 demonstrates the detectivity improvement obtained by a treatment. The sequence, from which fig 14 is obtained, is transformed with an algorithm of first order moment quotient. It consists in calculating for each point M (x,y), and at each instant t, the following quantity :

$$Q(x,y,t) = \{\Sigma_0^t [\ T(x,y,t) - T(x,y,repos)\ ]\}/\ \{\Sigma_0^{end} [\ T(x,y,t) - T(x,y,repos)\ ]\}$$

T is the temperature.

Q can be easily associated with the Biot number. Thus it can lead to the determination of parameters such as the thermal diffusivity, or the external convection coefficient [6].

The integration improves the signal/noise ratio, while the normalization (division) attenuates the influence of surface effects (emissivity variations, heat flux deposit variations). This character is put forward on figure 15 : the trace of the laser impact has nearly disappeared.

Figure 16 gives the contrasts between the four internal defective zones and a sound area. The curves are raw, which means that they have not been filtered or smoothed, in order to improve their aspects. They are remarkable. The following table indicates the positions and values of maximum contrasts.

	Value of the max of contrast	Instant of the max of contrast
Defect 1 Theoretical depth 150 microns	0.074705	0.88 s
Defect 2 Theoretical depth 450 microns	0.027441	1.20 s
Defect 3 Theoretical depth 750 microns	0.008302	1.32 s
Defect 4 Theoretical depth 1050 microns	0.001891	1.44 s

The deepest defect is not visible on a single image, and the contrast it induces is only detectable by the eye when using the recorder (spatio-temporel integration effect). The instant and value of the maximum of contrast lead to the determination of the depth and the importance of the defect. It is a classical proceeding.

156

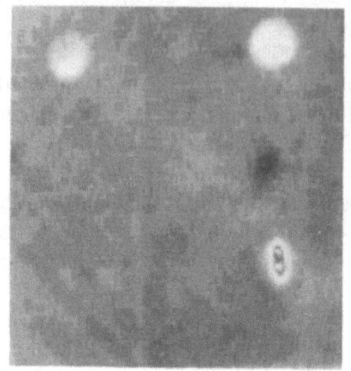 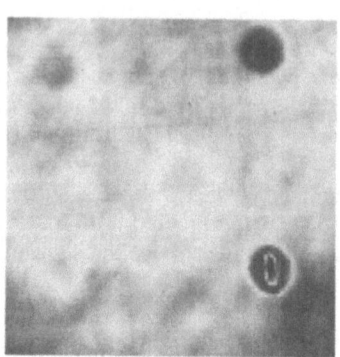

Fig 13 Floating point recorder. Temperature image (left).
Fig 15 Floating point recorder. Moment image (right).

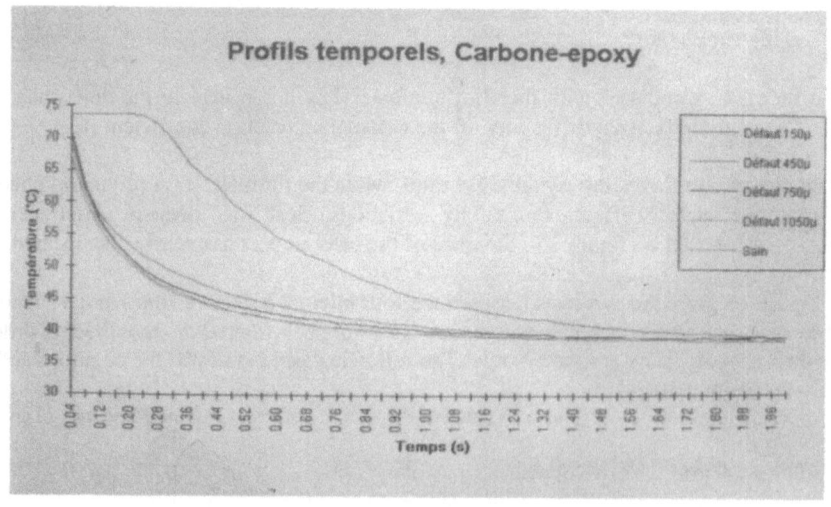

Fig 14   Temporal curves obtained on fig 13.

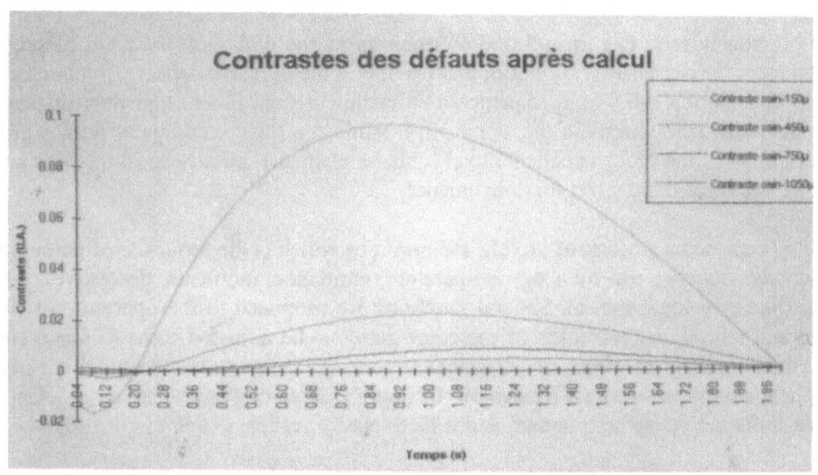

Fig 16   Contrast vs. time curves obtained on fig 15.

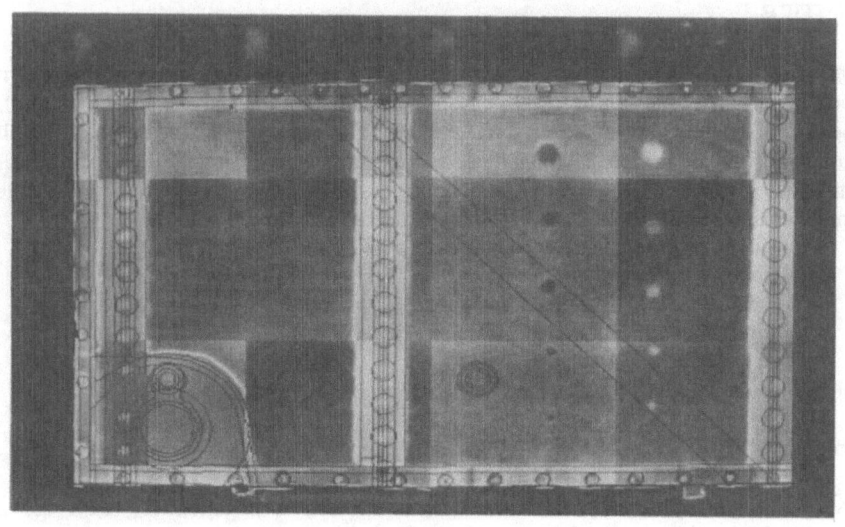

Fig 21   First result of the future software CAD/NDT

4.4. VOXEL MODE DISPLAY

We have previously seen that, in infrared thermography, the data took the form of sequences. The digital recorder is a visualization tool that privileges the dynamic aspect. Another technique consists in considering the whole sequence as a tridimensional block, and then displaying it. The two base dimensions remain the same (x,y), while the third component is played by the time. An element of the block is called Voxel (volume element). Strictly speaking, it is an abuse of language, but we use the word for convenience.

Figure 17 represents a block of voxels, the origin of which is the sequence of paragraph 3.3. Of course, any sequence can be used (temperature, emittance, moments, derivative), whatever its format (linear or logarithmic). Several functions are proposed to the operator. At first, the maximum and minimum thresholds of the color map can be adjusted so as to make elements invisible, thus creating an effect of "digging" (figure 18). Sometimes the block can become hollow. Then, complete temporal planes can be "torn", simulating an advanc or a delay (figure 19). At least, the block can be oriented, and it faces cut by vertical planes (figure 20)

Of course, it is always possible to associate an optimum color map to a block of voxels.

The use of voxels is apparently simple. In reality, a lot of precautions must be taken in order to avoid any false or unneeded conclusion.

4.5. FUTURE

A complementary software will soon be available. It will allow the superimposition of results obtained by any imaging non destructive testing system (not exclusively thermography), with tridimensional drawings generated by CAD. Figure 21 is a first result. The operator will thus be able to make the difference between real defects and natural singularities (reinforcements, overthicknesses, etc..). The control being done, the CAD data bases will be modified by adjunction of new parameters. The objectives of such a system are :
- to follow, in time, the evolution of a damaged part
- to incite a reaction, at the design bureau, if too much structure
  linked defects are discovered
- to improve the productivity

This project is supported by the French Ministery for Research and Space.

5. Conclusion

The results we have presented are issued from researches engaged 10 years ago at the Laboratoire d'Energétique et d'Optique of the University of Reims. Mecica is in charge of transferring them to the industry. The more recent progress concerns without any doubt the computer possibilities. In 1985, the capacity of storage hard-disks went rarely over 10 or 20 megabytes. Now, a 500 megabyte hard disk costs only about $ 1000. In parallel, new langages (C and C++) as well as new compilers allow to accelerate treatments, and also to implement on PC-compatibles functions that were previously reserved to larger computers : floating point calculations , FFT, etc..Thermography systems have also evolved, but their improvement can be compared to the considerable performance increases of computers

Non destructive control by thermography has now yielded its adult life. Its notoriety is high, and its performances acknowledged. However, the technique is still largely under employed. To

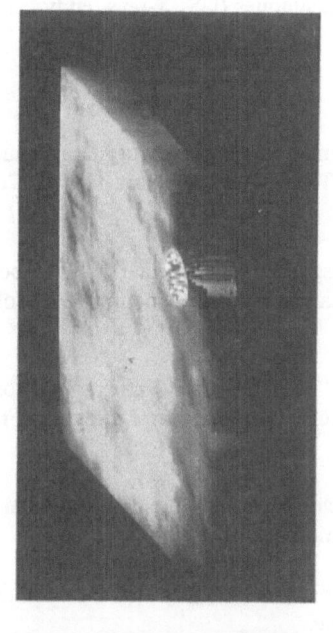

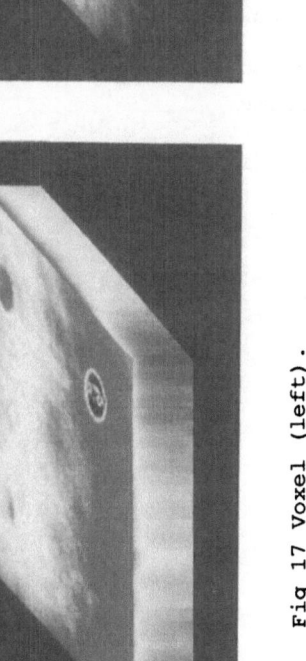

Fig 17 Voxel (left).
Fig 18 Voxel. Effect of a color threshold (right).

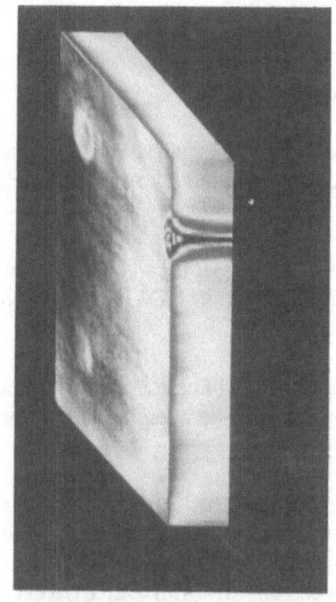

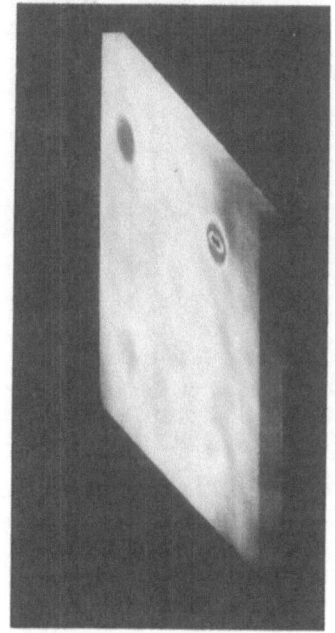

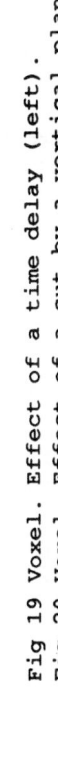

Fig 19 Voxel. Effect of a time delay (left).
Fig 20 Voxel. Effect of a cut by a vertical plane (right).

change this situation, a first step would probably be the setting up of international operators and procedure certifications, similar to what exists for other techniques (US, X-rays, eddy currents, etc..).

Bibliography

[1] Bekesho N.A and Upadyshev A.B. Use of thermal method for observing defects such as breaks in continuity in metallic and non metallic objects. The Soviet Journal of NDT, 1970, vol 1, p20-24.

[2] Vavilov V. and Taylor R. Theoretical and practical aspects of the thermal NDT of bonded structures. Res. Techn. in NDT. Edited by Sharpe R., Academic Press, London 1982, Vol 5, p 239-280.

[3] C. Martin. PhD Thesis. Contribution to the use of infrared thermography, for the measurement of flux, temperatures and various parameters. University of Limoges - France. Nov 1978.

[4] see proceedings of the last International Topical Meetings on Photothermal and Photoacoustical phenomena (Montreal, Heidelberg, Baltimore, etc..)

[5] see proceedings of QIRT 92. EUROTHERM 27. Chatenay Malabry - France. Edited by Editions Européennes Thermique et Industrie, Paris, France.

[6] David Crowther. Ph D Thesis. University of Reims. March 1990.

ADVANCES IN SIGNAL INVERSION WITH APPLICATIONS TO INFRARED THERMOGRAPHY

V.P.VAVILOV
Tomsk Polytechnic University
Institute of Introscopy
634028 Tomsk, Savinykh, 3
Russia

ABSTRACT. Approaches to performing the thermal characterization of defects in solids are discussed including phenomenological treatment of experimental data, use of classical heat transfer solutions and minimization of functionals. Simple analytical expressions proposed by various authors in order to estimate defect size, depth and thickness are verified by using well defined reference temperature data.

1. General remarks

A lot of cross-correlations could be referred to terminology "nondestructive testing", "technical diagnostics", "materials evaluation", *etc.* Historically, the first nondestructive techniques have been targeted to simple detection of defects without determining their parameters. The Russian term for such technique was "defectoscopy", *i.e.*, "seeing defects". Another Russian term nearly unknown in the West is "introscopy", *i.e.*, "seeing inside". Firstly, nondestructive testing (NDT) was considered as purely applied area without specific academic interest. Mathematically, this situation involved solutions of the so called "direct" problems where modeling of internal specimen structure allowed to determine its response to applied physical field or particles beam. Situation changed with worldwide spreading of super-sophisticated systems, like nuclear power stations, space shuttles, *etc.*, where payment for technical fault became too expensive in human and economical sense. Now NDT agenda includes trend prediction for processes occurring in systems like mentioned above. It requires to solve the so called "inverse" problems of technical diagnostics which represent the area where academic researchers could verify their algorithms. In Russia this new approach is called "defectometriya", *i.e.*, "measuring defects" with analog term in the West to be "defect characterization".

After these philosophic remarks we pass to thermal characterization of defects inside solids. Researchers who will try to solve inverse heat transfer problem in general 3D consideration (arbitrary defect in arbitrary body) will encounter "a riddle wrapped in a mystery inside an enigma" (as the Russian I can not to apply this famous words told by Winston Churchill about Russia to Thermal NDT). Fortunately, the most practical problems could be referred to more simple geometry shown in *Fig.1*. A specimen with thickness much less than its lateral size is heated on the surface with Dirac pulse, square-pulse, periodic or continuous heat beam which could be uniformly distributed or focused into a spot or line.

There is no need to discuss here the Thermal NDT (TNDT) terminology and mathematical description of arising direct heat transfer problems (for this we refer the reader to some fundamental works devoted to Thermal NDT techniques [1-5])

X. P. V. Malague (ed.), Advances in Signal Processing for Nondestructive Evaluation of Materials, 161–184.
© 1994 *Kluwer Academic Publishers.*

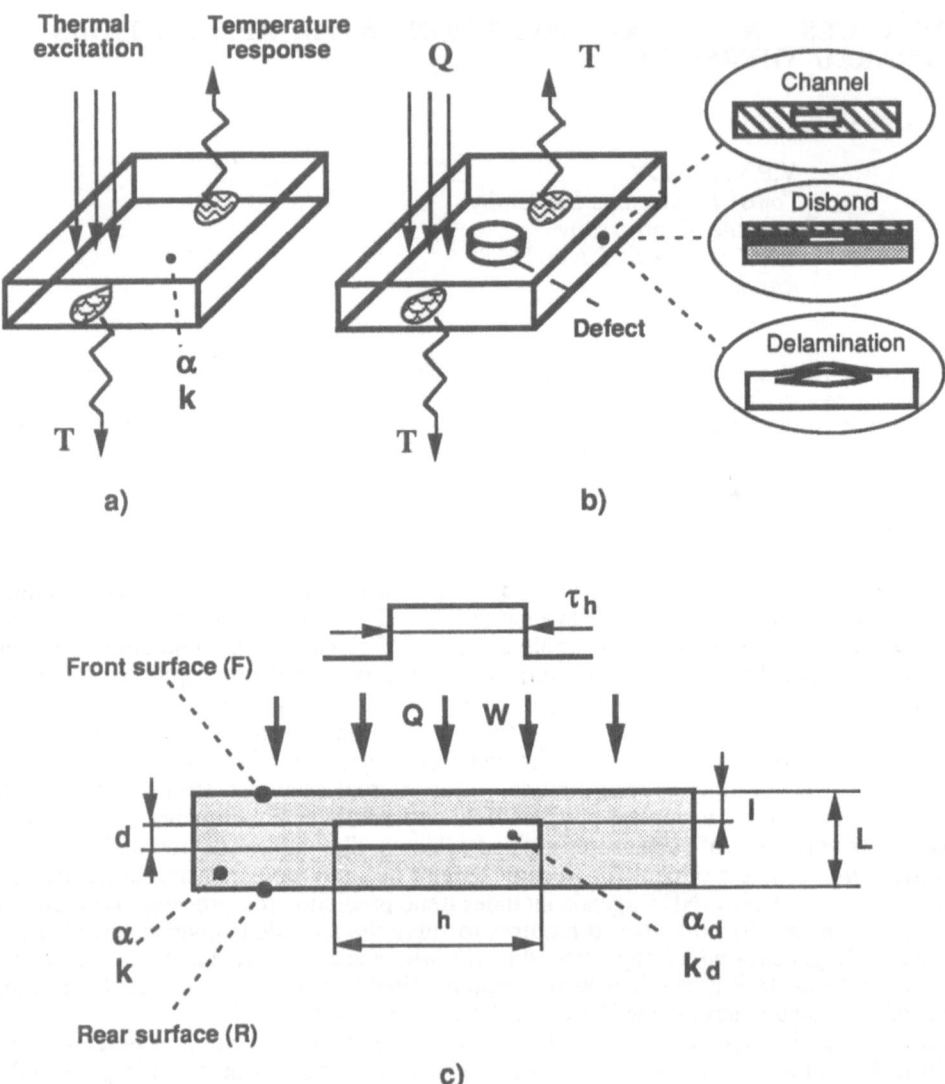

Fig.1. Thermal Characterization of Solids:

a-thermal properties measurement,
b-defects' detection,
c-specification of involved parameters

It is important to state that after performing thermal test we shall have set of 3D temperature data $T(x,y,\tau)$, (x,y-surface coordinates; τ-time) which will serve as input data to solve inverse TNDT problem. Then either thermal properties of a specimen (α-diffusivity; K-conductivity) or defect parameters (h-in-plane size; d-thickness; l-depth) are to be derived as the output parameters. Measurement of thermal properties is highly developed area deserving special discussion in NDT applications. In this paper we focus the attention onto defects' characterization. This area is also under intensive investigation (see Proceedings of the annual meetings "Thermosense" and "Review of Progress in Quantitative NDE", U.S.A, and the QIRT-92 Seminar, France).

2. Features of the Inverse Solutions for Thermal NDT and Algebraic Inversion in the Laplace Region

As we stated above, even simple NDT problems involve at least three unknown defect parameters: h, d, l. In many cases defects which are to be detected thermally have lateral size much larger than their thickness or depth. It follows from the fact that defects located parallel to main surface represent strong resistance to in-depth thermal flux and hence their shape and size could be determined by their surface "prints". So we concentrate ourselves on two unknown defect parameters: thickness d and depth l. There is temptation to treat heat transfer solutions as algebraic equations. Then we need only two linearly independent equations to determine d and l. The problem is that we are not free in obtaining independent temperature equations. If to forget about peculiarities in spatial distribution of temperature because of neglecting the lateral defect size, we are able to vary only recording time (not mentioning input energy which obviously affects the temperature values in a trivial form). Unfortunately, neither temperature functions in TNDT are non-algebraic, nor temperature read-outs taken at two times are not independent. There were some hints to use different time points to analyze the responses of different internal layers [6] or even to distinguish one defect located under another [7]. But so far our experimental attempts to implement these rather theoretical proposals are far from being excellent because of noise.

Simple and very illustrative general approach to problem discussed has been developed by Degiovanni, Maillet *et al.* who used quadripole method to get algebraic expressions in Laplace region [8]. For three-layer plate, where external layers are identical and the middle layer represents defect with thermal resistance R_c, the corresponding solutions which represent difference between defect and non-defect areas are as follows:

$$\Delta\theta^F = \frac{R_c^*\sinh^2[a(1-x^*)]}{\sinh(a)\{\sinh(a)+R_c^*a\sinh(ax^*)\sinh[a(1-x^*)]\}};$$

$$\Delta\theta^R = -\frac{R_c^*\sinh(ax^*)\sinh[a(1-x^*)]}{\sinh(a)\{\sinh(a)+R_c^*a\sinh(ax^*)\sinh^2[a(1-x^*)]\}}, \tag{1}$$

where $\Delta\theta$ is Laplace transform of $\Delta T/T_m$; F, R relate to front and rear surface; $T_m=W/C\rho L$; $x^*=l/L$; $R_c^*=R_c/R$; $R=L/K$; $a=\sqrt{p^*}$; $p^*=L^2p/\alpha$; p-Laplace transform of time τ.

Real functions $\Delta T(\tau)$ could be determined using numeric inversion algorithm. But it is important that the inversion procedure for defects parameters could be performed in Laplace region. Sensibility analysis of *Eqs.(1)* proved analytically the fact which is theoretically and experimentally confirmed by many researchers, namely:

*front surface (F) procedure is more sensible to defect depth;
*rear surface (R) procedure is more sensible to defect thickness.

In the R-procedure the thermal resistance of defect could be directly determined in particular case when two recording times τ_1, τ_2 are chosen in a such way that $p_2 = 4p_1$:

$$R_c = \frac{M_1^2[1 + a_2\sinh(a_2)M_2]\sinh(a_1)\tanh(a_1)}{[1 + M_1a_1\sinh(a_1)][-M_1 + M_2\cosh(a_1) - 1.5M_1M_2a_1\sinh(a_2)]},$$

where $M_i = \Delta\theta^R(p_i^*, R_c^*)$ for $i = 1, 2$.

In the F-procedure the defect depth value is "hidden" into transcendent equation but for thin defects it could be found directly from:

$$x^* = 1 - \frac{1}{a_1}\ln[(\frac{M_2}{M_1})^{1/2}\cosh(a_1) + (\frac{M_2}{M_1}\cosh^2(a_1) - 1)^{1/2}].$$

Resistive defects used in the solutions above fit well thin air gaps; in a case of solid inclusions the capacitive defects must be introduced.

Physically, it is worth to define the difficulties of inverse Thermal NDT problems solutions as follows:

*heat transfer solutions are dubious in sense that both defect depth and thickness influence temperature surface "prints"; one of simple solutions to separate these functions is to use one-side and two-side procedures;
*dependencies of temperature signals *vs.* defect depth and thickness represent functions with saturation, *i.e.*, for large defect depth and thickness small variation in measured temperature will lead to large variations in the obtained data (qualitatively this remark is illustrated with *Fig.2* which needs no further explanation).

3. Approaches to Solutions of the Inverse Thermal NDT Problems

Classification of inverse problems in Thermal NDT is not established. Those working in mathematical physics use more "academic" terminology, *e.g.*, detection of defects could be related by them to "geometrical" inverse problems which involve the determination of external and internal borders of geometrical regions where heat transfer occurs (see also the paper of Cassab and Hsieh [9]).

Not pretending for basic classification we relate more to published works which include the following approaches:

*Phenomenological (researcher estimates the apparent features of his particular experiment with less if any theoretical basis);

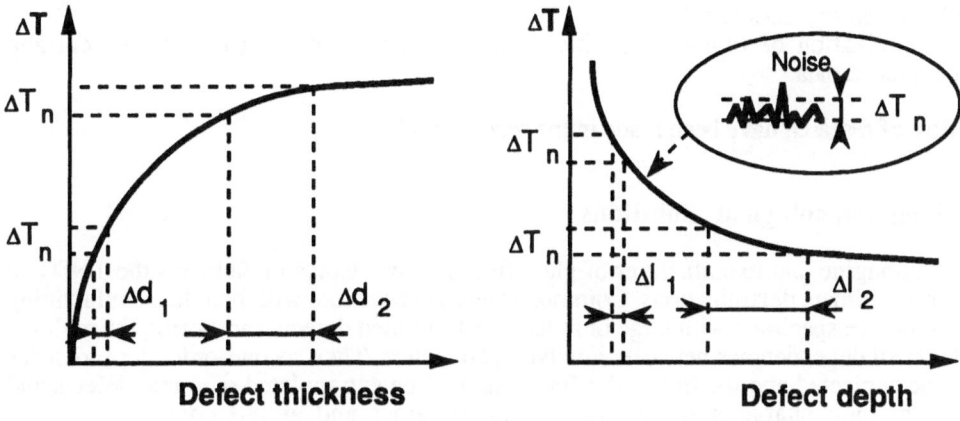

Fig.2. Why it is difficult to measure thickness of thicker defect and depth of deeper defect on the background of noise ?

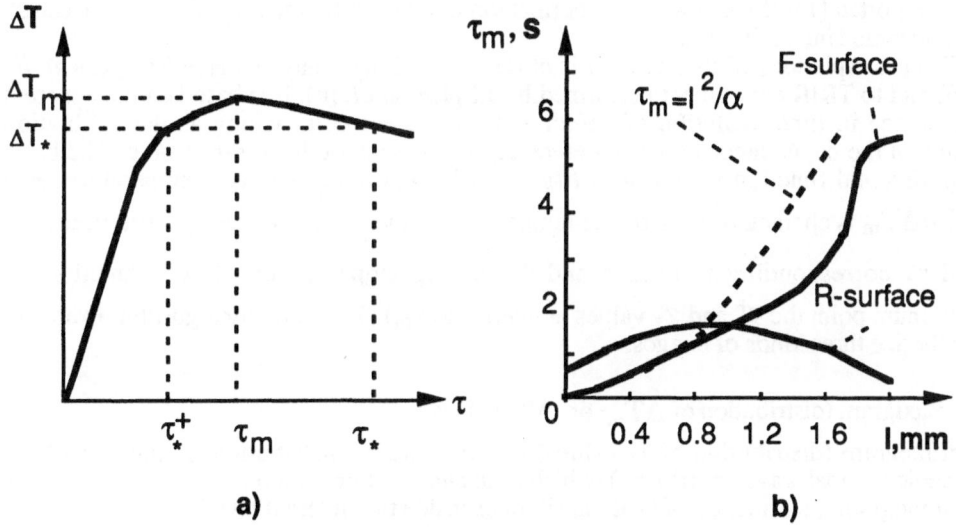

a)   **b)**

Fig.3. Basic Functions in Thermal Tomography:

a-maximum temperature signal over defect *vs.* time,
b-specific transit time *vs.* defect depth in one-side
(F-surface) and two-side (R-surface) inspection
of air delamination in carbon epoxy plastic of
thickness 2mm

*Inversion of classical solutions;

*Minimization of functionals which represent difference between theoretical and experimental data.

Most of research have been made in the second field.

4. Phenomenological Solutions

Let us imagine that manufacturer of glass fiber rocket casings implements the TNDT in order to estimate delaminations. Common research procedure will include manufacturing the reference specimen with large or at least well-defined defects and storing the multiple data on all dependencies between involved parameters. Then second-order dependencies will be neglected and the first-order functions will be extrapolated onto real defects and used for their characterization. Tables, nomogrames and graphs could be used for inversion of results. This simple but reliable for particular products procedure could satisfy practitioners.

More sophisticated version of this approach is the so called "dynamic thermal tomography" which was proposed by Vavilov and Shiryaev in 1985 [10]. The term "infrared tomography" was also used by Cassab and Hsieh but it related to the steady-state thermal process [9]. Since that time some theoretical and experimental results have been reported [11-15]. Commercial thermal tomography software is available from Tomsk Polytechnic University, Russia.

The principle idea of this method is close to effusivity analysis made Balageas et. al. [16] and to TRIR experiment performed by J.Spicer et al. [6]. It is based on the apparent difference in time evolution of defect and non-defect point taking as the calibration function the dependence between defect depth and specific heat transit time. The basic experimental function is shown in Fig.3a. It is important that each temperature level $\Delta T_*/\Delta T_m$ is characterized with one temperature value ΔT_* and two specific times τ_*^+ and τ_* corresponding to arising and decreasing slopes of the $\Delta T(\tau)$ function (at extremum point the τ_*^+ and τ_* values convert into τ_m). Tomography algorithm is able to synthesize three kinds of images:

*maxigram (distribution of ΔT_m or ΔT_* values);

*timegram (distribution of τ_* values) or phasegram (distribution of phase shift in periodic thermal wave experiment) which is an analog of timegram ;
*tomogram (distribution of heat transit times inside thin internal layer).

Tomograms could be easily obtained from timegrams or depth images. Inversion function "depth vs. heat transit time" is shown in Fig. 3b for carbon epoxy plastic. This function was calculated numerically. Its simple approximation (shown dotted in Fig.3b) is well known in the TNDT theory:

$$\tau_m = l^2/\alpha. \qquad (2)$$

To complete considerations of the previous paragraph the same function $\tau_m(l)$ is also presented in *Fig.3b* for R-surface illustrating clearly why two-side procedure can not be used in the inversion of defect depth.

Examples of thermal tomograms of carbon epoxy plastic are presented in *Fig.4*.

5. Inversion of the Classical Solutions

When using term "classical solutions" we mostly refer to two well known books (Carslow and Jaeger [17], Luikov[18]) as well as some 1D multi-layer solutions which are intensively discussed now [6,8,19]. Recent developments in the TNDT theory are more related to numeric algorithms.

5.1. MEASUREMENT OF THERMAL PROPERTIES

Measurement of conductivity and diffusivity or effusivity of solids is the excellent example of heat transfer inverse problems. This area is widely discussed by the community of specialists working on thermal properties measurement and we touch it here in order to show that detection of internal defect puts even more challenging problem.

Dirac-pulse adiabatic heating of semi-infinite body is described with very well known expression:

$$T(\tau) = \frac{W}{\sqrt{KC\rho}} \frac{1}{\sqrt{\pi\tau}}. \tag{3}$$

Values of T and τ are measured in thermal test, hence there are four unknown parameters: W, K, C, ρ.

Obviously it is impossible to distinguish K, C, ρ values in this model, then only two parameters (energy density W and effusivity $e=\sqrt{KC\rho}$) could be determined. To do this we need two independent equations which could not be obtained by recording temperature at two times. Usually energy W is measured independently or only a ratio of effusivities is determined.

Another example is provided by solution for temperature on the rear surface of a plate which is adiabatically heated with Dirac-pilse:

$$T(\tau) = \frac{W\alpha}{KL}(1 - 2e^{-\pi^2\frac{\alpha\tau}{L^2}}),$$

where the first member of infinite seria is taken for simplicity. Here the specimen thickness as new unknown parameter appears. This solution could be easily reduced to known Parker's formula which connects diffusivity, thickness and experimentally measured half-arise time $\tau_{0.5}$ as follows:

$$\alpha = \frac{0.139L^2}{\tau_{0.5}}.$$

168

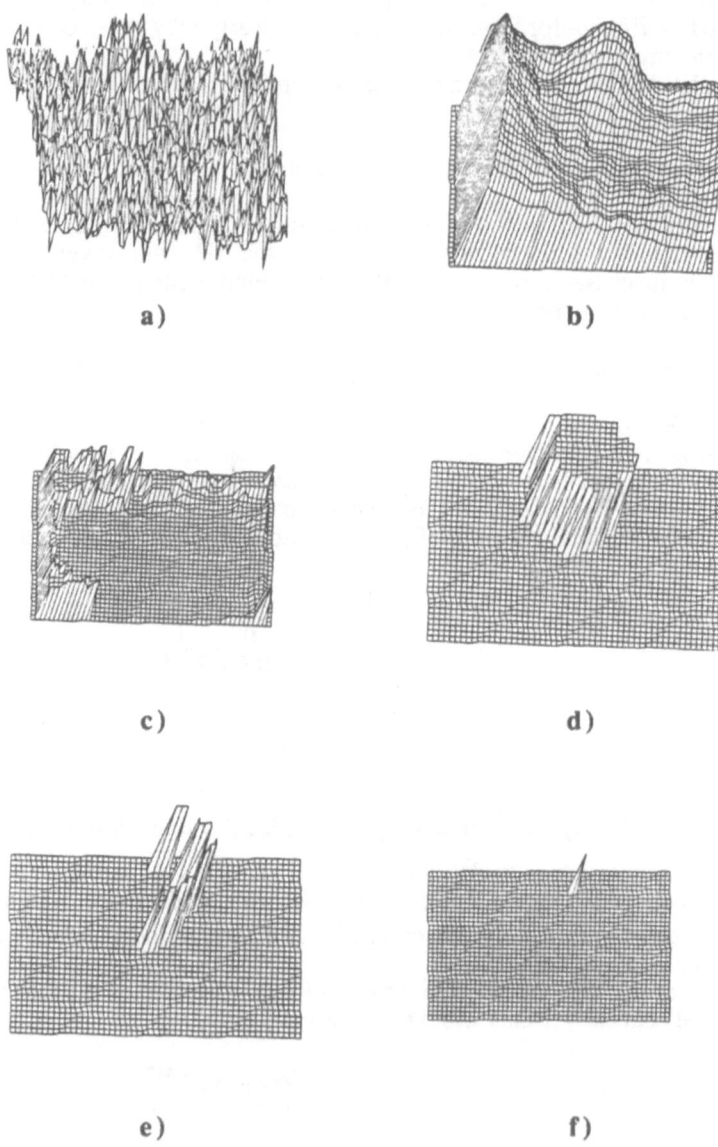

Fig.4. Dynamic Thermal Tomography of Teflon Insert in
Carbon Epoxy Specimen:

a-input IR image (from IR camera),
b-smoothed image,
c-timegram,
d-tomogram of the layer 0.8-1.5 mm (with insert),
e-tomogram of the layer 1.4-1.8 mm (with artifact),
f-tomogram of the layer 1.8-2 mm (with noise)

Notice that diffusivity does not include energy units and does not require to measure the incident energy.

These simple solutions show that one-side thermal test tends to be connected to effusivity measurement and two-side one-to diffusivity measurement. Both solutions are independent but they do not allow to solve the inverse problem without measurement of W because of appearance of new unknown parameter: L.

The situation aggravates essentially when we deal with localized internal variations of thermal properties, *i.e.*, with defects.

5.2. SIMPLE APPROXIMATION OF 1D FUNCTIONS

Storozhenko, Vavilov and Volchek performed the experimental research with glass fiber specimens and found that in a limited range of involved parameters the main TNDT dependencies could be approximated with simple expressions [20]:

$$\Delta T = A d \, Exp(-Bl) \quad \text{or}$$
$$\Delta T = (l/L)^m (d/L)^n (h/L)^s,$$

where A,B,m,n,s-constant values which are simply found experimentally or determined by the least square method comparing theoretical and experimental data.

Theoreticians could have much of criticism to such solution of inverse problem but it is close to what is asked by practitioners.

5.3. THERMAL WAVE APPROACH

Here we spread the term "thermal wave" into periodic, Dirac-pulse or step-function heating. Mathematically, solutions in two latter cases produce the same conclusions because Dirac pulse is just the derivative of step-function. Unlike sine-wave technique the pulsed heating generates a wide spectrum of frequencies which penetrate the solids to different depths, *i.e.*, low frequencies provide information about deeper layers for longer recording time and high frequencies are more responsible for information about shallow defects for shorter time. Physically it means that to make "in-depth scanning" we would need a lot of periodic experiments with varying frequency. Flash experiment will make it at once (to be correct we have to mention that periodic procedure enhances signal-to-noise ratio).

5.3.1. Single pulse- and step-heating. Reflection of thermal wave from defect. Most of researchers follow the assumption that the front of thermal wave has to reach defect and after being reflected to come back to F-surface. Conception of thermal fronts travelling between surface and defect follows from infinite series which are inherent to common heat transfer solutions. This approach involves defect reflectivity (or thermal mismatch) Γ as:

$$\Gamma = \frac{1-b}{1+b};$$
$$b = \sqrt{\frac{(KC\rho)_2}{(KC\rho)_1}}.$$

Lau, Almond and Patel proposed the following simple expression for temperature signal over large-size defect [21]:

$$\Delta T(\tau) = \frac{W}{2\sqrt{\pi K C \rho \tau}} \sum_{n=1}^{\infty} \Gamma^n Exp(\frac{-n^2 l^2}{\alpha \tau}). \tag{4}$$

For the first reverberation and $\Gamma = 1$ the last expression was transformed for disc-shaped defect of radius r_d [22]:

$$T(\tau) = \frac{W}{\sqrt{\pi K C \rho \tau}} \{ Exp[-\frac{l^2}{\alpha \tau}] - Exp[-\frac{(l + \sqrt{r_d^2 + l^2})^2}{4\alpha \tau}] \}. \tag{5}$$

Similar approach to thermal waves generated by step-function heating is developing by Aamodt, Maclachlan Spicer and Murphy [6] who called it TRIR (Time-Resolved Infrared Radiometry) procedure and spread it onto multi-layer structures using the artificial construction of heat packets. According to [6] each packet makes its prescribed number of passages between surface and layer boundary and then (and only then) contributes to the surface temperature and "dies" afterwards.

Important bridge between results of different authors is that all they agree about existence of specific heat transit time which determine the surface response of internal layers. The problem is to propose well verified expressions to calculate these times in various TNDT procedures and to use them in practice. Unfortunately, the up-to-date results show that in a case of real multi-layer structure the amplitude of heat packets is considerable only for the first layer (e.g., coating on a substrate). Even worse is the situation when one tries to see defect "through" another defect located above because the upper defect overshadows the deeper one [7] (but by reasonable remark of D.Burleigh at "Thermosense-XIV" there is no need to seek for deeper defect if serious shallow fault is once detected).

5.3.2. *Harmonic waves.* There are two large areas of practical application of phenomena which occur in solids under periodic heating:

*Analysis of thin layers and coatings using stimulated thermal waves with frequency in a Hz to MHz range;
* Building and terrestrial applications (including IR reconnaissance) using super-slow day-night variations in solar irradiation.

In the framework of the first area Alzofon demonstrated analogy between thermal and optical processes [23]. Rosencwaig and Busse paid more attention to thermoacoustic signals which always occur in materials with non-zero expansion [24,25]. Experimental results on measurement of sub-surface distribution of effusivity (determined mostly by wave amplitude) and diffusivity (determined mostly by wave phase) in graphite specimen were recently reported by Bein, Gu, Mensing *et.al.* [26]. Use of periodic thermal waves for detecting the subsurface defects is based on the fact that wave of particular frequency f will penetrate into solid up to depth which is in the range of length of thermal diffusion

$\mu=\sqrt{\alpha/\pi f}$. Hence, if we deal with a spectrum of temperature response we have to separate the signal which is within narrow frequency band, *i.e.*, corresponding to defect at particular depth. Contemporary thermal wave analysis involves the conception of the point spread function (PSF) which considers the point-like thermal inhomogeneity. It allows to represent the signal deviation S_{dev} in defect area from non-defect signal S_0 by using the multiple convolution integral:

$$S_{dev}=S_0[W**F**G], \tag{6}$$

where $W(x,y)$-heating beam profile; $F(x,y)$-shape of defect located at depth l; $G(x,y,\tau)$-PSF. The expression for G is given in [27] in the following form:

$$G(r)=\frac{\pi}{2}\sigma^2\frac{1}{\sqrt{r^2+l^2}}Exp[-\sigma(1+\sqrt{r^2+l^2})][\frac{Al}{\sqrt{r^2+l^2}}(1+\frac{1}{\sigma\sqrt{r^2+l^2}})+B],$$

where $A=\Delta K/K_{nd}$; $B=\Delta(C\rho)/(C\rho)_{nd}$ -defect thermal strength parameters; $\sigma^2=2i/\mu^2$; $\sigma=(1+i)/2$-wave vector; r,l-lateral and vertical offsets between point defect and heating beam (notice that $G(r)$ does not contain the time variable).

Natural thermal waves stimulated in solids under solar irradiation have been used to make inspection of roofs, roads and walls [28], as well as to detect the objects buried in soil [29, 30]. Software "Thermo.Tom-Build" of Tomsk Polytechnic University allows to synthesize positive and negative maxigrams and "phasegrams" (analogs of timegrams) from sets of IR images taken during day-night period.

5.4. ANALYTICAL INVERSION OF DEFECT RESISTANCE AND DEPTH

General approach to inversion of 1D inverse TNDT problems in the Laplace region has been described in p.2. In some extreme cases simple analytical expressions are available.

5.4.1. *Apparent effusivity method.* It is assumed that in one-side procedure the specimen under test could be considered as semi-infinite body. Specimen effusivity could be determined by expression which follows from *Eq.(3)*:

$$e=W/T^F(\tau)\sqrt{\pi\tau}.$$

It is evident that uniform body will be characterized with constant value of effusivity e independently on measurement time. Any internal variation in thermal properties will cause deviation in curve $e(\tau)$ from the straight line. The deeper is defect the later this deviation will occur and the less will be its amplitude. This procedure is illustrated with graph in *Fig.5*. Obviously, the effusivity approach is equivalent to dynamic thermal tomography (see p.4) with values of $(e/e_{nd})_{min}$ being related to maxigram and values of Fo_{dmin} -to timegram.

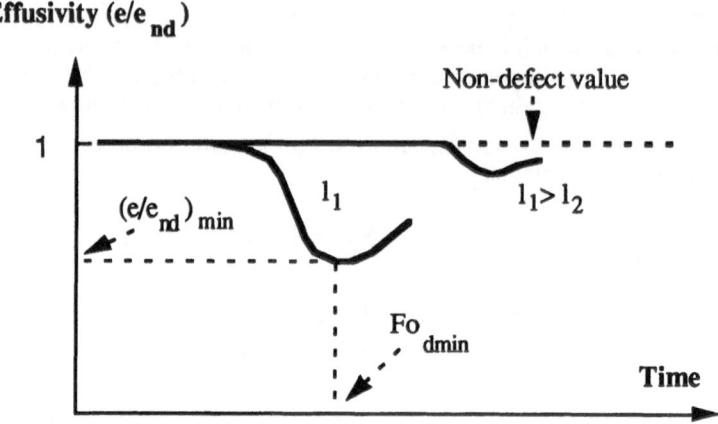

Fig. 5. Measurement of apparent effusivity

Balageas, Deom and Boscher proposed the following identification formulas for situation of *Fig.5* [16]:

$$l=\sqrt{\alpha\tau_{min}}(e/e_{nd})_{min}^{0.95};$$
$$R=(1.61/e_{nd})\sqrt{\tau_{min}}(e/e_{nd})_{min}^{0.95}[(e/e_{nd})_{min}^{-1.89}-1]. \qquad (7)$$

This method was used to synthesize images of defect depth and thermal resistance in carbon epoxy specimen [16]. It is limited by shallow defects because for deeper defects the approximation of semi-infinite body is not valid.

5.5. ISOTHERMAL MODEL

Simple identification method has been developed by Delpech *et al.* for thermal resistive layer between two layers of high-conductive specimen [31]. The inversion expressions for uniform specimen heated with square pulse of duration τ_h are as follows:

$$l=\frac{L}{1+\dfrac{s\tau_h[T^F(\tau=0)/T_m]}{Exp[\dfrac{\tau_h}{RKCl(1-l/L)}]-1}};$$
$$R=\frac{1}{sKCl(1-l/L)}. \qquad (8)$$

Here $T_m=W/KCL$; T^F ($\tau=0$)-artificial point obtained by crossing the straight line "Log T vs. τ" the ordinate axe; S-slope of the mentioned straight line.

This model was used to estimate parameters of glue layer between metallic and non-metallis plates and air defects in carbon epoxy specimen with divergence in defect depth values of 10-14 % and in defect thickness 10-95 % [31].

5.6. INVERSION OF THE TNDT SENSIBILITY TABLES

As we mentioned above, rather simple analytical expressions to identify defect parameters could be derived in some extreme cases for uniform bodies. When real milti-layer structures are involved the analytical formulas become too bulky to be easily calculated and interpreted.

Vavilov and Finkelstein analyzed solution for step heating a plate with 1D defect [32]. Mathematical problem was solved twice: firstly, in non-defect area which could also consist of three layers and, secondly, in defect area. Subtraction of two solutions allowed to calculate maximum signals over defect and optimum time for their recording. As a result the TNDT sensibility tables have been compiled. By now these tables include four particular cases: glass fiber-air; Teflon-air, steel-air and tungsten-air. Being compiled in dimensionless form for some specimen thickness these tables are helpful in understanding the TNDT limits. They could be also used for inversion purposes.

The similar tables have been compiled by Ivanov for two-side TNDT of air defects in carbon epoxy plastic and tungsten specimens heated with Dirac and square pulses [33].

These two kinds of tables have been compiled for 1D defects. More real results are based on solution of 2D problems. Vavilov compiled the corresponding TNDT sensibility tables for glass fiber, carbon epoxy, steel and aluminum specimens with thickness 1 mm (defect lateral size 0.2-0.8 mm) and 20 mm (defect size 2-8 mm) [3]. Manual use of the TNDT tables in the inversion purposes could be troublesome but using the computer database could make this approach valid.

5.6. ADAPTIVE INVERSION ALGORITHM (ADAPTIVE THERMAL TOMOGRAPHY

This algorithm assumes that [34]:

*Experimental "temperature vs. time" function is measured with $\Delta\tau$ interval which is sufficiently small;
*Thermal properties of sound material are known;
*Penetration of heat flux into specimen during the first temporal step is small, specimen is non-defect and behaves as semi-infinite body;
*Experimental values are selected step by step and compared with theoretical calculation according to explicit numeric scheme;
*Calculated temperature is modified to fit experimental value by changing the diffusivity of each coming layer.

Limit of this method was found in the used explicit scheme which requires the specific stability condition. Increasing thickness of deeper layers leads to divergence in values. Nevertheless, being applied to experimental IR images of carbon epoxy specimen the adaptive algorithm succeeded in "slicing" the specimen for four layers. Results are presented in *Fig.6*. Location of two rows of ply-cut-out defects in 8-ply specimen is

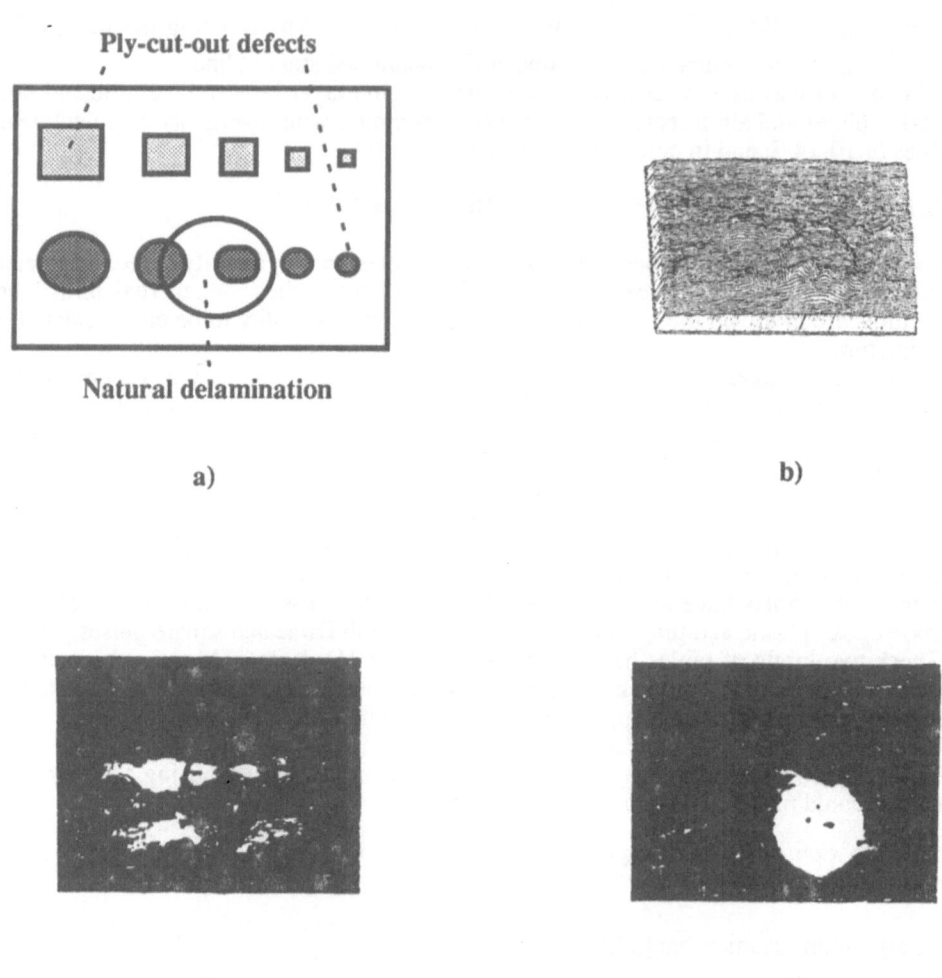

a)

b)

c)

d)

Fig. 6. Adaptive Thermal Tomography of Carbon Epoxy 8-Ply Specimen with Ply-Cut-Out and Ply-Added Artificial Defetcs and Large Natural Delamination:

a-defect's location;
b-smoothed IR image;
c-tomogram of layer 0-0.3 mm;
d-tomogram of layer 0.3-0.62 mm

shown in *Fig.6a*, smoothed IR image-in *Fig.6b*. Thermal tomogram of *Fig.6c* was obtained for the first layer of thickness 0.3 mm showing artificial ply-cut-out defects. The next layer of thickness 0.32 mm (*Fig.6d*) revealed the large size natural delamination.

6. Inversion by Minimizing Functional

The functional we are discussing here represents in fact the difference between experimental and theoretical data which is iteratively minimized by changing required values of defect size, depth and thickness, *i.e.*, $T^{exp}(\tau,x,y)-T^{theor}(\tau,h,l,d)\to\min$

In 1984-86 Cassab and Hsieh discussed steady-state thermal tomography by using very close approach [9]. Proposed equations allow to locate the irregular shape of the back specimen surface. Nevertheless, the gap which is to be filled between this research and modern tendencies in thermal characterization of defects is probably too wide.

In 1988 Kushch, Rapoport and Budadin developed algorithm which required firstly 56 iterations and up to 168 min of CPU time with ancient Russian computer EC-1055 to calculate defect thickness and depth in 2D model with accuracy 20 %. Later this troublesome computation time was reduced up to 3-5 min by introducing some preliminary data on expected defects [35].

The general approach to characterization of 1D defects which took into account the pulse duration, finite optical penetration of heating radiation and heat loss is described by Krapez and Cielo [36]. Errors in the best estimated values have been reported as follows: 5% in defect depth, 40 % in defect thickness, 40 % in heat loss. Experimental research using minimization algorithm has been made by the same authors together with Maldague [37]. Non-linear minimization of functional using 2D model was recently reported by Krapez, Boscher and Delpech [38] and Bonnet, Maigre and Manaa [39].

Minimization algorithms could be viewed as "correct" solutions of "incorrect" inverse problems. It is expected that large CPU time which is necessary to perform multiple iterations will be soon reduced with astonishing progress in computer technique.

7. Estimating the Lateral Size of Defects

In the previous paragraphs we discussed mostly 1D defects. Disbonds, delaminations, voids, *etc.*, are well approximated with 2D disc-shaped defect (cracks could require to introduce 3D models). Estimating the lateral size and shape of internal defects by their surface "prints" could be considered as single inversion problem.

Phenomenologically, it is believed that "print" of defect is larger than defect itself because of lateral heat diffusion. Nevertheless, this estimate depends on the accepted threshold in $\Delta T(\tau)$ value and dynamic behavior of this signal makes the existence of such stable threshold questionable. Bekeshko and Popov demonstrated in experiment with spot welding joints that the surface isotherm around defect could reproduce the size and shape of internal defect at particular time[40].

Later Vavilov and Shiryaev showed that extremums in the first surface derivative correspond exactly to the defect edges independently on time, defect depth and thermal properties [41]. This conclusion has received the later development with wide use of Laplace spatial filter which underlines the image edge. The problem here is that derivation technique also enhances the noise. Usually, it could be applied to signals not less than 2-5 °C.

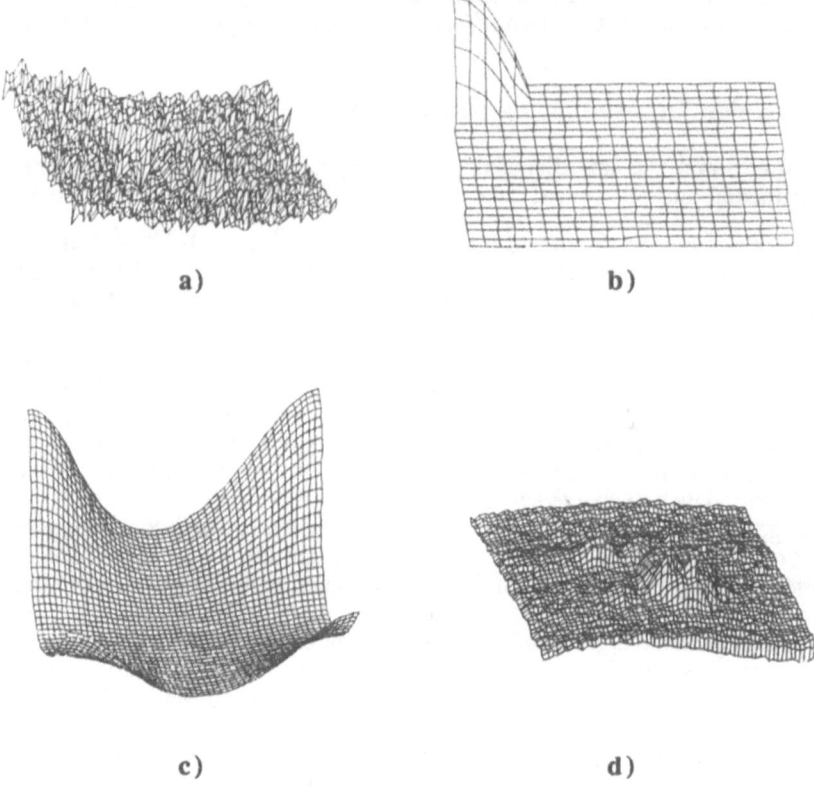

a)

b)

c)

d)

**Fig.7. Inversion of Defect Shape and Size in Carbon Epoxy Specimen
Using Green Function and Fast Fourier Transform:**

a-raw image,
b-Green function (PSF),
c-Fourier transform of Green function,
d-resulting image

Krapez, Maldague and Cielo proposed to estimate defect size at half-maximum level $\Delta T_m(\tau)/2$ [37]. Implementation of this simple technique was made by Vavilov, Grinzato, Bison *et al.* in the framework of morphological processing of thermal tomograms being resulted in suppression of the "crown-shaped" artifacts arising in timegrams around defect [14].

Using our previous classification the algorithms above could be viewed as phenomenological. More basic approach was recently developed by Thomas, Favro, Crowther and Kuo [42]. Inversion algorithm involves Green function determined for planar defect of shape $F(x,y)$ located at depth l. This function was derived using the first order Born approximation in the following form [42]:

$$G(x,y,\tau) = -\frac{A}{2\pi}\left(\frac{1}{\pi\alpha\tau}\right)^{1/2}\frac{\partial}{\partial l}\frac{Exp\{\frac{[(x^2+y^2+l^2)^{1/2}+l]^2}{4\alpha\tau}\}}{(x^2+y^2+l^2)^{1/2}}.$$

Green function is in fact the PSF for point-like defect in a solid heated with Dirac uniform pulse (notice that $G(x,y,\tau)$ depends on time). Then *Eq.(6)* could be reduced into more simple one:

$$\Delta T(x,y,\tau) = G(x,y,\tau)**F(x,y).$$

Inversion procedure includes the fast Fourier transform (FFT) which is now commercially available in order to replace the convolution with arithmetic operation. In a frequency region:

$$\Phi(v,\xi) = \theta(v,\xi,\tau)/\Gamma(v,\xi,\tau), \tag{9}$$

where Φ, θ, Γ-Fourier transforms of $F, \Delta T, G$ functions; v, ξ-spatial frequencies. Inverse FFT of Φ function will provide defect shape function.

We applied this algorithm to carbon epoxy specimen described in *Fig.6*. Example of raw IR image is shown in *Fig.7a*. As the Green function we accepted the temperature response from small defect (radius 0.2 mm, depth 0.95 mm) calculated with "Termo.Heat" software. This "PSF" is shown in *Fig.7b* with its FFT transform in *Fig.7c*. Resulting image presented in *Fig.7d* shows defects more clear than the input one.

By its final result the algorithm described works similarly to common filtration. Another problem in its correct use is that the function $\Phi(v,\xi)$ in the *Eq.(9)* which is result of division of two time-dependent functions must not depend on time itself. It requires careful choice of time in the Green function.

8. Verification of the Inverse Algorithms and the Reference TNDT Data

Experimental verification of the algorithms described above is difficult because of inevitable additional errors, especially in temperature values. We propose to develop the theoretical reference set of temperature data which will be well defined quantitatively and will be tentatively used by researchers. We believe that such reference model should be described numerically. Of course, the reference data themselves must be verified. It could be easily done in non-defect area with known classical solutions. The possible verification in defect area could be performed by using 1D analytical solutions for three-layer plate as well as analyzing reference numeric solutions in the extreme cases, *e.g.*, making defects infinite in lateral direction or reducing defect conductivity to zero value, *etc.*

As an example of such reference data we developed 2D cylindrical model with disc-shaped defect in anisotropic carbon epoxy specimen heated uniformly with square pulse which simulates the thermal excitation with flash tubes which are widely used Geometry of the specimen and all input parameters are presented in *Fig.8*.

To our opinion the reference data have to include:

*Time evolution of temperature in non-defect area;
*Time evolution of temperature in defect area for defects of various size and depth;
*Temperature profile over defect at particular time.

Time evolution data for model by *Fig.8* are presented in *Table 1*.

Table 1

Reference Temperature Data for Carbon Epoxy Plastic
(model by *Fig.8*; parameters of numeric uniform grid:
number of spatial steps r x z=50 x 100; time step 0.5 ms)*

Time,s	Temperature over non-defect area, oC	Temperature over defect area, oC	
		r_d=4.75 mm	r_d=0.475 mm
0.005	88.559	88.559	88.559
0.010	37.632	37.632	37.632
0.050	14.612	14.614	14.614
0.100	10.190	10.316	10.264
0.200	7.165	7.769	7.379
0.300	5.860	6.666	6.061
0.400	5.136	5.951	5.290
0.500	4.710	5.454	4.820
0.600	4.455	5.105	4.533
0.700	4.303	4.859	4.357
0.800	4.212	4.686	4.250
0.900	4.157	4.564	4.185
1.000	4.124	4.477	4.145

*Maximum temperature signal over defect by radius 4.75 mm:

ΔT_m=0.826 oC at τ=0.35 s.
Maximum temperature signal over defect by radius 0.475 mm:

ΔT_m=0.218 oC at τ=0.22 s.

The data of *Table 1* have been used to verify simple inversion formulas discussed above. Summarized results are presented in *Table 2*.

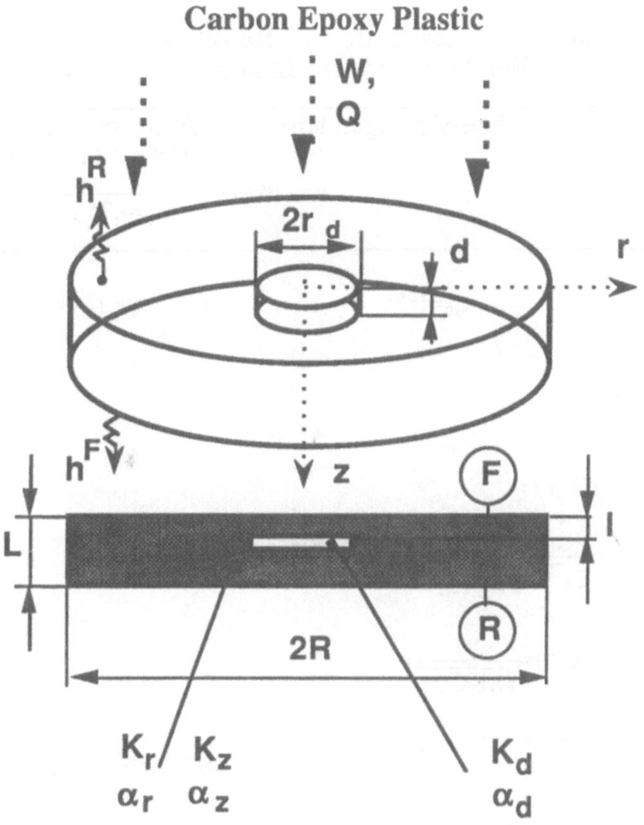

Specimen:

radius R=23.75 mm

thickness L=1mm

conductivity κ_r =1.28 W/m.K

k_z =0.64 W/m.K

diffusivity α_r=10.4x10-7 m^2 /s

a_z =5.2x10^{-7} m^2 /s

Defect:

radius r_d =4.75 mm; 0.475 mm

thickness d=0.05 mm

depth l=0.475 mm

conductivity k=0.07 W/m.K

diffusivity a=5.8x10 $^{-5}$m^2 /s

Uniform heating:

Power density Q=1 000 000 W/m^2

Pulse duration τ_h =5 ms

Heat loss:

Front surface h=0

Rear surface h=0

Fig. 8. Thermal NDT Reference Model and Input Parameters

Table 2

Thermal Characterization of Defects in   Solids:
Simple Inversion Expression

Parameter to be estimated	Inversion algorithm*		Accuracy when applying to the Reference Data of *Table 1*, %	
			r_d=4.75 mm	r_d=0.475 mm
Optimum recording time	$\tau_m = l^2/\alpha$	(2)	-24.0 **	-97.3
Temperature signal over defect	$\Delta T(\tau) =$			
	$= \dfrac{W}{2\sqrt{\pi K C \rho \tau}} \sum\limits_{n=1}^{\infty} \Gamma^n Exp(\dfrac{-n^2 l^2}{\alpha \tau})$	(4)	-17.8	-319.4
	$\Delta T(\tau) = \dfrac{W}{\sqrt{\pi K C \rho \tau}} x$			
	$x\{Exp[-\dfrac{l^2}{\alpha \tau}] - Exp[-\dfrac{(l+\sqrt{r_d^2+l^2})^2}{4\alpha \tau}]\}$	(5)	-87.9	-158.3
Defect depth	$l = \sqrt{\alpha \tau_{min}}(e/e_{nd})_{min}^{0.95}$	(7)	+20.9	+30.9
Defect thermal resistance (thickness)	$R = \dfrac{1}{sKCl(1-l/L)}$	(8)	-24.4	+10.2
	$R = (1.61/e_{nd})\sqrt{\tau_{min}}(e/e_{nd})_{min}^{0.95} x$ $x[(e/e_{nd})_{min}^{-1.89}-1]$	(7)	+59.7	+92.9

*Specification in formulas follows the one accepted in the main text.
**"+" and "-" mean that analytical formulas respectively decrease or increase the reference values.

9. Back-up Technique

We discussed above the inversion algorithms which are based on the features of heat transfer in solids. Implementation of these algorithms is usually accompanied with common image processing which could be performed in real time or/and as post-processing operation. Analysis of particular techniques deserves the special treatment which should be performed using signal-to-noise ratio as the comparison basis. Some recent information on the image processing in the TNDT could be found in [43]. Here we limit ourselves with some remarks.

The first kind of noise which operator encounters when analyzing IR images is the one presented by multiple "broken" areas which are formed by a few of pixels and clearly seen at high sensitivity spans of IR imager (sometimes it looks like "snow" noise which is well known in image processing). Suppression of this noise is usually made with 3x3 smoothing filters. Use of filters with window up to 12x12 is possible but such operation distorts the edges and may cause artifacts in resulting images. It is worth mentioning that filtering usually does not enhance the signal-to-noise ratio which is formed by signal over defect area related to signal over sound area. But filtered images look more friendly and fit better the next processing stages, e.g., image division.

Recently differentiation technique which is applied both in temporal and spatial coordinates became popular. The second spatial derivative is realized in the Laplace filter. The first temporal derivative was used to synthesize timegrams in thermal tomography [43]. The problem in applying this technique is essential increase in high-frequency noise.

Some special filters which have been calculated with use of heat transfer knowledge are also available although their universal character should be confirmed for various materials and heating technique [44].

Averaging the images is another commonly used operation. This is one of few techniques which is applied in real time in order to reduce the random noise. Averaging is highly effective in detecting the low amplitude defect signal in slow processes. In fast thermal processes the number of averaged images could not be large.

More powerful tools in image processing are supplied with Fourier transform and morphology technique. Common Fourier transform produces the results similar to filtration but its use with deeper physical basis could open new possibilities in detecting the hidden defect parameters (see p.7). Morphology being combined with some *a priori* knowledge about defects behavior in thermal images could be also very effective [13].

The modern image processing technique requires too much of operator involvement into pattern recognition and decision making. The coming new ideas in this area are based on the use of elements of artificial intelligence, e.g., in the form of neural networks [45.]

10. Conclusion

Thermal characterization of internal defects in solids is the rapidly developing area of the Thermal NDT. Starting from phenomenological treatment of apparent peculiarities in temperature data this method is now mostly based on the inversion of classical heat transfer solutions. Simple analytical expressions developed as a result of such approach allow to estimate defect depth and thickness with accuracy respectively in the range of 20-40 % and 20-100 % that could satisfy practitioners as the first step to general characterization of defects. More promising in this direction are numerical methods which could become common with growing computer performance and introducing to the NDT the elements of artificial intelligence.

Acknowledgment

I would like to thank Dr.X.Maldague (University Laval, Canada) for cooperation in obtaining the thermal tomography results, fruitful discussions and support.

References

1.Maldague X. Nondestructive Evaluation of Materials by Infrared Thermography. - Springer-Verlag, London,U.K., 1992. - 204 p.

2. Infrared Methodology and Technology.- Monograph Series: *Intern. Advances in NDT*, ed.by X.Maldague, Gordon & Breach Science Publisher, U.S.A., 1993 (to be published).

3. Vavilov V.P. Handbook of Thermal NDT. - Moscow, Mashinostroenie, 1991.-256 p.(*in Russian*).

4. Vavilov V.P., Taylor R. Theoretical and Practical Aspects of the Thermal NDT of Bonded Structures.- In: *Res. Techn. in NDT*, ed.by R.Sharpe, Academic Press, vol.5, p.239-280.

5. Balageas D. Le Controle Non Destructif par Methodes Thermiques.- *Rev. Gen. Therm. Fr.*, No.356-357, Aout-Septembre 1991, p.483-498.

6. Aamodt L.C., Maclachlan Spicer J.W., Murphy J.C. Analysis of Characteristic Thermal Transit Times for Time-Resolved Infrared Radiometry Studies of Multilayered Coatings.- *J.Appl.Phys.*, vol.68 (12), 15 December, 1990, p.6087-6098.

7. Bendada A., Maillet D., Degiovanni A. Nondestructive Transient Thermal Evaluation of Laminated Composites: Discrimination Between Delaminations, Thickness Variations and Multidelaminations.- In: *Proc. of the Eurotherm Seminar* "Quantitative InfraRed Thermography QIRT-92", July 7-9, 1992, Chatenay-Malabry, France, p.218-224.

8. Maillet D., Didierjean S., Houlbert A.S., Degiovanni A. Nondestructive Transient Thermal Evaluation of Delaminations Inside a Laminate: a Thermal Processing Technique of Thermal Images.-*See* [7], p.212--215.

9. Kassab A.G., Hsieh C.K. Application of Infrared Scanners and Inverse Heat Conduction Methods to Infrared Computerized Axial Tomography.- *Rev. Sci.Instrum.*, 58(1), January 1987, p.89-95.

11. Vavilov V.P., Shiryaev V.V. Thermal Tomograph.- *USSR Patent No.1,266,308* (appl. Jan.31, 1985).

12. Vavilov V.P., Ahmed T., Jin J., Thomas R.L., Favro L.D. Experimental Thermal Tomography of Solids Using Pulse One-Side Testing.- *Sov.J.NDT*, 1990, No.12, p.60-66.

13. Vavilov V.P.,Bison P., Bressan C., Marinetti S., Grinzato E. New Aspects of Dynamic Thermal Tomography.- *See* [7], p. 259-265.

14. Vavilov V.P., Maldague X., Picard J. *et al.* Dynamic Thermal Tomography: New NDE Technique to Reconstruct Inner Solds Structure Using Multiple IR Image Processing.- In: *Rev. of Progress in Quant. NDE*, ed.by D.Thompson and D.Chimenti, vol.11A, 1992, Plenum Press, N.Y., U.S.A., p.425-432.

15. Maldague X., Cote J., Poussart D., Vavilov V.P. Thermal Tomography for NDT of Industrial Materials. -*Canad.J.NDT,* May/June 1992, p.22-31.

16. Balageas D.L.,Deom A.A., Boscher D.M. Characterization and NDT of Carbon-Epoxy Composites by a Pulsed Photothermal Method.-*Mater. Evaluation,* April 1987, p.461-465.

17. Carslaw H.S., Jaeger T.S. Conduction of Heat in Solids.-1959, Oxford Univ.Press, Oxford, U.K.-580 p.

18. Luikov A.V. Heat Conductivity Theory.-Moscow, Vysshaya Shkola, 1967.-600 p. (*in Russian*).

19. Balageas D., Krapez J.C., Cielo P. Pulsed Photo-Thermal Modeling of Layered Materials.- *J.Appl.Physics,* vol.59, No.2, January 5, 1986, p.348-357.

20. Storozhenko V.A., Vavilov V.P., Volchek A.D. Nondestructive Testing of Industrial Products Quality Using the Active Thermal Method.-Kiev, Ukraine, Tekhnica Publisher, 1988 (*in Russian*).-168 p.

21. Lau S.K., Almond D.P., Patel P.M. Transient Thermal Wave Techniques for the Evaluation of Surface Coatings.-*J.Phys.D: Appl.Phys.,* vol.24(1991), p.428-436.

22. Lau S.K., Almond D.P., Milne J.M. A Quantitative Analysis of Pulsed Video Thermography.- *NDT & E. Intern.,* 4 August 1991, vol.24, No.4, p.195-202.

23. Alzofon F.E. The Optics of Heat Conduction in Solids.-In: *Proc.SPIE,* vol.780, "Thermosense-IX", 1987, p.215-233.

24. Rosencwaig A. Thermal Wave Physics in NDE.-In: *Rev. Progr. in Quant. NDE,* vol.5A, ed.by D.O.Thompson and D.E.Chimenti, Plenum Press, N.Y., p.429-437.

25. Wu D., Karpen W., Busse G. Lockin Thermography for Multiplex Photothermal Nondestructive Evaluation.-*See* [7], p.371-376.

26. Bein B.K., Gu J.H., Mensing A. *et al.* Modulated Infrared Radiometry of Rough Surfaces at High Temperatures.- *See* [7], p.393-398.

27. Walther H.G., Seidel U. Some Remarks on Definition, Resolution and Contrast in Photothermal Imaging.-*See* [7], p.251-258.

28. Grinzato E., Mazzoldi. Infrared Detection of Moist Areas in Monumental Buildings Based on Thermal Inertia Analysis.-In: *Proc. SPIE,* vol. 1467 "Thermosense-XIII", 3-5 April, 1991, Orlando, U.S.A., p.75-82.

29. Del Grande N. Buried Object Remote Detection Technology for Law Enforcement.-In: *Proc. SPIE,* vol.1479 "Surveillance Technology", 1-5 April, Orlando, U.S.A., p.234-239.

30. Weil G.J., Graf R.J. Infrared Thermographic Detection of Buried Grave Sides.-In : *Proc. SPIE,* vol. 1682 "Thermosense-XIV", 22-24 April, 1992, Orlando, U.S.A., p.347-353.

31. Delpesh Ph.M., Boscher D.M., Lepoutre F. et al. Time-Resolved Pulsed Stimulated Infrared Thermography Applied to Carbon/Carbon Nondestructive Evaluation.-*See* [7], p.201-205.

32. Vavilov V.P., Finkelstein S.V. Method of Estimation of Sensitivity Limits for Active Thermal NDT.- *Sov.J.NDT,* 1983, No.5, p.65-68.

33. Ivanov A.I. Development of Pulsed Thermal Testing of Thermal Properties and Defects in Multilayered Specimens with Improved Sensitivity and Productivity.- *PhD Thesis,* Tomsk Polytechnic Institute, 1989.- 220 p. (*in Russian*).

34. Vavilov V.P., Maldague X., Dufort B. Adaptive Thermal Tomography Algorithm.-In: *Proc.SPIE,* vol.1933 "Thermosense-XV", 14-16 April, 1993, Orlando, U.S.A., p. 166-173.

35. Kushch D.V., Rapoport D.A., Budadin O.N. Inverse Problem of Automatized Thermal Testing. - *Sov.J.NDT,* 1988, No.5, p.64-68.

36. Krapez J. C., Cielo P. Thermographic NDE: Data Inversion Procedure (Part I: 1D Analysis). - *Res.in NDE,* 1991, vol.3, No.2, p.69-100.

37. Krapez J.C., Maldague X., Cielo P. Thermograhic NDE: Data Inversion Procedure (Part II: 2D Analysis and Experimental Results).- *Res.in NDE*, vol.3, No.2, 1991, p.101-124.
38. Krapez J.C., Boscher D., Delpech Ph. *et al*. Time-Resolved Pulsed Stimulated Infrared Thermography Applied to Carbon-Epoxy Nondestructive Evaluation.-*See* [7], p.195-200.
39. Bonnet M., Maigre H., Manaa M. Numerical Reconstruction of Interfacial Defects and Interface Thermal Resistances Using Thermal Measurements.-*See* [7], p.266-271.
40. Bekeshko N.A., Popov Yu.A. Testing of Welding Zone Diameter in Spot Welded Joints Using Visualization of Thermal Fields.-*Sov.J.NDT*, 1972, No.6, p.86-92.
41. Vavilov V. P., Shiryaev V.V. Method of Determination of Sub-surface Defects In-Plane Size in Thermal NDT.-*Sov.J.NDT*, 1979, No.11, p.103-106.
42. Thomas R.L., Favro L.D.,Crowther D.J., Kuo P.K. Inversion of Thermal Wave Infrared Images.- *See* [7], p.278-282.
43. Vavilov V.P. Thermal Nondestructive Testing: Short History and State-of-Art.-See [7], p.179-194.
44. Winfree W.P., James P. Thermographic Detection of Disbonds.-In: *Proc. of the 35th Intern. Instrumentation Symp.*, 1989, p.183-189.
45. Prabhu D.R., Howell P.A., Syed H.I., Winfree W.P. Application of Artificial Neural Networks to Thermal Detection of Disbonds.-*See* [14], vol.11B, p.1331-1338.

USE OF FINITE DIFFERENCE METHOD FOR PREDICTING PROPAGATION OF ULTRASONIC WAVES IN SOLIDS[1]

D. KISHONI
College of William & Mary / NASA Langley Research Center
Hampton VA 23681-0001, USA

Successful interpretation and analysis of ultrasonic wave records depend on understanding the expected waveforms in the presence of material and structural defects. Measured signals can be compared to the predicted ones, and signal processing operations can use the predicted waveforms in various ways such as targets of filters, as teaching base for neural nets and expert systems and such. While exact elastic solution can usually be obtained only for simple geometries, finite difference method offers a flexible method that can accomodate defects and other discontinuities. In this paper, finite difference method is used to solve the elastic wave equation and obtain the wave propagation pattern. Particular attention is given to long distance wave propagation. Standard non-staggered methods produce spurious signals caused by numerical dispersion and require small grid size, resulting in a significant increase in computer memory requirements and in calculation time. Staggered grid, on the other hand, along with high-order method, reduce these requirements and improve the accuracy, thus, enabling practical solutions to these problems.

D. Kishoni, "Pulse shaping and extraction of Information from ultrasonic reflections in composite materials." NATO Advanced Research Workshop on Signal Processing and Pattern Recognition in Nondestructive Evaluation of Materials, Lac Beauport, Québec, Canada (August 19-22, 1987), C. H. Chen ed., Springer-Verlag, Berlin Heidelberg 1988, NATO ASI Series, Vol. F44, pp. 117-127.

K. J. Sun and D. Kishoni, "Ultrasonic investigation of disbonds in thin aluminum plates for aircraft applications." IEEE 1991 Ultrasonics Symposium, Orlando, FL. (December 8-11, 1991), pp. 859-862.

P. H. Johnston and D. Kishoni, "Evaluation of waveform mapping as a signal processing tool for quantitative ultrasonic NDE." Review of Progress in Quantitative NDE, University of California, San Diego, La Jolla, CA, (August 19-24, 1992), D. O. Thompson and D. E. Chimenti eds., Plenum Press, New York, 1993, Vol. 12A, pp. 711-718.

D. Kishoni and S. Ta'asan, "Improved finite difference method for long distance propagation of waves." Review of Progress in Quantitative NDE, University of California, San Diego, La Jolla, CA, (August 19-24, 1992), D. O. Thompson and D. E. Chimenti eds., Plenum Press, New York, 1993, Vol. 12A, pp. 139-146.

[1]this contribution: abstract only

X. P. V. Malague (ed.), Advances in Signal Processing for Nondestructive Evaluation of Materials, 185.
© 1994 *Kluwer Academic Publishers.*

INVERSE SCATTERING OF PULSED THERMAL WAVES

L.D. FAVRO, D.J. CROWTHER, P.K. KUO, AND R.L. THOMAS
Department of Physics and Institute for Manufacturing Research
Wayne State University, Detroit, MI, 48202, USA

ABSTRACT. Preliminary results on inversion of pulsed thermal wave images have been presented elsewhere [1-4]. In that work a technique was introduced for removing the blurring of thermal wave images of planar subsurface features such as delaminations. The purpose of the development of the technique was to provide an algorithm for inverting a *single* image obtained at some time after the interrogation of the sample with a thermal wave pulse, and to obtain the shape of the scatterer with a *single* calculation on that image. The method involves no tomographic reconstruction and does not rely on any kind of successive approximation scheme. Recently, we have extended the method to include thermally anisotropic materials and also multiple scattering effects. [5]

1. Mathematical Model

Here we will be concerned with modeling the propagation and scattering of pulsed thermal waves in such a way as to facilitate inverse scattering calculations. We will limit our considerations to planar scatterers which are parallel to the plane of the surface of the sample, and describe back-scattered rather than transmitted images. It is assumed that the incident thermal wave pulse initially has negligible duration, and is launched as a plane wave from the surface of the sample at time, $t = 0$. This corresponds to the common experimental situation in which the pulse is initiated by the absorption of the light from photographic flash lamps. The method can be applied to a variety of planar voids and inclusions in the sample, but here, for ease of description, we will imagine the scatterer to be the top surface of a flat-bottomed hole milled into the sample from the rear. A schematic drawing of such a sample is shown in Fig. 1a. In that drawing we have labeled the top surface of the sample (from which the thermal wave pulse is assumed to have been launched) as S_1, the top surface of the defect as S_3, and the infinite extension of the plane of S_3 has been labeled as S_2. Figure 1b shows a similar sample but with no hole. In this drawing, the surface corresponding to the combination of S_2 and S_3 is labeled S_4. The reason for this labeling scheme will become apparent as the description of the model progresses. The problem we wish to solve is the solution to the heat equation for the

187

X. P. V. Malague (ed.), Advances in Signal Processing for Nondestructive Evaluation of Materials, 187–191.
© 1994 *Kluwer Academic Publishers.*

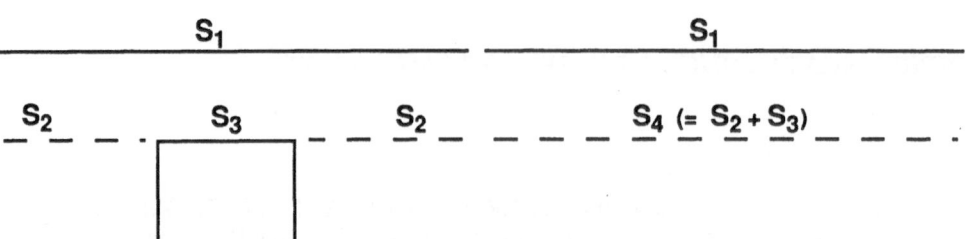

Fig. 1a Schematic diagram of a sample with a flat-bottomed hole milled into its rear surface

Fig. 1b Diagram of a sample like the one in Fig. 1a, but without the flat-bottomed hole.

geometry of Fig. 1a with a source which is uniformly distributed over the surface S_1, and which is a delta-function in time at $t = 0$. An equivalent problem is one in which there is no source, but in which it is specified that the initial temperature of the sample is a one-dimensional delta function in space centered on the surface of the sample. We will temporarily assume that both S_1 and S_3 are thermally insulated. If we take the diffusivity of the material to be α, and the inward normal direction to surface S_1 to be the positive z-direction, this problem can be stated as

$$\nabla'^2 T(\mathbf{r}',t') = \frac{1}{\alpha}\frac{\partial T(\mathbf{r}',t')}{\partial t'},$$

$$T(\mathbf{r}',0) = \delta(z'),$$

$$\text{and } \frac{\partial T}{\partial z'} = 0 \quad \text{on } S_1 \text{ and } S_3. \tag{1}$$

To facilitate the solution to this problem we will introduce a Green's function which is defined only within the region between the surfaces S_1 and $(S_2 + S_3)$,

$$\nabla'^2 G(\mathbf{r},\mathbf{r}',t-t') = \frac{-1}{\alpha}\frac{\partial G(\mathbf{r},\mathbf{r}',t-t')}{\partial t'},$$

$$G(\mathbf{r},\mathbf{r}',0) = \delta(\mathbf{r}-\mathbf{r}'),$$

$$\text{and } \frac{\partial G}{\partial z'} = 0 \quad \text{on } S_1. \tag{2}$$

At this point nothing is said about the boundary conditions on $G(\mathbf{r},\mathbf{r}',t)$ on the surfaces S_2 and S_3, nor is anything said about the behavior of G outside the region between the two specified surfaces. Thus there are infinitely many Green's functions which satisfy Eqs. 2. If one of these Green's functions and the solution to Eqs. 1 are inserted into Green's theorem, and the result is integrated over the region, V, between the two surfaces, and also integrated over t' from 0 to t, the result is,

$$T(\mathbf{r},t) = \int_V G(\mathbf{r},\mathbf{r}',t)T(\mathbf{r}',0)d^3\mathbf{r}' + \alpha \int_0^t dt' \int_{S_2+S_3} [G\frac{\partial}{\partial z'}T - T\frac{\partial}{\partial z'}G]\, dS' \tag{3}$$

A similar calculation for the sample described in Fig. 1b yields the result,

$$T_0(r,t) = \int_V G(r,r',t)T(r',0)d^3r' + \alpha \int_0^t dt' \int_{S_2+S_3} [G\frac{\partial}{\partial z'}T_0 - T_0\frac{\partial}{\partial z'}G]\,dS'$$

$$\tag{4}$$

where $T_0(r,t)$ is the (known) solution to the problem of Fig. 1b,

$$T_0(r,t) = \frac{1}{\sqrt{4\pi\alpha t}} \exp\left[-\frac{z^2}{4\alpha t} \right].$$

$$\tag{5}$$

When Eq. 4 is subtracted from Eq. 3, the result is an expression for the temperature contrast, $T(r,t) - T_0(r,t)$, which results from the presence of the hole in the sample,

$$T(r,t) - T_0(r,t) = \alpha \int_0^t dt' \int_{S_2+S_3} [G\frac{\partial}{\partial z'}(T-T_0) - (T-T_0)\frac{\partial}{\partial z'}G]\,dS'$$

$$\tag{6}$$

Equation 6 is, of course, not an expression for a solution to the problem, but only an integral equation for the temperature contrast. The success of the inversion scheme depends on an appropriate choice of Green's function, and more importantly, on the selection of a suitable approximate expression for T in the integral on the right hand side the equation. Our usual approximation for T is the one which is in almost universal use in optical diffraction calculations, namely that the incident wave is undisturbed up to the edge of the scatterer, and that on the scatterer it satisfies some simple law of reflection. In our case this translates into the statement that T and its gradient are undisturbed on S_2, i.e. that $(T-T_0)$ and its gradient are zero there, and that on S_3, the normal derivative of T vanishes. In the simplest approximation, one which corresponds to the inclusion of only a single scattering from the surface of S_3, the value of T on S_3 is taken to be twice the value of T_0. This choice is appropriate because the wave reflected from an insulating surface is in-phase with the incident wave and doubles its amplitude at the reflector. This choice of boundary approximation is then normally combined with a choice of the half-space Green's function for G in the integral equation. However, in many experimental situations the scatterer is so close to the surface of the sample that the thermal wave pulse makes several reflections back and forth between the scatterer and the surface, so that a single scattering approximation is not sufficient. In this situation one can achieve a reasonably good approximation to the multiple scattering situation by using a Green's function for a slab of the same thickness as the region between S_3 and S_1, and approximating T on S_3 by the one-dimensional solution for the same thickness. An advantage of these approximations is that they permit the time integration in the integral equation to be performed explicitly. The result of this process is a relatively simple expression for the temperature contrast,

$$T(r,t) - T_0(r,t) = \left(\frac{1}{4\pi\alpha t}\right)^{3/2} \iint_{S_3} dx'dy' \sum_{m=-\infty}^{\infty} \exp\left(\frac{-[u_m+d]^2}{4\alpha t}\right)\left[1 + \frac{d}{u_m}\right]$$

(7)

where

$$u_m = \left[(x-x')^2 + (y-y')^2 + (2m+1)^2 d^2\right]^{1/2}.$$

(7')

2. Inversion Algorithm

This can be rewritten as an integral over the entire x'-y' plane by introducing a "scatterer shape function", $f(x',y')$, which is defined as unity on the scatterer and zero elsewhere,

$$T(r,t) - T_0(r,t) = \left(\frac{1}{4\pi\alpha t}\right)^{3/2} \iint dx'dy' \sum_{m=-\infty}^{\infty} \exp\left(\frac{-[u_m+d]^2}{4\alpha t}\right)\left[1 + \frac{d}{u_m}\right] f(x',y')$$

(8)

The advantage of this form of the equation is that it is explicitly a convolution of a "point spread function" and the shape of the scatterer $f(x',y')$. Thus it readily lends itself to standard Fourier transform deconvolution techniques to perform the inverse scattering calculation. Our algorithm consists of the performance of a Fast Fourier Transform (FFT), followed by a deconvolution with the transform of the point spread function in Eq. 8. Either a Wiener filter or a Gaussian low-pass filter is used to suppress the effects of noise. An example of the result of such an inversion is shown in Fig. 2. In this case

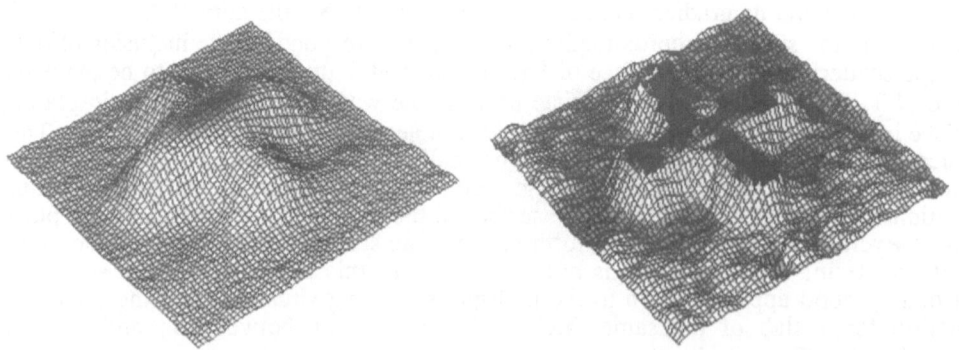

Fig. 2a A raw thermal wave image of a cross-shaped flat-bottomed hole milled into the rear side of an anisotropic graphite-epoxy slab.

Fig. 2a The result of inverting the image of Fig. 2a. Note the improvement in resolution of the cross and the appearance of "texture" in the image due to sample porosity which was unresolved in the original image

the sample consisted of an anisotropic graphite-epoxy slab with a cross- shaped flat-bottomed hole milled into its rear side. The anisotropy resulted from the fact that all of the fibers are in the same in-plane direction. In order to invert an image in such an anisotropic sample, one must, of course, replace the diffusivity α in Eq. 8 by an appropriately chosen diffusivity tensor. This causes no problems in the derivation which led to Eq. 8. In the raw image of the cross, shown in Fig. 2a, the two arms of the cross which are perpendicular to the high diffusivity direction (the direction of the fibers) have a lower intensity than the two arms which are parallel to the high diffusivity direction. In the inverted image, shown in Fig. 2b, the arms have all been restored to the same height, and the rounded contours of the raw image have been significantly sharpened. In addition, there is a distinct "texture" which appears in the inverted image, but not in the raw image. This texture seems to be associated with porosity in the sample which is a "home made" one. This porosity has been observed visually, and an attempt has been made to image it ultrasonically. However, the material is of such poor quality, that it was not possible to get an ultrasonic image it.

3. Acknowledgment

This work was sponsored by the Institute for Manufacturing Research, Wayne State University.

4. References

1.   "Inversion of pulsed thermal-wave images for defect sizing and shape recovery," L.D. Favro, D. Crowther, P.K. Kuo, and R.L. Thomas, Proc.SPIE Thermosense XIV Orlando, FL, April 20-24, 1992, Vol. 1682 Thermosense XIV (1992), pp. 178-181.

2.   "Inversion of Pulsed Thermal Wave Images for Recovery of the Shape of the Object," D.J. Crowther, L.D. Favro, P.K. Kuo, and R.L.Thomas, accepted for Proc. 7th Int. Topical Meeting on Photoacoustic and Photothermal Phenomena, Doowerth, The Netherlands, August 26-30, 1991, Photoacoustic and Photothermal Phenomena III, D. Bicanic (Ed.) Springer Series in Optical Sciences 69, 544-546, Springer-Verlag, Berlin, Heidelberg (1992).

3.   "Application of the Inverse Born Approximation to Experimental Thermal Wave Images," L.D. Favro, D.J. Crowther, P.K. Kuo, and R.L. Thomas, Proc. 14th International Congress on Acoustics, Beijing, Sept. 3-10, 1992, A13-5 (to be published).

4.   "An Inverse Scattering Algorithm Applied to Infrared Thermal Wave Images," D.J. Crowther, L.D.Favro, P.K.Kuo and R.L.Thomas (submitted for publication to J. Appl. Phys.).

5."Inversion of Pulse-Echo Thermal Wave Images," L.D. Favro, D. Crowther, P.K. Kuo, and R.L. Thomas, Proc.SPIE Thermosense XV Orlando, FL, April 12-16, 1993, (to be published).

INFORMATIVE PARAMETERS AND NOISE IN TRANSIENT THERMAL NDT

V. VAVILOV*, P.G. BISON, C. BRESSAN, E. GRINZATO, S. MARINETTI
CNR-ITEF, Corso Stati Uniti, 4
35100, Padova, Italy
* Tomsk Polytechnical University,
Russia, 634028, Tomsk, Savinykh, 3

ABSTRACT Detail description of physical and derived informative parameters used in transient thermal NDT is presented. Features of temperature-domain and time-domain parameters are discussed. Expressions for signal-to-noise ratio are specified and efficiency of spatial filters applied to IR thermograms with defects is discussed. Spatial frequency spectrums are obtained in non-defect and defect areas using fast Fourier transform.

1. Measured informative parameters

Result of any measurement is a set of numeric values organised in one of possible ways (image, profile, table, single values, *etc.*). These values represent a combination of useful signal and noise which have to be deciphered by an operator. Under term "informative parameters" we shall specify those ones which are accepted for processing in particular testing procedure. A lot of sophisticated informative parameters could be devised for decision making. Usually maximising signal-to-noise ratio is the comparison criterion for these parameters.

Firstly, we introduce only those parameters which are really measured in a physical experiment. In thermal non-destructive testing (TNDT) the only source of information about specimen's properties is the heat flux. In practice IR devices measure thermal radiation which depends not only on defect presence but on more wide range of optical and thermal properties. Estimation of these properties could be the goal of thermal test also (*e.g.*, we refer to analysis of emissivity distribution which is sometimes crucial for decision making). But for later use we just state that TNDT IR equipment is calibrated in black or grey body temperature. Hence, specimen's transient temperature $T(x,y,\tau)$ where x and y define specimen's surface, is the main measured parameter. Knowledge of ambient temperature T_a allows to introduce excess temperature of heating $t_h = T(x,y,\tau) - T_a$ which is usually used in theoretical analysis. Finally, in computerised IR systems it is also possible to measure the excess temperature signal upon defect $\Delta T(x,y,\tau) = T(x,y,\tau) - T(x_{ref}, y_{ref}, \tau)$ determining (in real time or not) the difference between reference point or area and each point or area of total image.

Amplitude temperature values could not always be regarded as reliable ones for decision making. Modern TNDT philosophy tends to be based on measurements of the so called heat transit times or intervals τ_γ specified in

193

X. P. V. Malague (ed.), Advances in Signal Processing for Nondestructive Evaluation of Materials, 193–207.
© 1994 *Kluwer Academic Publishers.*

temperature evolution curve for each pixel in image. Origin of temporal informative criteria could be traced to method of flash diffusivity measurement proposed by Parker *et al.* in 1961 [1]. Thermal properties values (diffusivity or/and effusivity) could be identified by τ_y values directly. Properties variations and internal defects are characterised by their variations $\Delta\tau_y$. In photothermal technique the most reliable analysis could be done in phase domain which in fact is the direct analogue of time domain in single pulse procedure [2].

Using amplitude temperature parameters (first of all ΔT) it is possible to synthesise the space-domain images (common thermograms). Temporal parameters could be reduced to time-domain or phase-domain images ("timegrams" and "phasegrams").

2. Derived informative parameters

Amplitude and temporal informative parameters allow to combine them in various ways in order to obtain new criteria which are more stable against the noise.

Most important amplitude parameter is dynamic temperature contrast defined as $C(x,y,\tau)=\Delta T(x,y,\tau)/T(x,y,\tau)$ (some authors use this term for temperature signal ΔT that contradicts to dimensionless character of contrast). Sometimes temperature contrast is used in form: $C^{(1)}=T/T_{ref}$ or $C^{(1)}=1+C$. Analysing temporal behaviour of temperature contrast it is possible to distinguish surface and sub-surface effects [3].

By analogy temporal contrast $\Delta\tau_y/\tau_y$ could be introduced also although its advantage, if to compare it with more simple τ_y or $\Delta\tau_y$ values, is questionable.

Considering not single temperature read-outs but thermal images the amplitude informative parameters discussed above could be easily extracted from image histograms. Thus, average temperatures in non-defect or defect areas correspond to particular peaks in respective distributions. Distance between peaks could be regarded as differential temperature signals. This situation involves signal-to-noise consideration and decision making rules.

3. Noise: general remarks

Any measurement would be much more simple without noise which interferes results and converts any procedure into statistical one. Noise is generated inside measuring tool, inside specimen under test and in the environment. These three sources of noise have to be taken into account in any practical test although their contribution into resulting signal could vary. Respectively we shall specify the apparatus, structural and external noise. In TNDT procedures each kind of noise could be described in terms of temperature, temperature contrast or any other informative parameter. General signal-to-noise ratio s could be written in the following form:

$$s = \frac{\Delta T(\tau)}{\sqrt{(\Delta\overline{T}_{ap})^2 + (\Delta\overline{T}_{str})^2 + (\Delta\overline{T}_{ext})^2}},$$

where $\Delta T(\tau)$ specifies signal upon defect and subscripts *ap*, *str*, and *ext* respectively relate to apparatus, structural and external noise. It is supposed that particular kinds of noise do not correlate each with other.

4. Additive apparatus noise

The only noise created by "good" equipment is that one produced by a sensor. In IR equipment this is IR detector noise which is specified by manufacturer as temperature resolution ΔT_{res}. This value for modern IR imagers working in common mode is in a range from 0.05 to 0.2 °C. Improvement of this parameter depends on IR detectors technology or processing algorithm. Most common algorithm is averaging N images that enhances resolution by $\sqrt{N}$ times. Practically N value is not more than 16-25 being limited with speed of image sampling. In some special cases (*e.g.* in equipment for stress pattern analysis by thermal emission SPATE [4]) averaging could be fulfilled over thousands of images resulting in essential improvement of temperature resolution.

It is important to state that no measurement (at least without sophisticated computer algorithms) can be done with resolution better than one determined by ΔT_{res} value. Modern transient thermal NDT tends to be done more in time domain than in temperature domain. Respectively term "resolution" is considered here in amplitude or temporal treatment. It means that minimum measured time interval is limited by two neighbouring amplitude values separated by ΔT_{res} value. In measurement of temporal informative parameters it is possible to introduce temporal resolution $\Delta \tau_{res}$ by analogy with amplitude resolution. Both values are connected by obvious expression which follows from consideration of general recorded function $T(\tau)$:

$$\Delta T_{res} = \frac{\partial T}{\partial \tau} \Delta \tau_{res}.$$

The last expression means that temporal measurements are more sensitive in such time intervals where the first temperature derivative is the steepest.

More considerations about peculiarities of acquisition technique and temporal resolution when fast changing temperature fields are scanned could be found in [5].

5. Additive background noise

This kind of noise is determined by environment where thermal test is done. Background radiation could be direct and reflected from the surface of tested object. There are plenty of natural and artificial sources which create essential thermal fluxes in a spectral range of IR detector: Sun, different kinds of heaters, lamps, radiators, human beings, *etc.*

Direct irradiation of IR device could be reduced in many cases with simple shields or blends which limit angle of view. Reflected radiation is more difficult to handle because it interferes inherent thermal flux of tested object. Its power depends on reflection coefficient. Hence, it is possible to reduce this flux by removing its source or covering surface under test with "black body" coating.

Accurate measurement of various components of general thermal flux could be done using thermal reference placed in the angle of view of IR imager. This technique complemented with automatic recognition of thermal reference image was described in [6].

Some errors of thermographic measurement of temperature in presence of operator and common electric bulb have been reported in [7]. These data presented in *Table 1* show that: 1) operator should stay at the distance not less than 1 m from object; 2) electric bulb placed far than 1 m from object does not influence test. It is usually recommended to place IR camera far from house radiators (or to cover heat sources with thermal shield).

Table 1: Relative methodical error of thermographic temperature measurement in presence of background radiation [7]

Distance between object and source of background radiation [m]	Error (in %) conditioned by thermal radiation of	
	human being	electric bulb
1	6.7	3.1
2	1.9	0.8
3	0.8	0.3
4	0.4	0.2

Outdoor radiation is generated by Sun, sky and Earth. Solar radiation reaches earth surface in a range of 0.3-3 μm providing heat power density of about 1350 W/m² outside atmosphere. This value reaching partially the Earth surface competes those thermal fluxes used in some TNDT tests. Moreover solar irradiation is used as the heater in some special cases. Atmosphere radiation includes the component conditioned by radiation of vapour, carbon dioxide and ozone as well as scattered solar radiation. These components produce similar thermal fluxes in a wavelength range of 3-4 μm. Atmosphere temperature is in range from 200 to 300 K. Hence maximum of radiation is around 10 μm. Cloudy sky emanates radiation as black body. Radiation of landscape elements consists of solar reflected component in daytime ($\lambda \leq 4$ μm) and inherent radiation of Earth itself which is close to black body radiation. It is clear that solar radiation is a main noise factor in outdoor thermographic measurement. Hence, it is highly recommended to fulfill tests during dawn or twilight hours (or during a night).

Additive background noise could be described with dispersion of equivalent temperature $\sqrt{\overline{\Delta T_b^2}}$ then value $\sqrt{\overline{\Delta T_b^2}} / \Delta T$ (with ΔT defect signal) could characterise TNDT sensitivity. By now there are plentiful data on spectral behaviour of background noise. Statistical description of noise has been mostly done in IR reconnaissance (see handbook [8]). As for industrial materials, a lack of sufficient information (*e.g.* on emissivity statistics on large surface) makes difficult reliable estimate of the TNDT sensitivity.

6. Multiplicative structural noise

This kind of noise multiplies useful signal (classic example: emissivity coefficient in M. Planck's law for grey body). It is inherently related to tested object. In fact the use of term "noise" should be careful in this case because defect signal could be often considered as a part of this noise. Conditionally, structural noise could be classified for surface (absorbtivity-emissivity) component and volumetric (thermal properties) component.

Multiplicative noise which depends on signal level is usually described in terms of contrast. More details on this subject are discussed below.

7. Phenomenological approach to noise estimation

In practical TNDT applications it is difficult to distinguish various noise sources. Noise signals could be estimated in a phenomenological way by analysing specimen's areas which are presumably non-defect. A simple way to do this is to apply standard image processing software. Non-defect area is determined either by operator or using alternative technique. Then variations of standard deviation and average temperature (or standard deviation of temperature contrast at once) are analysed in a set of recorded IR images. Two parameters are important in this process: size of chosen non-defect area and time. Usually noise is increasing with larger analysed area because of non-uniform heating, edge influence and non-stable convection. Study of time evolution of contrast could provide information about structure of noise. For instance, if noise temperature contrast is time-independent it means that main contribution is done by surface effects. Increase of noise signals for lower average temperatures could be explained by essential contribution of detector noise. Slow variations of noise in time could be conditioned by inner structure of a specimen, *etc*. Using this methodology it is easy to evaluate efficiency of image processing technique. Example of this is described below.

Noise estimation in carbon fiber reinforced plastic (CFRP)

The CFRP specimen by size 50x50 mm was heated by 2.5 kW quartz IR lamp. An AGEMA 782SW IR imager linked to thermal inspection station [3] was used to store the multiple images. Noise statistics was analysed by an operator in non-defect area using temperature profiles and/or 2D images. No special measures to subdue the non-uniform imager sensitivity have been done so this kind of noise complemented structural noise. In order to exclude the effects of back scattered radiation the specimen's surface was black-painted.

Statistical treatment of the obtained data included the determination of averaged value, standard deviation, histogram shape and spatial spectrum. The results are presented in *Table 2*.

Smoothing technique by using the 3x3 filter is common in the image processing. Applying it to IR image of CFRP it seemed possible to decrease the noise signals from 3.2 to 2.6 % in contrast terms (*Table 2*). It means that the threshold for decision making for this particular material could be recommended at the level of 5.2 % (2 standard deviations provide the false alarm probability of 95 %). High frequency filtering increases the noise up to 6.8 %.

Table 2: Standard deviations of the noise temperature contrast in the IR image of CFRP specimen.

Image description	Standard deviation, %
Raw image	3.2
Smoothed with filter 3x3	2.6
High-frequency filtered	6.8
Processed with Winfree's filter	2.7
Positive deviation in the raw image are taken into account	2.1
Positive deviation in the smoothed image are taken into account	1.8

Image processing with the Winfree's filter which is supposed to subdue the non-uniformity of heating-cooling process [9] produced the same result as smoothing. The minimum noise was obtained when only positive deviations of the temperature over the average level have been considered because it was assumed that possible defects are warmer than background (see the last line in *Table 2*).

8. Signal-to-noise ratio and sensitivity in TNDT

8.1 BASIC EXPRESSIONS

In IR systems signal-to-noise ratio is treated in terms of IR flux energy. We assume that TNDT devices are calibrated in temperature and introduce signal-to-noise ratio s in terms of absolute temperature and/or temporal values or their contrasts. Sensitivity of TNDT is treated as minimum variation of tested parameter which is detected with required probability of correct detection at pre-determined false alarm level.

We assume that result of any thermal test is represented by temperature function $T(q_i^d, q_i^n, \tau)$, where q_i^d describes any defect parameter (depth, size, thermal resistance, *etc.*) and q_i^n is any noise parameter (emissivity, background radiation, minor variation of thermal properties which are not considered as defects, *etc.*). In this treatment it is important that temperature and time represent equally ranked parameters which could be measured in order to estimate q_i^d values. Hence, measurement of any parameter q_i^d is based on general equation:

$$\Delta T = \Delta q_i^d \frac{\partial T}{\partial q_i^d} + \Delta \tau \frac{\partial T}{\partial \tau},$$

which connects temperature and temporal criteria. When TNDT sensitivity is evaluated by temperature amplitude the relative measured q_i^d value is as follows:

$$\frac{\Delta q_i^d}{q_i^d} = \frac{\Delta T}{q_i^d} \bigg/ \frac{\partial T}{\partial q_i^d}.$$

For temporal parameter it could be obtained respectively:

$$\frac{\Delta q_i^d}{q_i^d} = -\frac{\Delta \tau}{q_i^d} \frac{\partial T}{\partial \tau} \bigg/ \frac{\partial T}{\partial q_i^d}.$$

Practical application of the last two expressions for measurement of diffusivity and thickness was discussed in [10] concluding that each TNDT transient test could be specified with particular inspection time when signal-to-noise ratio is maximum. Basically the same approach but using another terminology was successfully applied to analysis of multi-layered plate in Laplace domain resulting in algebraic expressions for defect depth and thermal resistance [11].

Signal-to-noise ratio could be defined in this case as follows:

$$\Delta s = \frac{\Delta T(\tau)}{\sqrt{\sum_{j=1}^{N} \left[\frac{\partial T(\tau)}{\partial q_{ij}^n} \Delta q_j^n \right]^2}}.$$

The last expression could be modified in terms of contrast in the following form:

$$\Delta s = \frac{C(\tau)}{\sqrt{\sum_{j=1}^{N} \left[\frac{1}{T(\tau)} \frac{\partial T(\tau)}{\partial q_{ij}^n} \Delta q_j^n \right]^2}}.$$

By analogy the expressions for temporal informative criteria could be also specified.

8.2. TNDT SENSITIVITY IS LIMITED BY APPARATUS NOISE

In the ideal case when results of any thermal measurement depend only on apparatus noise the corresponding expression for s is as follows:

$$s = \frac{\Delta T(\tau)}{\Delta T_{res}},$$

where apparatus noise is reduced to temperature resolution ΔT_{res}.

8.3. TNDT SENSITIVITY IS LIMITED BY STRUCTURAL NOISE

Here we specify simple but practical case when structural noise in thermal test is described in phenomenological way with experimental value of temperature standard deviation and respective standard deviation of temperature contrast:

$$\sigma_n^C = \frac{\sqrt{\Delta \overline{T}_n^2}}{T(\tau)}.$$

Technique to obtain this value is discussed below. Here we state that signal-to-noise could be found as:

$$s = \frac{C}{k\sigma_n^C},$$

where constant k depends on the accepted false alarm level. Usually only gaussian noise is considered assuming $k=2$ (false alarm probability $P_{f.a.}=5\%$), i.e. noise of any specimen will be characterised with value of noise contrast $C_n=2\sigma_n^C$.

By analogy contrast temporal noise could be introduced as follows:

$$\sigma_n^\tau = \frac{\sqrt{\overline{\tau}_n^2}}{\tau}.$$

Enhancing signal-to-noise ratio by filtering

Example of non-optimum raw IR image of plastic specimen with cross-shaped defect by size 15x15 mm located at the depth 1.8 mm is shown in *Fig. 1a* (more details about specimen and measuring technique could be found in [12]). To estimate signal-to-noise ratio the IR image with maximum contrast over defect was chosen. Average temperature has been calculated over defect and presumably non-defect area both. Temperature signal from defect ΔT was found as difference between these averaged temperatures. Then standard deviation σ_{nd} in non-defect area was calculated. Respectively signal-to-noise ratio was defined as $s=\Delta T/2\sigma_{nd}$. Results are presented in *Table 3* for standard spatial filters. It is seen that the most effective smoothing filter 7x7 is able to increase s value by 30 %. Notice that this conclusion is valid for particular and rather large defect. Signals from small defects could be suppressed by filtering as noise signals and the gain in signal-to-noise ratio could be minimum.

Table 3: Signal-to-noise ratio in plastic specimen using spatial filtering.

	Filter	s
Raw image smoothing		12.1
	3x3	14.1
	5x5	14.9
	7x7	15.7
Gaussian	3x3	13.9
	5x5	14.4
	7x7	14.9

Data of *Table 3* illustrate assumption that low frequency noise dominates in TNDT when large areas are inspected. This conclusion will be confirmed below using Fourier analysis (see *Fig. 1b* and discussion below).

8.4. SIGNAL-TO-NOISE RATIO IN TERMS OF IR FLUX ENERGY

In IR systems the problem of optimum spectral range is often arising. We consider signal-to-noise ratio in a case when "pure' temperature signal $\Delta T(\tau)$ is detected on the background of emissivity ε noise.

With accuracy of constant coefficient thermal flux recorded by IR detector is $J(T,\varepsilon)$, where J is spectral integral $J = \int_{\lambda 1}^{\lambda 2} R(\lambda,T)\nu(\lambda)\varepsilon(\lambda)d\lambda$ (where $R(\lambda,T)$ is Planck's function, $\nu(\lambda)$ is the relative spectral sensitivity of IR detector, $\varepsilon(\lambda)$ is the spectral emissivity, (λ_1, λ_2) is used spectral interval). Obvious expression for s is as follows:

$$s = Ct_h \frac{\partial J / \partial T}{(\partial J / \partial \varepsilon)\Delta \varepsilon},$$

where $Ct_h=\Delta T$ (we remind that with t_h we specify excess temperature of an object so the temperature contrast $C=\Delta T/t_h$; absolute temperature of an object in this treatment is $T=T_a+t_h$). For grey bodies:

$$s = \frac{\Delta J / J}{\Delta \varepsilon / \varepsilon}.$$

Values of J, ΔJ for common IR detectors and black body are given in [13]. It was also shown that for t_h value in a range from 0 to 100 °C optimum spectral filtration increases signal-to-noise value up to 1.5-2 times for grey bodies. For non-grey bodies efficiency of filtration depends strongly on particular spectral behaviour of emissivity.

For monochromatic radiation:

$$s = 5C \frac{\lambda_m}{\lambda} \frac{1}{1+T_{amb} / t_h} / \frac{\Delta \varepsilon}{\varepsilon},$$

where T_{amb} is expressed in K, t_h - in °C or °K, λ_m-maximum wavelength determined by $\lambda_m [\mu m]=3000/T$ [K].

The most simple expression is valid for whole spectral range following Stephan-Boltzman law:

$$s = 4C \frac{1}{1+T_{amb} / t_h} / \frac{\Delta \varepsilon}{\varepsilon}.$$

Three last expressions illustrate dependence of TNDT sensitivity on emissivity fluctuations. Let us assume as an example that $T_{amb}=300K$; $t_h=10$ °C and $\Delta \varepsilon/\varepsilon=1\%$. If IR device is working in wide wavelength region the last expression will give contrast value $C=7.75\%$ to provide $s=1$. In its turn it means that differential temperature signal from defect must not be less than $\Delta T=Ct_h=0.775$ °C. The last value is essentially more than possible temperature resolution of modern IR equipment illustrating how emissivity fluctuations could limit the detection of defects.

9. Fourier analysis in TNDT

Fourier analysis is the powerful tool to understand "hidden" features in spatial-temporal temperature data. It is especially useful to analyse the process of defects detection on noise background. By now most of researchers limit themselves with more simple spatial filtering technique. Recent review of standard image processing has been done in [14].

Nevertheless new inversion algorithms using Fourier transformation of IR images are appearing now. They are based on fact that convolution in original domain means simple multiplication in Fourier domain. With fast Fourier transformation (FFT) being available some inversion procedures could be easily done with clear physical sense. Let us introduce: $W(x,y,\tau)$ heating function profile, $F(x,y)$ function describing extended defect of a complicated shape, $G(x,y,\tau,l)$ point-defect response function depending on defect depth. The last function is borrowed from optics being called as "point spread function (PSF)" [15]. Defect signal is the convolution of these three functions:

$$\Delta T(x,y,\tau) = W(x,y,\tau)*F(x,y)*G(x,y,\tau,l).$$

Hence, shape of unknown defect could be restored by two consecutive operations:
1) determining Fourier-image of F:

$$F^*(v,\xi) = \Delta T^*(v,\xi,\tau)/[W^*(v,\xi,\tau)G^*(v,\xi,\tau,l)];$$

2) making inverse Fourier transform:

$$F(x,y) = [F^*(v,\xi)]^{-*}.$$

Here "*" means 2D Fourier transform and v, ξ are spatial frequencies.

Algorithm described above was applied to defects in carbon plastic specimen. Raw IR thermogram is shown in *Fig. 2a*. Temperature image of point spread function $G^*(v,\xi)$ (which could be called as "point-defect response function") computed numerically for small size defect with TURBO.HEAT program [3] is shown in *Fig. 2b*. Final $F(x,y)$ image is presented in *Fig. 2c,d* showing clearly two rows of artificial defects (upper row defects are of square-shape and lower row defects are of round shape) and one large natural delamination of round shape.

Earlier this algorithm using theoretical approximation of PSF was described in [16].

In equations above the presence of time and defect depth is important making the inversion procedure not very stable. Generally in the approach above Fourier transform is done by coordinates for inversion purposes. We investigated use of FFT also instead of spatial filtering introducing spatial spectrums of defect signals and noise.

For simulating defect signal we used simple cylindrical model which describes gaussian shape of temperature signal over disc-shaped defect by radius r_d not specifying dependence of this signal on defect depth, thickness, time, *etc.* :

$$\Delta T / \Delta T_m = \exp\left(-\frac{x^2 + y^2}{2r_d^2}\right).$$

It was reported that this model provides accuracy in a range of 40 % for round- or square-shaped defects [8].

There are simulated signals for defects by radius 10 and 1 mm and their Fourier spectrums in *Fig. 3*. For large defect (*Fig. 3a*) most of signal energy is concentrated in low frequency range (about 1.5 mm^{-1}, see scale in *Fig. 3b* and notice that zero spatial frequency point is located in the center of image). Small gaussian defect of *Fig. 3c* is characterised by more wide spectrum with some noticeable energy in a range up to 30 mm^{-1} (*Fig. 3d*) although main part of energy is still located in the range up to 8 mm^{-1}. It is worth mentioning here that theoretically gaussian signal has gaussian Fourier spectrum but the use of FFT for finite size images produces some zero frequency points especially for small defects (compare spectrums of *Fig. 3b* and *Fig. 3d*).

The next step was to analyse experimental Fourier spectrums of non-defect specimen under transient TNDT test. Raw IR image of plastic specimen is presented in *Fig. 4* with its Fourier image. Effects of non-uniform heating, edges influence and digital noise are clearly seen in *Fig. 4a*. Respectively spectrum of this image is wide enough with most of energy concentrated in a range up to 30 mm^{-1} (notice that scales in *Fig. 3* and *Fig. 4* are different). Nevertheless, the main conclusion after comparison of theoretical and experimental data is that spectrums essentially cross each other. It makes reliable distinguishing of defect and noise difficult.

Finally, we simulated defect with size close to cross-shaped defect in *Fig. 1* and found its spatial spectrum as above. Then this spectrum was used to mask experimental spectrum of image in *Fig. 1a*. Inverse FFT produced the image of *Fig. 1b* where high frequency components are essentially suppressed making this image more comfortable for operator than original image of *Fig. 1a* (cross shape of defect could be followed better also in *Fig. 1b*). It is worth mentioning that the same quality image could be produced using smoothing filters. Hence, the use of FFT for filtering purposes is probably not convenient disposal of this potentially powerful technique.

10. Conclusion

With research described in the present paper we attempt to focus the attention of researchers onto problem which is rarely discussed in current publications: importance of noise description in transient TNDT as the background for reliable decision making. We investigated standard image processing technique being applied to IR thermograms and found that typical defects and noise are characterised by similar features in amplitude and frequency domain both. It converts recognition of small defects into challenging task which needs more deep treatment in future.

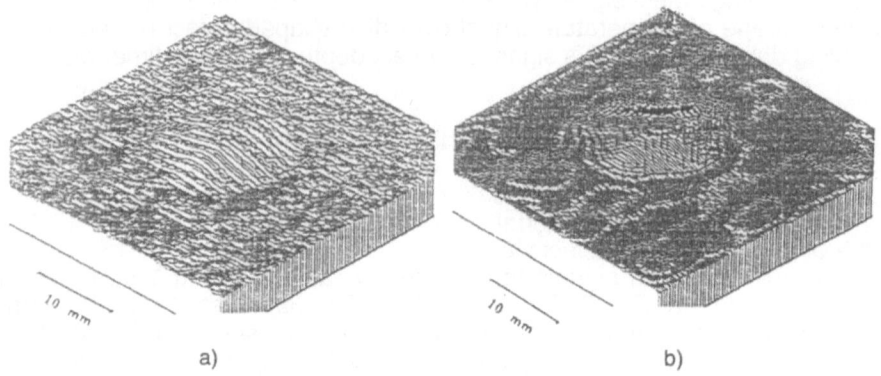

Fig. 1: Thermal NDT of cross-shaped air defect by size 15x15 mm at depth 1.8 mm in plastic specimen: a) raw IR Thermogram, b) same as a) after processing in Fourier domain.

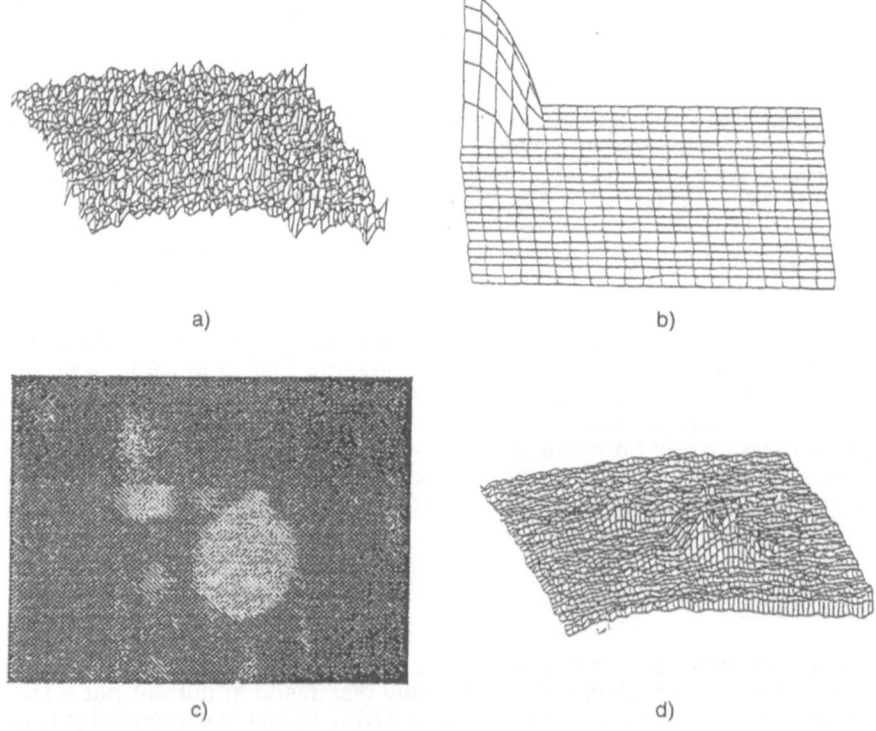

Fig. 2: Restoration of defect size and shape in carbon plastic 8-ply specimen using numerically simulated point-defect response function: a) raw IR thermogram, b) point-defect response function (one quadrant is shown), c) processed image, d) same as c) in 3D presentation.

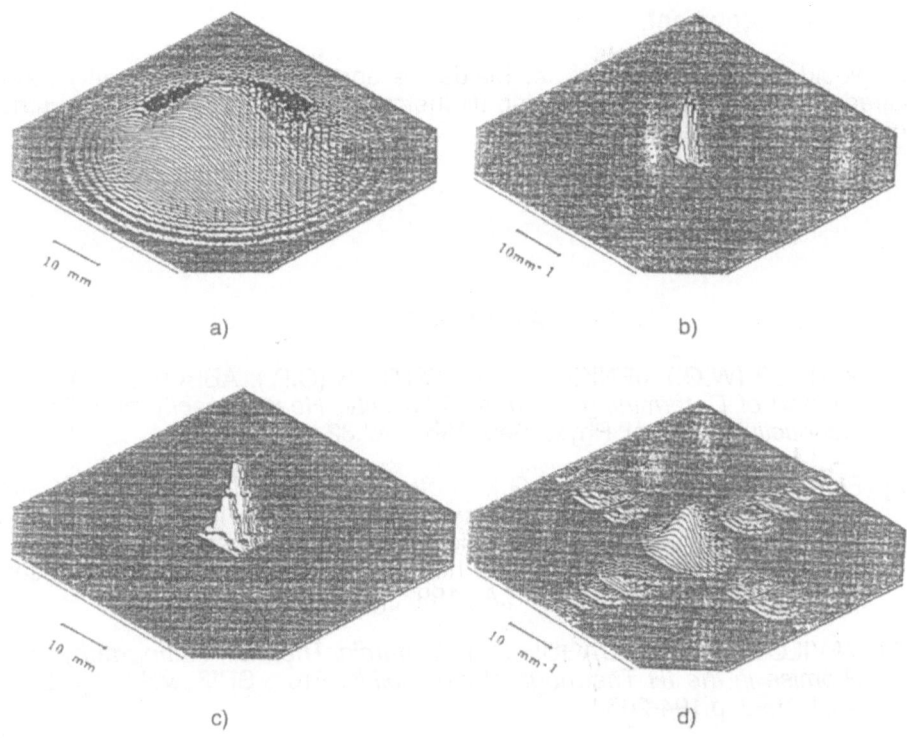

Fig. 3: Simulated temperature signals over defects and their Fourier images: a) defect by radius 10 mm, b) Fourier image of a), c) defect by radius 1 mm, d) Fourier image of c).

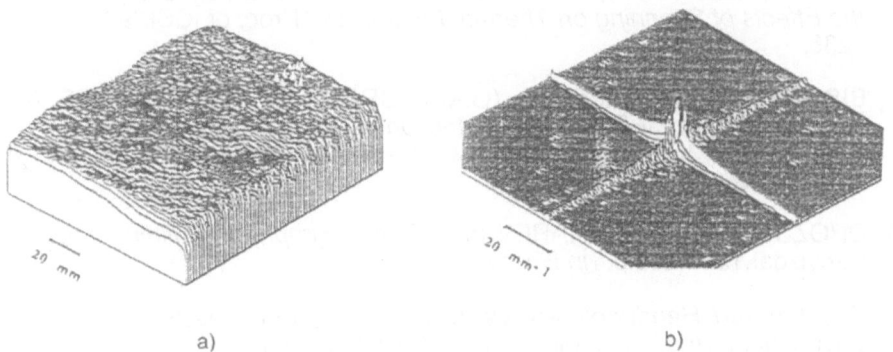

Fig. 4: Raw IR thermogram of non-defect plastic specimen a) and its Fourier image b).

11. Acknowledgement

We would like to thank Prof. X. Maldague and B. Dufort (University Laval, Canada) for assistance in obtaining IR thermograms of CFRP, producing their Fourier images and useful discussion.

REFERENCES

[1] PARKER (W.G.), JENKINS (R.J.), BUTLER (C.P.), ABBOT (G.L.), *Flash Method of Determining Thermal Diffusivity, Heat Capacity and Thermal Conductivity.* J. Appl. Phys., Sept.1961, vol.32, p.1679-1684.

[2] BUSSE (G.), WALTHER (H.G.), *Photothermal Non-destructive Evaluation of Materials with Thermal Waves.* In:"Progress in Photothermal & Photoacoustic Science & Technology (vol.1 "Principles & Perspectives of Photothermal & Photoacoustic Phenomena"), ed.by A.Mandelis, Elsevir Science Publish. Co., Inc., U.S.A., 1992, p.206-298.

[3] VAVILOV (V.), MALDAGUE (X.), *Dynamic Thermal Tomography: New Promise in the IR Thermography of Solids.* Proc. SPIE, vol.1682, 22-24 April, 1992, p.194-206.

[4] TENEK (L.H.), HENNEKE (E.G.), Flaw Dynamics and Vibrothermographic-Thermoelastic NDE of Advanced Composite Materials. Proc. SPIE, vol.1467, "Thermosense-XIII", 3-5 April, 1991, Orlando, U.S.A., p.252-263.

[5] CORTELAZZO (G.M.), MANDUCHI (R.), GRINZATO (E.), BISON (P.), *On the Effects of Scanning on Thermal Images.* In: "Proc. of ICSP'90", p.1233-1236.

[6] BISON (P.), CORTELAZZO (G.M.), GRINZATO (E.), SVAIZER (P.), *Automatic Thermal Reference Detection in Thermographic Images.* Proc. SPIE, vol.1313, "Thermosense-XII", 18-20 April, 1990, Orlando, U.S.A., p.269-277.

[7] DROZDOV (V.A.), SUKHAREV (V.I), *Thermography in Building.* Moscow, Stroyizdat,1987,238 p. (*in Russian*).

[8] *The Infrared Handbook*, ed. by W.L.Wolfe and G.J.Zissis, IRIA Center, Environment Res.Inst.of Michigan,USA,1978, 2520 p.

[9] WINFREE (W.P.) and JAMES (P.), *Thermographic Detection of Disbonds.* Proc. of the 35th Intern. Instrumentation Symp., 1989, p.183-188.

[10] VAVILOV (V.P.), Informativity of Thermal Fields in the Active Thermal NDT Problems. Sov.J.NDT, 1987, No.3, p.67-77.

[11] MAILLET (D.), DIDIERJEAN (S.), DEGIOVANNI (A.), *Identification of Thin Films Thermal Properties by the Heat Pulse Method Using a Sensitivity Analysis.* In: "Proc. of the 3rd Annual Inverse Problem in Engineering Seminar, Michigan State Univ., East Lancing, U.S.A., 25 June, 1990.

[12] VAVILOV (V.P.), BISON (P.G.), BRESSAN (C.), GRINZATO (E.), MARINETTI (S.), *Some New Ideas in Dynamic Thermal Tomography.* Proc.of QUIRT Seminar No.27, 7-9 July, 1992, Paris, France.

[13] VAVILOV (V.), Enhancement of Signal-to-Noise Ratio in Thermal NDT by Using Spectral Filtration. Sov.J.NDT,1985, No.8, p.65-71.

[14] PATEL (P.M.), LAU (S.K.), ALMOND (D.P.), *A Review of Image Analysis Techniques Applied in Transient Thermographic NDT.* Nondestr. Test. Eval., vol.6, 1992, Gordon & Breach S.A., p.343-364.

[15] WALTHER (H.G.), SEIDEL (U.), Some Theoretical and Experimental Aspects of Photothermal Imaging. In :" Abstracts of QUIRT No.27", 7-9 July,1992, Paris, France, p.150-158.

[16] FAVRO (L.D.), CROWTHER (D.J.), KUO (P.K.) and THOMAS (R.L.), *Inversion of Pulsed Thermal-Wave Images for Defect Sizing and Shape Recovery.*, See [3], p.178-181.

ON METHODS FOR SHAPE CORRECTION AND RECONSTRUCTION IN THERMOGRAPHIC NDE

X. MALDAGUE, E. BARKER, A. NOUAH, E. BOISVERT,
B. DUFORT, L. FORTIN, D. LAURENDEAU
Electrical Engineering Dept., Université Laval
Québec, Qc
Canada G1K 7P4
ph (418) 656-2962, fax (418) 656-3594, e-mail: maldagx@gel.ulaval.ca

ABSTRACT. In this paper, various methods are proposed to correct distorted infrared images recorded over curved specimens. This problem is particularly important in thermal inspection applications, in order to avoid misinterpretation of the results. In the paper four methods are discussed: point source heating, video thermal stereovision, direct IR image calibration and shape from heating. Particular emphasis is given on the last two methods. Details of the methods are given and results presented for each method along with particular advantages and drawbacks.

1.   Introduction

Thermographic NonDestructive Testing or TNDT is now one of the many methods internationally accepted for NDT. On this respect, it is interesting to note that level III thermographer certification started to be offered by the American Society of NonDestructive Testing (ASNT) just recently (in Summer 1993). This trend towards standardization now imposes more rigor in the processing of TNDT results.

In this paper, TNDT is defined as an active technique where the part to inspect is briefly thermally heated (pulse stimulation), while the temperature decay recorded on the front surface reveals the subsurface structure of the material inspected. Due to their different thermal properties with respect to surrounding sound material, defects exhibit a different thermal behavior (hotter, cooler), after a time which depends on their depths. Good discussion of this method can be found in [1,2,3,4]. Variations of this technique where the part is cyclically [5] or continuously [6] heated are possible as well providing the specimen is not overheated and there is no problem of reflection of thermal noise from the heating apparatus into the infrared (IR) camera objective.

The reader is referred to [1] for an exhaustive discussion of the main problems afflict-

X. P. V. Malague (ed.), Advances in Signal Processing for Nondestructive Evaluation of Materials, 209–224.
© 1994 *Kluwer Academic Publishers.*

ing TNDT, in this paper, we will address the problem of curved surfaces inspection. Fig. 1 illustrates the problem. On left, a visible image shows part of the tail section of an aircraft which is inspected by TNDT. On right, the corresponding IR image is misleading. First, thickness variation of the internal structure of the airplane is visible (for instance, notice the two plies of aluminium which are cooler on left). Moreover, due to the aircraft geometry, a discontinuous heating pattern is observed and not having Fig. 1-a at hand, a misinterpretation on thickness evaluation of the various layers in the IR image of Fig. 1-a could have resulted.

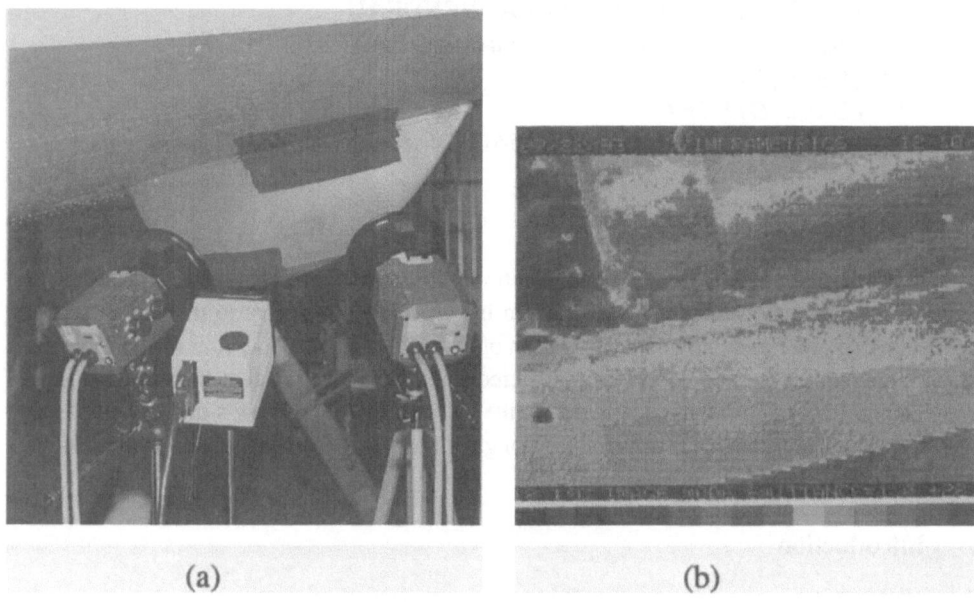

(a)   (b)

Fig. 1   Illustration of the problem of shape curvature in TNDT. (a) visible image of rear part of
tail section of a DASH-8 aircraft. (b) corresponding thermogram recorded 150 ms after
flash and showing all subsurface structure (flash duration is 15 ms). Color scale: 18.1 to
28.1 °C.

This qualitative discussion shows the importance to have mechanisms for shape correction while inspecting curved specimens. In the TNDT community itself (*e.g.* annual *SPIE Thermosense* conferences in Orlando, Fl.), not much authors have addressed this problem [7, 8]. In visible industrial inspection, this problem has already found several solutions [9,10,11].

In this paper, we will discuss four different methods enabling correction for shape-distorted thermograms: point source heating, video thermal stereovision, direct IR image calibration and shape from heating. Particular emphasis will be given on the last two methods.

2. Thermal stimulation background

Thermographers have experimented several ways to stimulate samples, from snow, water bags, liquid nitrogen to more conventional heating devices such as lamps. There is a wide variety of lamps, the more simple being probably the common household bulb whose heating distribution can be expressed by the well known $1/R^2$ relationship:

$$\Delta T_p \sim \varepsilon \frac{P\cos\theta\Delta t}{4\pi R^2 \rho C_p dz} \; [°C] \tag{1}$$

where:

R	=	distance between point heating source and sample surface [m],
θ	=	angle between normal to sample surface and incident stimulating rays [rad],
P	=	heating power of the point source [W],
ρ	=	specimen density [kg/m^3],
Δt	=	thermal pulse length [s],
dz	=	depth of penetration of heating front during Δt [m],
C_p	=	specific heat [J/kg °C],
δ	=	thermal diffusivity [m^2/s],
ε	=	spectral emissivity.

and where dz is approximately given by:

$$dz \sim \sqrt{\delta\Delta t} \tag{2}$$

As seen in eq. (1), two parameters (R and $\cos\theta$) are particularly involved in the inspection of curved objects and contribute to distort thermograms. The correction mechanism thus implies to find a way to derive these parameters to have the actual correction taking place.

Other heating devices behave differently. For instance, if a parabolic reflector is placed behind a point source, thus concentrating the heating rays, dependance on distance R may either disappear or may become linear (on restricted distances), thus simplifying the correction process. In this paper, a bank of six thirty centimeter long 1200 W quartz infrared lamps was used as a stimulation device. A parabolic reflector is mounted at the back of the unit while a shutter door is operated in order to obtain "square heating pulses" (Fig. 2). Minimum duration of heating pulse is about 150 ms. Such a long heating pulses are well suited for studies on low diffusivity materials (such as graphite epoxy or plastic specimens). For high diffusivity materials (such as metal and aluminum specimens), shorter heating pulses are required. In this case, high power flashes are preferred (for instance, the two units depicted on Fig. 1-a deliver 6.4 KJ in 10 ms).

The objective of the correction process is to suppress inhomogeneities due to shape curvature from the thermograms, thus improving reliability of the subsequent interpretation. If θ and R are found independently of the TNDT procedure itself, all images recorded can be corrected (a 3-D ranging camera can be used for such purpose [12]). However, in order to restrict the equipment used to a minimum, some methods (discussed below) rely only on the first IR image recorded to extract the correction parameters. In such case depth of penetration dz (eq. 2) becomes an important factor to consider: the first image must be recorded as early as possible, even before the appearance of the most shallow suspected defects to prevent the correction process to reduce visibility of these defects as it does for shaped sound material.

Fig. 2   Pictures of the heating unit with six 30 cm long quartz infrared lamps and a mechanical shutter. Notice IR camera on top of the device.

3. Point source heating correction

This method has been described in details in [7], here we will only recall its main features. This method is directly derived from eq. (1) and on dependance of the rising temperature on parameters R and θ. In this case an high power bulb (1000 W) is used as a thermal stimulation device, it is mounted on top of the IR camera (Fig. 3). In a first step, a calibration phase is necessary. This calibration procedure consists to record an early temperature image (ETI) in the tridimensional inspection volume located in front of the IR camera. As discussed above, the ETI is the earliest image recorded just after the thermal stimulation is completed (IR image recorded before shallow defects start to perturb surface temperature distribution). Typically, in the calibration phase, the ETIs are recorded on a plane covered with a high emissivity paint (same paint as for the specimens) at various distances in front of the IR camera enabling to constitute a complete data bank of the expected temperature distribution within the inspection volume. The method exposed in [8] is also based on a similar calibration technique.

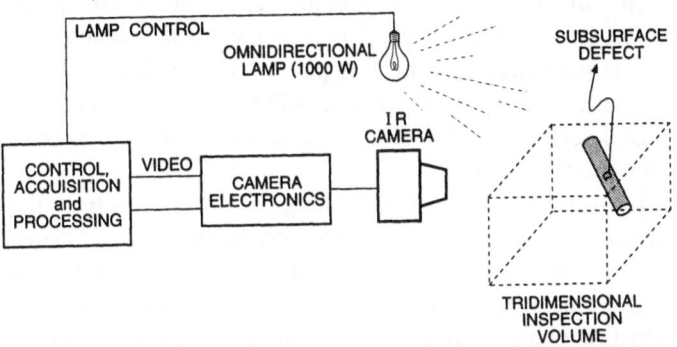

Fig. 3   Schematic diagram of the experimental set-up needed for the point source heating correction method.

Once the data base is completed, the actual inspection proceeds and the shape correction consists in matching specimen ETI with ETI of the database on a pixel per pixel basis. This matching produces a range image of the inspected scene which can be further used to correct subsequent thermograms from the specimen. Fig. 4 shows result of such (RTI -> range) image conversion.

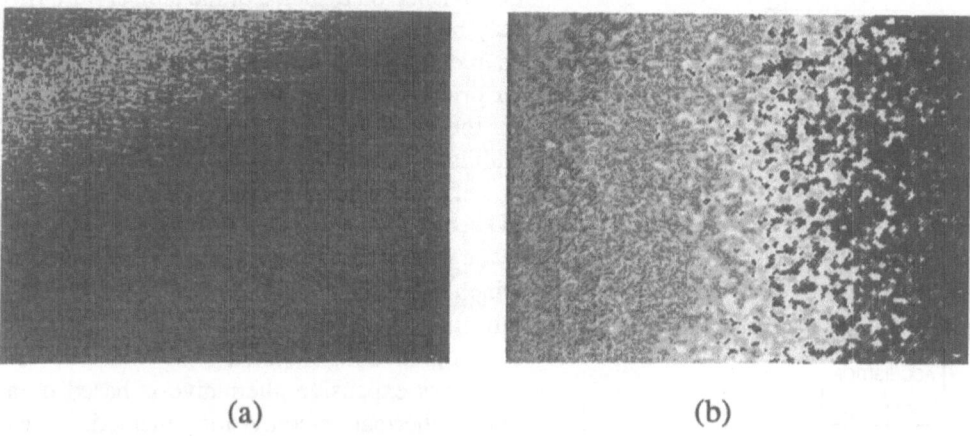

(a)   (b)

Fig. 4   Results of the point source heating correction method. (a) raw thermogram recorded on a tilted plastic plane, notice the heating trend with top left corner hotter than bottom right corner. (b) processed distance image where corrected trend appears (closer on left and further on right). Arbitrary color units.

Drawbacks of this technique are two fold. An high power source is required and only a fraction of the available power is used due to the inherent spreading of the radiation in the whole space, consequently the inspection volume is restricted. Experiments described in [7] reported a maximum workable distance of about 20 cm in front of the IR camera. Moreover, the calibration procedure is slow if the calibration plane is to be tested at many distances in front of the camera. A possibility discussed in [8] is to use a tilted plane covering the whole distance span in front of the IR camera. In such case one experiment may be sufficient (however this is not valid if a point-source-like heating device is used due to the nonuniform heating pattern of such device). Moreover, the database has to be periodically recalibrated as any calibration-based technique. Finally, since the method neglects parameter $\cos\theta$ of eq. 1, it is limited to objects of restricted curvature ($\cos\theta \sim 1$ for $\theta = 90 \pm 20°$).

On the other hand the technique is interesting since it allows from the thermograms, to extract information which would not be available otherwise and besides the calibrated database, it does not require extra hardware with respect to the standard TNDT procedure. Moreover, if the geometry of the inspected object is known (a reasonable hypothesis), parameter $\cos\theta$ becomes available for a more accurate correction.

4. Video thermal stereovision

One of the main problem of the previous correction technique is related to the point source like heating device which provides non uniform heating patterns and for which lot of the heating power is lost, thus reducing amplitude of the available thermal contrasts. In most case, it is better to rely on a uniform heating device. If such device is used, the dependance on distance of the temperature rise may be unclear, for instance if more than one source is used with each one having a particular orientation. In such a case, a better choice for thermograms correction is to rely on some additional hardware.

Different hardware can be envisaged such a 3D sensor (for example the Jupiter 3D camera made by Servo-Robot, Boucherville, Canada) which can serve to provide direct ranging information enabling object reconstruction and thermogram correction [12].

A less expensive alternative is based on a video thermal stereovision method. Such method has been described in details in [13], here we will only recall its main features. To apply this technique, in addition to the standard TNDT apparatus (heating device and IR

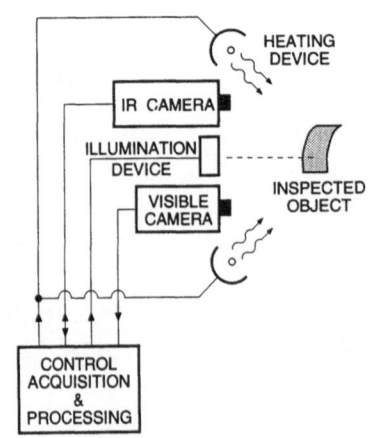

Fig. 5 Schematic diagram of the experimental set-up needed for the video thermal stereo correction method.

camera), a video camera and an illumination device are required along with usual computer control and frame grabbing facilities (Fig. 5).

The reflectance information of the scene present in a visible image makes possible to determine θ, this method is based on the work of Nandhakumar *et al.* [14] and assumes that observed surfaces are Lambertian and objects opaque (these hypotheses are valid in most TNDT applications). In these conditions:

$$L = K\cos\theta + C \tag{2}$$

where: L is the digitized intensity value of the visible image,
         K and C are the overall calibration constants of the visible imaging system.

K and C are specific to a given experimental apparatus, they are found experimentally using the following procedure. A visible image of a scene with two surfaces of known orientation is digitized. Substitution of known orientations and digitized values for the two surfaces in eq. (2) allows to write a two-equation system which solves for K and C. Once K and C have been calibrated for a given apparatus, the orientation image $\cos\theta$ can be computed with eq. (2) from the visible image of the inspected scene.

Before to proceed to curvature correction of the thermograms, it is necessary to align visible and IR image formats. This is essential to have a direct pixel to pixel correspondence between the two images. This registration step is based on the following transformation between the coordinates (x, y) and (u, v) for infrared and visible images respectively. For the u coordinate:

$$u = \sum_{i=0}^{N} \sum_{j=0}^{N-i} a_{ij} x^i y^j , \tag{3}$$

with: N degree of the polynomial (usually 2),
       a_{ij} polynomial coefficients.

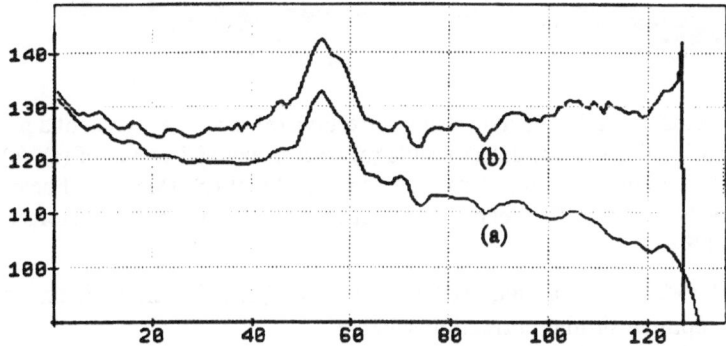

Fig. 6   Results of the video thermal stereo correction method profile extracted from a thermogram of a plastic tilted plane (a) before and (b) after correction. Arbitrary vertical units, horizontal scales: pixel coordinates.

The v coordinate is computed in a similar fashion.

A calibration step is required to compute the polynomial coefficients. A set of common points between two images of the same scene, visible and infrared, are selected and eq. (3) is solved. Finally curvature correction of thermograms is done by dividing successive thermograms by $\cos\theta$ following eq. (1).

This double image correction rises several problems. First of all, it is necessary to cover specimen surface with a paint having both good pseudo-Lambertian properties in the visible spectrum and acceptable emissivity value in the sensibility band of the IR camera. Second the illumination device must have limited dependance with parameter R for eq. (2) to be valid. Besides these difficulties, this method is an effective correction tool as illustrated by Fig. 6 (correction process performed in the case of the plastic tilted plane of Fig. 4).

5. Direct thermogram correction

The idea of the direct thermogram correction is to suppress the extra hardware (illumina-

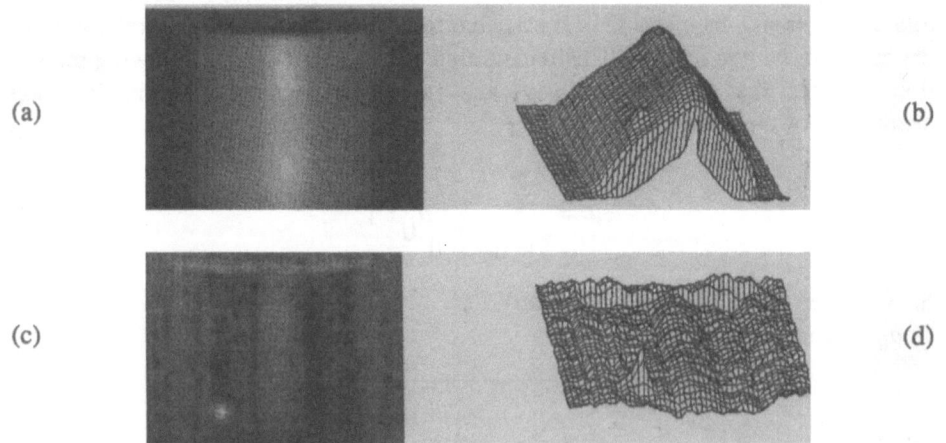

(a)

(b)

(c)

(d)

Fig. 7   Results of the direct thermogram shape correction. (a) raw thermogram of a plastic half-cylinder with a subsurface defect visible (b) corresponding 3-D plot of (a). (c) thermogram (a) after correction. (d) corresponding 3-D plot of (c). Duration of heating was 2 s, observation time is 10 s, depth of defect is about 2 mm under the surface (heating device of Fig. 2).

tion device and video camera) required in the videothermal stereo vision correction method while keeping a similar correction scheme.

The idea is as followed. For opaque bodies at close to room temperature, one of the main difference between visible and IR images lies in their formation process, visible images are formed due to a reflection phenomenon while IR images are formed by an emission mechanism. Nevertheless, if the heating device provides little dependance on distance R, eq. (2) can be adapted to TNDT applications with L being the digitized value of the thermogram, K and C being the overall calibration constants of the IR imaging system and θ being the angle between normal to surface patch (for a given pixel set x,y) and direction of observation with the IR camera. Eq. (2) thus provides a reasonable estimation of local surface orientation θ once parameters K and C have been calibrated using a similar technique as the one exposed in the previous section (using planes of various orientation).

The orientation correction process is next completed using an early recorded IR image after thermal pulse. With calibration parameters K and C and eq. (2), the orientation image is formed and all the subsequent thermograms are divided by this orientation image. Fig. 7 shows some results of this method. An important point is that of course thermal pulse specifications (duration, power) must be the same for both calibration and correction steps.

Drawback of this method is that heating distance (parameter R) is assumed sufficiently large so that local depth variations on the specimen do not affect the IR emission process. This hypothesis is not always valid. This can be seen for instance on Fig. 7-d where the central band still exhibits shape distortion after processing. Nevertheless, there is an obvious interest for such a simple correction process, at least for defect detection purposes.

6. Shape from heating

The last method we propose is originally based on shape from shading theory although substantially modified for TNDT applications [15]. The idea is, as in section 5, to rely solely on the standard TNDT apparatus to perform shape correction, using only an early recorded IR image (before apparition of subsurface defects) as stated before.

In order to use this correction technique, the following hypothesis must be valid. The heating flux must be orthogonal to the specimen, if considered flat. In the case of the heating apparatus of Fig. 2, this is justified since the thermal profile obtained on a tilted plane is a straight line and also since the experimental curve temperature rise versus distance R is linear on TNDT working distances.

The early thermogram of a specimen is analyzed row by row. Each row is divided in segments which are classified as whether or not they are linear. Two cases may occur.

First case: linear segments

If a segment is linear, this implies that variation of the distance R (heating device to specimen distance) is greater than orientation angle θ. This is the case for objects with flat surfaces. Fig. 8 depicts this case. From this Fig. 8 and with eq. (1), assuming that initial before heating specimen surface temperature is T_i, we can write:

$$T_2 - T_i = (T_1 - T_i) \cos\theta, \tag{4}$$
$$T_3 - T_i = (T_4 - T_i) \cos\theta,$$

218

from which we obtain:

$$\cos\theta = \frac{T_2 - T_3}{T_1 - T_4} \quad .$$ (5)

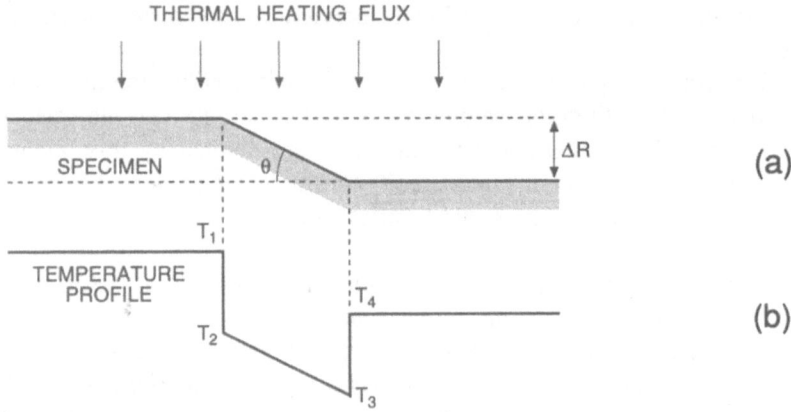

Fig. 8   Heating analysis of a flat surface specimen (a) experimental configuration. (b) corresponding thermal profile.

Notice T_1 and T_4 values are not always available, think to a tilted plate for instance. In such case, a variant procedure is to be used as seen on next page.

Providing the experimental curve temperature rise versus distance R is known, slope S becomes a known parameter with the following definition (on TNDT workable distances):

$$S = \frac{\Delta T}{\Delta R}$$ (6)

with respect to eqs. (5, 6):

$$\Delta R = \frac{(T_1 - T_4)}{S} \quad ,$$ (7)

and finally with eqs (4, 7):

$$\Delta R = \frac{(T_2 - T_3)}{S \cos\theta} = \frac{\Delta T}{S \cos\theta} \quad .$$ (8)

Using eq. (5), it is possible to compute surface orientation θ, while eq. (8) relates both orientation to height variation together (in an orthogonal projection).

Fig. 9 depicts the case of a single plane segment. From Fig. 9, the following relationship is obtained:

$$\Delta R = d\tan\theta. \tag{9}$$

where d is the length of the projection and is measured by the number of pixels N associated with the segment, $d = N\,Step$, with $Step$ being the horizontal distance per pixel in the field of view, at focus. This relationship is only valid for orthogonal projections. Combining eqs. (8) and (9), we obtain:

$$\sin\theta = \frac{\Delta T}{dS} \tag{10}$$

which allows to determine orientation θ for cases where T_1 and T_4 are not available.

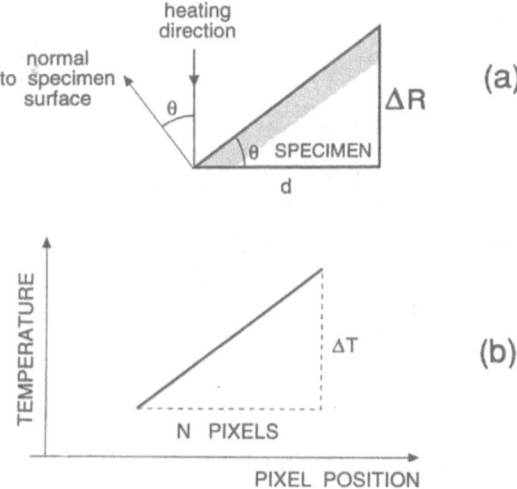

Fig. 9   Heating analysis for a tilted plane (a) experimental configuration. (b) corresponding thermal profile.

To summarize, for linear row segments, orientation θ is first computed using eq. (5). Next, height variations ΔR can be found either with eq. (7) or (8) if the rising temperature versus distance curve is available. Otherwise an approximative solution of ΔR is given by eq. (9) and θ is found using eq. (10).

Second case: non-linear segments

In these cases, variations of distance R are much less than variations due to orientation θ. It is then assumed that, for small segments, temperature variations is only proportional to orientation θ:

$$T - T_i = (T_{max} - T_i) \cos\theta. \tag{11}$$

where T_{max} is the local maximum temperature close to pixel of interest which is at temperature T. Knowing T_i (before heating initial temperature), it becomes possible to compute local surface orientation. Fig. 10 illustrates the studied geometry.

Following this scheme, non-linear segments are divided into elementary segments for which local orientation are found with eq. (11). Next segment size is derived using eq. (9) with $d = Step$ for one-pixel-segment size. Complete shape reconstruction proceeds by merging all segments one after one. Image correction can proceed as exposed in section 5 once the orientation image is generated.

Some results are presented in Fig. 11. Fig. 11-a depicts the experimental set-up: a

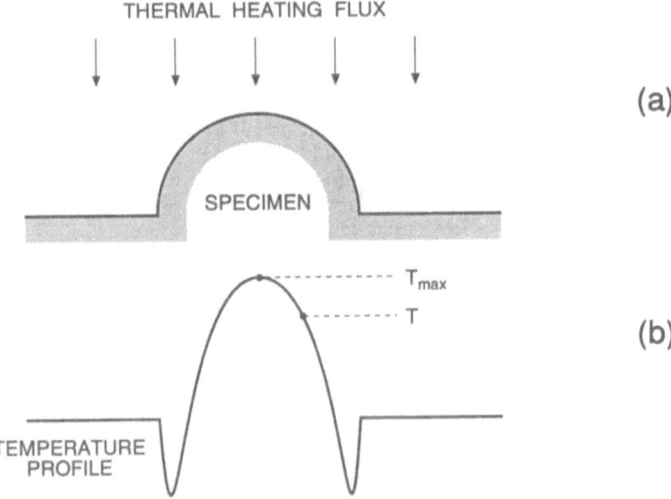

Fig. 10 Heating analysis of a curved specimen (a) experimental configuration. (b) corresponding thermal profile.

plastic part constituted of a plane and a half cylinder is tilted by 30° and is heated orthogonally. Fig. 11-b shows the corresponding raw temperature image of this part geometry. Fig. 11-c shows profile of the middle row of thermogram of Fig. 11-b. Due to high level of noise, the thermogram must be smoothed before processing (using a 5 x 5 averaging filter), Fig. 11-d shows the corresponding smoothed profile. Fig. 11-e presents the result of the shape reconstruction: line 1 is the reconstructed profile superimposed on the exact profile (line 2). Maximum reported error is 2° on global orientation while the two shapes (exact and reconstructed) match quite well.

In shape from shading, the question to determine whether or not a surface is convex or concave is a problem. The same problem arises here although the known direction of heating helps, moreover, it is possible to have a priori information on object shape to solve this ambiguity.

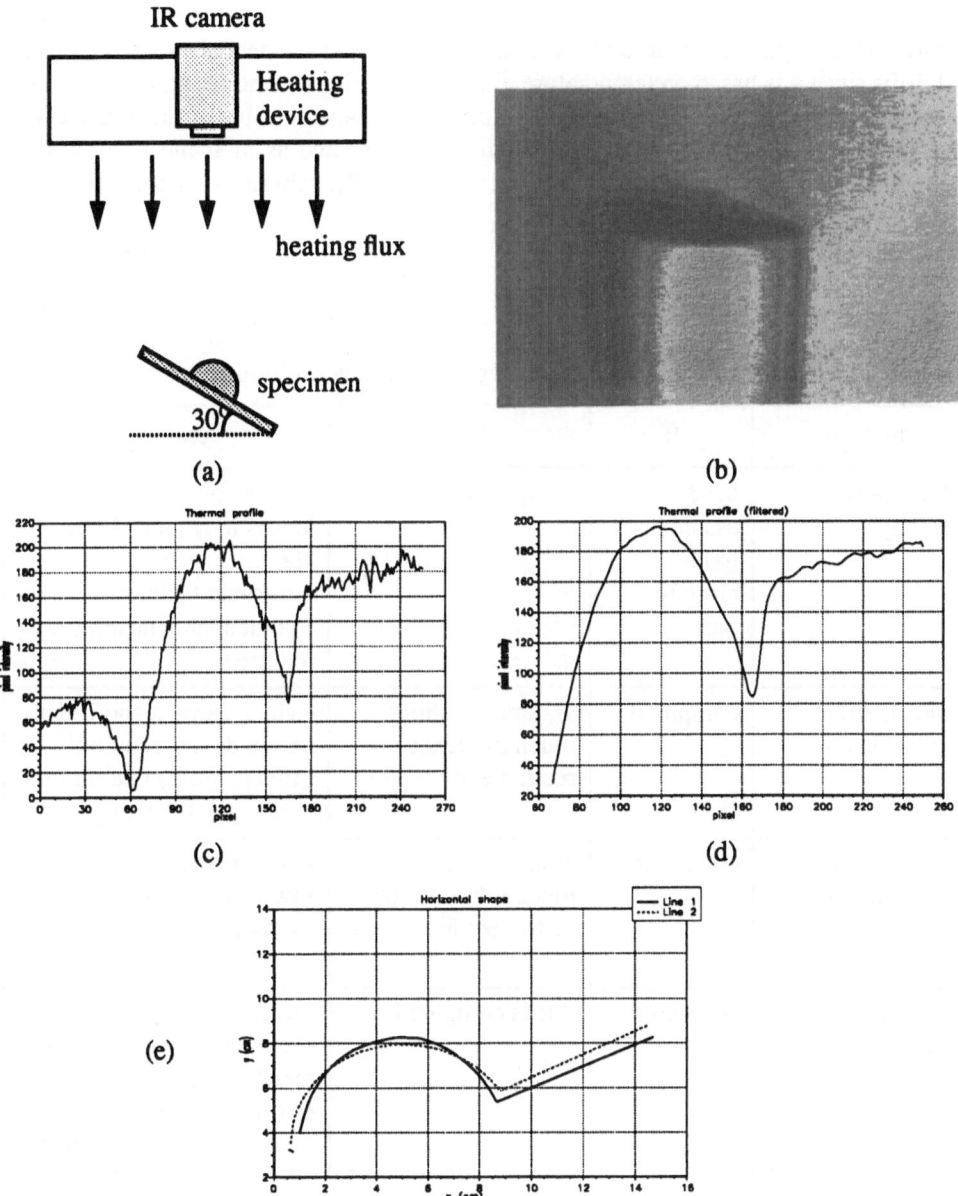

Fig. 11   Results of the shape from heating reconstruction method. (a) experimental configuration. (b) early recorded thermogram, heating pulse (heating device of Fig. 2) for t = 0 to 125 ms, observation at t = 167 ms. (c) raw extracted profile: middle row of image (b). (d) profile (c) after smoothing. (e) result of the shape reconstruction: line 1 is reconstructed profile while line 2 is exact specimen shape.

Advantage of this reconstruction process is that it only requires a single early recorded thermogram while it is more robust than absolute temperature method to thermal drifts since it is based on temperature differences within the thermograms (this is not the case of point source correction method described above and based on absolute values). On the other hand, only relative heights are computed as in shape from shading technique, hence ΔR becomes available, but not R. Finally, calibration is not mandatory.

7. Conclusion

In this paper, four different techniques were described for shape reconstruction and ther-

Table 1: Characteristics of discussed methods

method	θ	R	comments
point source heating	cosθ assumed to be close to one	computed	• works on limited distances in front of IR camera • absolute calibration • point source heating required
video thermal stereo vision	computed	assumed without much influence on restricted distances	• video camera and illumination device required • two parameter calibration • orthogonal heating
direct thermogram correction	computed	assumed without much influence on restricted distances	• no extra hardware • two parameter calibration • orthogonal heating
shape from heating	computed	ΔR is computed	• no extra hardware • relative calibration (not mandatory) • orthogonal heating • more complex processing

mogram correction in TNDT applications. Main advantages and drawbacks were dis cussed for each method. Table 1 summarizes characteristics of the four methods discussed. From this analysis, it appears that shape from heating is probably the most attractive method although it requires more complex processing than the other methods, especially for the segmentation task.

Attractive fields of application of theses methods are in the aerospace industry and also for pipe inspection. Further work is now needed to link these restoration processes to the inverse procedures for extraction of subsurface defect parameters such as depths, size, thermal resistance.

8. Acknowledgments

Collaboration of Mr. B. Tanguay from Air Alliance and his team is acknowledged. Support of Natural Sciences and Engineering Research Council of Canada and of IRIS is acknowledged.

9. References

1. X. Maldague, *Nondestructive Evaluation of Materials by Infrared Thermography*, Springer-Verlag, 241 p., 1993.

2. V. P. Vavilov, R. Taylor, "Theoretical and Practical Aspects of the Thermal NDT of Bonded Structures," in *Res. Techn. in NDT*, **5**, R. Sharpe (ed.), Academic Press, p. 239-280, 1982.

3. D.L. Balageas, D. Bosher, A. Déom, J. Fournier, "La thermographie infrarouge: un outil quantitatif à la disposition du thermicien," *Rev. Gén. Therm. Fr.*, no. 322, p. 501-510, oct. 1988. [in French]

4. S.K. Lau, D.P. Almond, J. Milne, "A quantitative analysis of pulsed video thermography," *NDT & E International*, 24 [4], p. 195-202, 1991.

5. T. Ahmed, Z.J. Feng, P.K. Kuo, J. Hartikainen, J. Jaarinen, "Characterization of Plasma Sprayed Coatings Using Themal Wave Infrared Video Imaging, *J. of NDE*, 6[4], p. 169-175, 1987.

6. J.W.M. Spicer, W.D. Kerns, L.C. Aamodt, R. Osiander, J.C. Murphy, "Time-Resolved Infrared Radiometry (TRIR) using a Focal Plane Array for Characterization of Hidden Corrosion," in: L.R. Allen (ed.), *Thermosense XV*, SPIE Proc., vol. **1933**, p. 148-159, 1993.

7. X. Maldague, L. Fortin, J. Picard, "Applications of tridimensional heat calibration to a thermographic NDE station," in: G. S. Baird (ed.), *Thermosense XIII*, SPIE Proc., vol. **1467**, p. 239-251, 1991.

8. H. Kiiskinen, P. Pakarinen, M. Luontama, A.-L. Laitinen, "Using infrared thermography as a tool to analyze curling and cockling of paper," in: J.K. Eklund (ed.), *Thermosense XIV*, SPIE Proc., vol. **1682**, p. 134-141, 1992.

9. A. Blake, A. Zissermaan, *Visual Reconstruction*, MIT Press, Cambridge, Mass., 1987.

10. H. T. Tanaka , F. Kishino, "Visual Reconstruction with Adaptive and Arbitrarily Oriented Meshes," *Proc. 11th Int'l Conf. Pattern Recogni.*, pp. 775-779, 1992.

11. P. Hébert, D. Laurendeau, D. Poussart, "From 3-D scattered data to geometric signal description: invariant stable recovery of straight line segments," *SPIE Proc. Appli-*

cations of Artificial Intelligence, vol. **1964**, pp. 135-146, 1993.

12. L. Pitre, X. Maldague, D. Laurendeau, "Range Finder Based Guidance of Puma Robot for the Infrared Inspection of Non-Planar Material," in: V. Bhargava (ed.), *Proc. Can. Conf. Electrical Comput. Eng.*, 1993. [in press]

13. A. Nouah, X. Maldague, F. Robitaille, "Shape correction in transient thermography inspection of non-planar components," in: D. Balageas, G. Busse, G.M. Carlomagno (eds.), *Proc. Quantitative Infrared Thermography*, Eurotherm Seminar **27**, p. 224-228, 1992.

14. Nandhakumar N, Aggarwal JK. "Integrated Analysis of Thermal and Visual Images for Scenes Interpretation," *IEEE Trans on Pattern Analysis and Machine Intelligence*, **10** [4], p. 469-481, July 1988.

15. A.M. Bruckstein, "On shape from shading," *Comput. Vision, Graph., and Image Processing*, **44**, p. 139-154, 1988.

USE OF NEURAL NETWORKS AND EXPERT SYSTEMS FOR EVALUATION

Professor Kevin Warwick
Department of Cybernetics
University of Reading
Reading, RG6 2AY, UK

ABSTRACT. This paper considers recent developments in the techniques for evaluation, in particular those which are useful, as alternatives to complex mathematical expressions, for modelling non-trivial, non-linear relationships. Neural Networks and Expert Systems, with the ability to learn, are considered in depth and a number of practical examples are included to show the techniques in operation.

1. INTRODUCTION

The last decade has seen a rapid increase in the nature and number of "intelligent" systems being investigated and/or employed for data evaluation and control. This has come about principally because of the considerable performance improvement in on-line computing power. It is therefore now possible for such systems to learn from experience and organise their own behaviour in order to adapt to an unknown and/or changing situation.

Perhaps the most widely encountered intelligent system in practice is an expert system, which is usually a rule/knowledge based system in which the rules have been obtained not only from expert input but also from basic physical laws, specific plant data and experimental or testing data. At present fixed rule expert systems are rather restricted in their performance, although nevertheless more and more application areas are being found in which their employment is seen to be of benefit [1]. It may be however that the use of an expert system is fairly simple and straightforward. and is along the lines of rules such as: IF the Temperature is too high THEN close the valve. The overlap between logic, simple rules and expert systems is therefore clear, with expert systems taking a more heuristic role, i.e. a role gained more from experience than basic theory. It is however also possible for an Expert System to be originated with a programmed set of rules, and then to learn itself how these rules should be modified or increased in number [2]. The specific problem of how such systems learn and improve will be dealt with, by means of a power systems example, in section 2.

A more recently introduced intelligent system procedure is the use of neural networks,

X. P. V. Malague (ed.), Advances in Signal Processing for Nondestructive Evaluation of Materials, 225–240.

both analog and digital approaches being possible [3]. Such networks are fundamentally constructed as networks of neurons which are simple, approximations to brain neurons, and in this respect the technique is a sensible "bottom up" approach to artificial intelligence [4]. It is worth emphasising however that the artificial neurons are generally very much simpler than their human counterparts, a rigorous, well defined, structure is usually assumed, interactions between neurons are restricted to a certain class of behaviour and the total number of artificial neurons in a network is very, very much less than the number in the average human brain. As an example of this latter point, artificial neural networks can be put together consisting of less than 20 neurons whereas in the human brain there are, something in the order of 20 billion neurons [5]. In short therefore, in many ways artificial neural networks still have some way to go before they are compatible with human brains. One common element however is that both artificial and human neural networks can learn and adapt by a variety of means, one approach for artificial neural networks is by the use of Genetic Algorithms [6], these also being applicable for expert system learning.

This paper contains a look at some different types of neural networks, both those with a single layer of neurons and those with multi-layers, and also expert systems and their use for information extraction and processing. Further, some practical examples are given to show how such methods can be employed.

2. ADAPTIVE EXPERT SYSTEMS

Adaptive Expert Systems are often referred to as "Learning Classifier Systems" [7], the originating rule-base being a population of rules which can either fire individually or in a chain, the latter being more usual. This can be viewed as:

Rule 1: IF A AND B AND C THEN X
Rule 2: IF A AND D AND E THEN Y

Here we have a population of 2 base rules, in which A, B and C must all be true for Rule 1 to fire and hence X to become true.

It may be for a particular problem that if A, B and C are all true it would actually be better if we took another action, say W; or if A,B and D are true then an action Z is taken. From the initialised rule base of 2 the system must either be given, via input, new rules in order that W and Z can be achieved, or it must learn for itself. This paper is concerned with the latter case.

One distinct problem with setting up an Expert System rule base is the experts themselves. For a specific application problem (a) experts can disagree - each has their own problem solution, (b) it can be difficult to extract the desired information from an expert, and (c) an expert's opinion can often differ with that suggested by data sheets or common sense information. Further, if rules are not regularly updated, then the rule base can quickly get out of date, i.e. the system itself may change and unless the rules change in sympathy, wrong decisions may be made. In practice however it is very difficult to keep checking that the rules employed are still acceptable. All of these points indicate that it is often sensible to firstly seed an expert system with a population of originating rules and then to use some form of critic to adapt the rule base.

An adaptive expert system consists of a standard originating rule base and an adaptation procedure via a critical analysis i.e. a critic. The population size is however important as it is not sensible to simply add more and more rules to the rule base - although more eventualities can be accounted for by employing more rules, actually working through rules at a particular time can be problematic, especially when conflict occurs, i.e. when several rules have fired a decision must be taken as to which rule has precedence. It is therefore normal for a fixed population size to be set, this population consisting of the "best" set of rules, where "best" is defined in an appropriate way for a particular problem. The introduction of any new member into the population does however mean that a certain member of the original population must be removed/ sacrificed, thereby creating a space. The weakest member is the obvious choice.

As a practical example of an adaptive expert system, an alarm processing scheme designed for use on the U.K National Grid network is detailed, this application making use of a Learning Classifier System [8].

The National Grid is a high power transmission network of considerable complexity, consisting of interconnected high power components. Overall grid control is organised in a hierarchical fashion with power components being grouped together in substations, these substations are then under the control of Area Control Centres (ACCs), of which there are five in the U.K, and a National Control Centre oversees the total. It is at the ACC level however that operational control of the network is handled, to the extent of realising and diagnosing problems on the Grid and deciding on a course of action.

When a fault occurs, automatic switches act to protect the system by isolating the area about the fault, thereby minimising its effect on the remainder of the network. The substations receive direct information on the actions taken in terms of switch openings and closures, and this information is held until the substation is polled by an ACC. At the ACC the switch messages, along with supporting analog information, can be used to diagnose the position and nature of the originating fault. The ACC interpretation of the message stream depends on network topology and current state, but is not distinctly time dependant due to the substation polling action. Switching information occurs very rapidly, possibly several hundred relevant switch operations within a minute, and further the fault can be either permanent or transient. All of this means that the problem is most definitely non-trivial..

Historically a grid control engineer has faced the task of carrying out the fault diagnosis and suggesting remedial action. However the sheer quantity of switch messages now available, principally through improved communication links, means that the engineer may become overloaded at peak times, e.g. in stormy weather.

A Learning Classifier System (LCS) has been developed, as an aid to the engineer, to tackle the on-line fault diagnosis problem. Although essentially a rule-based system, it differs from standard expert systems in that the diagnostic rules are not pre-programmed into the system, but rather the rules are learnt through training by example via a fault simulator. In this sense a large section of the grid network has been respectably simulated in terms of switch operation on control centre messages. A fault can be entered to the simulation and from the switch messages that occur the LCS must learn which fault causes that particular sequence of messages. Genetic Algorithms [2,7] are used to adapt the system to enable the generation of new rule sequences. Fig. 1 shows a schematic of the LCS with the Alarm Handling and Fault Analysis (AHFA) simulator. Once the LCS has been trained on the simulator, it can be operated on the actual grid network, on the basis that if the AHFA simulation is reliable, so the LCS conclusions will also be reliable.

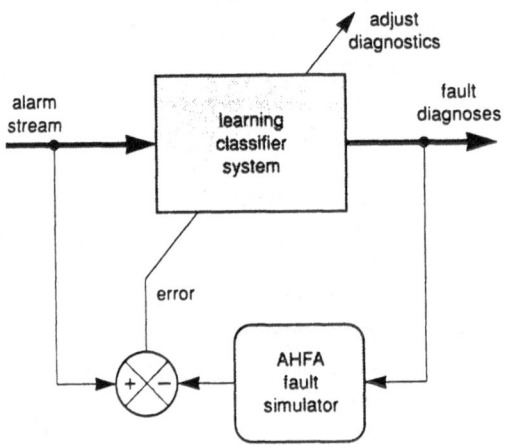

Fig. 1 Adaptive system for fault diagnosis on the electrical
power transmission network

Although described through one specific example, the LCS operates on a continuous stream of data, with the help of further domain specific supporting evidence. In the power transmission example, a further (positive) constraint is that only a time window of data need be considered for the analysis without degrading performance. Essentially however the LCS can competently cope with any data driven system in which a certain number of events, or data pieces, will realise a particular conclusion. The LCS can readily learn this sequence such that when these events occur in the future, the LCS can draw the same conclusion.

The LCS uses blackboards to look at an alarm stream and various hypotheses are developed as the stream is received, such that at any one time, one particular hypothesis is favourite. However a considerable amount of the alarm values can be termed noise through switches dithering, opening and closing or being affected by unrelated transients on the network. All of these values must be processed in some way, but frequently repeating alarms can usually be ignored. Any completed hypotheses are passed to a higher level of hypothesis testing which, if it recognises the hypothesis fingerprint, will put forward the suggestion to the diagnostic critic, as being a full diagnosis of the fault. This can be seen in Fig. 2.

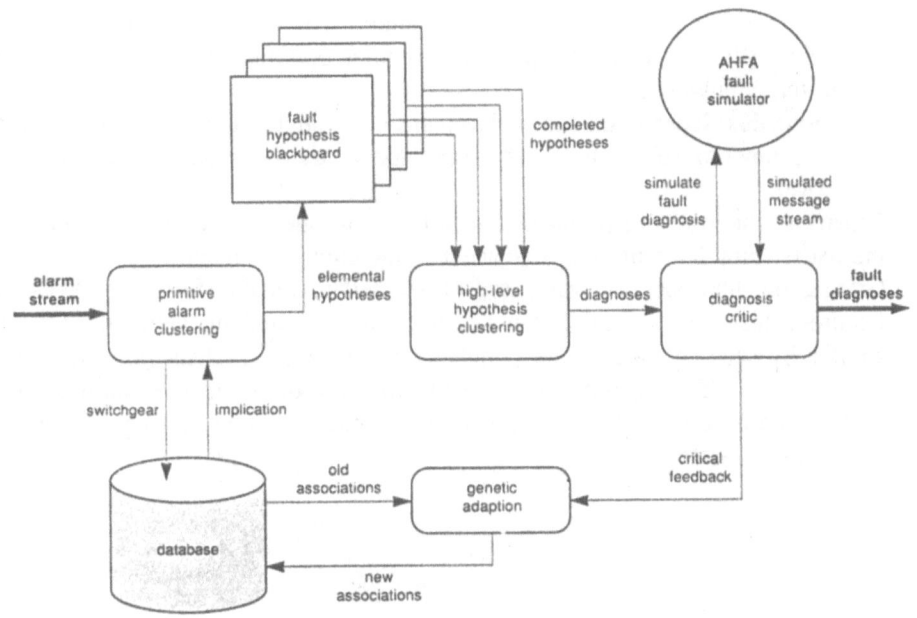

Fig. 2 A schematic of the LCS for alarm analysis and diagnosis

The diagnosis critic makes good use of the AHFA simulator. For a suggested fault diagnosis the critic initiates this fault in the simulator, thereby realising a simulated message stream. A closeness of fit test is then carried out between the actual and simulated message streams, in which if the two streams are very different it is taken that the diagnosis is poor, with the converse also holding.

Genetic operators are then employed in the feedback path, as shown in Figure 2, to recombine the population of fault association listings, the overall aim for the population being a bias towards the fittest strings, and it is these that are put forward for reproduction. Re-combination by the genetic operators means that the weakest of the fault sets are replaced and the new population defines the desired mapping more accurately.

Two standard genetic operators form the basis of the genetic operations, these being crossover and mutation as shown in Fig. 3. These two methods assume no knowledge of the strings being manipulated, merely operating on the selected, and therefore successful, strings. Further heuristic genetic operators namely gravitation and smearing have however been introduced, as complementary procedures, to improve the optimization process.

The four genetic operators employed, and shown in Fig. 3 can then be described as follows:

(i)   Crossover: this allows parts of optimal solutions to be developed via implicit parallelism [9]. This operator performs a course optimisation of the associations within the population.

(ii)   Mutation: this depends on the replacement of a single element in the list with another, random selection. This operator is used far less frequently than Crossover.

(iii)   Smearing: this operator produces a new list from one element in the parent list. the list is compiled from items adjacent to the chosen element.

(iv)   Gravitation: this operator was based on the self-organising feature of Kohonen feature maps [10]. Gravitation has the effect of making neighbouring lists similar by seeding part of the populations with adjacent lists (akin to cross-pollination). This operator was specifically introduced to the National Grid problem because of an heuristic pointing to proximal faults causing similar sets of reported alarms.

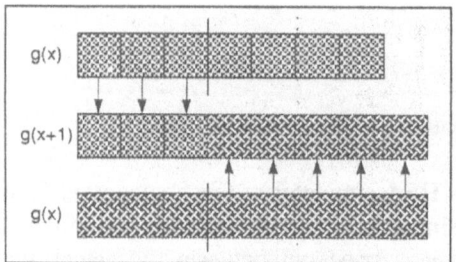

Fig. 3a The crossover

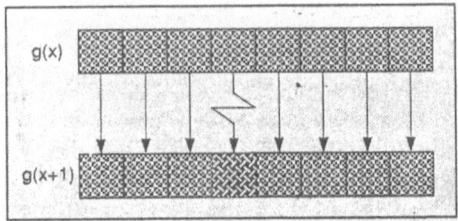

Fig. 3b The mutation operator

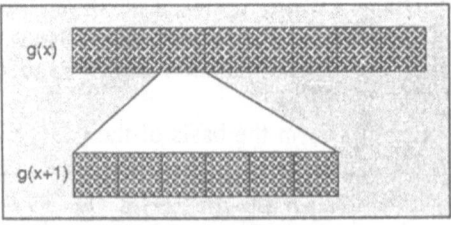

Fig. 3c The smearing operator

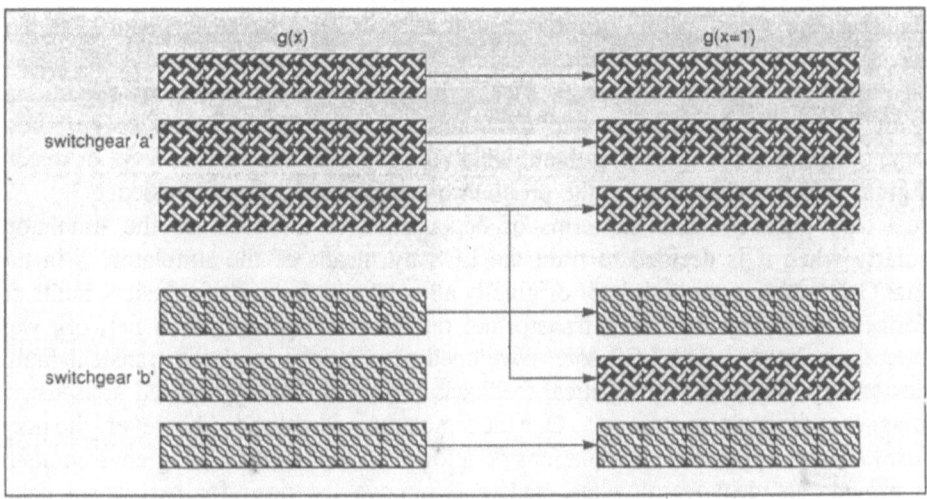

Fig. 3d The gravitation operator

Control of the application of the genetic operators is achieved through selection of the probability of their application, such probabilities are however very problem dependent and apart from initial seeding values, can only be selected through heuristics and subsequently trial and error.

Any new strings produced by the genetic operators will subsequently live or die through their own usefulness.   If they prove to be successful then they in turn will be used as a parent string, whereas if they prove to be unsuccessful they will surely be removed from the population before too long.

Overall, the process of developing hypotheses, which are evaluated and subsequently rule adaption, leads the system to improved performance in fault diagnosis.   the hypothesis processing section is however static in that the processing characteristics do not change, it is merely the low-level associations that do through genetic operations.

Operational results from the LCS when operating on the power transmission network have been reported extensively [8] and are not seen to be appropriate here.   Some important elements from the general nature of the experiments are however worth recording, and these are as follows.

Critical analysis of each fault diagnosis can take up to about 4 seconds, which is due solely to the speed of the AHFA simulator.   This time delay means that operation of the critic, credit assignment to the successful strings etc., and hence genetic adaption must, in practice, be performed in an off-line fashion when little switchgear activity is apparent.   This is in fact rather appropriate in that the system will perform in a uniform way when a flow of switchgear messages are apparent, through a series of faults, and yet will re-configure during quiet periods.

The feedback loop shown in Figure 2 certainly need not be due simply to simulation and computer based critical assessment or matching of strings.   It is quite possible for the control engineer (the expert) to feed back a critical assessment independently from the simulator, the overall nature of the system then however becomes dependent on that engineer's viewpoint and hence before too long the LCS takes on the operational nature of that particular engineer - whereas usually it develops an independent operational nature of its own. Indeed one additional result is that the LCS can suggest new analyses, or new ways of achieving them, which the engineer can either accept or reject, i.e. it gives a fresh view point on the problem based on hard, factual evidence.

Great care must be taken in terms of dependence of the LCS on the simulator, particularly when it is decided to train the LCS by means of the simulator. In the National Grid problem the simulator originally allowed simulations of transient faults on transformers when any fault on a transformer on the actual transmission network was permanently isolated. The LCS soon developed associations implying transient faults on transformers, because of the critical feedback loop, and thereby realised nonsensical fault diagnoses from alarm streams. One the simulator was upgraded however, the poor diagnoses soon disappeared. Simulations do allow self-contained training environments which provide an ideal learning test bed, and remove the need for real-world data, however the LCS used for real-world problems should really receive extra training in its real-world environment, or at least using real-world data, before it is considered to be competent.

In this section the concept of an intelligent expert system has been introduced, i.e. a rule based system that learns its behaviour through experience. The structure and adaptive nature of the system are operator selected however the rules, and hence response, are learnt by the system. Such a system can operate on any data streams in which from a certain sequence of data elements, a conclusion can be drawn. The overall LCS model can therefore be seen, from the outside, as one in which a data stream is entered as input, and a conclusion or result is presented as output, i.e. in the form of a logical model.

3. NEURAL NETWORKS

Over the last few years the implementation of neural networks has been a highly productive research area. The general field of study brings together a number of closely related areas such as connectionism, parallel distributed processing [11] and neural computing as exhibited by biological neural systems. One useful definition of the topic [12] states that it is a study of networks of adaptable nodes which, through a process of learning from task examples, store experiential knowledge and make it available for use.

A key aspect of present day artificial networks is the use of fairly simple processing elements which are models of neurons in the brain. These elements are then connected together in a well structured fashion. One important property of such networks is their ability to infer and induce from possibly incomplete or inexact information, along with a learning capability in particular where real-world data problems are apparent. The network can be taught particular data patterns when they are presented and can subsequently not only recognise these patterns when they occur again, but also recognise similar patterns when presented by means of generalisation. It is such properties as generalisation and learning that distinguish neural networks from conventional algorithm based computers.

3.1   N-TUPLE NETWORKS

N-tuple nets are also known as digital networks in that [3] each node element makes use of binary data via a digital memory chip. Inputs to the neuron are memory address lines whilst the neuron output is the data value stored at that address, as depicted in Fig. 4.

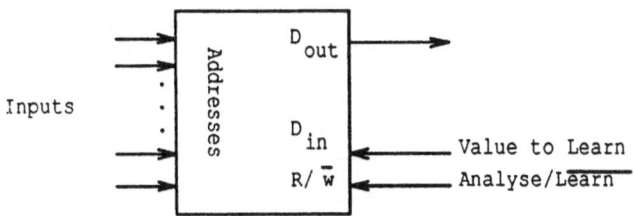

Fig. 4 RAM Neuron

When in learning mode, the pattern or data being taught is placed on the memory address lines and the value is stored. In terms of Fig. 4, the pattern is a set of 0 or 1 binary signals and the Random Access Memory Output is set to a 1 for that particular pattern.   The term tuple is then used to describe how many inputs/address lines are employed, e.g. if 4 lines are employed then this represents a 4-tuple, i.e. 4 bits of data are considered as input to each neuron. When in analysis mode however, a pattern is placed on the neuron input lines, and if it is the same pattern as that learnt, a 1 will be output, if it is a different pattern however, a 0 will be output.

One neuron/RAM chip will merely deal with 4 or 8 pattern data values and if it is considered that a large binary image is to be learnt, then a large number of neurons are required, as shown in Fig. 5. Input patterns are fed into the single layer bank of neurons in a pseudo random fashion, such that all pixels in the image, or data values in the pattern, are applied as an input to a neuron.   The first n-tuple is used to address the first neuron, the second n-tuple the second neuron and so on. For a particular input pattern to be learnt, when this is presented to the neurons in binary form, all of the neuron outputs are set to a 1.

The n-tuple network operates as follows. Once a pattern has been learnt, as described above, if the identical pattern is presented again then all of the neuron outputs will show a 1. In practice, however, it is unlikely that all of the input signals will be exactly the same, because of noise or corruption, e.g. in a visual image, changes in lighting or shadow readily change the input data. So the neuron outputs are summed and if the summation shows 100% then this denotes a perfect image or pattern presented at input, i.e. this is exactly the same pattern as was learnt earlier. If however 90% or more of the outputs are 1 then most likely it is the same input pattern in a practical sense,

234

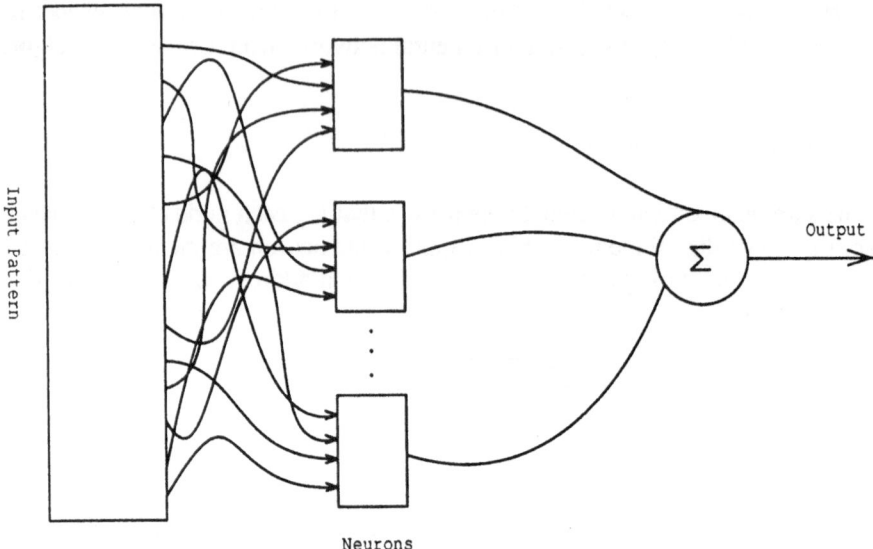

Input Pattern

Output

Σ

Neurons

Fig. 5 n-tuple network

accounting for noise etc. A certainty factor can therefore be placed on the analysis, which is dependent on the particular problem and our own accepted margin of error.

The network can readily be made to generalise, simply by teaching it not one but a number of "similar" input patterns, i.e. set the neuron outputs to one for a number of input patterns. Then when an input pattern is presented it merely has to fit within an image range such that the network recognises it. Obviously it is possible to teach a network many input patterns of a dissimilar nature. The problem with this is that when in analysis mode, the network will recognise just about everything.

In practice, it is best to teach the network basically one input pattern along with any corrupted values or different views of the same pattern. As an example the input pattern could be a binary 0, 1 map from the visual image of a cup and the network is required to recognise the cup when it sees it again. The cup can be shown to the network in a range of upright positions, with the handle providing dissimilarity between images. When subsequently the network is in analysis mode, the cup image can be shown to the network in any upright position and it will immediately be recognised by the network as the cup. Further, any other object of a similar appearance, e.g. a slightly bigger or smaller cup, will also be recognised - this depending on the difference in size and shape and the confidence limits placed on acceptance percentage [4].

If it is desired for an overall network to learn very different data inputs then effectively a number of networks can be employed, each one being termed a discriminator. Each discriminator learns one particular input pattern, such that with a bank of discriminators a range of input patterns can be recognised.

This type of network is particularly suitable for binary data processing and hence it is directly useful on thresholded image data. Further, because RAM chips are used in a single layer fashion, they can operate very rapidly, which means that when such a network is linked to a thresholded camera image, the network can produce decisions at camera frame rate, i.e. 50 frames or decisions per second. This type of network is therefore particularly well suited to high speed visual decision making tasks, such as quality control on a high speed production line [13] or for a robot vision system.

3.2   GENERAL NEURON MODEL

The n-tuple neuron described in the previous section is fairly exceptional in the study of neural networks not only because of its binary nature but also because firstly each of the inputs are afforded equal weighting and secondly the output is a Boolean function of the inputs.

The general development of neural networks has concentrated on an analog form of network [11,14], this being assumed as a more representative model of a biological neuron. One commonly encountered model is a form of the McCullock and Pitts neuron [3], and this is shown in Fig. 6.

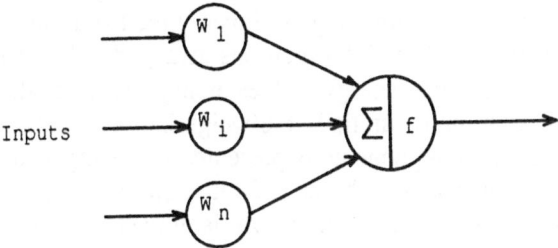

Fig. 6 Basic neuron model

The inputs to the neuron in Fig. 6 are essentially analog in nature and are either from the real/outside world or from other network elements, i.e. from the outputs of neurons in a previous layer. The output of the neuron is then found as a function of the summed weighted strength inputs.

The neuron output can subsequently be directly made use of or it can be processed by a further functional action, for example thresholding, one popular approach is to employ a sigmoid function. So in some networks the neuron outputs are employed directly whereas in others they must be subject to a further filtering action or will be high or low depending on whether a weighted sum exceeds a particular preset value.

In general, neural networks consist of a large number of neurons, either arranged in the form of a single layer, as in the case of the n-tuple network, or in terms of several layers, the outputs from neurons in one layer being connected through to some or all of the neuron inputs in the next layer. The weightings placed on the neuron inputs will then define the overall functional operation of that network, given its fixed structure.

The values of weightings placed on the neuron inputs within a network are selected as the network learns its method of operation and desired response. The learning procedure can use a block of data such that the network learns its function, the weightings are fixed, and then it can operate on further new data in the fashion expected.

Conversely learning can continue throughout the network's lifetime, more usually with a faster learning rate in the initial time frame with a limited amount of adaptability later on. The learning can be supervised or unsupervised, dependent on whether an external critic is present.

3.3   KOHONEN NETWORKS

Kohonen networks [10] are essentially single layer networks, which were originally employed for speech recognition, but are in fact useful for general classification issues. In terms of the speech analysis problem however, a speech signal is looked at in terms of each phoneme, whereby the energy content of the phoneme across the frequency range is considered. The phoneme energy is divided into frequency components by means of bandpass filters, the energy output of each filter acting as an input to the Kohonen network.

An individual phoneme is represented at the network input in terms of its energy input pattern, such that the frequency range 0-250 Hz has an analog value associated with it, as does the range 250-500 Hz etc, up to approximately the band range 3,500 Hz-3,750 Hz. Typically the network could total 400 neurons arranged "physically" in a 20 x 20 matrix, with all 17 analog input energy values being input to all neurons with a weighting associated to each input. Initially the weightings on each neuron are pseudo randomly selected, hence when a phoneme is presented at the input, the neuron outputs will present a variety of values across the 20 x 20 neuron mapping.

One neuron output is selected as that which is linked to a particular phoneme utterance, possibly the neuron whose output is strongest following the initial random weighting.   That neuron and the surrounding neurons then have their weights adjusted, the amount of adjustment being dependent on their distance from the selected neuron, as shown if Fig. 7. By this means the utterance of a particular phoneme is shown to coincide with an area on the Kohonen map, with a centre about the selected neuron, whose outputs are actively high.

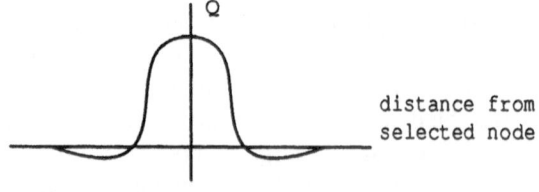

Fig.7 Mexican hat

A second phoneme is then uttered, another neuron selected as centre and the weights of neighbouring neurons are adjusted as before, for the new phoneme. By inputting different phonemes, so different small regions of the map are selected, by adjustment of the neuron input weights, to coincide to particular phonemes.

Having been trained on a particular set of phonemes, the network is now ready, with fixed input weightings, to be used in practice. When a new phoneme is then uttered, this will be recognisable on the Kohonen map by a small area of the neuron outputs being active, when another new phoneme is uttered so another area will be active. Speech signals are made up of a number of phonemes, such that when a word or sentence is uttered this will correspond to a moving area (or ball) on the map as it moves from one phoneme to the next. The uttered speech can, therefore, be picked out in terms of its phoneme representation on the Kohonen map.

Problems in the Kohonen map for speech processing arise when (a) a different speaker inputs to the map, thereby inputting a different energy spectra, (b) the same speaker utters a learnt phoneme in a different way, e.g. after a late night party (!), or (c) a phoneme which has not been learnt is used as input. In each case the map will give a best guess approach, even if this means that several areas of the map are active, to some extent at least. This latter point therefore raises the need for a further level of intelligence or decision making, which can sit on top of the Kohonen network map, deciding on conflicts and making use of phoneme trajectories [15]. Natural language information can then be used to suggest whether the map's solution is sensible at a particular instant, given the limitations and rules of the language.

Kohonen networks have been described here through the example of speech processing. Such networks are however useful for the classification of any data types, such as failure patterns, when containing measurement values which indicate that a known failure type has occurred or is likely to occur. The discussion has also concentrated on a single dimensional network, i.e. one input type is used. It is however quite possible to employ multi-dimensional maps for which numerous input types are used, one example being in production machine monitoring in which information is available on such as product flow, belt speed, tension, motor speed etc. The Kohonen map can then show, by clustering of the network outputs, likely regions on the multi-dimensional map, within which failures can be expected or conversely good machine running can be achieved.

3.4   FEEDFORWARD NETWORKS

Feedforward neural networks are perhaps the most commonly encountered at the present time and in general consist of analog neurons of the form shown in Fig. 6. These neurons are arranged in layers, starting with an input layer, the neurons of which take in real-world signals as input, and the network concludes with an output layer, in which the outputs of its constituent neurons deliver signals to the real-world. The number of neurons in the input layer and output layer is therefore very much dependent on the real-world problem being dealt with. Between the input and output layers exists one or more hidden layers, the number of neurons in each hidden layer being dependent on the complexity of the overall network requirements. Inside the feedforward network, the neuron outputs from one layer are fed as the weighted inputs to the next layer.

A basic neural network non-linear approximator can be described as follows:
The input-output relationship of the ith network node in the jth layer is given by·

$$z_i(j) = \sum_{k=1}^{n_{j-1}} \eta_{ik}(j).x_k(j - 1)\mu_i(j) \tag{1}$$

$$x_i(j) = a\big(z_i(j)\big) \tag{2}$$

in which $\eta_{ik}(j)$ and $\mu_i(j)$ are respectively the connection weights of that node and the node threshold, whereas a(.) is the activation function for that node.

For non-linear modelling, the input to the network is given by x(t), and the activation function is chosen, for example as:

$$a(z) = \frac{1}{1 + \exp(-z)} \tag{3}$$

although several other respectable choices of a(.) can be employed. Further, the output nodes usually do not themselves contain a threshold parameter and in fact often employ linear association functions, i.e.

$$\hat{y}_i = x_i(m) = \sum_{k=1}^{n_{m-1}} \eta_{ik}(m)x_k(m - 1) \tag{4}$$

Weight learning can then be carried out by means of such as a prediction error algorithm employing

$$\epsilon(t) = y(t) - \hat{y}(t) \tag{5}$$

in which y(t) is the actual plant output at time t whereas $\hat{y}(t)$ is the output of the network, dependent on the weights $\eta_{ik}(t)$. These weights are then adjusted to minimize an appropriate function of $\varepsilon(t)$[16].

By this means a feedforward neural network can be used to approximate any non-linear functional relationship between input and output data, the network weights being selected for each particular function. One common form of learning employed for such networks is back propagation [11] in which the error $\varepsilon(t)$ in each layer is respectively minimized by selection of weights, until the input layer is reached.

It has been shown [17] how neural networks can be used extensively for non-linear modelling. In fact when a priori information about a model is held, it is possible to use one neural network to learn this information, with a further neural network acting in parallel to learn initially unknown characteristics.

4. CONCLUSIONS

In this paper a look has been taken at some artificial intelligence approaches to function evaluation, in particular expert systems and neural networks. Such techniques are, when learning characteristics are required, still in their infancy in terms of both conclusive research results and practical usage.

The type of intelligent procedure to be used for a particular problem is highly dependent on the problem to be addressed. For example if a decision is taken by an intelligent system or a conclusion is drawn, and a human operator desires to know why that conclusion was drawn or they wish to verify the results, then an expert, rule based system must be used in order that one can detect what rules fired in the case in question. A neural network technique does not in general afford this luxury, although there are methods of extracting rules from neural networks [18] that can work in some cases.

ACKNOWLEDGEMENTS

The author wishes to thank the UK National Grid Company for their financial support of part of the work described.

REFERENCES

1. I.M. Graham and R.W. Milne (eds.), "Research and development in expert systems", Cambridge University Press, 1991.

2. D.E. Goldberg, "Genetic algorithms in search optimization and machine learning", Addison Wesley, 1989.

3. I. Aleksander (ed.), "Neural computing architectures", North Oxford Academic Publishers, 1989.

4. K. Warwick, G.W. Irwin and K.J. Hunt (eds.), "Neural networks for control and systems", Peter Peregrinus Ltd., 1992.

5. A.C. Stauggard Jr., "Robotics and AI", Prentice Hall Inc., 1987.

6. E. Alba, J.F. Aldana and J.M. Troya, "Genetic algorithms as heuristics for optimizing ANN design", Proc. Int. Conference on Artificial Neural Networks and Genetic Algorithms, Innsbruck, pp. 683-690, 1993.

7. L. Davis (ed.), "Handbook of genetic algorithms", Van Nostrand Reinhold, 1991.

8. L. Kiernan and K. Warwick, "Adaptive alarm processor for fault diagnosis on power transmission networks", Information Systems Engineering, Vol. 2, pp. 25-37, 1993.

9. J.H. Holland, "Adaption in natural and artificial systems", University of Michigan Press, Ann Arbor, 1975.

10. T. Kohonen, "Self organising and associative memory", Springer-Verlag, 1989.

11. D.E. Rumelhart and J.L. McClelland, "Parallel distributed processing", MIT Press, Vols. 1 and 2, 1986.

12. I. Aleksander and H. Morton, "An introduction to neural computing", Chapman and Hall, 1990.

13. R. Gelaky, M.J. Usher and K. Warwick, "The implementation of a low-cost production line inspection system", Computer Aided Engineering Journal, Vol. 7, No. 6, pp. 180-184, December 1990.

14. Y-H. Pao, "Adaptive pattern recognition and neural networks", Addison-Wesley, 1989.

15. P. Wu and K. Warwick, "A structure for neural networks with artificial intelligence", in Research and Development in Expert Systems, I.M. Graham and R.W. Milne (eds.), Cambridge University Press, pp. 104-111, 1991.

16. S. Chen and S.A. Billings, "Neural networks for non-linear dynamic system modelling and identification", Int. Journal of Control, Vol. 56, pp. 319-346, 1992.

17. M.A. Sartori and P.J. Antsaklis, "Implementation of learning control systems using neural networks", IEEE Control Systems Magazine, pp. 49-57, 1992.

18. R.J. Mitchell, J.M. Bishop and W. Low, "Using a genetic algorithm to find the rules of a neural network", Proc. Int. Conference on Artificial Neural Nets and Genetic Algorithms, Innsbruck, pp. 664-669, 1993.

APPLICATIONS OF ARTIFICIAL INTELLIGENCE IN NONDESTRUCTIVE EVALUATION

C.H. CHEN
University of Massachusetts Dartmouth
Electrical and Computer Engineering Dept.
Old Westport Road
N. Dartmouth, MA 02747-2300 USA

ABSTRACT. To meet the increased demand for reliable inspection in complex NDE tasks, artificial intelligence (AI) can provide new and effective approaches to many problems. AI expert systems (often rule-based) offer solutions to problems for which numerical algorithms may not be suitable and for which nonnumeric information is important. Expert systems allow data fusion to utilize information from different NDE sensors, and can deal with uncertainty more effectively. Another advance in AI perhaps is in artificial neural networks which have been increasingly used in NDE. These and other AI capabilities have been applied successfully to several NDE systems which will be discussed in this paper.

1. Introduction

The progress in artificial intelligence (AI) has generated much interest to build intelligent systems for nondestructive evaluation (NDE) of materials. A fully intelligent system has yet to be seen in any application. However the concepts and techniques in AI have been very useful to enhance an NDE system's capability. Artificial intelligence obviously is much needed in various steps to achieve more autonomous systems in NDE. Potential applications of AI in NDE include signal and knowledge representation, expert systems artificial neural networks (hereafter simply called neural networks), fuzzy logic, sensor fusion, and computer vision. The traditional pattern recognition approaches have been used in NDE, but will not be discussed in this paper. For a recent survey of pattern recognition in NDE, see Ref. 1.

Traditional signal processing deals with numerical analysis. Nonnumerical information can be just as important which usually is expressed in relations and rules. To fully characterize the material being tested and the inspection process, it is necessary to provide both numeric and nonnumeric information. The signal and knowledge representation encompasses all techniques to best present such information. The expert systems, especially the rule-based expert systems, are particularly useful to perform specific nondestructive inspection tasks

X. P. V. Malague (ed.), Advances in Signal Processing for Nondestructive Evaluation of Materials, 241–249.
© 1994 *Kluwer Academic Publishers.*

which require expert knowledge expressed in rules. Neural networks represent a different approach which learn from the training data about the characteristics of the system, and then perform classification, estimation and control functions. Fuzzy logic is a way to represent the uncertainty that uses, instead of a probability, a grade of class membership. This concept is useful in NDE as most often a precise probabilistic description of an event is not possible. Sensor fusion refers to ways to merge information from different sensors, which again is important in NDE as often information from a single sensor is not adequate and several sensors must be used. Perhaps computer vision is the most well developed area in AI, and for NDE images, computer vision techniques can be used to extract more information from the data than simple image processing techniques.

The capabilities of AI as presented above can find many different uses in NDE. They will be examined in detail in this paper.

2. Signal and Knowledge Representation

Signal processing allows us to filter out noise and interferences in the data and enhance certain desirable characteristics which help us to extract certain information from the NDE signals (including waveforms and images). A complete representation of the waveforms for example requires high resolution time domain, frequency domain and joint time-frequency domain information. Fig. 1a is an example of a set of ultrasonic signals taken from a thin graphite epoxy specimen. The variation from no defect to defect area is not evident, but is a lot more clear from using the Gabor transform as shown in Fig. 1b, which is also called time-frequency discrete series (TFDS)[2]. Often there are a large volume of data in inspection. Effective signal representation is quite essential to extract essential information from data.

In NDE there is no substitute for better sensors. For example, the synthetic aperture focusing technique (SAFT) ultrasonic inspection system scans an area of material in nuclear reactor and produces a three-dimensional view of the entire volume of material tested. Such data has a lot more information than simple ultrasonic testing signals. A SAFT image interpretation assistant (SIIA) [3] has been developed as a prototype knowledge based system designed to assist in making the system more reliable and efficient. The knowledge related to the SAFT inspection problem is of three types: procedural, inferential and structural. The sequence of activities that must be performed for a SAFT inspection may be represented procedurally. Such procedural knowledge is represented as a sequential computer program. Then there is a set of decisions to be made such as on what type of SAFT processing is to be used. This type of decision making may be performed through logical inference using production rules. The physical objects (and relations between them) that are part of a problem are an example of structural knowledge. A common way to represent such knowledge is through the use of object oriented programming techniques. For complex inspection problems the three types of knowledge described above must be integrated to develop an effective solution. In general human inspectors can be

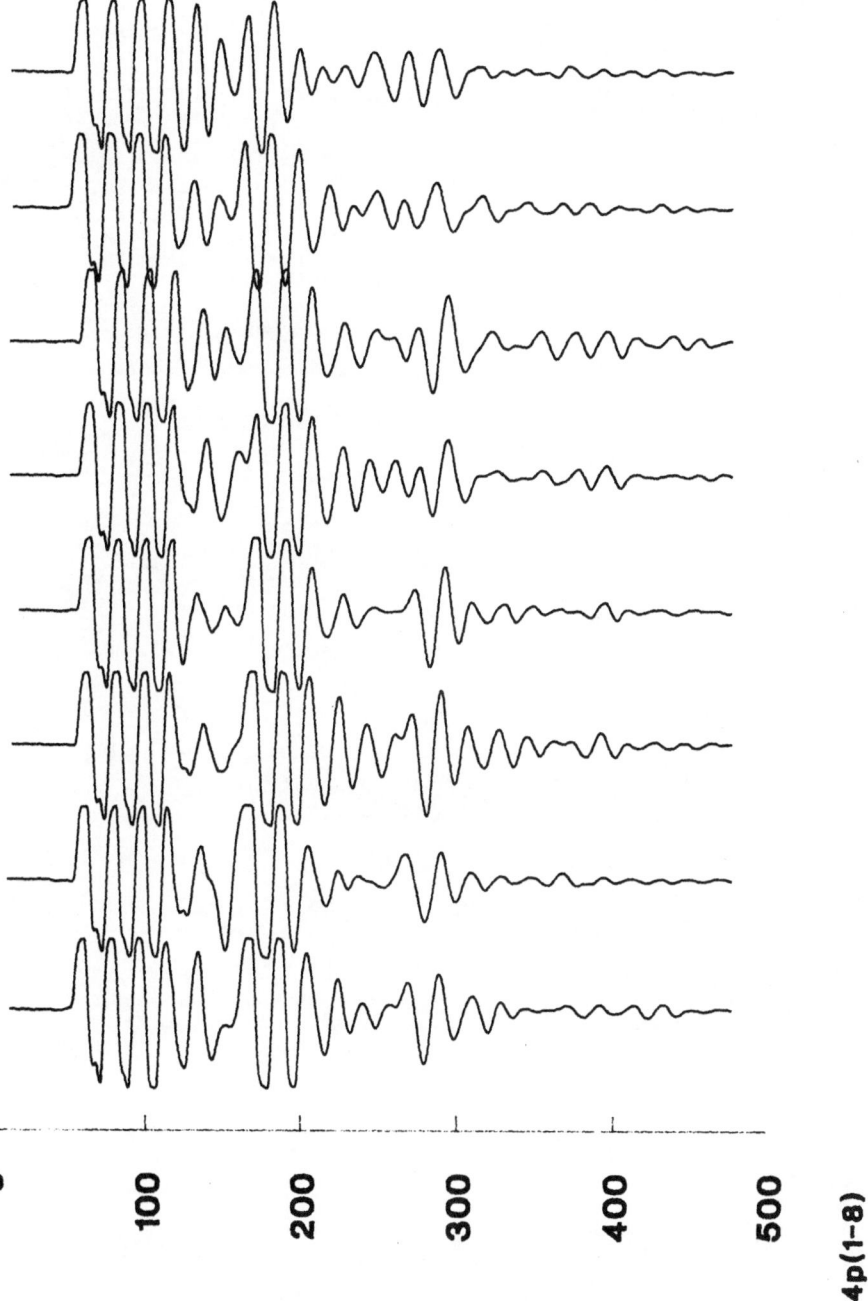

0   100   200   300   400   500

4p(1-8)

Fig. 1a A set of ultrasonic B-scan signals from composite material.

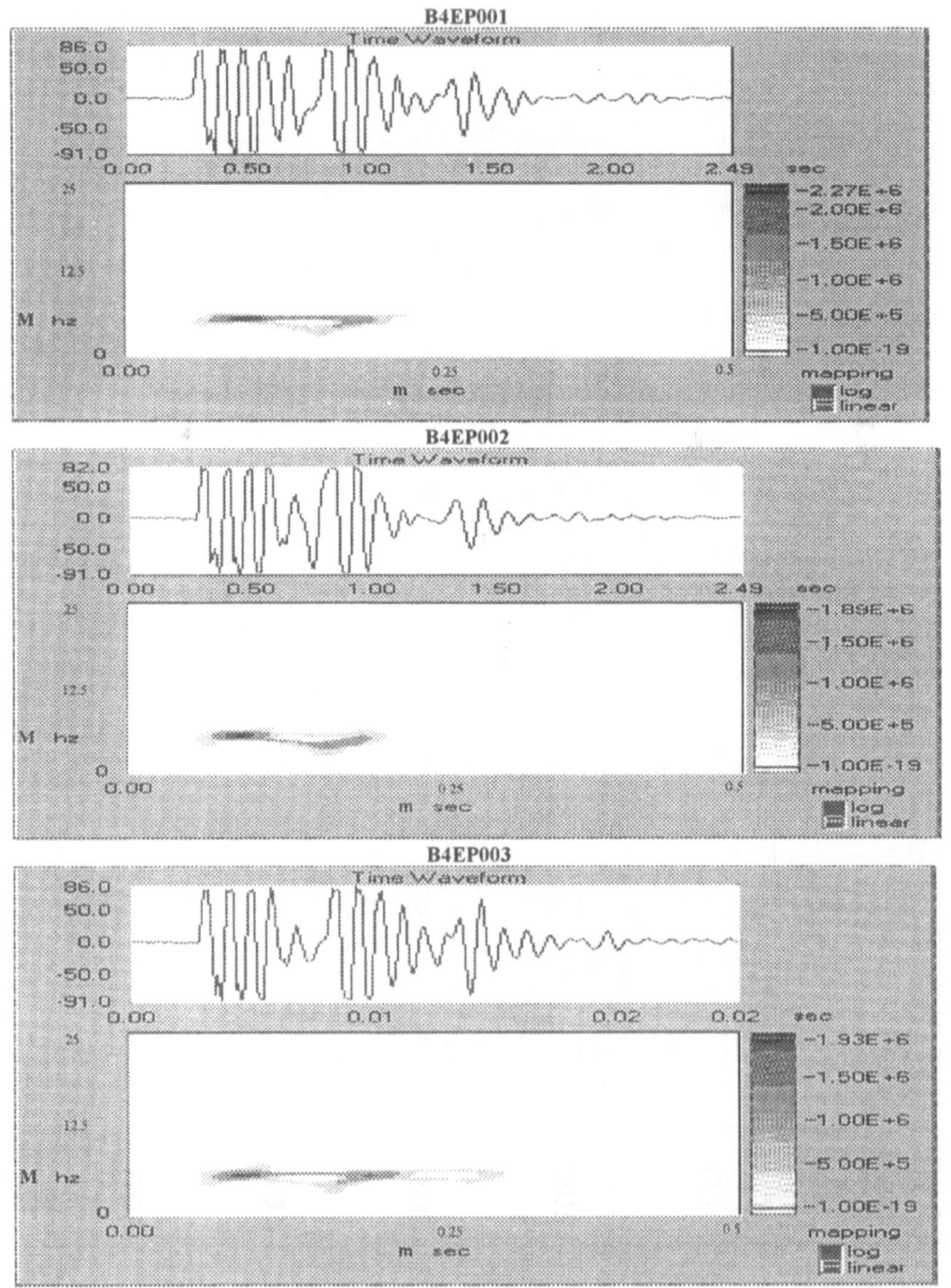

Fig. 1b The Gabor transform of several signals shown in Fig. 1a.

inconsistent. Knowledge based system can assist with consistent solution to many inspection problems.

3. Expert Systems

An expert system is a computing system which implements organized knowledge on some specific area of human expertise. Basically the expert system consists of knowledge (data, rules) and control strategy that determines how and in what order the rules are applied. Rule-based expert systems encode the knowledge into IF....THEN....rules. The use of expert systems, especially the rule-based expert systems [4] is probably the single most important application of AI in NDE. In almost all inspection tasks, there is certain expert knowledge which can be represented by rules. Also there are expert system shells available that may be used for different inspection tasks which differ only by the expert knowledge. A good example of an expert system is the SIIA system mentioned above. Another example is an expert system for automatic analysis of eddy current signals provided by the multifrequency control of steam generator tubing [5]. The TestPro system of Informetrics [6] has certain expert system capability. An expert system that interprets the results of nondestructive testing is much needed.

4. Artificial Neural Networks

Instead of encoding expert knowledge into production rules as in expert systems, neural networks are trained to learn through experience represented by the training data set. Once trained, the network can perform classification, estimation, prediction and control functions. The feedforward types of networks are particularly suitable for classification while the Hopfield types of networks may be more suitable for image reconstruction. Both are very useful for many tasks in NDE including defect classification (see e.g. [7] [8] [9] and characterization, depth estimation [10], eddy current image restoration [11], leak location [12], etc. Empirically it has been proven that neural networks often offer better classification and estimation performance than traditional techniques. For example for the composite material testing signals as shown in Fig. 1, Table 1 is a comparison between the classification of backpropagation trained network and the traditional nearest neighbor decision rule. The neural network performance is clearly superior. It is also noted that signals taken from materials may vary in statistical characteristics, and the number of signals available may be very small. Neural networks can be more suitable in such cases. Neural networks have been integrated into some NDE systems (see e.g. [7]) though a dedicated neural network based NDE system is yet to be seen for real-time operation. It may not be long however before such system is available in view of the rapid progress in neural network software and hardware development.

Nondestructive Evaluation Results for Composite Materials

	B3EP		B3EN		B4EP		B4EN	
No.	NNDR	BPNN	NNDR	BPNN	NNDR	BPNN	NNDR	BPNN
1	55.00	78.75	62.50	77.50	48.75	68.75	51.25	72.50
2	67.50	77.50	60.00	77.50	50.00	66.25	62.50	75.00
3	66.25	78.75	51.25	77.50	58.75	70.00	56.25	73.75
4	75.00	80.00	57.50	76.25	60.00	67.50	53.75	75.00
5	67.50	78.75	63.75	77.50	47.50	71.25	58.75	73.75
6	63.75	78.75	66.25	78.75	46.25	70.00	47.50	75.00
7	56.25	78.75	70.00	78.75	61.25	68.75	53.75	73.75
8	63.75	76.25	68.75	77.50	53.75	70.00	57.50	71.25
9	68.75	82.50	67.50	77.50	43.75	70.00	52.50	73.75
10	60.00	77.50	60.00	78.75	55.00	67.50	56.25	73.75
11	75.00	78.75	60.00	76.25	45.00	72.50	51.25	75.00
12	68.75	78.75	66.25	80.00	51.25	66.25	55.00	71.25
13	68.75	81.25	66.25	77.50	47.50	68.75	58.75	76.25
14	75.00	80.00	68.75	75.00	42.50	65.00	61.25	72.50
15	62.50	81.25	60.00	77.50	56.25	68.75	66.25	76.25
16	60.00	78.75	62.50	78.75	48.75	68.75	63.75	76.25
17	68.75	78.75	65.00	80.00	48.75	70.00	51.25	73.75
18	52.50	77.50	56.25	77.50	65.00	65.00	72.50	71.25
19	65.00	80.00	63.75	77.50	57.50	71.25	58.75	68.75
20	65.00	77.50	60.00	69.00	51.25	68.75	55.00	75.00
average:	65.25	79.00	62.81	77.33	51.94	68.75	57.19	73.69
sta-dev:	6.17	1.46	4.64	2.23	6.09	1.98	5.78	1.92

Table 1 A comparison of classification performance between the
backpropagation trained neural network and the nearest
neighbor decision rule for the B-scan signals of the
composite material considered in Fig. 1a.

5. Fuzzy Logic

In dealing with uncertainty, the use of probability theory is still the best numerical approach from theoretical point of view. This area of AI is called numerical reasoning or Bayesian inferences [13] [14] [15]. In a less restrictive form the Dempster-Shafer theory [16] [17] has been quite popular. However fuzzy logic not only has sound theoretical basis but also has been useful in practice to design products more human and user friendly (see e.g. [18]). A large percentage of all microcontroller chips will incorporate fuzzy logic in their design in a few years because of performance improvements over conventional microcontrollers. Fuzzy logic undoubtedly will be used to improve existing NDE systems in different ways.

6. Sensor Fusion

There are a number of sensors in NDE. For some inspection problems, more than one sensor may be needed to provide us enough information to make a sound judgment. The information gathered in different sensors must be integrated in order to derive the best NDE solution for a given task. The sensor fusion problem has been very popular in other areas such as radar sonar and optical systems. Both expert systems and neural networks can perform data or sensor fusion. However the strategy for fusion may depend on the problem considered. For example in nondestructive inspection of fiber reinforced composite material, a simple fusion technique [19] was used that takes the X-ray and C-scan images and the location distribution histogram of events from acoustic emission to separate the real defects from the spurious components resulting from the image segmentation, so that a more complete defect map of the specimen can be obtained.

7. Computer Vision

Images taken from radiography, ultrasonic imaging and visual inspection can be digitized and processed by computer. Image processing and low-level computer vision provide many effective procedures to enhance and segment the images, and to extract useful feature. Another major application is in computed tomography for which many computation intensive algorithms have been developed (see e.g. [20]). A number of publications are available in both areas. In using middle and high level computer vision for which computer interpretation is a major objective (see e.g. [21]), there has been very little progress in the NDE community however.

8. Concluding Remarks

The application of AI concepts and techniques in NDE already has provided a lot of improvements in NDE capabilities. The progress is fairly slow as compared to many other disciplines, largely because the computer technology is yet to be fully integrated into many NDE systems and components. On the other hand, the process of building an intelligent and autonomous NDE system can be very long. Any incremental steps of improving the existing system using AI thus can be significant and we are looking for continued incremental progress that eventually will result in a highly intelligent NDE system.

REFERENCES

1. C.H. Chen, "Pattern recognition in nondestructive evaluation of materials" in "Handbook of Pattern Recognition and Computer Vision", eds. C.H. Chen, L.F. Pau and P.S.P. Wang; World Scientific Publishing 1993.

2. S. Qian and D. Chen, "Discrete Gabor transform", IEEE Trans. on Signal Processing, July 1993.

3. R.B. Melton, "Knowledge based systems in nondestructive evaluation", in "Signal Processing and Pattern Recognition in Nondestructive Evaluation of Materials", ed. C.H. Chen; Springer-Verlag, 1988.

4. G.F. Luger and W.A. Stubblefield, "Rule-based expert systems", in "Computer Engineering Handbook", ed. C.H. Chen; McGraw-Hill 1992.

5. B. Benoist, et. al., "Increasing reliability of defect characterization on SG tubings using a combination of signal processing and expert system" in Proc. of 12th World Conference on Non-Destructive Testing, eds. J. Boogaard and G.M. van Dijk, Elsevier Science Publishers, 1989.

6. TestPro AI enhancement brochure from Informetrics, 1987.

7. C.H. Chen, "High resolution ultrasonic spectroscopy system for NDE", Proc. of NASA Tech 2001 Conference, Dec. 1991.

8. L.M. Brown, J.S. Lin and R.W. Newman, "Signature classification development system", ASME, June 1993.

9. L. Udpa and S.S. Udpa, "Eddy current defect characterization using neural networks", Material Evaluation, vol. 48, pp. 342-353, March 1990.

10. M. Takadoya, M. Kitahara, et. al., "Depth estimation of inclined surface-breaking crack by a neural network", presented at the Review of QNDE Meeting, Brunswick, ME, Aug. 4, 1993.

11. T. Stepinski, "Restoration of eddy current images with neural networks", presented at the Review of QNDE Meeting, Brunswick, MD, Aug. 4, 1993.

12. M.A. Friesel, et.al., "Acoustic emissions applications on the NASA space station", in "Review of Progress in QNDE", vol. 11A, eds. D.O. Thompson and D.E. Chimenti, Plenum Press, pp. 725-732, 1992.

13. J. Pearl, "Probabilistic Reasoning in Intelligent Systems", Morgan Kaufman, 1988.

14. P. Cheeseman, "An inquiry into computer understanding", Computational intelligence, vol. 4, 1988.

15. R.E. Neapolitan, "Probabilistic Reasoning in Expert Systems, Theory and Algorithms", Wiley 1990.

16. A.P. Dempster, "A generalization of Bayesian inference", J. Royal Stat. Society, Sec. B, vol. 30, 1968.

17. G. Shafer, "A Mathematical Theory of Evidence", Princeton Univ. Press 1976.

18. J. R. Adamski, "Fuzzy logic for appliance control", Electronic Progress, vol. 32, no. 1, Raytheon Co., 1992.

19. A.K. Jain, et. al., "Multi-sensor fusion for nondestructive inspection of fiber reinforced composite materials", Proc. of the American Society for Composites Conference, Albany, NY, Oct. 1991.

20. K.C. Tam, "Limited-angle image reconstruction in non-destructive evaluation", in "Signal Processing and Pattern Recognition in Nondestructive Evaluation of Materials", ed. C.H. Chen; Springer-Verlag, 1988.

21. G. Garibotto and S. Masciangelo, "Computer vision principles and applications", in "Computer Engineering Handbook", ed. C.H. Chen; McGraw-Hill, 1992.

FUZZY MIN-MAX NEURAL NETWORK BASED CLASSIFICATION OF UNDERWATER LAYERED MEDIA DUE TO ATTENUATION EFFECTS.

FERIAL EL-HAWARY AND SAEED SETAYESHI

Technical University of Nova Scotia
P.O.Box 1000, Halifax, Nova Scotia,
B3J 2X4, Canada.

ABSTRACT. A feed-forward fuzzy dual-connected neural network classifier that uses min-max amplitude ranges to define classes is designed and evaluated for underwater layered media recognition based on a computer simulation of synthetic data. A supervised fuzzy min-max learning rule updates the weights corresponding to the input data sets used for training. A simple nonlinear structure modeling underwater layered media response to acoustic inputs accounting for time delay in layers and the exponential decay of the output signals' amplitude due to attenuation effects is employed for simulations used to test the classifier. Following the training stage, the classifier is tested with different test data sets. The results suggest that the application of fuzzy min-max neural networks in pattern recognition will enable automatic classification of the layered media with reasonable accuracy.

1. Introduction

Sound propagation in underwater layered media depends on the structure and properties of the layers traversed. A pressure pulse introduced into the water column undergoes successive transmissions and reflections as it travels towards the seabed [1-2]. Information collected by hydrophones from reflected waves are used to classify, and identify the subsurface layers. The propagation pattern for the case of six layers is shown in Figure (1).

In sonar applications it is required to evaluate attenuation as a function of frequency with good accuracy. A model of underwater layered media accounting for attenuation is used as the basis to classify and identify the layered media structure. F. El-Hawary points out in [3] that accounting for attenuation effects adds more complexity to the estimation process than is the case neglecting attenuation.

The recent work of [4] treats classifying sonar signals in underwater layered media by applying a back-propagation algorithm in a multilayer perceptron-based neural network [5-6] to identify the media structure. The proposed network identified the layers with a success rate of close to 90 %. In [7], an associative memory neural network [8] is evaluated in solving the problem. The network is trained using attenuation factors. The network

X. P. V. Malague (ed.), Advances in Signal Processing for Nondestructive Evaluation of Materials, 251-267.

252

classifies the minimum and maximum points of the amplitude of peaks to identify the layers with good success. This work is part of a recently completed research [9].

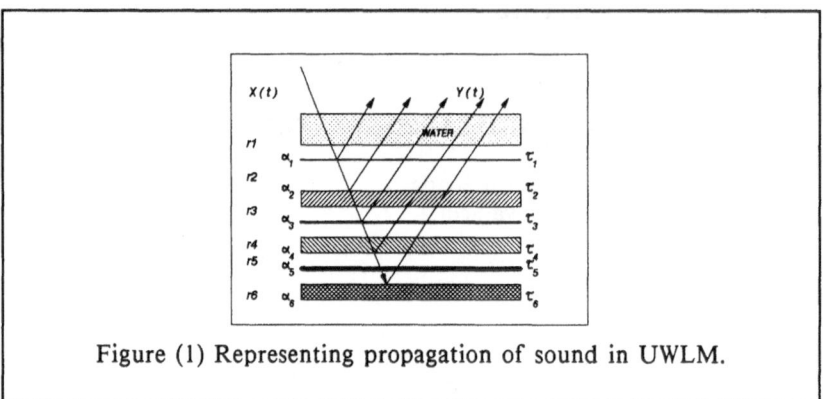

Figure (1) Representing propagation of sound in UWLM.

Zadeh [10] introduced the idea of fuzzy sets in 1965, to allow representing and manipulating inexact concepts and data that is not precise, but rather vague. This includes, for example, linguistic constructs such as "many," "few," "often," and "sometimes" which mean different things depending on the situation and the observer.

The synergism of neural networks and fuzzy sets seems natural. The former compare an input pattern with a set of stored code, or reference vectors. The closer the input matches the data stored in the network, the higher the output values of the corresponding processing element [11].

Assume that subsets of the processing elements in a neural network represent separate categories, such as the output processing elements of an adaptive resonance network. The closer an input pattern A fits within a category, the higher the corresponding value of that processing element. By viewing each category as a fuzzy set and identifying the processing element transfer function with the membership function, a direct relationship between fuzzy sets and the pattern classification is realized.

Data sets tend to have inherent ambiguities, and there never seem to be enough data to describe completely all the possible patterns that will be seen in a real-world applications. By employing fuzzy sets as pattern classifiers, it is possible to describe the degree to which a pattern belongs to one class or another. This relationship was realized very early in the development of fuzzy sets and is the basis for fuzzy pattern classifiers.

In this paper, the concept of feed-forward fuzzy dual-connected neural network classifier that uses min-max amplitude ranges to define classes [11-13] is examined. A classifier is designed and evaluated for underwater layered media recognition based on a computer simulation of synthetic data. Following the training stage, the classifier is tested with different test data sets. In this paper the relationship between fuzzy sets and pattern classification, the fuzzy min-max neural network model implementation, learning and recalling algorithms, and results of the modeling are given.

2. Attenuation Model of UWLM

Accounting for primary reflections only, the return pressure signal $y(t)$ is expressed as the sum of delayed replicas of the source signal $x(t)$

$$y(t) = \sum_{i=1}^{N} A_i x(t - \tau_i) \qquad (1)$$

N is the number of layers, and the A_i are the medium amplitude scale factors whose values depend on each layer's thickness and density, and are also frequency dependent due to attenuation effects.

Attenuation causes the amplitude of a plane sinusoidal wave in a linear medium to fall off exponentially with the distance travelled. Thus, during propagation in one layer at a given frequency f, the amplitude of the sound pressure wave at distance d from the source is given by:

$$P = P_0 e^{-\alpha(f) \cdot d} \qquad (2)$$

where P is the sound pressure wave amplitude,
   P_0 is the sound pressure wave amplitude at $d = 0$,
   $\alpha(f)$ is the attenuation rate, and
   d is the distance travelled [14-16].

The damping factor D_i is used. It depends on the thickness of the layer and is proportional to the attenuation ratio $\alpha(f)$. The medium amplitude scale factors A_i in terms of frequency and damping ratio are expressed as:

$$A_i = \alpha_i e^{-2\pi f_i D_i} \qquad (3)$$

where α_i are the frequency independent amplitude scale factors which related to the layers' reflection coefficients r_i.

Substituting (3) in (1) results in:

$$y(t) = \sum_{i=1}^{N} \alpha_i e^{-2\pi f_i D_i} x(t - \tau_i) \qquad (4)$$

where τ_i are time delays for propagation in layers.

3. Fuzzy Sets and Neural Networks

A fuzzy set A is a subset of the universe of discourse, X, that ranges from no membership (the empty set) to full membership. A membership function, $\mu_A(X)$, is used to describe the degree to which the object x belongs to the set A. A value $\mu_A(x) = 0$ represents no membership, and $\mu_A(x) = 1$ represents full membership. A fuzzy set A admits partial membership, and is represented by the ordered pair $A = \{x, \mu_A(x)\}$, with $x \in X$ & $0 \leq \mu_A(x) \leq 1$.

In the same manner as in traditional set theory, there is a whole set of operations identified with fuzzy sets to allow their manipulation. Basic operations of interest in our work are those of union, intersection, and complementation. Assuming that $A \in X$ & $B \in X$, the following are the original definitions:

For union:

$$A \cup B = C \ni \mu_{A \cup B}(x) = \max(\mu_A(x), \mu_B(x)) \quad \forall x \in X, \qquad (5)$$

254

For intersection:

$$A \cap B = C \ni \mu_{A \cap B}(x) = \min(\mu_A(x), \mu_B(x)) \quad \forall x \in X, \tag{6}$$

For complement:

$$complement(A) = \overline{A} \ni \mu_{\overline{A}}(x) = 1 - \mu_A(x) \quad \forall x \in X. \tag{7}$$

It is also useful in many instances to identify the size of a fuzzy set. The sigma-count, the sum of all the set membership values, provides one measure. Given the fuzzy set A, the sigma-count of A in defined as:

$$\sum_{count}(A) = \sum_{x \in X} \mu_A(x) \quad \forall x \in X \tag{8}$$

It is important to note that the above definitions constitute the original way in which crisp set theoretic concepts were extended to fuzzy sets. There many alternative ways to define the basic operations mentioned above [17-18].

3.1. FUZZY PATTERNS

The membership values of an n-element ordered fuzzy set are used to construct an n-dimensional fuzzy pattern. The fuzzy pattern for the ordered fuzzy set $Y = (y_1, y_2, ..., y_n)$ is considered to be a vector constructed from the membership values of each of the n corresponding fuzzy set elements.

A histogram with the bars replaced by a point at the maximum value of each bar is called a histograph and is the preferred means of representing higher dimensional data, since it makes the visualization of the data much easier. Here the fuzzy set elements are represented as points along the abscissa of the histograph and the degree of membership for each set element along the ordinate of the histograph [19].

3.2. NEURAL NETWORKS

A neural or a connectionist network is a group of simple, highly interconnected processing units acting collectively to perform computable functions [5]. A network consists of a number of simple processors called nodes or units, each connected to many others through weighted links. A unit in a neural network contains a numerical activation value which it sends to other units in the network. Activating a particular unit affects the activation of other units depending on the strength of the connection between the two, and information processing occurs as these activities flow to other units via the weighted interconnections. The values of designated input units can be set to correspond to the presence or absence of features in the problem domain such as sensor reading or visual scene descriptions. Designated output units contain activity values which may again represent features in an environment such as physical objects in a scene.

A neural network is a distributed processing system that utilizes only local information to carry out useful information processing tasks. Local processing, in the neural network sense, means that all the information a processing element needs to compute an output value is available at the adjacent connections.

Neural networks can exhibit properties of associative memory, recognition, learning and computation. Other special and very useful features of these networks include an ability to perform general mappings, a resistance to noisy data, and massive parallelism [20].

In supervised learning, we try to adapt an artificial neural network so that its actual outputs B come close to the class of the input pattern which contains n features. The goal is to adapt the connections of the network so that it performs well for patterns from outside the training set. There are two ingredients involved in supervised learning [21]:

1- defining the connections and equations for a network with inputs a_k and outputs b_j.

2- the relation between the inputs and outputs must depend on a set of weights (parameters) w so as to make the actual outputs b_j close to the input class.

The simplest form of such a learning procedure involves a two layered network whose input layer contains k units, and a layer that contains j units at the output. In the network, the total input, y_j to units j at the output layer is a linear function of the outputs, a_k of the units that are connected to k and of the weights, w_{jk} on these connections

$$y_j = \sum_k a_k w_{jk} \tag{9}$$

A unit has a real-valued output, b_j which is a nonlinear function (sigmoid function) of its total input

$$b_j = \frac{1}{1 + e^{-y_j}} \tag{10}$$

The weights may be adjusted using weight correction rules.

4. Fuzzy Min-Max Pattern Classifiers

Given an input pattern, the object of a pattern classifier is to establish the class to which this pattern belongs. The classification problem requires defining boundaries as shown in Figure (2). Traditional pattern classification is usually performed by collecting data, extracting features, normalizing the features, and then attempting to find a set of class boundaries that minimizes the classification error. In a majority of classifiers, the decision boundaries are determined by attempting to minimize the intraclass distance, x_1 and x_2, and maximize the interclass distance, y.

Adaptive non-parametric neural network classifiers work well for many real world problems. These classifiers frequently provide reduced error rates when compared to more conventional Bayesian approaches and they also provide selection of differing characteristics.

Applying a pattern classifier requires first selecting features that must be tailored separately for each problem domain. The selected features should contain information required to distinguish between classes, be insensitive to irrelevant variability in the input, and also be limited in number to permit efficient computation of discriminating functions and to limit the amount of training data required.

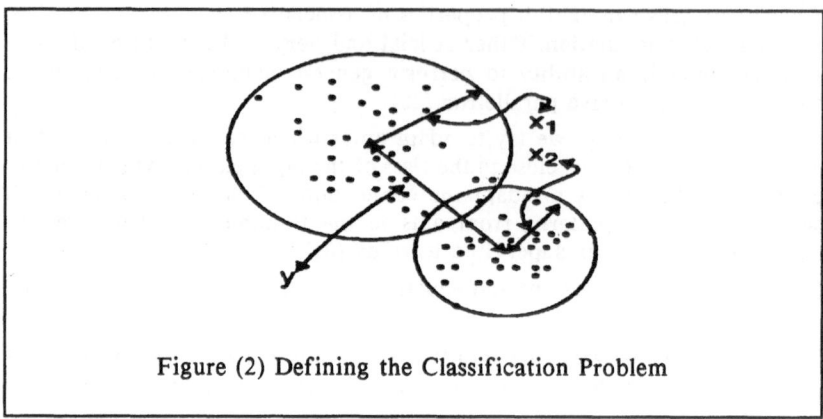

Figure (2) Defining the Classification Problem

Fuzzy pattern classification offers an appealing approach, by seeking to find the subset of the universe of discourse that best represents a given pattern class. Recall that conventional techniques to establish the boundaries find the mean and the variance of each class as shown in Figure (3-a). This approach results in a set of n-dimensional hyperspheres (or hyper-ellipsoids), one for each class.

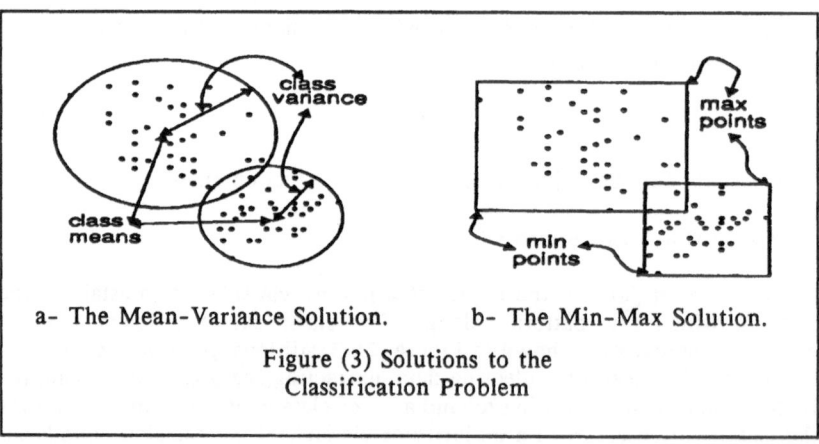

a- The Mean-Variance Solution.   b- The Min-Max Solution.

Figure (3) Solutions to the
Classification Problem

Fuzzy min-max classification works from a different perspective. Class boundaries are hyperboxes (boxes pack into cubes much better than spheres and ellipsoids). The min and max points of a hyperbox are all that are required to completely define its size and shape. Hence, fuzzy min-max classification searches for the min point and max point for each class as shown in Figure (3-b).

Each n-dimensional hyperbox represents a class in the n-dimensional hypercube. The hyperboxes, like the hyperspheres, can overlap. Traditional pattern classification seeks to minimize the misclassification between classes. The end result is a hard decision that represents a willingness to live with a fixed amount of misclassification. In fuzzy min-max classification, if a class overlaps, the data point that falls within the overlap belongs to both classes.

The motivation is that there are many areas in pattern classification where patterns are not strictly of one class or another, rather they are a mixture of two classes. The overlap between the classes represents that mixture.

4.1. MEASURING THE DEGREE OF CLASSIFICATION

Fuzzy min-max classes are represented by min and max points as shown in Figure (4-a), using the histograph representation. As shown in Figure (4-b), if an input pattern is entirely contained within the max and min, then the input is a member of the class with degree 1. Figure (4-c) shows that the input is completely outside the min-max boundaries, then the input is a member of the class with degree 0. If as shown in Figure (4-d), the input pattern is neither completely within nor completely outside of the min-max class boundaries, then the degree to which the input is a member of the class lies between 0 and 1.

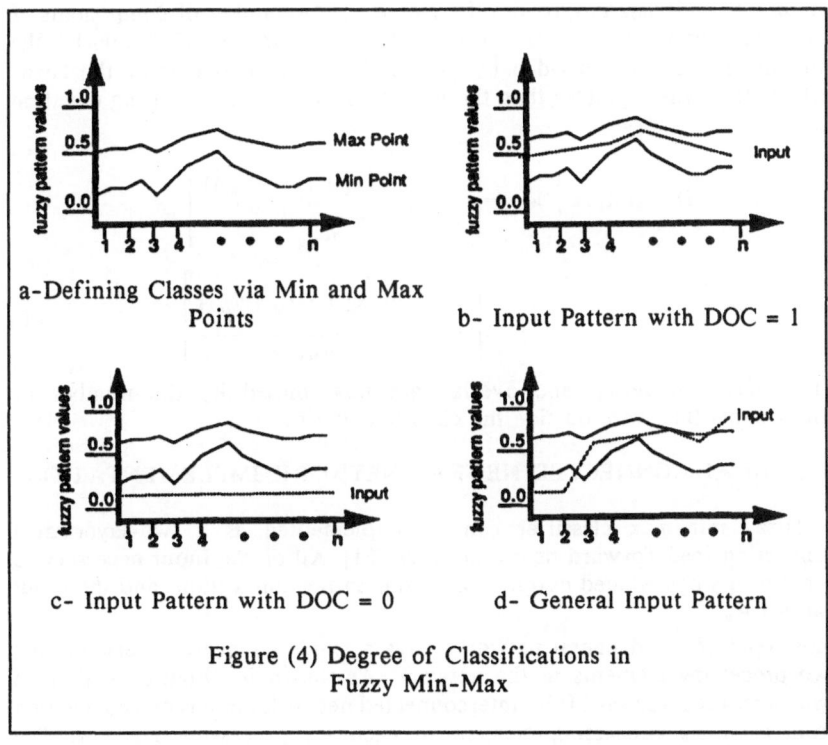

a-Defining Classes via Min and Max Points

b- Input Pattern with DOC = 1

c- Input Pattern with DOC = 0

d- General Input Pattern

Figure (4) Degree of Classifications in Fuzzy Min-Max

The degree of classification (DOC) is a measure of how well the kth input pattern, A_k, falls between the min point of the jth class, V_j, and the max point of the same class, W_j. Assume that:

$A_k = \{a_{ki}\}$ i= 1,2,....,n

$V_j = \{v_{ji}\}$ i= 1,2,....,n

$W_j = \{w_{ji}\}$ i= 1,2,....,n

Three measure have been proposed to describe the degree of classification for min-max classes [19]:

1- subsethood/supersethood,

2- average underlap/overlap, and

3- biased underlap/overlap.

The disadvantage of the subsethood/supersethood measure is that a large sized input pattern, i.e. with a large n, reduces the influence of the sum of the violations.

To overcome the bias created by the input pattern's size, the second classification measure is introduced by normalizing by the pattern dimensionality. For large dimension patterns, the average difference between the input and the class boundaries becomes very small. Hence the relative difference between the classification responses becomes very small. The resulting DOC values are close to unity most of the time, making it difficult to discriminate class membership.

The biased underlap/overlap measure is a compromise that calculates the amount of underlap and divides that value by the number of components of the fuzzy pattern that were less than the min point (n_{sub}). Similarly, the amount of overlap is divided by the number of components of the fuzzy pattern that where greater than the max point (n_{sup}). The resulting equation is:

$$DOC_3(A_k, V_j, W_j) = \left[1 - \frac{\sum_{i=1}^{n} \max[0, (v_{ji} - a_{ki})]}{n_{sub}} \right] x$$

$$\left[1 - \frac{\sum_{i=1}^{n} \max[0, (a_{ki} - w_{ji})]}{n_{sup}} \right] \tag{11}$$

The average underlap and overlap are now biased by the number of components that were outside the class boundaries.

4.2. DUAL CONNECTED NEURAL NETWORK IMPLEMENTATION

A fuzzy min-max classifier can be implemented as a two-layer dual connection feed-forward neural network [21]. All of the input necessary to compute a single valued output is available in the connections and the input pattern A_k.

Most feed-forward neural networks have a single connection between any two processing elements in the network. As shown in Figure (5-a), in a two-layer feed-forward fully interconnected network there is one connection from each F_A processing element to each F_B processing element. The connection strength between the ith F_A processing element to the jth F_B Processing element is w_{ji}.

Dual connection neural networks involve two connections between the processing elements. Figure (5-b) shows a two-layer feed-forward fully interconnected network with two connections from each F_A processing element to each F_B processing element. The two connections between the ith F_A and the jth F_B processing elements are v_{ji} and w_{ji}. These dual connections are used to store the min and max points of a class.

In this structure, the min point for a class is represented by the values on the V_j connections (min connections for the jth class), $V_j = v_{jl}, v_{j2}, ..., v_{jn}$, where v_{ji}, is the strength of connection from the ith F_A processing element to the jth F_B processing element in a two layer feed-forward neural network. The max point for a class is represented by the values on the W_j connections (max connections for the jth class), $W_j = w_{j1}, w_{j2}, ..., w_{jn}$, where w_{ji} is the strength of the connection from the ith F_A processing element to the jth F_B processing element in a two-layer feed-forward neural network.

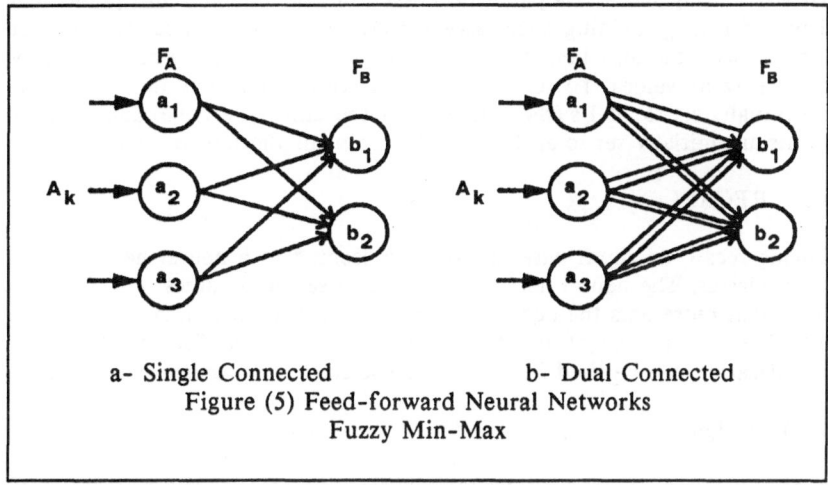

a- Single Connected   b- Dual Connected
Figure (5) Feed-forward Neural Networks
Fuzzy Min-Max

4.3.   ADAPTATION

The fuzzy min-max classifier employs supervised learning. Each fuzzy pattern is associated with a class. Only the min and max points of that class are adjusted during the presentation of that fuzzy pattern. The learning procedure utilizes the fuzzy union operation to adjust the max point and the fuzzy intersection operation to adjust the min point. Therefore, the min point represents the set of lowest values among all the fuzzy patterns associated with a given class. Similarly, the max point represents the set of highest values among all the fuzzy patterns associated with a given class. The min connections are adjusted using:

$$v_{ji}^{new} = v_{ji}^{old} \cap a_{ki} = \min(v_{ji}^{old}, a_{ki}) \qquad (12)$$

for all $i = 1, 2, ..., n$, where j is the class associated with the fuzzy input pattern $A_k = (a_{k1}, a_{k2}, ..., a_{kn})$, $k = 1, 2, ..., m$. The max connections are adjusted according to:

$$w_{ji}^{new} = w_{ji}^{old} \cup a_{ki} = \max(v_{ji}^{old}, a_{ki}) \qquad (13)$$

for all $i = 1, 2, ..., n$ and j is the class associated with the fuzzy input pattern A_k.

It is significant to observe that the adaptation does not require several presentations of the input patterns, rather the learning is immediate. Moreover, the learning equations are able to learn on-line. New data can be added to the system without complete retraining. These qualities are advantageous in applications where it is difficult to maintain a data set for retraining and a very short training time is required. The fuzzy min-max neural network classifier is powerful for pattern recognition, and dynamic modeling.

4.4. CLASS INITIALIZATION

Initially each class is empty. During the presentation of the first pattern for a class, the min and max points should assume the same values as the input pattern. During successive pattern presentations the min and max points will then separate (providing successive patterns are not identical to the first pattern) with the max boundary increasing its values and the min boundary decreasing its values. To accomplish this learning behavior, the min points are initially set to all 1's (the full point of the unit hypercube) and the max points are initially set to all 0's (the null point of the unit hypercube).

4.5. RECALL

During recall, an input pattern, A_k, is presented to a neural network that has p classes. The neural network produces a set of p values, one for each class, that represents the degree to which the input pattern, A_k, fits within the class j. Any one of the three equations described for the degree of classification can be used for recall. The selection is problem dependant.

5. Test Signals

A computer simulation program was developed to realize the mechanism of acoustic signals propagating through the underwater layered media accounting for attenuation effects. A seven layer underwater structure was assumed based on data given in [22]. The layers' composition is shown in Table (1), and the corresponding model parameters are listed in Table (2).

Table (1)
Layers' Composition

Layer #	Sand %	Silt %	Clay %	Sediment Identification
Layer 1	65	21.6	13.4	Silty Sand
Layer 2	88.1	6.3	7.1	Fine Sand
Layer 3	83.9	13	2.9	Very Fine Sand
Layer 4	32.6	41.2	26.1	Sand-Silt-Clayey
Layer 5	99.8	0.2	0.00	Medium Sand
Layer 6	100	0.00	0.00	Coarse Sand
Layer 7	6.1	59.3	34.8	Clayey Silt

Figure (6) displays a synthetic acoustic return signal. The waveform indicates that the amplitude of the output signal is affected by the attenuation parameter of each layer traversed. Portions of the figure are labelled with layers' identities which are considered as distinct classes.

Successive maxima and minima of the signal of Figure (6) were extracted. The result is used as a discretized form of the return signal displayed in Figure (7). In subsequent work, it is assumed that the minimum and maximum points of a pattern are all that are required to completely define its class.

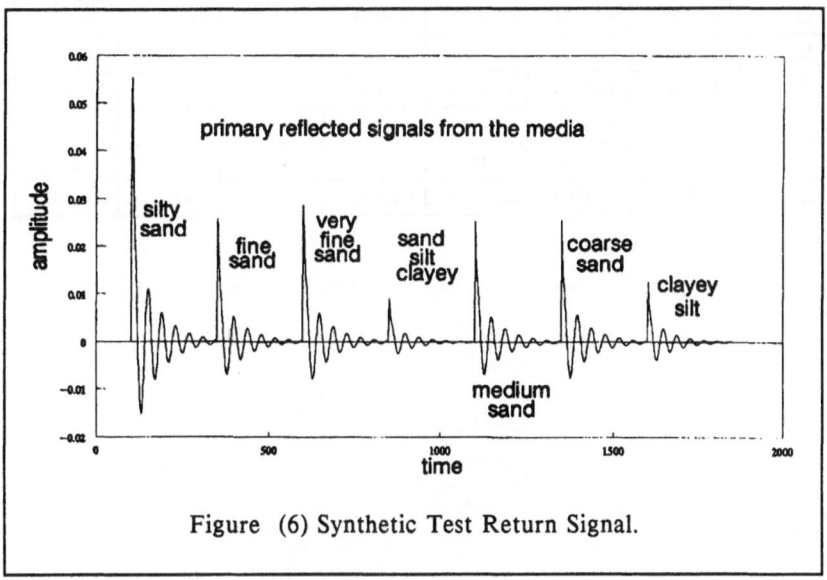

Figure (6) Synthetic Test Return Signal.

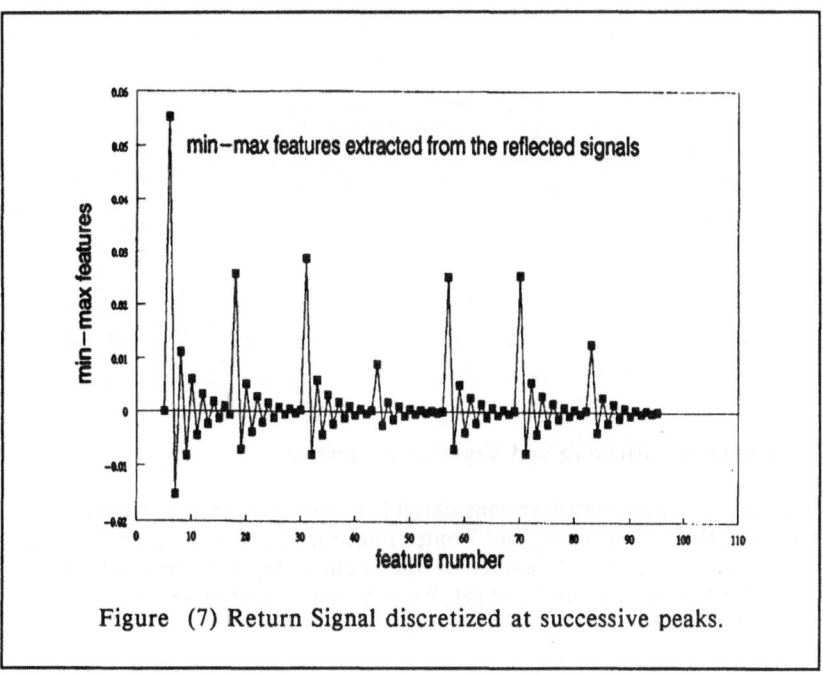

Figure (7) Return Signal discretized at successive peaks.

Table (2)
Material Characteristics

Layer #	Ref. Coef	Amp.	Thick.	Delay	Depth	Atn. Coef.	Decay
Layer 1	0.3228	0.06133	0.0001	0.0002	27	2.8	0.00504
Layer 2	0.3749	0.06381	0.00018	0.00056	45	7.0	0.021
Layer 3	0.3517	0.05145	0.00009	0.00074	48	4.5	0.0144
Layer 4	0.2504	0.03210	0.00003	0.0008	56	7.5	0.028
Layer 5	0.3835	0.04608	0.0003	0.0014	62	3.6	0.01488
Layer 6	0.4098	0.04200	0.0006	0.0026	64	2.6	0.01109
Layer 7	0.1767	0.01507	0.00001	0.00262	72	0.97	0.00465

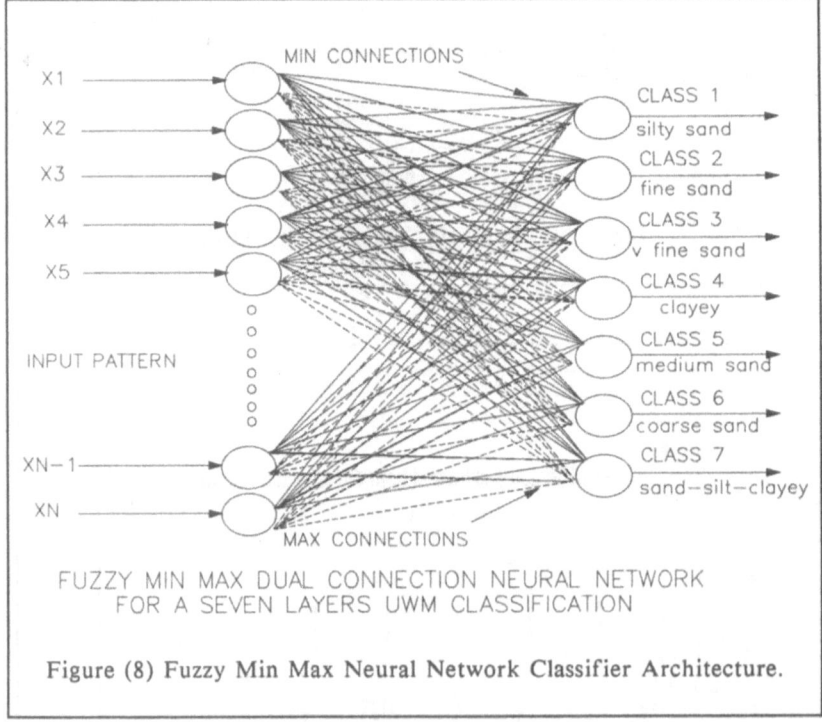

Figure (8) Fuzzy Min Max Neural Network Classifier Architecture.

6. Network Architecture and Training Procedure

The fuzzy min-max classifier considered is a two-layer feed-forward neural network with 13 input units, and 7 output units as shown in Figure (8). Each input unit represents one feature of the input data, and each output unit determines the class of the input set. We considered a set of seven 13-feature patterns associated with 7 classes of layers.

To store the patterns requires 13 x 7 dual connections. Through the learning procedure explained earlier the min and max connection strength were computed. Figures (9-10) show the final min and max weight matrices at the end of the training procedure in the first and second layers.

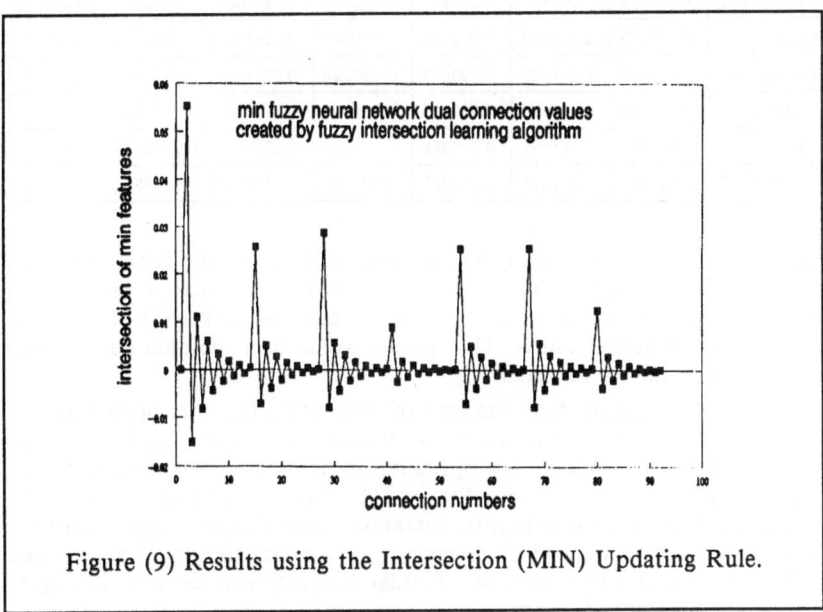

Figure (9) Results using the Intersection (MIN) Updating Rule.

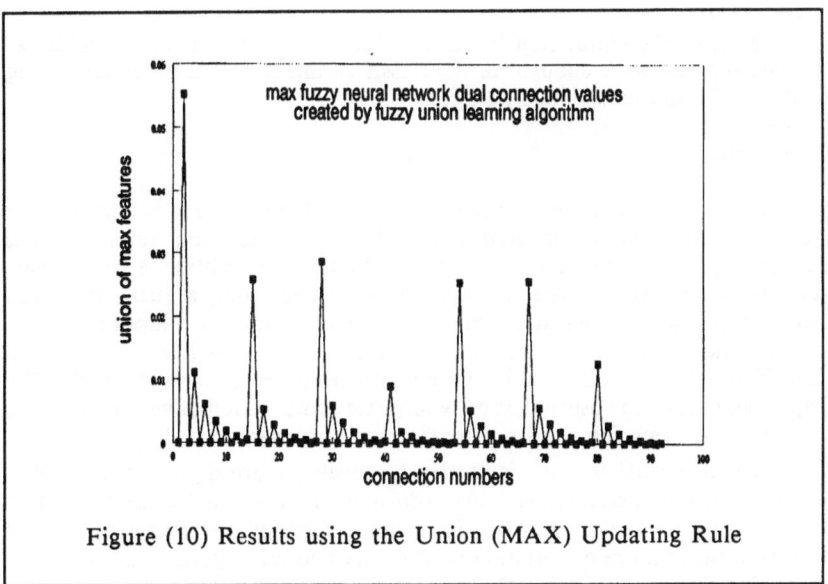

Figure (10) Results using the Union (MAX) Updating Rule

Table (3)
Output Unit Values During Tests for Signals Classification

Output #	Class 1	Class 2	Class 3	Class 4	Class 5	Class 6	Class 7
Output # 1	0.999998	0.999980	0.999983	0.999968	0.999980	0.999982	0.999972
Output # 2	0.999975	0.999997	0.999994	0.999991	0.999997	0.999995	0.999995
Output # 3	0.999983	0.999996	0.999998	0.999984	0.999995	0.999997	0.999987
Output # 4	0.999968	0.999989	0.999987	0.999999	0.999989	0.999987	0.999998
Output # 5	0.999975	0.999997	0.999994	0.999991	0.999997	0.999995	0.999995
Output # 6	0.999983	0.999996	0.999998	0.999984	0.999995	0.999997	0.999987
Output # 7	0.999968	0.999989	0.999987	0.99999	0.999989	0.999987	0.999998

The original pattern of Figure (7) has been used to test the fuzzy min-max classifier after training. Figure (11) shows the performance of each output unit in determining of the classes of the test data set. Table (3) displays the contents of the output units. The largest value in a column corresponds to the class of the tested signal.

Figure (11-a) corresponds to the case of presenting the first portion of the synthetic record of Figure (6) to the trained classifier. The classification rate value at class 1 is closest to the optimum, indicating a clear discrimination in this case. The situation with Figure (11-b) corresponding to data from the second data segment is slightly different. Here Classes 2 and 5 are very close. This can be explained by referring to the properties of the second and fifth layers which are very close. Similar findings can be seen in Figure (11-e). The measure of the degree of classification given in equation (11) is used to resolve the issue with results as indicated in the seven displays of Figure (11).

As a result of the simulation it was concluded that the maximum points of each output unit were enough for the classification of real data corresponding to their min and max points patterns.

7. Conclusions

The concept of a fuzzy min-max neural network classifier was employed in classifying underwater layered media based on acoustic return signals accounting for delay and attenuation effects. We applied a two layers feed-forward neural network, which was trained using a fuzzy min-max neural network classifier algorithm, to process the unknown input data. The performance of the fuzzy min-max neural network classifier algorithm in identifying classes and in reducing classification error was acceptable. The paper established the usefulness of synergistic combination of neural networks and fuzzy systems for pattern classification purposes.

It is clear that further work is needed to develop more discriminating DOC measures than is currently available. Additional features such as zero crossings may also prove useful. Finally, implementations of the fuzzy union and intersections that are different from the max and min operations may prove to be fruitful areas of research.

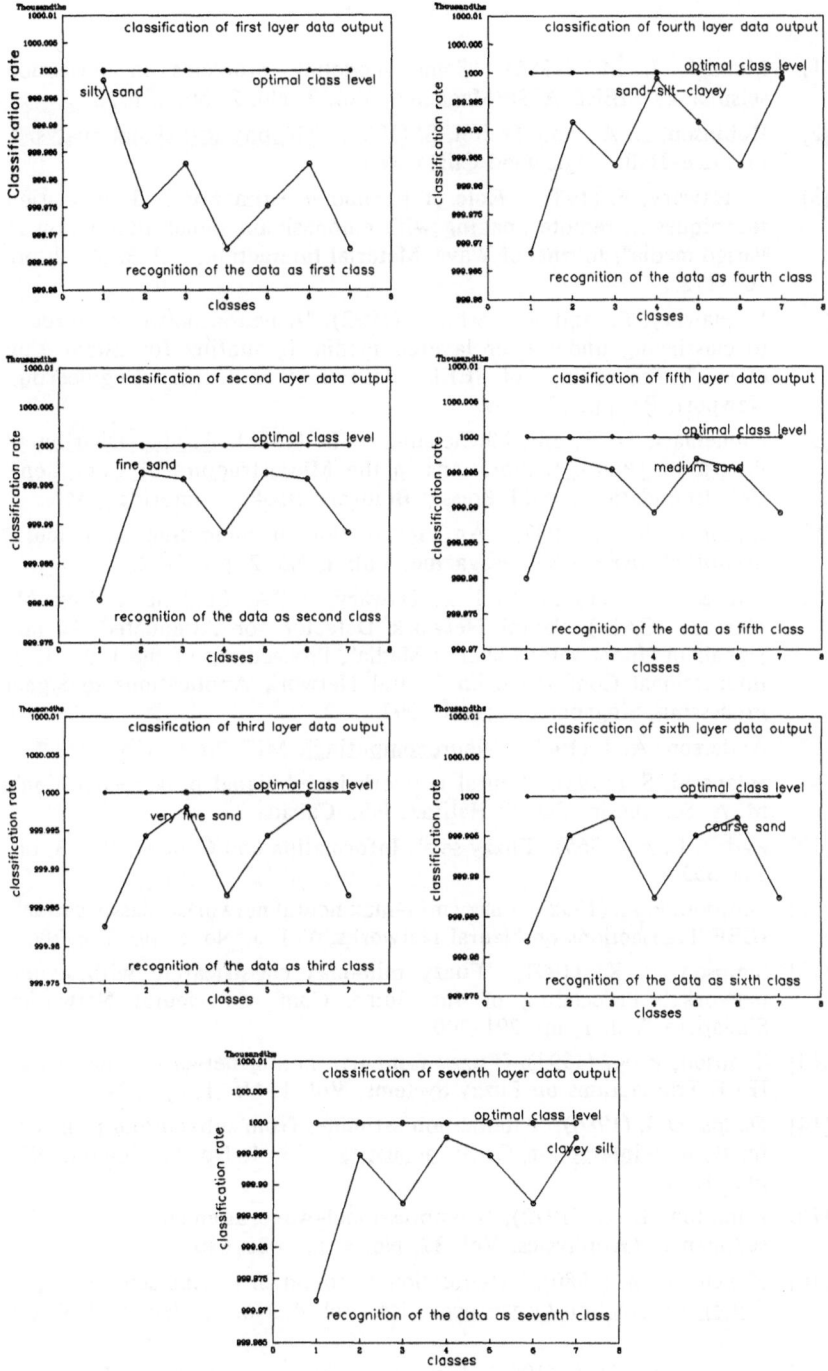

Figure (11) Min-Max Classification Results

REFERENCES

[1] Mendel, J. M. (1986), "Some modeling problems in reflection seismology", IEEE ASSP Magazine, Vol. 3, No. 2, pp. 4-17.

[2] Robinson, E. A., and Treitel, S. (1980), "Geophysical signal analysis". Prentice-Hall, Englewood Cliffs, N.J.

[3] El-Hawary, F. (1988), "Role of parameter estimation and correlation techniques in remote imaging with emphasis on signal attenuation in varied media", Journal of Wave-Material Interaction, Vol. 3, No. 2, pp. 163-172.

[4] El-Hawary, F., and Setayeshi, S. (1992), "A neural network approach to classifying underwater layered media accounting for attenuation effects", Proceedings of IEEE conference on Oceans Engineering, Newport, RI, pp. 134-139.

[5] Rumelhart, D .E. and McClelland, J. L. (1986). Parallel Distributed Processing (PDP): Explorations in the Microstructure of Cognition". Vol. 1:Fundations., MIT Press / Bradford Books, Cambridge, MA.

[6] Lippman, R. P. (1987). "An introduction to computing with neural networks". IEEE ASSP Magazine, Vol. 4, No. 2, pp. 4-22.

[7] Setayeshi, S. (1993), and El-Hawary, F.,"A Content-Addressable Memory (CAM) Neural Network Detector for Attenuated Seismic Signals in Underwater Layered Media", Proceedings of the 1993 IEEE International Conference on Neural Network Applications to Signal Processing, Singapore, August 1993.

[8] Anderson, A. J. (1987), "Neurocomputing". MIT Press, MA.

[9] Setayeshi, S. (1993). "Neural network based signal peak recognition". M. A. Sc.. thesis, TUNS, Halifax, NS., Canada.

[10] Zadeh, L. A. (1965). "Fuzzy sets". Information and Control, Vol. 8, pp. 338-353.

[11] Simpson, P. K. (1992). "Fuzzy min-max neural networks: classification". IEEE Transactions on Neural Networks, Vol. 3, No. 5, pp. 776-786.

[12] Simpson, P. K. (1991). "Fuzzy min-max classification with neural networks". Proceeding of Int. Joint. Conf. on Neural Networks, Singapore, Vol. 1, pp. 291-300.

[13] Simpson, P. K. (1993). "Fuzzy min-max neural networks: clustering". IEEE Transactions on Fuzzy Systems, Vol. 1, No. 1, pp. 32-45.

[14] Dodds, D. I. (1980), "Attenuation estimates from sub-bottom profiler", in: Bottom-interaction Ocean acoustics, Vol. 4, No. 5, Plenum, NY, pp. 183-191.

[15] Hamilton, E. L. (1972), "Compressional-wave attenuation in marine sediments", Geophysics, Vol. 37, No. 4, pp. 620-640.

[16] Hovem, J. M. (1980), "Attenuation of sound in marine sediments", in Bottom-interaction Ocean acoustics, Vol. 4, No. 5, Plenum, NY, pp. 1-13.

[17] Zimmerman, H-J, (1991), "Fuzzy Set Theory and its Applications", Second Edition, Boston, Kluwer Academic Publishers.

[18] Dubois, D., and Prade, H., 1980, "Fuzzy Sets and Systems, Theory and Applications", Academic Press, San Diego.

[19] Kosko, B., (1990), "Fuzziness vs. Probability", International Journal of General Systems, Vol. 17, pp. 211-240.

[20] Hecht-Nielson, R. (1991). <u>Neurocomputing</u>. Addison-Wesley Publishing Company, CA.

[21] Simpson, P. K. (1991). <u>Foundation of neural networks: Neural Networks.</u>, Chapter 1, Lau, C., Sanchez-Sinecio, Eds., IEEE Press: Piscataway, NJ.

[22] Hamilton, E. L., Shumway, G., Menard, H. W. , and Shipek, C. J., "Acoustic and Other Physical Properties of Shallow Water Sediments Off San Diego", Journal of the Acoustical Society of America, Volume 28, No. 1, pp. 1-15, Jan. 1956.

WAVELET TRANSFORM BASED TRANSIENT DETECTION

E. Ç. GÜLER[1], B. SANKUR[2], E. ANARIM[2], C. D. MENDİ[2], O. ALKIN[2],
Y. P. KAHYA[2], T. ENGİN[3]

[1]*Biomedical Eng. Institute, *[2]*Department of Electronics and Electrical Eng.,*
*Bogaziçi University, Bebek 80815, Istanbul, Turkey, *[3]*Faculty of Electronics*
and Electrical Eng., Technical University of Istanbul, Maslak, Istanbul, Turkey

ABSTRACT. In this work, the problem of detection of transients in noise is considered. A method for the detection and classification of transients based on the matched discrete orthonormal wavelet transform is proposed. The idea behind the method is to increase the transient / bacground ratio by expanding the signals in the time-scale space, identifying the bands containing transients and using various linear / nonlinear operations, such as the Teager's energy operator, for enhancement. Matched wavelet filters are designed from typical transient waveforms by using the Frazier-Kumar method in order to increase the correlation between the wavelet family and the transients to be detected. Transients are also analysed via the short-time Fourier transform, the Gabor representation and the Wigner distribution.

1. Introduction

The detection of transients in a stationary stochastic process is an important and well studied problem in many signal processing applications. The analysis of transients buried in a wideband signal by classical Fourier transform analysis is usually not satisfactory since the Fourier-basis functions infinitely extend along time whereas the duration of transients to be analysed are short with respect to the observation interval, however, these signals change the Fourier spectrum of the signal under analysis, but this change depends on the ratio of transient energy to noise energy, and if the effect of transients on the time and frequency axis is limited they cannot be separated from the background signal, thus the interest of this work is to analyse and detect transients which cannot be easily detected by standard Fourier techniques, and whose precise shapes and time of occurences are not well known.

Time-frequency or time-scale analysis of transients in the presence of noise seems to be a better solution to the detection problem. Four of the well known time-frequency or time-scale representations are the short-time Fourier transform, the Gabor representation, the Wigner distribution and the wavelet transform. Among them, the wavelet transform has been very popular in many applications including speech, music, seismic, sonar, biomedical because it provides a very robust time-scale resolution, hence, it is of special interest in this work.

The time-frequency expansion of a signal can be implemented via orthonormal wavelet transforms which decomposes an arbitrary signal into a succession of orthogonal subspaces at different scales. This decomposition is enacted using a single prototype

X. P. V. Malague (ed.), Advances in Signal Processing for Nondestructive Evaluation of Materials, 269–283.
© 1994 *Kluwer Academic Publishers.*

function called an "analysing" or a "mother" wavelet. Fine temporal analysis is done with the contracted versions of the mother wavelet while fine frequency analysis uses the dilated versions.

In this work, a method for the detection and classification of transients based on the matched discrete orthonormal wavelet transform is proposed and the transients are also analysed via the short-time Fourier transform, the Gabor representation and the Wigner distribution. Synthetically generated real exponentials and pathological respiratory sounds containing nonstationarities which are called crackles are used as transients. The outline of the paper is as follows. Section 2 gives a brief survey of previously mentioned time-frequency and time-scale representations. Details of the proposed analysis and detection method are given in Section 3. Section 4 is dedicated to the experimental results and conclusions.

2. A Survey of Several Time-Frequency and Time-Scale Representations

2.1. SHORT-TIME FOURIER TRANSFORM

The Fourier Transform (Fourier Spectrum) $X(f)$ of a signal $x(t)$

$$X(f) = \int_{-\infty}^{+\infty} x(t) \exp(-j2\pi ft) \, dt \tag{1}$$

characterizes the signal $x(t)$ uniquely, the Inverse Fourier Transform of $X(f)$ turns out one and only one signal which is $x(t)$. However, there is only one Fourier transform for the entire signal, so that a subtle change in the observed signal cannot be easily characterized and also it is impossible to find the time of occurrence of this change.

A popular method in analyzing transients is to introduce time localization of frequency components to Fourier spectrum, this is performed by pre-windowing the signal $x(t)$ and then by taking the Fourier transform. This method is called the short-time Fourier transform (STFT). In other words, the STFT consists of a separate Fourier transform for each instant in time. The STFT of a signal $x(t)$ is defined as

$$S_x(t, f) = \int_{-\infty}^{+\infty} x(\tau) w(t-\tau) \exp(-j2\pi f\tau) \, d\tau \tag{2}$$

where $w(t)$ is the sliding pre-windowing function. If the window function is a Gaussian function this transform is called the Gabor transform which is going to be handled in the following subsection. Windowing the signal causes a lack of resolution either in time or in frequency because both fine temporal analysis and fine frequency analysis is done with a fixed window function. The STFT can be seen as a filtered version of $x(t)$, the analysis window $w(t)$ being the impulse response of the filter, in order to see this, the value of f is fixed at f_0 and Eq. (2) is rewritten as

$$S_x(t, f_0) = \int_{-\infty}^{+\infty} [x(\tau) \exp(-j2\pi f_0\tau)] w(t-\tau) \, d\tau \,. \tag{3}$$

It can easily be noticed from Eq. (3) that this operation is in fact the modulation of $x(t)$ with $\exp(-j 2 \pi f_0 \tau)$ up to frequency f_0 and then the convolution of the resulted sequence with $w(t)$. The STFT is a linear transform, however, its energy and the phase distributions which are generally used to represent the signal, are nonlinear in nature [1], [2].

2.2. GABOR REPRESENTATION

The Gabor representation (GR) seems to be suitable for modelling transients, but the original Gabor window function which is a two-sided symmetric Gaussian function usually fails to represent one-sided, nonsymmetric transients. To use a one-sided exponential function resembling many physical transient phenomena may be more suitable for analysing transients by the Gabor representation. The jump discontinuity and the gradual decay of a one-sided exponential window can model a transient better than a classical Gabor window function [3]. The one-sided exponential function is first proposed and used in the detection phenomena by Friedlander and Porat [3].

The GR of a continuous time signal $x(t)$ is

$$x(t) = \sum_{m,n=-\infty}^{\infty} C_{m,n} w(t - n \alpha) \exp[j 2 \pi m \beta (t - n \alpha)] \tag{4}$$

where $w(t)$ is a nonnegative window function with unit area, $C_{m,n}$ are the Gabor coefficients, $\alpha > 0$, $\beta > 0$ and $\alpha\beta \leq 1$. If $\alpha\beta = 1$ Eq. (4) is rewritten as

$$x(t) = \sum_{m,n=-\infty}^{\infty} C_{m,n} w(t - n \alpha) \exp(j 2 \pi m \beta t). \tag{5}$$

The equation given above indicates that the GR is a sampled-parameter equivalent of the inverse STFT [3].

The important issues of the Gabor expansion are the completeness, linear independence and orthogonality of the Gabor basis. Completeness ($\alpha\beta \leq 1$) of the Gabor basis guarantees the exsistence of the representation. This condition is a bound on the time-frequency sampling grid [1].

The coefficients $C_{m,n}$ are given by

$$C_{m,n} = \int_0^1 \int_0^1 \exp[-j 2 \pi (m z + n w)] \frac{X(z,w)}{W(z,w)} dz \, dw \tag{6}$$

where $X(z,w)$ and $W(z,w)$ are Zak transforms of the signal $x(t)$ and the window function $w(t)$, respectively. Zak transform of an arbitrary signal $g(t)$ is defined by

$$Zak\{ g(t) \} = G(z,w) = \sum_{k=-\infty}^{\infty} g(z - k) \exp(-j 2 \pi k w). \tag{7}$$

Another way of computing the Gabor coefficients is via the biorthogonal function of w(t). The biorthogonal function $\omega(t)$ of the window function w(t) is given by the following relationship

$$\omega (z) = \int_0^1 \frac{dw}{W^* (z, w)} \qquad (8)$$

where $(\bullet)^*$ denotes the complex conjugation. The Gabor coefficients are computed in terms of $\omega(t)$ as follows

$$C_{m,n} = \int_{-\infty}^{\infty} \exp[-j 2 \pi m t] \omega^*(t) x(t + n) dt . \qquad (9)$$

The coefficients are related to the signal by an operation similar to the match filtering [3].
    The one-sided exponential window function used in this work is same as the one used in [3]; this function and its biorthogonal function are given by

$$w (t) = (2 \lambda)^{1/2} \exp(- \lambda t) u(t) , \qquad (10)$$

$$\omega (t) = [\exp(\lambda t) / (2 \lambda)^{1/2}] [- u(t + 1) + 2 u(t) - u(t - 1)] . \qquad (11)$$

where λ is a parameter that is used to control the width of the window, u(t) is the unit step function. Since $\omega(t)$ is zero outside the interval $-1 \leq t \leq 1$, Eq. (9) can be rewritten as

$$C_{m,n} = \int_{-1}^{1} \exp[-j 2 \pi m t] \omega^*(t) x(t + n) dt . \qquad (12)$$

A more convenient expression of (12) is given in [12] as follows

$$C_{m,n} = D_{m,n} - \exp(- \lambda) D_{m,n-1} \qquad (13)$$

where

$$D_{m,n} = \frac{1}{\sqrt{2 \lambda}} \int_0^1 \exp[-j 2 \pi m t] \exp(\lambda t) x(t + n) dt . \qquad (14)$$

If both the signal x(t) and w(t) are sampled at intervals $1 / L$ then $D_{m,n}$ becomes [3]

$$D_{m,n} \approx \frac{\exp(-j \pi m / L)}{L \sqrt{2 \lambda}} \sum_{l=0}^{L-1} \exp[-j \frac{2 \pi l m}{L}] \exp(\frac{\lambda l}{L} + \frac{\lambda}{2 L}) x(n + \frac{l}{L} + \frac{1}{2 L}). \qquad (15)$$

2.3. WIGNER DISTRIBUTION

The Wigner distribution (WD) of a signal $x(t)$ is given by

$$WD_x(t,f) = \int_{-\infty}^{\infty} x(t+\frac{\tau}{2}) x^*(t-\frac{\tau}{2}) \exp(-j2\pi f\tau) d\tau \qquad (16)$$

and in the frequency domain, the WD is

$$WD_x(t,f) = \int_{-\infty}^{\infty} X(f+\frac{\xi}{2}) X^*(f-\frac{\xi}{2}) \exp(j2\pi\xi t) d\xi \qquad (17)$$

where $X(f)$ is the Fourier transform of $x(t)$. The WD is a time-frequency representation that overcomes the time versus frequency resolution problem presented by the STFT or the GR. It can easily be noticed from Eq. (16) that the WD first computes the autocorrelation terms at each time instant t and then evaluates the power spectrum of the signal. The WD preserves time, provides frequency support, instantaneous frequency and group delay [2], these properties has made it popular in transient analysis, however, its bilinear nature produces cross-product interference terms that might be a nuisance in multiple data interpretation when compared with the linear transforms. It has been shown in [2] that the cross terms (i) occur at midtime and midfrequency, (ii) oscillate at a frequency proportional to the difference in frequency and time of the perpended signals, (iii) oscillate in the direction orthogonal to the line that connects the autocomponents and (iv) can have an amplitude twice as large as the product of the magnitudes of the WD of the signals under consideration.

The cross terms of bilinear time-frequency representations cannot be completely eliminated but they can be significantly suppressed by low-pass filtering the WD since their oscillations are at relatively high frequencies. The low-pass filtering operation causes broadening of autocomponents [4]. Filtering the WD by a function $c(t)$ introduces the pseudo-Wigner distribution (PWD) which can be expressed as

$$PWD_x(t,f) = \int_{-\infty}^{\infty} x(t+\frac{\tau}{2}) x^*(t-\frac{\tau}{2}) c(\frac{\tau}{2}) c^*(\frac{-\tau}{2}) \exp(-j2\pi f\tau) d\tau . \qquad (18)$$

2.4. WAVELET TRANSFORM

Wavelet transform is a tool to view signals at different scales. It decomposes an arbitrary signal $x(t)$ by expressing it as a superposition of projections onto a family of functions generated from a square integrable mother wavelet $\psi(t)$ by shifts and dilations. The WT coefficients $W_x(a,b)$ of the signal $x(t)$ are defined as

$$W_x(a,b) = <x(t),\psi_{a,b}(t)> = \frac{1}{a^{1/2}} \int_{-\infty}^{+\infty} x(t) \psi^*(\frac{t-b}{a}) dt \qquad (19)$$

where $\psi_{a,b}(t)$ is a complete orthogonal set of daughter wavelets generated from the mother wavelet by dilation and shift operations,

$$\psi_{a,b}(t) = (1/a^{1/2})\,\psi((t-b)/a)\,, a \in R^+, b \in R \tag{20}$$

where a and b are dilation and translation parameters, respectively, the normalization constant in Eq. (20) lowers a daughter wavelet by the factor $a^{1/2}$. Each daughter wavelet has a different length and location but their number of cycles and shapes are identical to the mother wavelet. The wavelet coefficients in Eq. (18) can be written in the frequency domain as

$$W_x(a,b) = \sqrt{a}\int_{-\infty}^{\infty} e^{j2\pi fb}\,X(f)\,\Psi^*(af)\,df \tag{21}$$

where $\Psi(f)$ and $X(f)$ are the Fourier transforms of the mother wavelet $\psi(t)$ and the signal $x(t)$, respectively, the Fourier transform of a daughter wavelet $\psi_{a,b}(t)$ is given by

$$\Psi_{a,b}(f) = \sqrt{a}\,e^{-j2\pi fb}\,\Psi(af) \tag{22}$$

The WT can also be viewed as the output of a filtering operation where the function $\psi_{a,b}(t)$ plays the role of the filter impulse response. If $\gamma_a(t) = \psi^*_{a,b}(-t/a)$ Eq. (19) be rewritten

$$W_x(a,b) = \int_{-\infty}^{+\infty} x(t)\,\gamma_a(b-t)\,dt \tag{23}$$

One of the major advantages of the WT can be noticed from Eqs. (21) and (23), it is the efficiency caused by the special kind of matched filter of identical shape which nevertheless permits the filter function to change its size, and its phase shift, and by the definition of Eq. (19), the WT is a correlation between the signal and the wavelet function. A mother wavelet must satisfy a certain condition so that the inverse WT exists:

$$\int_{-\infty}^{\infty} \frac{|\Psi(f)|^2}{|f|}\,df \; < \; \infty\,. \tag{24}$$

This condition implies that $|\Psi(f)|^2 = 0$ when $f = 0$ and then

$$\int_{-\infty}^{\infty} \psi(t)\,dt = 0\,. \tag{25}$$

Hence, a mother wavelet has a zero integrated area and a zero dc component.

If the WT is evaluated on a discrete grid on the time-scale plane, the redundancy caused by the continuous choice of parameters (a, b) can be avoided. The problem is to find such a grid where the set of wavelet functions forms an orthonormal basis. It is possible to design wavelet families such that functions $\psi(\cdot)$ are regular and the set of their dilated and translated versions forms an orthonormal basis [5], [6]. If the translation and dilation parameters of the wavelet given by (20) are discretized, the following expression is obtained:

$$\psi_{m,n}(t) = (a_0)^{-m/2}\,\psi((a_0)^{-m}\,t - nb_0)\,,\ m,\, n \in Z\,,\ a_0 > 0,\ b_0 \neq 0 \qquad (26)$$

which corresponds to $a = (a_0)^m$ and $b = n(a_0)^m b_0$. In this case functions $\psi(\cdot)$ can be constructed so that the set $\{\,\psi_{m,n}(\cdot)\,\}$ is orthonormal.

2.4.1. *The Discrete WT as a Bank of Bandpass Filters* . A wavelet can be viewed as a bandpass filter in signal processing applications, for example, it is an octave band filter if a_0 and b_0 given in Eq. (26) are chosen as 2 and 1, respectively. In the discrete case, WT can be regarded as a mathematically formalized subband coder and usually implemented as a bank of bandpass filters [5].

Let us assume that the set of functions $\{\,\Phi(x-k)\,,\ k \in Z\,\}$ forms an orthonormal basis for the space V_0 which is the space of all these band-limited functions with frequencies in the interval ($-\pi$, π). In this case, we will call V_{-1} as the space of band-limited functions with frequencies in the interval (-2π , 2π). Therefore the set $\{\,(2)^{1/2}\Phi(2x-k)\,,\ k \in Z\,\}$ is an orthonormal basis for V_{-1} and also this space contains space V_0, and if a discrete time signal x(n) is in V_0 then x(2n) $\in$ V_{-1}. The space of bandpass functions which is an orthogonal complement in V_{-1} of V_0, and with frequencies in the interval (-2π , $-\pi$) $\cup$ (π , 2π) will be named as W_0. These can be generalized for the space V_i, the space of band-limited functions with frequencies in the interval ($-2^{-i}\pi$, $2^{-i}\pi$), and for the space W_i, the space of band-limited functions with frequencies in the interval ($-2^{-i+1}\pi$, $-2^{-i}\pi$) $\cup$ ($2^{-i}\pi$, $2^{-i+1}\pi$), thus, following equations are obtained:

$$\dots V_{i+2} \subset V_{i+1} \subset V_i \subset V_{i-1} \subset V_{i-2} \dots \qquad (27)$$

$$V_i = W_{i+1} \oplus W_{i+2} \oplus W_{i+3} \ \dots\ . \qquad (28)$$

When the set $\{\,\Phi(x-k)\,,\ k \in Z\,\}$ constitutes a basis for V_0, $\{\,(2)^{1/2}\,\Phi(2x-k)$, $k \in Z\,\}$ constitutes a basis for V_{-1}. In the sampled version of V_{-1}, $\Phi(x)$ is given by the perfect halfband lowpass filter with the impulse response $(2)^{1/2}\,c_n$ where $\|\,c_n\,\| = 2^{1/2}$. Since $\Phi(x)$ is the interpolation, by $\Phi(2x)$, of the perfect halfband lowpass filter, it can be written as

$$\Phi(x) = \sum_{n=-\infty}^{n=+\infty} c_n\,\Phi(2x-n). \qquad (29)$$

$\Phi(x)$ is called a scaling function because it derives an approximation in V_0 of signals in V_{-1}. The detail corresponding to W_0 is given by the halfband highpass functions $\psi(x)$, $\psi(x)$ is given by

$$\psi(x) = \sum_{n=-\infty}^{n=+\infty} (-1)^n c_{-n+1} \Phi(2x-n). \qquad (30)$$

Let us now consider discrete time square summable sequences $x(n)$ and a finite impulse response (FIR) lowpass filter having the impulse response $[h_L(0), h_L(1),...,h_L(L-1)]$ with $h_L(n) = c_n / 2^{1/2}$ [5]. If the original sequence $x(n)$ is filtered with $h_L(n)$ and then the filtered output is downsampled by 2, this procedure corresponds to a coarse half resolution approximation to the infinite signal vector x and it can be represented by matrix multiplication of x by

$$H_L = \begin{bmatrix} & \cdots & & \cdots & & \cdots & & \cdots & & \cdots & & \cdots & \\ \cdots\ h_L(L-1) & h_L(L-2) & \cdots & & & h_L(0) & 0 & 0 & \cdots \\ \cdots & 0 & & 0 & h_L(L-1) & \cdots & h_L(2) & h_L(1) & h_L(0) \cdots \\ & \cdots & & \cdots & & \cdots & \cdots & \cdots & \end{bmatrix}. \qquad (31)$$

Let us further assume $h_L(n)$ and its even versions form an orthonormal basis for V_0:

$$< h_L(n-2l), h_L(n-2k) > = \partial_{kl}, \quad k, l \in Z \quad \text{or} \quad H_L H_L^* = I. \qquad (32)$$

Therefore $H_L H_L^* x$ is the projection of $x(n)$ onto the subspace spanned by the rows of H_L and multiplication by H_L^* corresponds to upsampling by 2 followed by a convolution with a filter having the impulse response $\tilde{h}_L(n) = [h_L(L-1) \ \ \cdots \ \ h_L(0)]$. The filter with the impulse response

$$h_H(n) = (-1)^n h_L(L-1-n) \qquad (33)$$

and its even shifted versions form an orthonormal basis for W_0. Therefore,

$$< h_H(n-2l), h_L(n-2k) > = 0, \quad k, l \in Z \quad \text{or} \quad H_L H_H^* = 0 \qquad (34)$$

and

$$< h_H(n-2l), h_H(n-2k) > = \partial_{kl}, \quad k, l \in Z \quad \text{or} \quad H_H H_H^* = I. \qquad (35)$$

Note that, $H_H^* H_H x$ is the projection of $x(n)$ onto W_0 and the space V_0 is orthogonal to the space W_0. The subspace V_0 can be decomposed into V_1 and W_1 and so on using the same procedures. Fig. 1. shows the decomposition of V_{i-1} into V_i and W_i using multirate filters, and recombination to achieve perfect reconstruction. Since the decomposition of V_{-1} into W_0, W_1, etc, splits the original space in two and then splits one of the resulting half spaces in two, it is actually a WT. The discrete WT as a bank of filters is shown in Fig. 2.

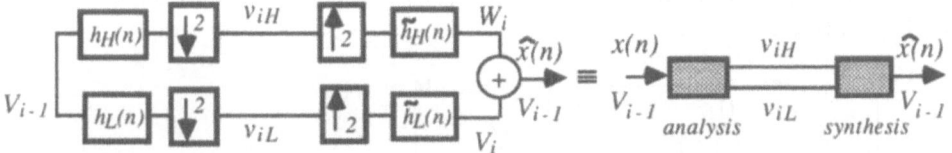

Fig. 1. Decomposition of V_{i-1} into V_i and W_i using multirate filters, and recombination to achieve perfect reconstruction.

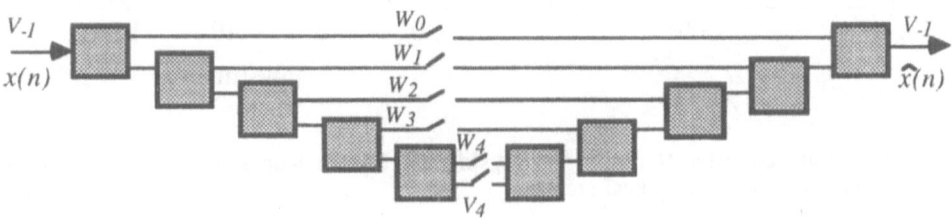

Fig. 2. The use of discrete wavelet transform as a bank of bandpass filters.

3. Transient Analysis and Detection Methods

In this work, a method based on the discrete orthonormal wavelet transform analysis for the detection of transients is proposed. The idea behind the method is to increase the transient / bacground ratio by expanding the signals in the time-scale space, identifying the bands containing transients and using various linear/nonlinear operations for enhancement. The method works as follows.

Firstly, the discrete wavelet transform of a length N signal is computed into octave bands, *see Figs. 1 and 2* by the help of wavelet filters. In other words, the signal space is decomposed into a succession of orthogonal subspaces, v_{LM}, v_{Hj}, $j = 0, 1, ..., M - 1$. The signal in v_{LM} represents blurred version of the original signal, on the other hand, the signal in v_{Hj} corresponds to details in the original signal, the choice of M depends on the original signal bandwidth. As mother wavelet filters, we have used the Daubechies wavelet with regularity six [6], and wavelets designed from transients using the Frazier-Kumar method [7]. In the latter method, portions of typical transient waveforms have been downsampled by considering the bandwidth of a transient and the sampling rate; long length filters may cause false alarms in detection, and then downsampled portions of transients have been mapped onto a wavelet space. The goal in Frazier-Kumar method is to find a wavelet filter $h_Y(n)$ which is closest to a length K sequence $t(n)$, in other words, $\| t(n) - h_H(n) \|^2$ is minimized. Steps of the wavelet filter $h_H(n)$ design procedure from a given sequence $t(n)$ are as follows:

(i) If $T(m) \neq T(m + K/2)$, ($T(m) = F\{t(n)\}$, $F\{ \bullet \}$: Fourier transform operator), Fourier transform coefficients $H_L(m)$ of $h_L(n)$ are given by

$$H_L(m) = \frac{2^{1/2} T(m)}{\sqrt{|T(m)|^2 + |T(m + K/2)|^2}} \quad , \quad m = 0, 1, ... , K - 1 . \tag{36}$$

(ii) $h_L(n)$ is calculated by taking the inverse discrete Fourier transform of $H_L(m)$ and then $h_H(n)$ is evaluated by Eq. (33). A matched wavelet designed by using the Frazier-

Kumar method and its reference function which is a typical crackle are shown in Fig. 3(a) and 3(b), respectively.

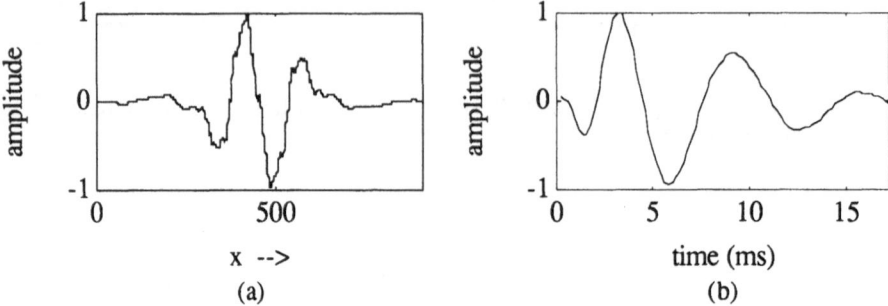

Fig. 3.(a) A matched wavelet designed by using the Frazier-Kumar method and, (b) its reference function which is a typical crackle.

In the second step full scale waveforms are reconstructed selectively from individual wavelet components. This represents in fact bandpass filtering operation in the scale space. This approach is used in order to consider the effect of filtering on transients at different scales (levels), and also avails us of the possibility to apply nonlinear background suppression algorithms in the individual bands thus M signals with lengths N in W_0, W_1, ... , W_{M-1} spaces are obtained. Teager's energy operator is then applied to each of M signals in order to enhance the transient / bacground ratio. Teager's energy operator is an energy function which is a very local property of the signal depending only on the signal and its first two derivatives [8], and it is defined as

$$T \{ x(n) \} = x^2(n) - x(n + 1) x(n - 1) . \tag{37}$$

In the third step an M-by-N matrix is constructed from those "projections" and the absolute value of this signal array is taken. Finally, signal array is treated as an image, I(x, y). Since transient phenomena like crackles are expected to show up on more than one scale, then vertical edges in this "image" are identified as crackles (transients). The center of gravity of the edge is an estimate for the epoch and scale of the crackle (transient). The peaks being vertical edges in the image with higher intensity than the background is found by applying an M-by-2 spatial mask. The spatial mask consists of two columns, of 1's and -1's, respectively. This mask is moved horizantally accross the image and the correlation between the image and the mask is evaluated at each step as follows:

$$C(i) = x_{i,M} + ... + x_{i,1} - x_{i+1,M} - ... - x_{i+1,1} . \tag{38}$$

So that, an (N - 1) length correlation vector is obtained and then the absolute of this vector is taken; when a value of an element of the resultant vector exceeds a certain predetermined empirical threshold C_{TH}, | C(i) | > C_{TH}, it is assumed that the mask just has passed an edge of a crackle (transient) for the signal under analysis.

4.   Experimental Results and Conclusions

This section illustrates and visualizes the results of application of analysis methods and the proposed detection scheme to synthetically generated real exponentials in additive Gaussian white noise (AGWN) and crackles in pathological respiratory sounds. The transients used in the analyses are generated by the following equation

$$x_i(k) = \text{Real}\{A_i \exp[-(\alpha_i + j\,2\,\pi\,f_i)(k-1)/L + j\,\phi_i]\}, \, k = 1, 2, ..., K \qquad (39)$$

where K is the length of a transient, i denotes the i th transient (i = 1, 2, ..., J), $\{A_i, \alpha_i, f_i, \phi_i\}$ are the amplitude, damping coefficient, frequency and phase of the i th transient, respectively, and L is the sampling rate. Fig. 4(a) shows a signal consisting of three transients (J = 3). They are neither well separated in time nor in frequency. In this signal, the amplitudes are 1, 1, 0.8; the damping coefficients of all three transients are 0.2; the frequencies are 6, 5, 3.5; their phases are 0; and the times of arrival are 4, 8, 10.5 , respectively, sampling rate is L = 64. Fig. 4(b) shows the same signal with zero mean AGWN having a standard deviation of 0.5. These two signals are analysed via the STFT,the GR,the PWD and the discrete orthonormal WT. On the other hand, the proposed detection scheme based on the discrete orthonormal WT is used to detect respiratory sound crackles. Crackles are discontinuous, adventitious non-musical respiratory sounds which may indicate an abnormality in the lungs. Crackles are characterised by their short duration and relatively high frequency content. Their effect on the time and frequency axes is limited. Their duration is usually not more than 70 ms, and they have a spectrum of frequencies between 100 and 2000 Hz. The timing, profusion and duration of crackles in a respiration cycle are significant clinically and physiologically. Estimation or detection of crackles by auscultation is subjective and most of the time is inaccurate since human ear can not quantitate exactly the events occuring in milliseconds. A respiratory sound signal containing 11 crackles is depicted in Fig. 5.

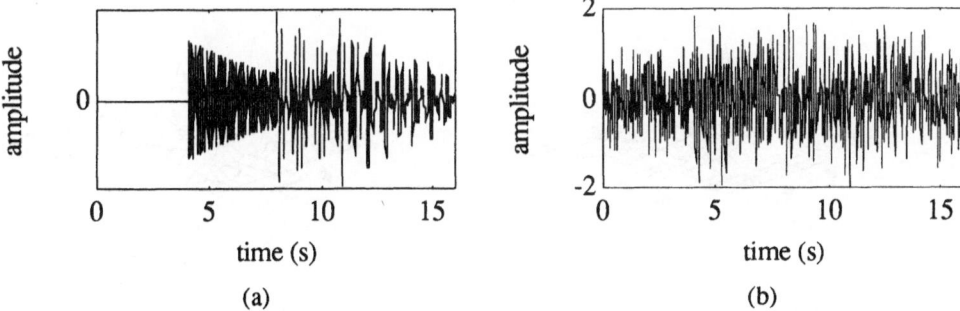

Fig. 4. (a) A signal consisted consisting of three transients, (b) the same signal with noise.

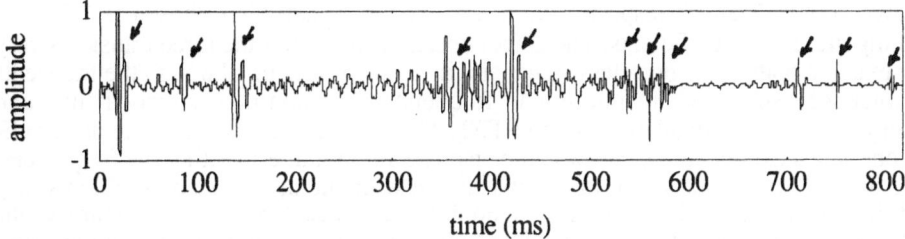

Fig. 5. A Respiratory sound signal containing crackles, crackles are indicated by arrows.

Fig. 6(a) and 6(b) show three-dimensional mesh plots of the absolute values of short-time Fourier transforms of signals given in Figs. 4(a) and 4(b), respectively. STFT analysis is performed with a window function which is a two-sided, length 64 Hann window having a unit energy. The main drawback of this approach is that the noninteger arrival time and frequency give rise to sidelobes which may mislead our further decision rule. The sidelobes resulting from noninteger arrival time are mainly along the time axis while the sidelobes resulting from noninteger frequency are in both axes. This problem may be overcome by using overlapping window functions. Another disadvantage of this technique is that it captures the damping rate of a transient along the time axis; this causes problems in finding time of arrivals of transients in noise especially when they overlap in time and their damping rates are slow. The unwanted effect of damping rate may be decreased by taking the derivative of the STFT coefficients over the frequency, we will call this method " the difference STFT " (Figs. 6(c)-(d)). As it is seen, the presence of the three transients is much more visible in the difference STFT than in the original STFT.

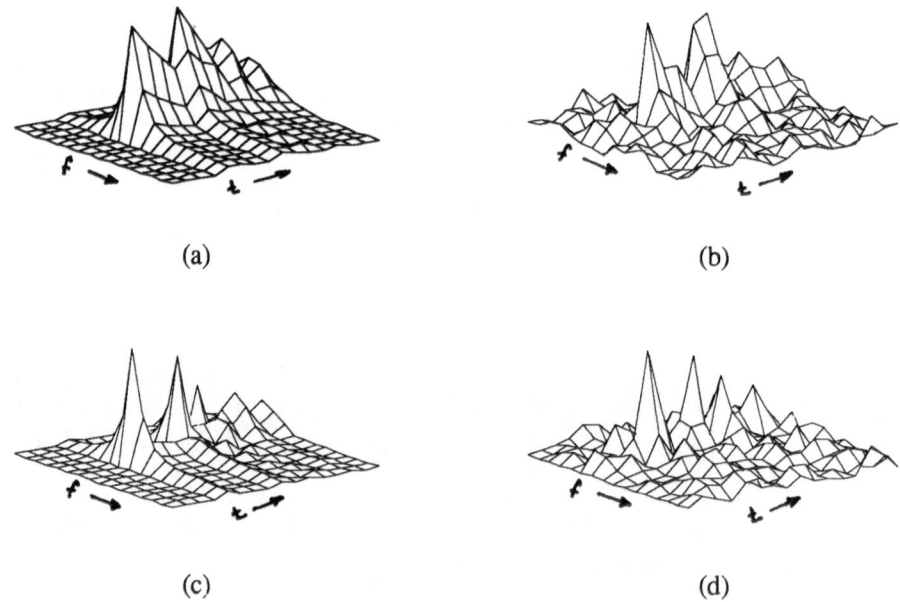

(a)

(b)

(c)

(d)

Fig. 6 (a) The STFT of the signal given in Fig 4.(a), (b) the STFT of the signal given in Fig 4.(b), (c) the difference STFT of (a), (d) the difference STFT of (b), $0 \le f \le 15$, $0 \le t \le 16$.

Absolute values of the Gabor coefficients of the signals depicted in Figs. 4(a) and 4(b) are shown in Figs. 7(a)-(b), respectively. The window parameter λ is chosen as 0.2 while computing the Gabor coefficients. The Gabor coefficients exhibit the three transients very clearly existing both in the noiseless signal and in its noisy version although the transients are neither well separated in time nor in frequency. The noninteger arrival time and frequency give rise to sidelobes as in the STFT. These examples show that the use of GR with a one-sided exponential window can be a powerful tool in estimating the frequencies and the time of arrivals of single tone transients in white noise, but some problems may arise if time of arrivals and frequencies are noninteger values. Unfortunately, many of the transients in physical world exist in a colored noise and usually they are not single tones.

The mismatch between the transient damping factor and the window parameter creates sidelobes along time and frequency axes but especially along the time axis. A smaller value of λ is more tolerant than a narrower window. In physical world, the mismatch between λ and α is important and α is usually not known a priori. This problem may be solved by estimating the damping factor of a transient using the Prony's method. It may help us to find a reasonable λ,but this time at least one transient has to be found and modeled, this does not guarantee that other transients in the signal have the same damping factors. We also note that the GR is less sensitive to noise than the STFT.

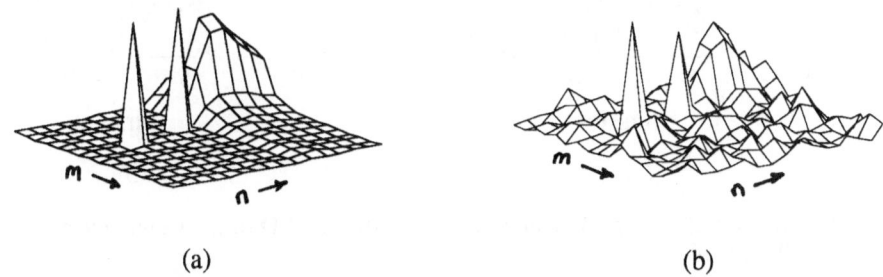

(a)   (b)

Fig. 7(a)-(b) Gabor coefficients of the signals given in Figs 4(a), (b), respectively, M=16, N=16, L=64, m, n ≥ 0.

Figs. 8(a) and 8(b) show the contour plots of the pseudo Wigner distributions of signals given in Figs. 4(a) and 4(b), respectively. PWD of the noiseless signal gives sufficient information for the detection of the transients, however, the transients are hardly discernible in the PWD of the noisy signal, therefore it requires more complex decision rules. Main drawbacks of the PWD are the nonlinear behaivour which also leads to cross terms between the overlapping time intervals and frequency bandwidths and computational load when compared to other time-frequency representations.

Many of the disadvantages presented by time-frequency representations mentioned in the previous sections can be overcome by using the wavelet transform. The signals shown in Figs. 4(a) and 4(b) are also analysed by the orthonormal wavelet transform which decomposes a signal into a succession of orthogonal subspaces at different scales, *see Fig. 2*. number of scales M is chosen as five, the reason is related to the signal bandwidth (32 Hz), thus the signal is analysed at five bands: 16-32 Hz (level 1), 8-16 Hz (level 2), 4-8 Hz (level 3), 2-4 Hz (level 4), 1-2 Hz (level 5). The wavelet coefficients at the third and fourth different scales of the signal depicted in Fig. 4(b) are shown in Figs. 9(a) and 9(b), respectively. As it is seen, the wavelet coefficients still exhibit the three transients at their own bands in the existence of noise although they overlap both in time and frequency, but the effect of overlapping cannot be eliminated completely. In order to obtain a more clear image, the sampling rate could be increased or continuous wavelet transform approximation could be considered in the analysis. Teager's energy operator is applied to these wavelet components in order to suppress the overlapping effects and background noise (Figs. 9(c) and 9(d)). Daubechies wavelet filter with the regularity degree six has been used in the analysis since it was the best one matching with the three transients.

The transient detection method proposed in this work is applied to the respiratory sound signal waveform which is shown in Fig. (5). M is again five since the signal bandwidth is 2500 Hz; thus the signal is analysed at five bands: 1250-2500 Hz, 625-1250 Hz, 312.5-625 Hz, 156.25-312.5 Hz, 78.125-156.25 Hz. The matched wavelet filter, *see Fig. 3* , which is designed by using the Frazier-Kumar method from a crackle waveform is used as an analysing wavelet in multiresolution decomposition of the signal. The use of a matched wavelet filter in the wavelet transform increases the correlation between crackles and the wavelet family. The threshold C_{TH} is found as 0.12, empirically. Fig. 10 and Fig. 11

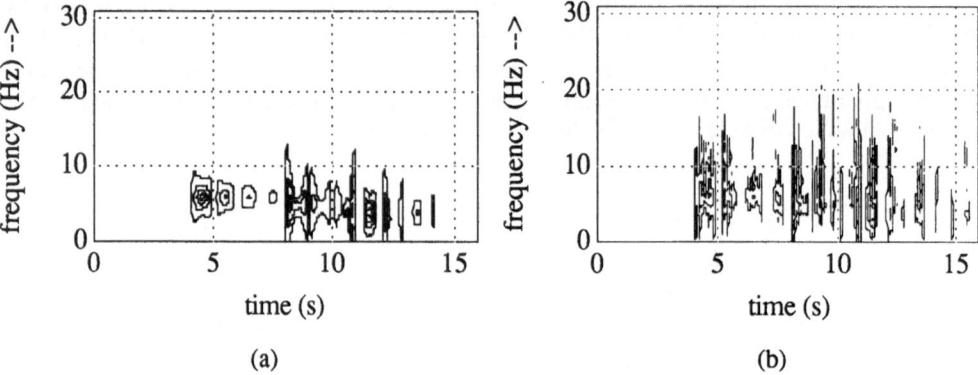

Fig. 8 (a) The WD of the signal given in Fig 4.(a), (b) the WD of the signal given in Fig 4.(b), $0 \le f \le 31, 0 \le t \le 16$.

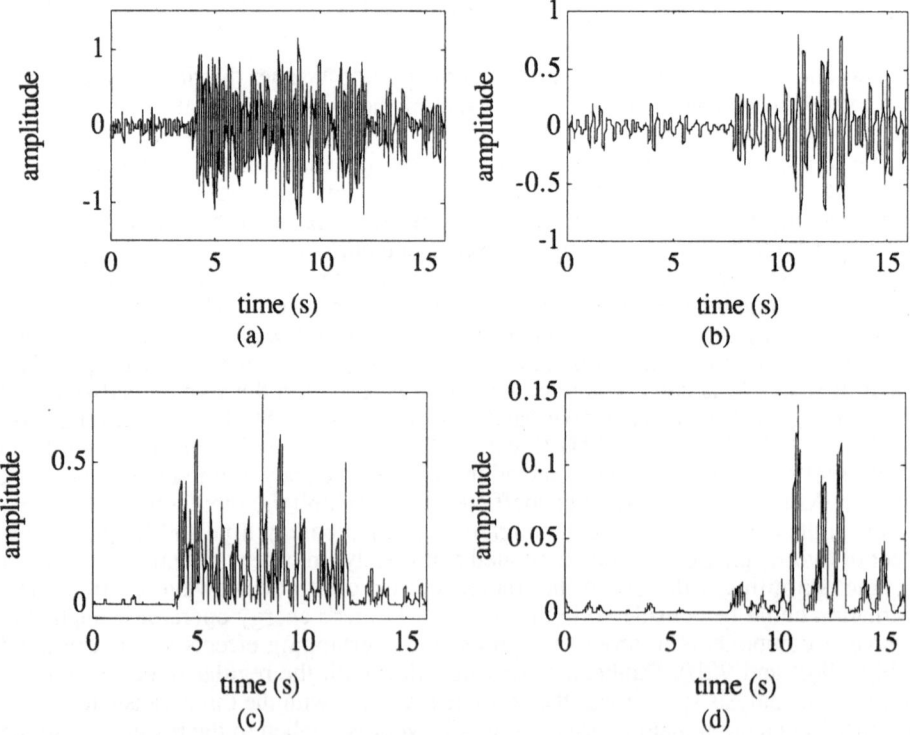

Fig. 9(a)-(b) The wavelet coefficients at scales three (4-8 Hz band) and four, (2-4 Hz Band) respectively, (c)-(d) The wavelet coefficients at the same scales after the Teager's energy operator is applied.

show the image I(x, y) obtained from the wavelet coefficients after the Teager's energy operator has been applied and the output of the detector, respectively. As it can be noticed from Fig. 10 that each crackle is seen at its own scale, this scale information may be useful in the classification of crackles.

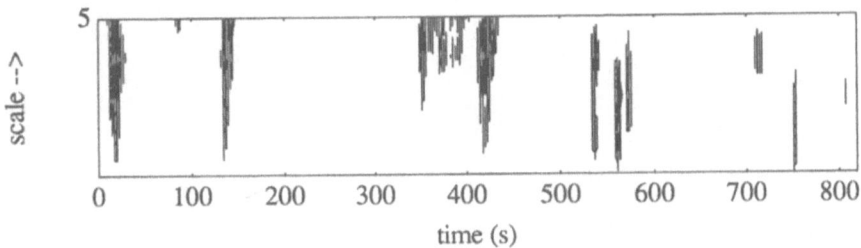

Fig. 10. The image I(x, y) which is obtained from the wavelet coefficients after the Teager's energy operator has been applied.

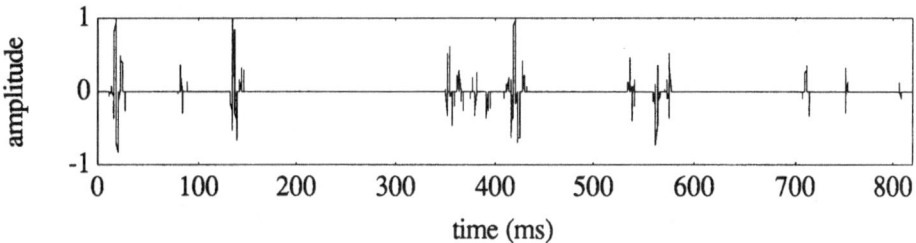

Fig. 11. The output of the detector, all crackles are correctly detected in this example.

References

[1] F. Hlawatsch, G. F. Boudreaux-Bartels, "Linear and quadratic time-frequency representations", IEEE Signal Processing Magazine, pp. 21-67, Apr. 1992.

[2] S. Kadambe, G. F. Boudreaux-Bartels, "A comparison of the existence of cross terms in the Wigner distribution and the squared magnitude of the wavelet transform and the short time Fourier transform", IEEE Trans. Signal Processing, vol. 40, no. 10, pp. 2498-2517, 1992.

[3] B. Friedlander, B. Porat, "Detection of transient signals by the Gabor Representation", IEEE Trans.Acoust., Speech, Signal Processing, vol. 37, no. 2, pp. 169-180, 1989.

[4] D. L. Jones, "A resolution comparison of several time-frequency representations", IEEE Trans. Signal Processing, vol. 40, no. 2, pp. 413-420, 1992.

[5] M. Vetterli, C. Herley, "Wavelets and filter banks: theory and design", IEEE Trans.Acoust., Speech, Signal Processing, vol. 40, no. 9, pp. 2207-2232, 1992.

[6] I. Daubechies, "Orthonormal bases of compactly supported wavelets", Comm. Pure and Appl. Mathematics, Vol. XLI, pp. 909-996, 1988.

[7] M. Frazier, A. Kumar, "The dicrete orthonormal wavelet transform: an introduction", Tech. Report, Wisconsin University, Sept. 1990.

[8] J. Kaiser, "Some useful properties of Teager's Energy operators", in Proc. IEEE-ICASSP-93 (Minneapolis, MN), pp. 149-152, 1993.

THE ROLE OF MODELING IN THE DETERMINATION
OF PROBABILITY OF DETECTION

D. O. THOMPSON
L. W. SCHMERR, JR.
Center for NDE and Dept. of Aerospace
Engineering and Engineering Mechanics
Iowa State University
Ames, Iowa 50011

ABSTRACT. The reliability of nondestructive evaluation (NDE) methods has traditionally been determined through costly round-robin test block trials. Here, an alternative model-based approach is described, where computer simulations are used to predict the probability of detection (POD) versus flaw size curves for components. Uses of such POD curves are shown to include estimating NDE inspection performance, setting of inspection parameters, evaluating the influence of inspections on component reliability, and designing parts with inspectability requirements.

1. Introduction

In many industries, nondestructive evaluation (NDE) methods now play key roles in guaranteeing the safety and reliability of high performance systems and components. These same NDE methods, when combined with modern failure analysis techniques such as fracture mechanics, also can serve as tools for extending life and reducing unnecessary costs. A recent example of this type of application of NDE was the U.S. Air Force Retirement-for-Cause program [1].

The ability of NDE methods to significantly affect the reliability of components is in a large measure a function of the capabilities and reliability of the NDE techniques themselves. Quantifying the sources of variability in NDE measurements and the influence of these variabilities on the evaluation process, however, has been a challenge. Most attempts have relied on "round-robin" tests where samples are made containing "typical" defects and then evaluated by different teams of inspectors. The PVRC program in the United States to evaluate thick section steel parts [2] and its off-shoot, the PISC I/II programs in England [3], the "have-cracks-will-travel" program of the USAF [4] to evaluate aircraft structures, and the more recent UKAEA defect detection trials that simulated aspects of the inspection of pressurized water reactors [5], [6], are all examples of such round-robin trials.

Round-robin studies typically involve considerable complexity and cost. For example, twenty one air force bases participated in the have-cracks program while PISC I involved thirty four teams from ten European countries [7]. Manufacturing and sectioning a single test block containing 15 defects for such studies was estimated by Silk et al. [7] to cost over $100,000 (in 1987 U.S. dollars). As a consequence, these test block exercises have a number of significant operational disadvantages including 1) their inability to vary more than a few of the important test parameters (as Silk et. al. [7] point out, even as few as 3 parameters, each having 5 values results in 243 possible combinations), 2) their lack

X. P. V. Malague (ed.), Advances in Signal Processing for Nondestructive Evaluation of Materials, 285–301.
© 1994 *Kluwer Academic Publishers.*

of transferability to other problem areas, and 3) their inability to effectively compare different test methods (ultrasonic, eddy current, x-ray, etc.).

Recent advances in modeling NDE measurement systems, however, now offer an alternative to empirical test block exercises. For example, Coffey et. al. [8], Ogilvy and Temple [9], Haines et. al. [10], and Thompson and Achenbach [11], all describe the uses of models of ultrasonic systems to evaluate NDE performance, while Segal et. al. [12] has used a similar modeling approach for x-rays. Gray et. al. [13] have described a comprehensive set of models that cover ultrasonic, eddy current, and x-ray NDE systems. In formulating such modeling approaches, two components are essential if the models are to address the capability and reliability of the NDE method being considered. First, the models must be capable of simulating the physics of the measurement processes to the point where actual measured signals received can be predicted, i.e. complete "measurement models" [13] must be available. Second, the important sources of variability in the measurements must be included and used to estimate the probability of detection (POD) of defects. In an NDE experiment, these variability sources include operators, defect character, techniques and procedures, equipment, component geometry and surface properties, and others.

Operator dependencies are obviously a primary source of variability in NDE tests and have been a major focus of round-robin test block exercises. However, the other sources of variability are also important. For example, in reporting variability of echo amplitudes measured in ultrasonic weld tests, Forli [14] quoted 5 dB differences due to operators and 3 dB differences due to characteristics of defects of the same type (in this case lack of root penetration and lack of side wall fusion being the two flaw types being considered). These other (than operator) sources of variability can now be measured and/or modeled in many cases and combined with NDE measurement models to simulate the POD characteristics of automated NDE measurement systems. Such POD models, in fact, have already been developed in a unified fashion for ultrasonics, eddy currents, and x-rays [13].

Here, we will demonstrate the role that POD models can now play in improving the capability and reliability of NDE measurements, in guaranteeing the reliability of the components to which they are applied, and in designing new components with built in inspectability requirements. First, a description will be given of the elements that go into the formulation of a POD model. Then, a discussion will be given of how POD versus flaw size curves, obtained from such POD models, can be used to define NDE system capabilities and determine the influence of inspections on component reliability. Finally, a discussion will be given of how POD models can also now be used even in the very early design stage of components.

2. POD Concepts

Probability of detection for a component is, by definition, the ratio of the number of flaws detected by a given technique to the total number of flaws in the inspected components. POD is a well established measure of inspection performance that is directly related to important issues such as accept-reject criteria, frequency and quality of inspection, cost of failure and repair, etc. [15]. Mathematically, POD can be expressed as the integral of a conditional probability. For example, in a 1-D model of a detection process, where a rectified signal amplitude, y, received from a flaw of size a, is compared to a given threshold level, y_{th}, we have:

$$POD(a) = \int_{y_{th}}^{\infty} P(y|a)\,dy \tag{1}$$

where P(y | a) is the probability that a signal amplitude, y, is measured, given that a flaw of size a is present. This type of situation is common, for example, in ultrasonics, where y is the amplitude of a voltage signal on an oscilloscope screen and y_{th} is a voltage threshold set on the detecting equipment. Varying the flaw size, a, and plotting POD(a) then gives a POD curve (Fig. 1) which typically has a sigmoidal shape. In round-robin test block tests, the POD curve is often given by an assumed distribution function that can be characterized by a few free parameters. For example, Berens [16] used a 'log odds model' to express POD(a) as

$$POD(a) = \frac{\exp(\alpha + \beta \ln a)}{1 + \exp(\alpha + \beta \ln a)} \tag{2}$$

which has the general form of Fig. 1. However, if the POD curve is obtained through the use of models, no such strong a priori assumed shapes are necessary. This is important since the exact shape of the POD curve determines quantities such as false accepts (FA) and false rejects (FR), which are areas under the POD curve above and below a preset "critical" flaw size, a_c (Fig. 1). Models can estimate POD(a) directly provided that they can predict the signals received in a deterministic (i.e. non-probabilistic) measurement process, and measure and/or model the uncertainties and variability present in the signal. For example, in the ultrasonic case,

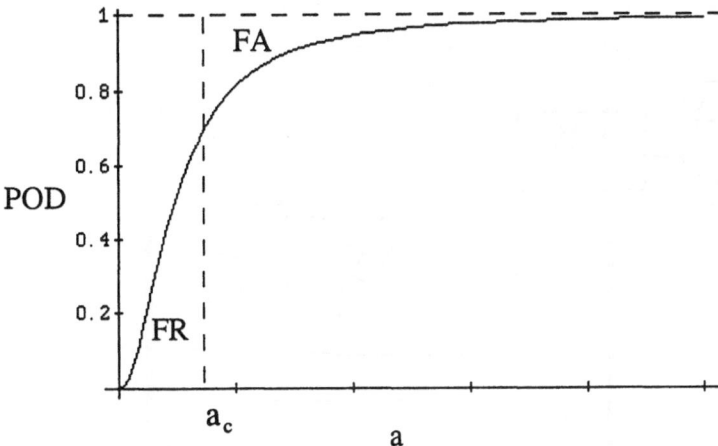

Fig. 1   POD versus flaw size curve.

288

using an approach developed by Rice [13], [17] for the detection of a narrow band rectified signal of amplitude, y, in noise, the probability of detection is given by

$$POD(a) = \int_{r=y_{th}}^{r=\infty} \int_{x_1} ... \int_{x_n} r/N^2 \exp[(r - y(a, x_1, ... x_n))^2 / 2N^2]$$

$$i_0[r\, y(a, x_1, ... x_n)/N^2]p(x_1)... p(x_n)dx_1...dx_n dr$$

(3)

where N^2 is the total power of the noise, $i_0(z) = \exp(-z)I_0(z)$, $I_0(z)$ being a modified Bessel function of the first kind of order zero, and the $p(x_i)$ (i = 1,2,... n) are the probability densities associated with the various measurement parameters x_i other than flaw size such as, for example, flaw location, orientation, type, etc. If these probability densities are measured or estimated and a model is present to calculate y , i.e. obtain the relationship

$$y = y(a, x_1, ... x_n)$$

(4)

then Eq. (3) can be integrated to yield a predicted POD(a) . Gray et. al. [13], describe just such a complete POD modeling process for ultrasonics and similar methods for x-rays and eddy currents. The resulting POD models have been used in a variety of applications, several of which we will describe later. Regardless of the NDE method being considered, however, a crucial element in the above process is obtaining Eq.(4) explicitly for a given NDE method, i.e. having a " measurement model" that can predict the measured signals in an NDE test. These measurement models have, in fact, been developed for ultrasonics , x-ray, and eddy current testing, each of which we will describe briefly.

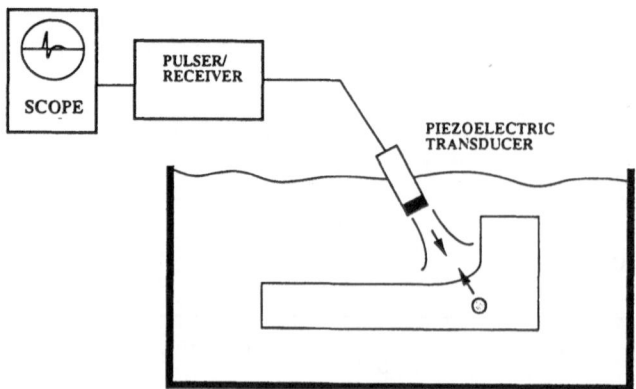

Fig. 2 Ultrasonic NDE measurement system.

3. Measurement Models

3.1 ULTRASONIC MEASUREMENT MODEL

In ultrasonic NDE, the output of a typical measurement is in the form of an rf voltage versus time waveform, V(t), (flaw signal) on an oscilloscope screen (Fig. 2). The frequency components of this waveform, $V(\omega)$, can be obtained through the use of Fourier analysis, using Fast Fourier Transforms, and, conversely, a knowledge of $V(\omega)$ is all that is needed to recover V(t). Using a general form of the electromechanical reciprocity relations [18], it is possible to model these frequency components as:

$$V(\omega) = \beta(\omega) \int_{S_f} (v_b \cdot \sigma_a - v_a \cdot \sigma_b) \cdot n \, dS \tag{5}$$

where v_a and σ_a are the velocity and stress fields, respectively, produced in the presence of the flaw and v_b and σ_b are the velocity and stresses produced in the material of the component when the flaw is absent. $\beta(\omega)$ is an "efficiency factor " that relates the conversion of electrical energy to mechanical motion and vice versa, during the transmission and reception processes. The integration in Eq.(5) is over any closed surface S_f that encloses the flaw. Although Eq.(5) appears simple in form, its practical implementation requires the extensive use of models combined with experiments. These models include models of 1) the ultrasound beam generated by the transducer, 2) the interaction of that beam with the boundaries of the part, 3) the propagation and attenuation of sound in the part material, and 4) the additional scattered waves generated by the flaw. Fortunately, such models have been developed recently and integrated into an explicit measurement model form [19] that makes the application of Eq.(5) possible for many important situations. That form is :

$$V(\omega) = \beta(\omega) P_a T_a C_a P_b T_b C_b [2 i A \rho_1 v_1 / k_1 r^2 \rho_0 v_0] \tag{6}$$

where P_a and P_b are propagation terms given by

$$P_{a,b} = \exp[\sum_n (i k_n d_n - \alpha_n d_n)] \tag{7}$$

with d_n and α_n being the propagation distances and material attenuation coefficients, respectively, associated with the waves traveling from the transducer to the flaw and back. Similarly, $T_a C_a$ and $T_b C_b$ are the product of plane wave transmission coefficients (T's) and diffraction coefficients (C's) associated with the waves traveling to the flaw and back. The quantity A is the plane wave far-field scattering amplitude of the flaw, r is the transducer

290

radius, and $\rho_1 v_1, \rho_0 v_0$ are the acoustic impedances of the component and surrounding fluid media, respectively. Finally, k_1 in Eq.(6) is the wavenumber associated with the received wave mode. Thompson and Gray, have shown [19] how all the terms in Eq (6) can be modeled, except for the distances d_n, attenuation coefficients α_n, and the efficiency factor, $\beta(\omega)$, which can be obtained experimentally through measurements taken either in the flaw setup of Fig. 2, or in separate reference/calibration experiments.

When Eq. (6) is Fourier transformed back into a received voltage versus time signal, $V(t)$, whose peak rectified amplitude, V_0, is taken, an explicit measurement model prediction of the form indicated in Eq.(4) is found for the ultrasonic case, with $y = V_0$. When these measurement model results are subsequently placed back into Eq. (3), a complete POD model for the ultrasonic case is then obtained.

The quality of such POD calculations, of course, is strongly dependent on the ability of the ultrasonic measurement model to accurately predict the features of an experimental ultrasonic signal. Figures (3) and (4) show a simulated (Fig 3) and experimental (Fig. 4) waveform produced by a 1 mm diameter glass sphere placed in a powder metal 304 stainless steel sample. Comparing the major amplitudes of these two signals, it is evident that the ultrasonic model does indeed model well this measurement process.

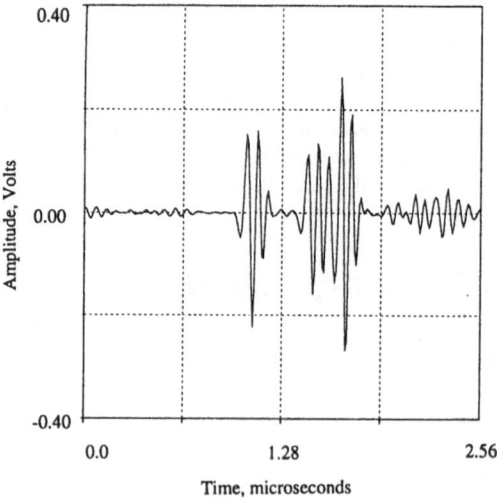

Fig. 3 Simulated waveform from a 1 mm glass sphere in powder metal 304 stainless steel specimen.

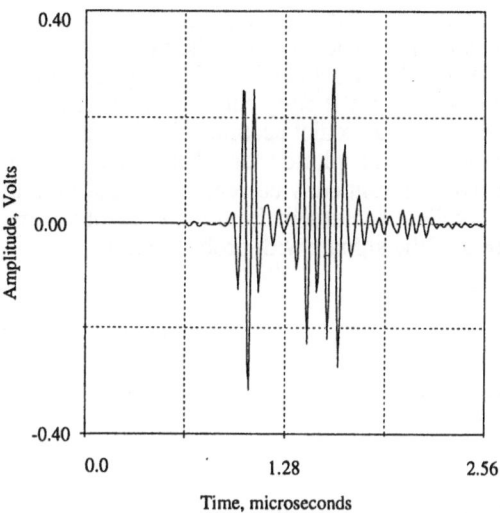

Fig. 4   Measured waveform from a 1 mm glass sphere in
powder metal 304 stainless steel specimen.

3.2 EDDY CURRENT MEASUREMENT MODEL

In eddy current testing, a relationship similar to the ultrasound case can be developed as
[13]:

$$\Delta Z = 1 / I^2 \int_{S_f} (E_a \cdot H_b - E_b \cdot H_a) \cdot n\, dS \qquad (8)$$

which relates the complex electrical impedance produced by a flaw, ΔZ, whose real and
imaginary parts, $\text{Re}\,\Delta Z$ and $\text{Im}\,\Delta Z$, respectively, are typically displayed on a storage
oscilloscope screen (Fig. 5) called an eddyscope. In this case, E_a and H_a are the electrical
and magnetic fields generated in the presence of the flaw by the transducer (a coil carrying
an a.c. current I) and E_b and H_b are fields generated when the flaw is absent. Again, S_f
is an arbitrary closed surface that surrounds the flaw. As in the ultrasound case, detailed
models are needed of the fields generated and their interaction with the part and the flaw.
Currently, this modelling is done numerically using the Boundary Element technique. Since
eddy current inspection is primarily a surface inspection (eddy currents in general do not
penetrate deeply into the material being inspected), the Boundary Element method, which
requires the modelling of the fields only on the surfaces of the part and flaws, is
particularly efficient.
        For the eddy current case , the output signal ΔZ = y in Eq. (4). Since ΔZ is
complex, the POD can be determined by using a two-dimensional detection criterion and
introducing the sources of variability (liftoff, material noise, etc.) [13].

As in the ultrasound case, the eddy current model must be able to simulate the signals received from an actual experimental setup if the POD calculations are to be valid.The amplitude of these signals, of course, is highly dependent on the eddy current probe itself. Fig. 6 shows the ability of the eddy current measurement model to take such probe characteristics into account and predict accurate signals. In Fig. 6 the magnitude of the impedance change versus position is plotted for a 1mm deep crack in Ti6Al4V as modeled and measured for five different eddy current coils. It is clear that the model and experimental results are in excellent agreement, both in the absolute amplitude of the signals measured and in the variation of the signals with probe position.

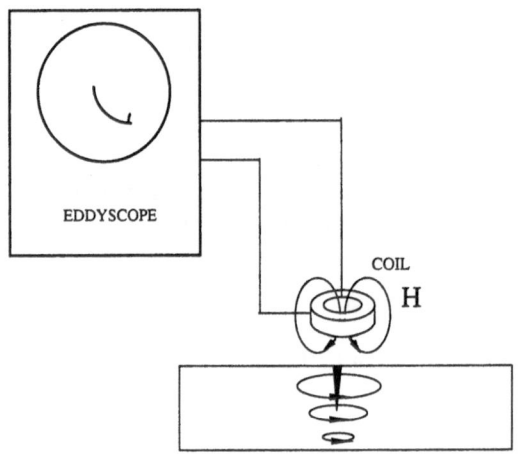

Fig. 5 Elements of an eddy current measurement system.

3.3 X-RAY MEASUREMENT MODEL

A measurement model has also been developed for a general x-ray inspection setup (Fig. 7). In this case the basic governing relation is [13]:

$$I(x,y,E) = I_0(E) \int_{Source} \{\exp[-\mu(x,y,E)\rho(x,y)] / r^2(x,y)\} dS \tag{9}$$

where I is the radiation intensity at the surface of the film (or other detector), E is the energy of the source, I_0 is the intensity produced at the x-ray generator, μ is the linear absorption coefficient, ρ is the x-ray path length through the sample, and r is the

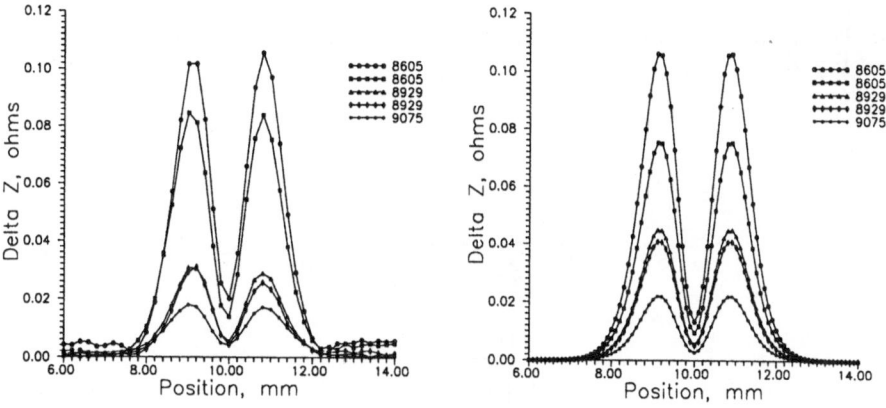

Fig. 6 Comparison of (a) measured and (b) calculated flaw
signals for a 1-mm deep crack in Ti6Al4V using five
different eddy current probes.

distance from the source to the detector. The integration is over the area of the source.

Since the received x-rays are typically recorded as an image, a model must also be
available for the image-forming process. This model for film is:

$$D(x, y, E) = D_0[1 - \exp\{\sigma(1 + \eta)I(x, y, E)t + \delta\}] \tag{10}$$

where D is the film density, σ is the interaction cross section of an x-ray with a film grain,
η is the coefficient of x-ray scattering, t is the time of film exposure, δ is the natural film

fog density, and D_0 is the maximum film density.

Implementation of this x-ray inspection model typically involves the use of ray
tracing. For complex geometries, this tracing is done directly off a CAD solid model.

Determining a POD for x-rays is somewhat different from the ultrasonic and eddy
current cases because the data is in the form of an image on film whose density differences
are detected by the human eye . Thus, the POD must be determined from a knowledge of
human vision capabilities [13].

To illustrate the capabilities of the x-ray measurement model to quantitatively predict
the changes present in a simulated image, Fig 8 shows film contrast versus thickness of an
aluminum alloy shim as predicted and measured for a specific industrial x-ray generator. As
can be seen, the agreement is quite good. One of the added advantages of using the model
approach, however, is the ability to also change many of the input parameters of the
measurement process and see immediately the effect on the output. For example, if the

294

aluminum alloy sample (which contains 4% copper) were replaced by pure aluminum instead, the resulting contrast versus thickness curve would be as shown by the dot-dash curve of Fig. 8.

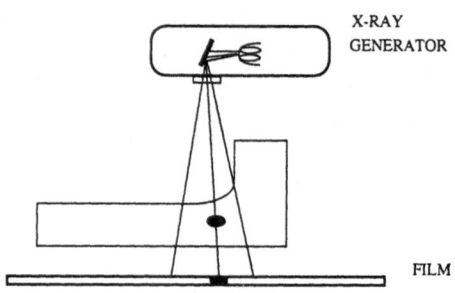

Fig. 7 Elements of an x-ray measurement system.

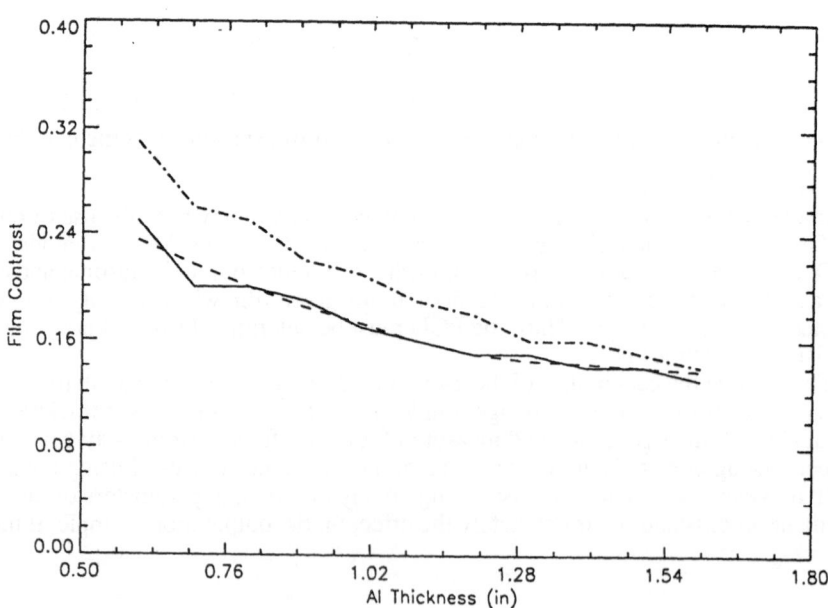

Fig. 8 Measured (dashed line) and modeled (solid line) contrast
curves for varying thickness of aluminum. Contrast curve
(dot-dash line) for pure aluminum.

4. POD and NDE System Performance

As an example of how POD models can be used to evaluate and validate NDE system performance and help in the setting of calibration thresholds, etc., consider the

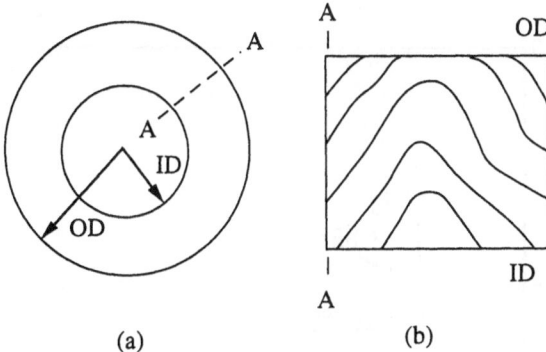

Fig. 9 (a) Geometry of a forged steel annular ring (b) Cross-sectional view
showing forging flow lines.

simulated ultrasonic inspection of a steel annular forging in the shape of a ring (Fig. 9a). Here, the defect that is simulated is a "pancake-like" silica inclusion. A defect of this type tends to align along the forging flow lines (Fig. 9b) whose precise directions are unknown. The simulated inspection system is one where the probe, a 10 MHz planar immersion transducer, can be indexed either vertically or radially and the flaw signal, detected in back scatter, is rectified and compared to a fixed (voltage) threshold. The threshold is set here at a 100% level corresponding to the signal received from a #1 flat-bottom hole at normal incidence to the transducer beam. The inspection requirement is to detect reliably (POD = 0.9) a 0.635 mm diameter flaw. Figure 10 shows the ultrasonic POD model predictions of the POD versus flaw size curve for this type of inspection as a function of the choice of the threshold. The dotted vertical line in Fig. 10 is placed at the target flaw size mentioned previously and the dotted horizontal line at the specified POD value. It can be seen from Fig. 10 that the 50% and 85% thresholds both give adequate detectability, but the 100% choice is inadequate.

As another indication of how inspection parameters can be set with such POD models, Fig. 11 shows the results of varying both scan index spacing (distance between successive transducer locations where data are taken) and threshold. Continuous rotational scan spacings of .05, .10, and .25 inches were simulated, and the thresholds chosen to have POD = 0.9 at the specified defect size. It can be seen that the coarse scan setting/low threshold combination gives a less steep POD curve, resulting in higher false rejects, but that there is no significant difference between the moderate scan spacing/threshold settings and the very fine/high threshold.

Finally, in Fig. 12 we show the effects that flaw characteristics can have on the POD curves. In that figure, the orientation of the transducer is held fixed relative to the surface of the forging in a continuous rotational scan, with scan index of 0.10 inches and threshold set at 100%, but the mean angle of the flaw relative to the probe beam is varied. It can be seen that the POD curves in Fig. 12 change considerably between mean tilts of 0-30 degrees. This indicates that a single fixed probe angle is inadequate to guarantee good detectability in this geometry.

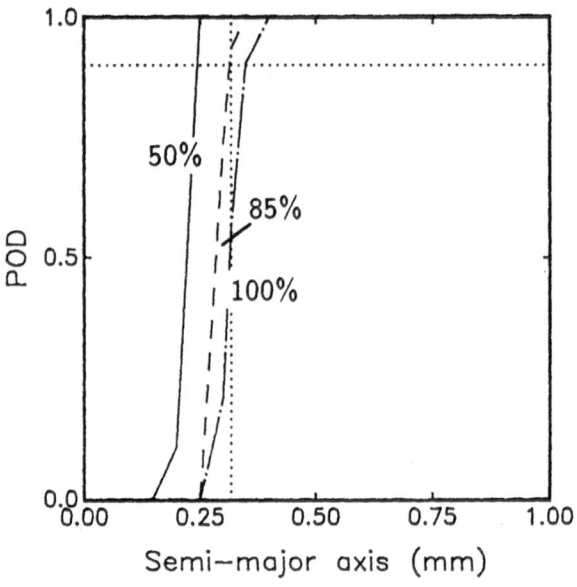

Fig. 10 Influence of threshold setting on the POD curve.

These examples demonstrate some of the power of POD models to contribute to quantitatively improving NDE inspection performance and reliability. POD models can also be used to evaluate the influence of inspections on the reliability of the component being inspected itself, as shown in the next section.

5. POD and Reliability

In many applications, NDE techniques are applied to metal components that are subject to progressive failure mechanisms such as fatigue. In these cases, inspection information can be used to help extend life and guarantee safe performance. To join NDE inspection capability and component reliability, however, it is not adequate to compare detectability at a single size, as done in the previous examples. Instead, the POD curve and entire flaw growth must be considered together. A recent study [20], for example, has combined inspection capabilities and fatigue crack propagation information to produce a reliability model that can quantitatively predict the effects of inspections on component failure. In that model, the measure of reliability was taken as the hazard rate function for a component, which is essentially the probability of the component failing in the next cycle. Garrigoux [20] developed an expression of the form:

$$\lambda_i(t) = \lambda_{NI}(t) \prod_{m=1}^{m=i} [1 - P_m(t)] f_i(t) \tag{11}$$

where λ_i is the hazard rate with i inspections, λ_{NI} is the hazard rate with no inspections, and $P_m(t)$ is the probability of detection of a flaw at inspection m of a part that will fail at

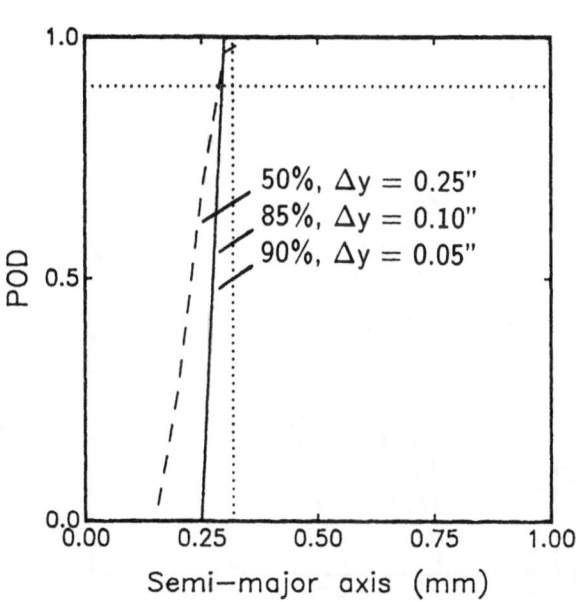

Fig. 11 Influence of scan spacing and threshold settings on POD.

time t, or equivalently, $P_m(t) = POD[a_m(a_c,t)]$ where a_m is the size of crack that will grow to a critical crack size, a_c, at time t. The function $f_i(t)$ is a function of a number of terms involving failure and detection probabilities [20].

The hazard rate function is a fundamental reliability function that can be used to determine a variety of important figures of merit such as cumulative probability of failure, mean time between failures, etc. Thus, Eq. (11) gives a methodology of calculating those important parameters and the influence of inspections on them. Also, if cost tradeoff information is available, Eq. (11) is an important input to life cycle cost estimates.

298

If one examines Eq.(11) carefully, it follows that the basic inputs needed to calculate the hazard rate function (with inspections) come from materials information (crack growth behavior a = a(t) or a = a(N) where N is the number of cycles, and an initial flaw size distribution at t=0), failure information (the definition of a critical flaw size, a_c), and inspectability information (the POD(a) curve). Thus, POD curves also serve as a direct input to reliability evaluations. For example, Fig. 13 shows the hazard function versus number of cycles (without inspections) calculated for the fatigue of a "pin" component analyzed in the NIST/ISU/NU program at the Iowa State University Center for NDE [21]. As expected, this function simply grows monotonically with number of cycles. Figure 14 shows the hazard function curve behavior when ultrasonic inspections are performed at inspection opportunities 11, 13, and 15. It is seen that each inspection in this case produces a significant drop in the hazard rate function, but that because of the rapid crack growth process in this example, the hazard function "recovers" before the next inspection. Having quantitative information of this type, one can set the recommended inspection intervals to guarantee a specified level of reliability.

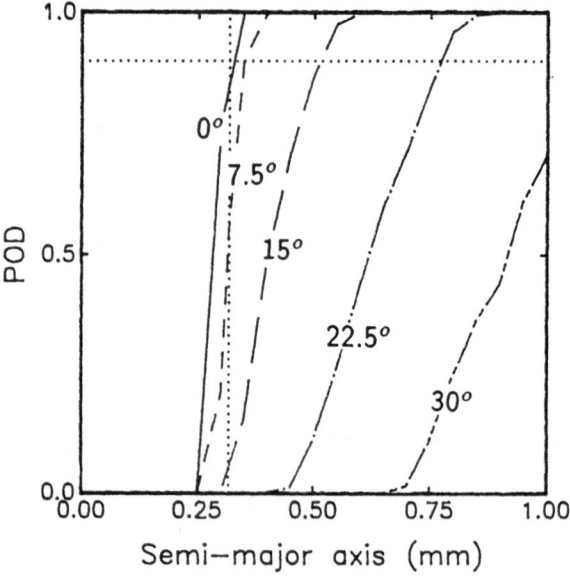

Fig. 12 Influence of flaw orientation on POD

6. POD and Design

Another application of POD and measurement models is in the area of design. NDE has traditionally not played a role in the design of components because NDE evaluations normally have required the existence of prototype parts to evaluate. At that stage, designs are often difficult and expensive to change. However, with NDE models available, if those models are linked to the CAD geometry description of a part, and combined with stress, materials, and other design information, NDE issues can be considered, even at the very early design stages. If, in addition, reliability and associated life cycle cost models are integrated into the design process as well, a new unified life cycle engineering (ULCE)

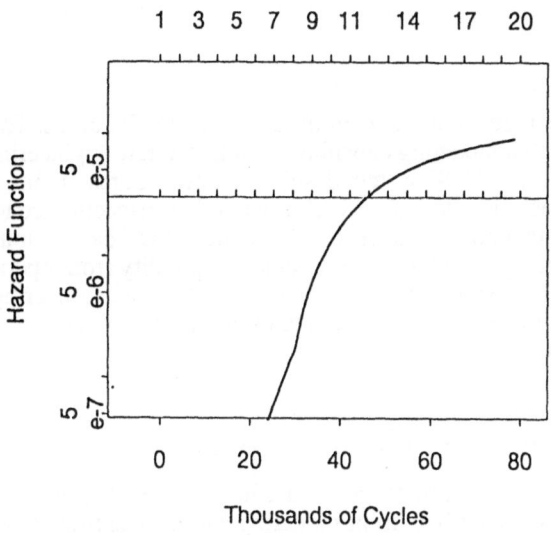

Fig. 13 Hazard function versus time (cycles) - no inspections.

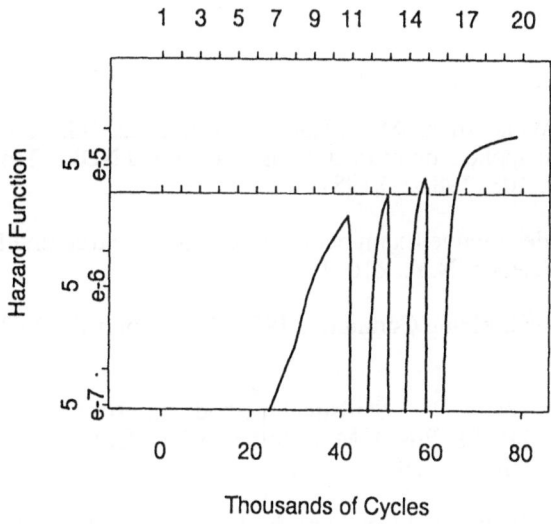

Fig. 14 Hazard function versus time with ultrasonic inspections at opportunities 11, 13, and 15.

approach can be taken to the design and analysis of components. Such an ULCE capability has in fact been recently developed in the NIST/ISU/NU program [21]. POD models and

POD calculations certainly play key roles in the extended design procedures of this ULCE environment. A description of those roles can be found in the article by Schmerr and Thompson [21].

7. Conclusions

We have described a number of the uses of model-based POD curves. There are, however, many more. Having such a modeling capability is in fact a new engineering technology that can be exploited to optimize NDE system designs and procedures, define NDE system performance capabilities, and allow transferability to other inspection conditions than initially assumed. This technology can also result in new NDE calibrations/standards, allow NDE system validation, separate NDE measurement capability from operator effects, and bring NDE into the early design process. A future use of these engineering tools could be in the design of NDE simulators for the training of operators and the solution of particular technical inspection problems.

8. Acknowledgements

We would especially like to acknowledge the use here of the ultrasonic results of Drs. T. A. Gray and R. Roberts, the eddy current modeling and measurements of Dr. N. Nakagawa and J. Moulder, the x-ray results of Dr. J. Gray, and the reliability calculations of Prof. B. Meeker and Dr. C. Garrigoux. The reliability calculations in particular also rely on auxiliary design, stress, and materials information which was contributed by a number of faculty and students in the Center for NDE and at Northwestern University.

9. References

1. Annis, C. G., Van Wanderham, M.C., Harris, D. L. Sims, "Gas turbine engine disk retirement-for-cause: an application of fracture mechanics and NDE," Trans. ASME, Journ. Eng. for Power, 103, 198-200,1981.

2. Chockie, L. J., Nondestructive Examination in Relation to Structural Integrity, Ed. R. W. Nichols, Applied Science, N.Y., 116, 1978.

3. PISC Plate Inspection Steering Committee 1979 CEC Report EUR 6371, vol. I-V, 1979.

4. Berens, A. P. and Hovey, P. W., "Statistical methods for estimating crack detection probabilities," ASTM STP 798, Eds. J. M. Bloom and J. C. Eckvall, Am. Soc. for Testing and Materials, Phil., Pa., 79-94,1983.

5. Watkins, B., Ervine, R. W., and Cowburn, K. J., "The UKAEA defect detection trials," Br. J. NDT, 25, 179-185, 1983.

6. Watkins, B., Lock D., Cowburn, K. J., and Ervine, R. W., "The UKAEA defect detection trials on test pieces 3 and 4," Br. J. NDT, 26, 97-105, 1984.

7. Silk, M. G., Stoneham, A.M., and J. A. G. Temple, The Reliability of Non-Destructive Inspection, Adam Hilger, Bristol, England, 1987.

8. Coffey, J. M., Chapman, R. K., and Hanstock, D. J. ,"The ultrasonic detectability of a postulated 'worst case' flaw in a PWR pressure vessel," CEGB Report NW/SSD/82/0045/R April, 1982, CEGB, London, England, 1982.

9. Ogilvy, J. A. and Temple, J. A. G.," Diffraction of elastic waves by cracks: application to time-of-flight inspection," Ultrasonics, 21, 259-269,1983.

10. Haines, N. F., Langston, D. B., Green, A. J., and Wilson, R., "An assessment of the reliability of ultrasonic inspection methods," in Periodic Inspection of Pressurized Components, Institute of Mechanical Engineering, London, England, 239-255, 1982.

11. Thompson, R. B.., and Achenbach, J. D., EPRI Report NP-3822,1985.

12. Segal, Y., Notea, A., and Segal E., Research Techniques in NDT, Vol. 3, Chapter 9, Ed. R. S. Sharpe, Academic Press, London, England, 1977.

13. Gray, J. N., Gray, T. A., Nakagawa, N., and Thompson, R. B.," Models for predicting NDE reliability," Metals Handbook, Vol. 17, ASM International, Metals Park, Ohio, 702-715, 1989.

14. Forli, O.," Reliability of ultrasonic and radiographic weld testing," Proc. 9th World Conf. on NDT, Melbourne, Australia, 3A, 7, 1979.

15. Bray, D. E., and R. K. Stanley, Nondestructive Evaluation, McGraw-Hill Book Co., New York, 1989.

16. Berens, A. P.," NDE reliability analysis," Metals Handbook, Vol. 17, ASM International, Metals Park, Ohio, 689-701, 1989.

17. Rice, S. O., " Mathematical analysis of random noise," Bell Syst. Tech. J., 23, 282, 1944; 24,96,1945

18. Auld, B. A.," General electromechanical reciprocity relations applied to the calculation of elastic wave scattering coefficients," Wave Motion, 1, 3, 1979.

19. Thompson, R. B. and T. A. Gray, " A model relating ultrasonic scattering measurements through liquid-solid interfaces to unbounded medium scattering amplitudes," J. Acoust. Soc. Am., 74, 1279, 1983.

20. Garrigoux, C. ,"A reliability model for planning in-service inspections of components subject to fatigue failure," Ph.D. thesis, Iowa State University, 1992.

21. Schmerr, L. W., and D. O. Thompson, " Incorporating inspectability into design: the new role of NDE in concurrent engineering," Proc. ASME/NDE Division Topical Conference, " NDE's Role in Concurrent Engineering," San Antonio, Texas, April 21-23,1992.

EARLY DETECTION BY STIMULATED INFRARED THERMOGRAPHY. COMPARISON WITH ULTRASONICS AND HOLO/SHEAROGRAPHY

––––––––––

J.C. KRAPEZ, D. BALAGEAS, A. DEOM and F. LEPOUTRE
ONERA/L3C, B.P.79, F92322 CHATILLON Cedex, FRANCE

––––––––––

Summary

Infra-red stimulated thermography is a fast and global method. Among the recently emerging NDE methods, it is probably the less intrusive one since it really needs no contact at all with the tested structure. A new inversion, using an early detection of the contrast, is presented to demonstrate how it recovers the depth of the defects with accuracy and removes partially the loss of resolution produced by the lateral heat diffusion.

Ultrasonics methods are well established NDT techniques and start to become remote and contactless with the development of laser ultrasonics. Holography / shearography are global methods which produce images of rather large areas with a good spatial resolution.

A second goal of this paper is to compare the results obtained in thermography with these different new techniques. We have mainly used an artificial sample of carbon epoxy composite containing teflon inserts and a sample of composite with impact damages. The comparison of the results obtained with the first sample was exploited in order to determine what are the practical interests of each method. The second comparison aims mainly to evaluate the precision of ultrasonics and infra-red thermography in the determination of the spatial extension of multidelaminations.

303

X. P. V. Malague (ed.), Advances in Signal Processing for Nondestructive Evaluation of Materials, 303–321.
© 1994 *Kluwer Academic Publishers.*

EARLY DETECTION BY STIMULATED INFRARED THERMOGRAPHY. COMPARISON WITH ULTRASONICS AND HOLO/SHEAROGRAPHY

1. INTRODUCTION

A modern Non Destructive Testing (NDT) should satisfy three criteria to be considered as an industrial tool. It must be :
- fast and low cost,
- contactless,
- quantitative.

These criteria are valid not only for applications to cheap and primary goods but also for expansive and sophisticated systems as the ones developped in aeronautics. A good demonstration of this necessity is given by the control of the aging aircrafts for which the main part of the cost of inspections is the duration of the operations. With the standard methods used, the immobilization time for a C-type inspection (heavy control), which is of the order of 10 days, represents 80% of the price of the control /1/! This very long duration is essentially due to the step by step character of inspections, for instance the control of the rivets should be performed at each point with a local eddy current probe.

A solution is obviously the developpement of global contactless methods able to give fast images of large areas. For this reason, during these last ten years, the NDT community has proposed a few new methods with the challenge of a complete fulfil of the above mentionned criteria.

The L3C/ONERA contribution, in this domain, lays on the techniques of the Stimulated Infrared Thermography (TIS) which consists in a pulse uniform illumination of the surface of the studied specimen, followed by the analysis of its InfraRed (IR) emission as a function of time /2/. The second section of this paper is devoted to the recent improvements brought by our group to this analysis and which allows :
- to find with a good precision the in-depth position of the subsurface defects,
- and, with lesser importance, to remove the effects of the lateral heat diffusion.
The advantages of this new inversion will be checked, in an absolute way, by using artificial defects prepared in a composite material.

It will be then shown that the method still suffers some inconvenients :
- the defects localised at more than half the material thickness are difficult to be detected,
- the three dimensional (3D) heat diffusion modifies the shape of the deep defects, and their resolution which fastly decreases with the depth.

For these reasons, it is interesting to test, with the same artificial samples, other new global methods (laser ultrasonics, holo/shearography) to evaluate the solutions that they provide to the fundamental problem of inversion. A third part of this paper devoted to this study, aims to precise the field of application of these three different optical methods (or able to become fully optical) with respect to the TIS method. In a first step, this evaluation, managed in an absolute way by using the artificial samples tested by the TIS method, will lead to a selection of the most satisfying method. And finally, the TIS and this selected method will be compared during a common test of an actual impact damage in a composite.

2. EARLY DETECTION IN TIS INSPECTION

2.1. The industrial set-up developed at ONERA/L3C

The schematic set-up represented in figure 1, and used in the experiments described this paper, corresponds to the front surface configuration, i.e. illumination and detection on the same surface.

The pulsed sources available in our system can be either flashes (pulse duration 4 ms) or continuous IR lamps (48 kW), the pulse being achieved, in that case, by the opening/closing of mechanical shutters electromechanically piloted (the smallest pulse duration which can be obtained with this system is typically 200 ms). The flashes are only used to study very shallow defects, or defects in high conductive materials such as metals /3/. With the samples used in this paper, the pulse durations which are necessary, are of the order of a few seconds (we chose 2 s) and, thus, the experiments have been performed with the IR lamps.

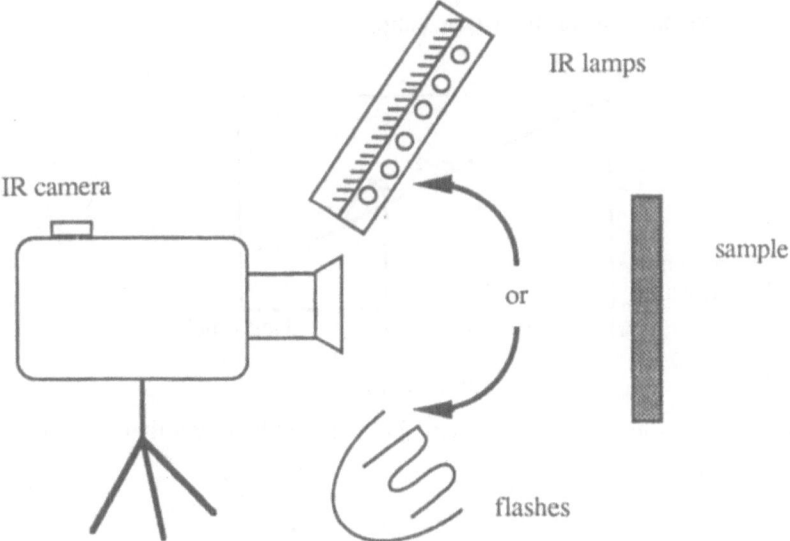

Figure 1
Schematic view of the TIS set-up.
Two kinds of illuminators are available : flashes for very fast materials and IR lamps with shutters. The camera is an AGEMA 880 long wave. The image are digitized (12 bits) and recorded by a microcomputer.

The sources placed at 40 cm from the sample surface, allow a power deposition of about 0.9 W/cm^2.

An AGEMA 880LW records the surface temperature as a function of time during periods of times ranging from 1 to 10 mn after the pulsed illumination. It is placed at 1.2 m from the sample surface which allows to have a field of view of 20 cm and a spatial resolution of 2 mm with the objective of 12° used for these experiments.

2.2. Inversion using the early detection

The detection and the characterization of the resistive subsurfacic defects is achieved by the research of the local emergence of a thermal contrast after the pulse illumination, i.e. an increase of the local temperature above the defect with respect to the one in a sound region, (see figure 2a). Generally, the resistance and the depth of the defect are deduced from the value and the time of the maximum contrast /4/.

This research is made pixel per pixel and finally leads to two images, respectively for the local resistance and for the depth of the subsurface defects. An alternative consists in the detection of the time corresponding to half the development of the contrast /5/. But, in both the cases, deeper are the defects, longer are these times, with important lateral heat diffusions as consequences. Under these conditions, the result of this inversion is an undersetimation of both the depths and the thermal resistances of the detected defects.

Figure 2
Maximum contrast and early detections.

a/ schematic explanations of the two detections

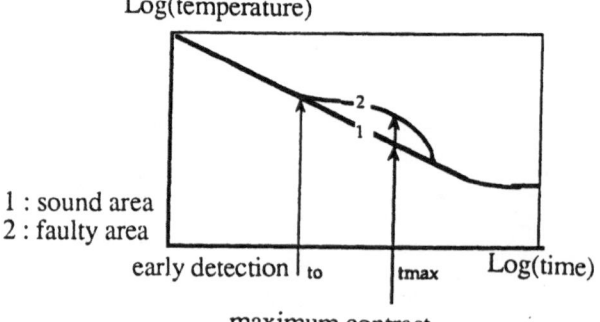

b/ definition of the threshold of detection. 2 ∂T should be larger than the noise.

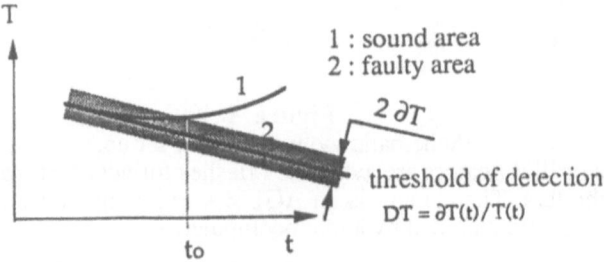

In order to reduce the importance of the 3D diffusion, it is obvious that the detection of the thermal contrast should be achieved as early as possible. The threshold of detection is depending upon the noise level of the experiment. Let us express this threshold, DT, as a percentage of the temperature above a sound region (more precisely the *increase* of temperature with respect to the temperature before the illumination) :

$$DT = \frac{\partial T(t)}{T_{ref}(t)}$$

(1)

This percentage is chosen very weak, for instance 1, 3, or 5%. For a given threshold, the time variation of the temperature of each pixel, $T(t)$, is then analysed and compared with the time variation of the temperature of an area known to be sound, $T_{ref}(t)$. When $T(t)$ exceeds $T_{ref}(t)$ of more than $DT * T_{ref}(t)$, the pixel is recognised to be above a faulty area (see figure 2b).

A very simple relation, between the depth z of the defect, the threshold DT, the time t_0 at which the threshold is reached and the thermal diffusivity κ of the material, has been established:

$$z = \sqrt{t_0 \, \kappa \, Ln(\frac{2}{DT})}$$

(2)

Actually, it is difficult to decide of an a priori value for DT : the best method consists in trying successive decreasing values of DT, down to a threshold for which the depth image starts to become ramdom, proving that the detection level is, at that moment, below the noise. The procedure is then started again with the immediately larger threshold value.

This early detection is insensitive to the thermal resistance since it has been demonstrated /6/ that all the curves corresponding to different resistances, for a given depth, are merged at the early times. It is thus necessary to deduce the thermal resistance image from the detection of the maximum contrast. This image is of course more perturbed by the 3D effects than the in-depth (z) image in spite of the use of a more accurate value for z.

2.3. Artificial defects

In order to demonstrate the advantages of the early detection of deep defects, two artificial samples (called respectively 1 and 2) were prepared :
- the sample 1 (see figure 3) is a 5.25 mm thick plate of carbon epoxy containing small inserts of teflon (thickness 80 µm) at depths varying from 0.4 mm to 4 mm. This sample allows to make an absolute test of the lateral and in-depth resolutions and also of the ability of the method to recontruct the shape of superimposed defects.
- the sample 2 (see figure 4) is a disc of polyester in the center of which concentric cavities of different depths are machined. This sample is used to probe the reconstruction of the contours, problem of crucial interest in the analysis of multidelaminations.

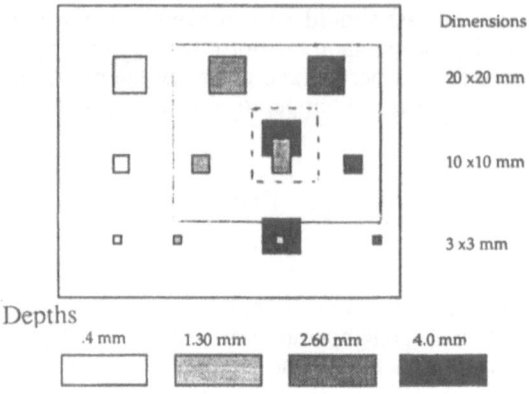

Figure 3
Carbon epoxy plate
The plate is 5.25 mm thick and contains inserts of teflon, 80 µm thick, at depths above-mentionned. Continuous line : field of view of figure 5. Dashed line : field of view of figure 7.

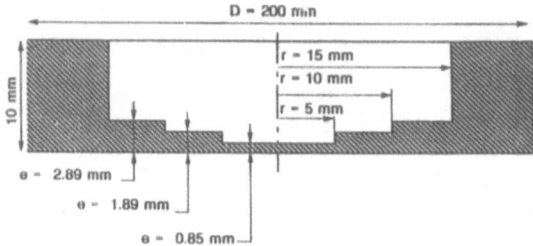

Figure 4
Concentric holes machined in a polyester plate.

As explained in the previous section, the comparison between the results deduced from the maximum contrast and the early detection is interesting only for the depth determination since almost no improvement is obtained in the resistance measurements.

For two different fields of view of the IR camera (see figures 5 and 7), the two in-depth images are quantitatively discussed. In figure 5, the relative improvement obtained with the early detection for the z determination of the deepest defects (z = 4mm) is of about 35 %.

A general comparison between the results of both the methods is given in figure 6 where the depths, identified in the centers of the defects, are plotted as functions of the true depths. It is also interesting to note the overestimation of z for the shallowest defects (z = 0.4 mm). This error is related to a bad use of the early detection. Indeed, for these shallow defects, the pulse duration (2s) is not negligible with respect to the early time of detection (5s). To obtain a better value, a shorter pulse (500 ms) should have been necessary.

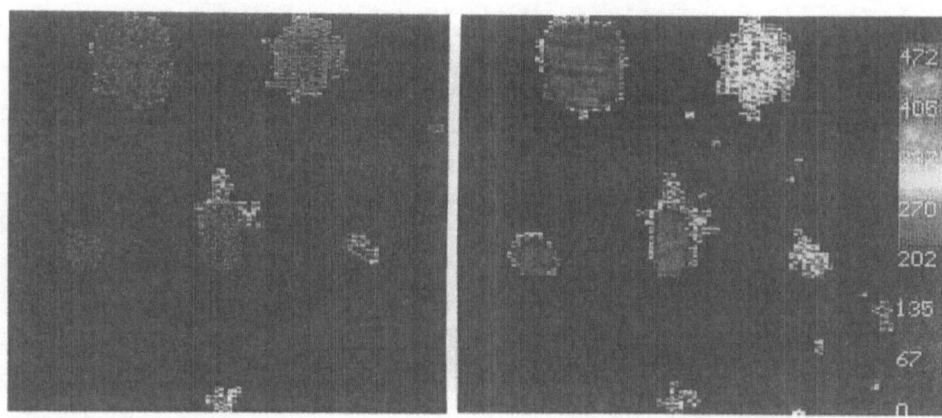

Figure 5
TIS in-depth images
In-depth images of the carbon epoxy plate (large field of view) obtained by the
method of the maximum contrast (left side) and the method of the early contrast
(right side), unit of the scale : 10 µm.

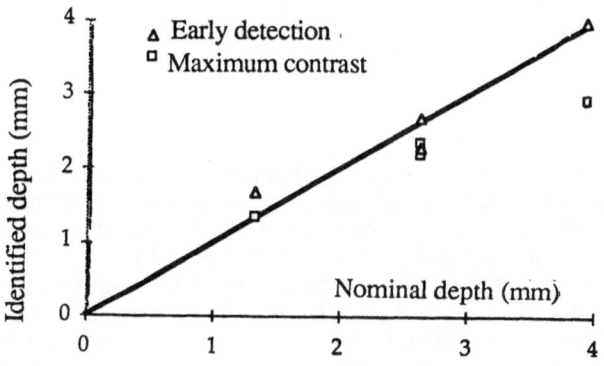

Figure 6
Comparison between the depths deduced by the early detection and the
maximum contrast detection.

The comparison between the images of the superimposed defects (see figure 7)
shows that the 3D effects which strongly alter the contours of the defects in the
method of the maximum contrast are also reduced with the early detection : the
detected surface of the partially hidden defect at z = 4mm is almost doubled.

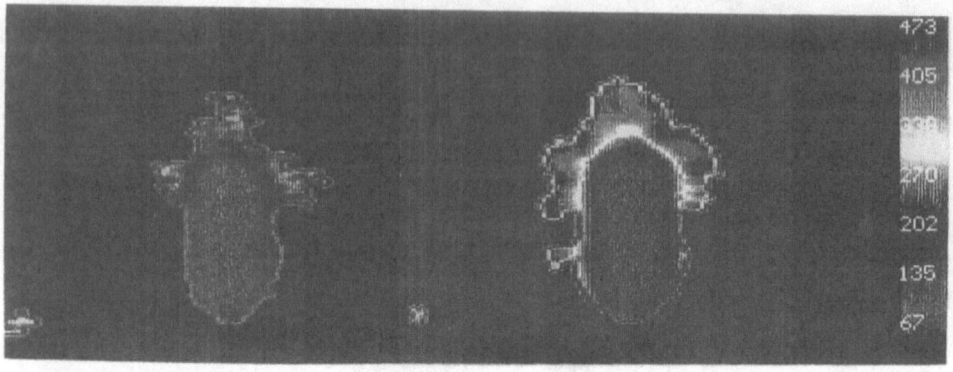

Figure 7
TIS in-depth images
In-depth images of the carbon epoxy plate (small field of view) obtained by the
method of the maximum contrast (left side) and the method of the early contrast
(right side), unit of the scale : 10 μm.

This result is even more obvious in figure 8 obtained with the sample 2 : the
profiles of the concentric cavities are much more precise in the case of the early
detection provide that the threshold is kept above the noise level (a deterioration of
the results is observed when the threshold is taken smaller than 3%).

Nevertheless, these remarkable improvements are not totally satisfying. The
images of the defects located at more than half the thickness are still strongly
distorted by the 3D diffusion. New efforts to achieve a fast deconvolution of the
surface temperature in order to remove the destruction of the information by the
heat diffusion are in progress in several laboratories and promising results were
presented during this workshop by Prof. Favro of the Wayne State University /7/.
But, the inversion of a true defect will remain always a difficult problem to solve.
It is thus now interesting to have a glance at the performances of the other new
methods.

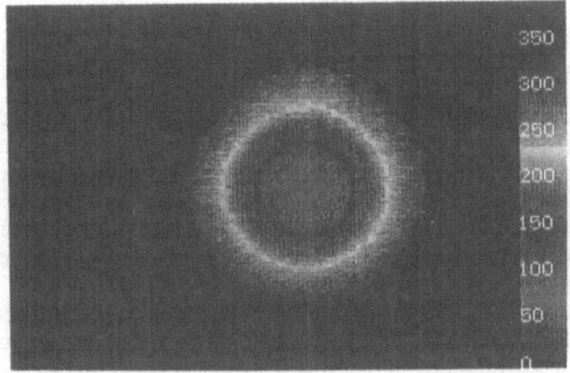

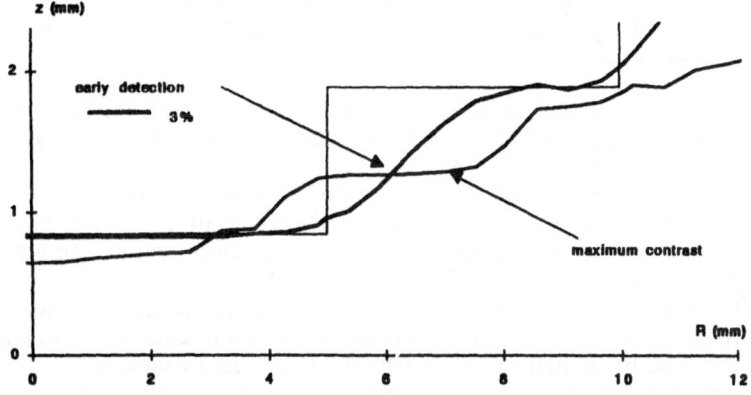

Figure 8
TIS image of the concentric holes.
The in-depth image (upper view) is obtained by the early detection with a threshold equal to 3%, unit of the scale : 10 μm. The profiles given in the lower view compare the results of this early detection and the ones deduced from the maximun contrast.

3. OPTICAL METHODS

3.1. General principles

It is well known that the interference of two beams of coherent light leads to a resulting intensity depending upon the difference of their optical pathes, from the region where they are produced to the one where they are mixed. The optical methods of NDT make use of this very simple idea to detect the anomaleous sample surface deformation above a subsurface defect after, or during, the application of a perturbation (generally a mechanical excitation) to the specimen.

The measurements can be performed in two different ways :

- **locally** : in that case, the interference is produced between a reference beam which is internal (i.e. independant of the sample) and a probe beam reflected by the sample surface during or after its excitation. This method is used in laser ultrasonics (US) and in vibrometry. Good descriptions of these technics are given in reference /8/.

- **globally** : an interference pattern is then realised between two surface waves diffused by the sample surface. The reference one, diffused by the surface at rest, and the probe one, diffused by the sample when submitted to the excitation. Such a method is used in holography and in shearography. Good descriptions of these technics are given in reference /9/.

If the detection consists simply in a counting of the fringes, the sensitivity to the surface displacement is a fraction of the wavelength of the laser used to produce the interference, i.e. typically 1000 Å.

This value can be too small for applications in which the excitation is weak. It is especially the case for the US generated by the absorption of a laser pulse. In the thermoelastic regime (non destructive), the surface deformations are generally of the order of a few tens Å.

To achieve such a sensitivity, it is necessary to read the phase of the sinusoidal interference pattern. This measurement is impossible directly with accuracy. The technique used is the heterodynage which consists in a frequency shift from the light domain (10^{13} Hz) to the MHz domain /9/. The heterodynage is well mastered in the two beams interferometers like the ones developed by the company BMI and presented in figure 9. Practically, the internal reference beam is mixed with an ultrasonic wave in an acousto-optic crystal (see figure 9). This technique is now currently used for laser US detection, i.e. for local detection, but, up to now, it is not the case for the global detections since it is then necessary to frequency-shift a very complex reference wave (the probe wave diffused by the sample surface at rest).

As a conclusion of this general approach, it can be said that the optical methods are divided in two parts:

- *local detections with very high sensitivities.* It is the domain of the laser US. Their main inconvenient comes from their local aspect : they are slow.

- *global detection with medium sensitivities*. It is the domain of holo/shearography. Their main inconvenient is the necessity of using rather hard excitations, but they are very fast.

Figure 9

Principle of the heterodyne interferometric detection of ultrasounds used by the BMI company.

The He Ne laser (frequency f_L) is seperated in two beams by a beam splitter. The reference beam (upper beam in the Dove prism) stays at the frequency f_L, while the measuring beam (sent to the sample surface) is shifted at f_L+f_B by an acousto-optic crystal excited at f_B. After the reflection on the sample surface, the two beams are mixed and their sum exhibits, around f_B, three frequency components. The intensity of the lateral bands are proportional to the surface displacement (amplitude of the ultrasound) at the frequency f of the ultrasound.

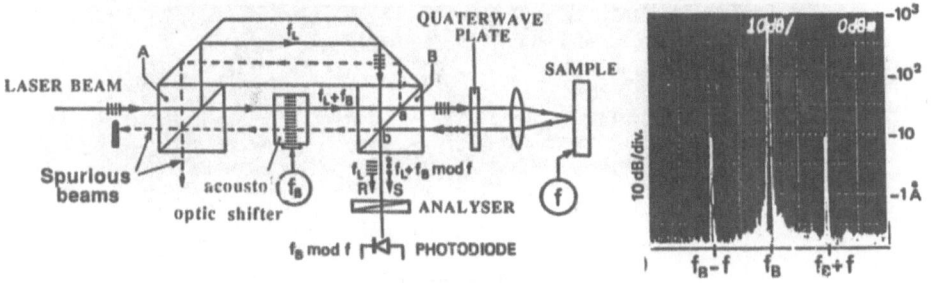

3.2. Artificial defects

3.2.1. Local measurements. Ultrasonics.

The system that we plan to use in a next future is the BMI apparatus described in figure 9. It makes use of a pulse YAG laser for the generation and of an heterodyne Mac Zenhder interferometer for the interferometric detection. The principle of the apparatus can be found in /9/. D. Royer /10/ with the same system has obtained a sensitivity of about 10^{-4} Å.

The first experiments, that were performed in our group with this apparatus, aimed at defining the power and the directivity of the ultrasounds generated by the laser. Typically with an energy of 10mJ, a pulse duration of 15 ns, the shear waves detected at a few MHz correspond to a displacement of about 10 Å.

The main problem encountered with this system, and still not solved, comes from the detection which needs to be adjusted at each point of the sample surface if it is not perfectly reflecting. For this reason, the results which are presented in this section, were not obtained with the laser US apparatus, but with conventional piezoelectric sensors approaching the conditions of the laser generation.

The US images of figures 10 to 12 are obtained by the pulse-echo method, in total immersion, with a unique piezoelectric sensor working at 10 MHz and focalised in the medium part of the sample. The focal area is of the order of 1 mm^2 and the measured displacement (longitudinal wave) is equal to about 10 Å (very similar to the one produced with the laser).

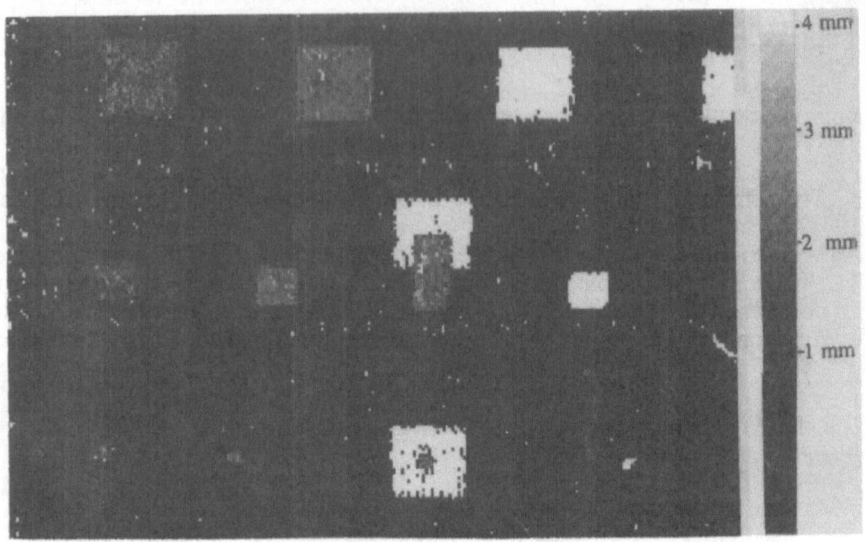

Figure 10
US image of the carbon epoxy plate.

All the inserts in the carbon epoxy plate are detected and the depth deduced from the time of flight is close to the nominal value except for the shallow defects which produce an echo too early with respect to the time delay of the electronic used. The general view of the plate clearly shows that the resolution is of the order of the diameter of the focal spot. It is also interesting to note that the superposition of two defects does not affect this resolution (see figures 11 for sample 1 and 12 for sample 2).

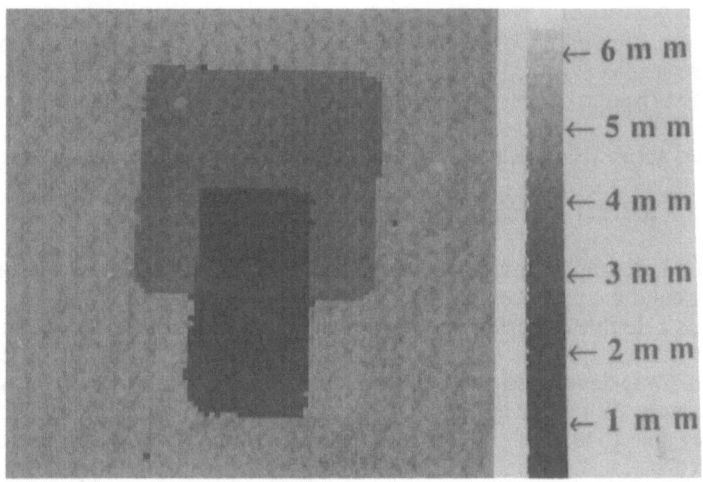

Figure 11
US image of the superimposed defects in the carbon epoxy plate.

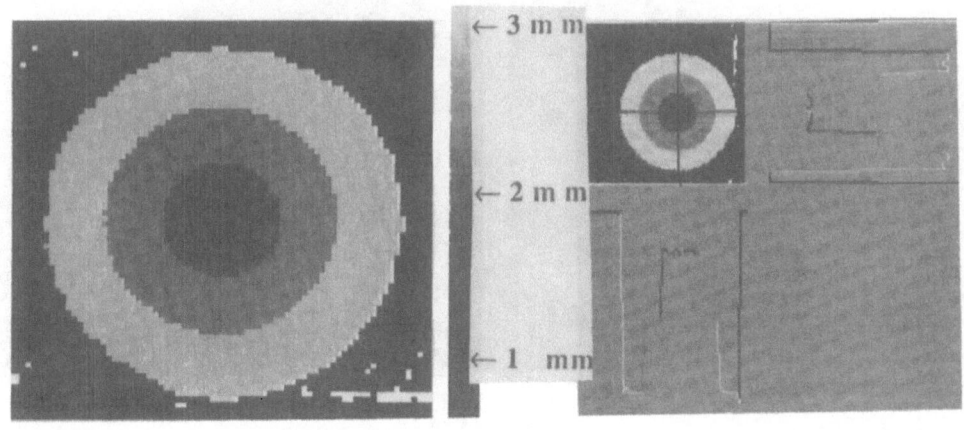

Figure 12
US image of the concentric holes in the sample 2.

3.2.2. Global methods

The carbon epoxy plate described in section 2.3., was tested by holography and shearography. In both the methods, the excitation used was a partial depressurisation of the whole sample placed in a vacuum chamber : as the pressure around the plate decreases, a small bump appears in the regions of the surface located above the inserts of teflon which are always accompanied of an air gap.

3.2.2.1. Holography

The holographic image presented in figure 13 was obtained at a pressure of 0.5 atm around the plate.

It is a double exposure photography /11/ : a first hologram is recorded with the wave diffused by the plate without any depressurisation in the chamber. The hologram is developed and placed exactly at the location where it was recorded, the sample staying always at the same place. This hologram illuminated by the laser is now able to restitute, at any moment, the coherent wave diffused by the sample surface at rest, i.e. what was called the reference wave in the previous section.

A depressurisation is then realised in the chamber while the sample and the hologram are still coherently illuminated. If the sample is observed through the hologram, an interference pattern is produced between the reference wave coming from the hologram and the wave diffused by the distorted sample. This interference pattern, which represents the local deformations of the sample under depressurisation, can be recorded by a visible camera as the pressure in the chamber is decreased, or photographied at a given pressure as it was done for the figure 13.

Figure 13
Holographic image of the carbon epoxy plate.

The restitution of the shape of the inserts is excellent at least for the shallow ones. The resolution is surpringly good : even the very small inserts (first column, third raw) are revealed. In addition this resolution is not stongly affected when two defects are superimposed.

It is also interesting to note the large fringes which cover the whole sample and which are related to a global deformation of the plate. This deformation depends clearly on the clamping of the plate (see the holder in the upper left side).

The deeper are the defects, the larger is the interfringe, but this interfringe depends also upon the lateral dimension of the insert : note for instance the fringes in the second column of defects much more spaced for the small defect (second raw) than for the large one (first raw). In our opinion, this problem is not easy to solve since it means that to be really quantitative, the method needs the inversion of a thermoelastic calculation already difficult in the case of artificial defects and probably impossible for true defects.

Nevertheless, there may be another approach in the observation of the movie of the generation of the fringes as the pressure decreases : it seems that it exists a clear relation between the depth of the defect and the pressure at which the first black fringe occurs.

3.2.2.2. Shearography

The principle of a shearographic camera /12/ is to produce an interference pattern by mixing the speckles of two images slighted sheared of the surface coherently illuminated. This interference pattern can be considered as a field of fringes which are very tight (invisible). The procedure of NDT with a shearographic camera is almost the same as the one described in the previous section for holography :

- a first shearographic image is recorded on a photographic plate when the sample is at rest.

- a second exposure of the same photographic plate is then made when the sample is deformed. The superposition of the two shearograms produces a moiré

which is related to the ratio $\frac{\partial z}{\partial x}$ where ∂z is the difference of the displacements between the two corresponding shear points, and ∂x is the shear displacement produced by the camera (the direction of the shear displacement is adjustable, for instance it could be y instead of x).

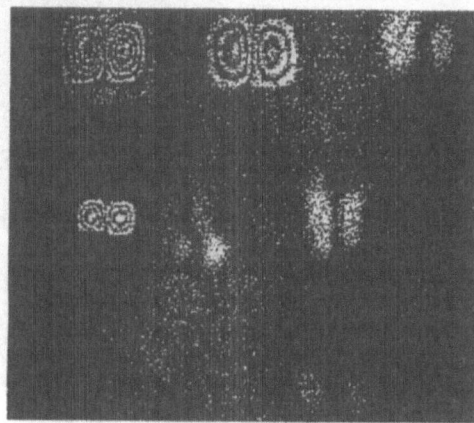

Figure 14
Shearographic image of the carbon epoxy plate.

The shearography of the carbon epoxy plate is presented in figure 14 : this image is composed of several partial views of the sample because the field of view of the camera used was not large enough to record the whole surface. The maximum depressurisation possible in the chamber was equal to 10 kPa below the atmospheric pressure. In the composite figure 14, the fringe patterns correspond to different values of the depressurisation for the different depths of the teflon inserts : 1 kPa for the depth of 0.4 mm, 4 kPa for the depth of 1.3 mm, and 10 kPa for the depth of 2.6 mm. The deepest defects were not detected, which is not surprising taking into account the weak depressurisation which was possible with that system.

The defects appear as "bull eyes" : the distance between the centers of the "eyes" of a given defect is directly related to the shear distance of the camera (it can be verified in the image that it is identical for all the inserts). The dimension of the interference pattern above a defect is equal to its lateral extension. When the number of fringes is large enough, the resolution is as good as for holography.

Unlike the holographic image (see figure 13), there are no fringes due to the global deformation of the whole sample. The shearographic image is only sensitive to the sharp local deformations. This should be related to a lesser sensitivity to a noisy environment of shearography than holography.

As in the case of holography, the main problem encountered here is the ambiguity of the relation between the value of the interfringe on the one hand and, on the other hand, the depth and the lateral extension of the defect. Here too, the problem of inversion seems really quite difficult to solve.

3.3. Conclusion

As a conclusion of these tests, it appears that, up to now, there is still not a new NDT method which fulfils all the crireria defined in the begining of this paper :
- the global methods are fast, low cost and contactless, but the quantitative information is not available for the moment,
- the laser US are contactless, quantitative but difficult to use during a scanning on an ordinary surface. Nevertheless, technical solutions start to become available /8/ though very expansive at that time.

Anyway, the quality (resolution, large signal to noise ratio, quantitativity) of the conventional US images seems so good that the laser US is probably the most attractive of the optical methods. For this reason, the selected method used in the next section is ultrasonics.

4. COMPARISON BETWEEN US AND TIS DETECTIONS OF IMPACT DAMAGES IN COMPOSITES. GENERAL CONCLUSION

4.1. Impact damage in composites

The goal of this section is to compare the performances of the TIS/early detection and the conventional US (guessing that it could be an anticipation of laser US) methods for the quantitative evaluation of the extension of an impact damage in a composite.

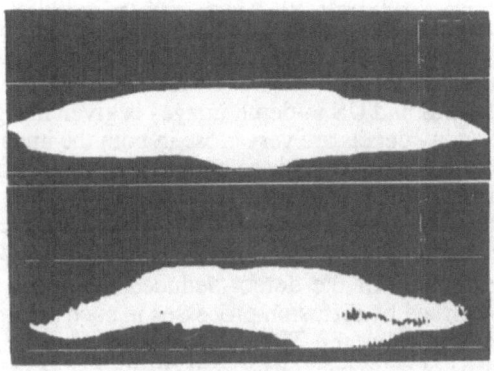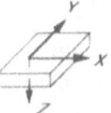

Figure 15
TIS measurement (early detection) of an impact damage.
Projection of the multidelamination on two perpendicular cross sections.

The sample used is an element of a composite wing which was impacted in several points with known impulses. In carbon epoxy composites, the defects produced by impacts are quite dangerous because, generally, they do not let any visible traces at the surface even when large multidelaminations are present in the thickness of the material /13/. The quantitative NDT problem to be solved here, is more the evaluation, as accurate as possible, of the extension of the internal dammages than the exact reconstruction of the shape of the delamination.

The early TIS detection is applied to such an impact with records after illuminations and measurements on the two faces of the sample. This double operation is necessary to detect the totality of the multidelamination while avoiding the too important perturbations produced, on the one hand by the 3D heat diffusion, and, on the other hand, by the rear surface /14/. It is then possible to reconstruct the total extension of the multidelamination in the thickness of the wing. The projections of this extension on two perpendicular cross sections of the sample are given in figure 15.

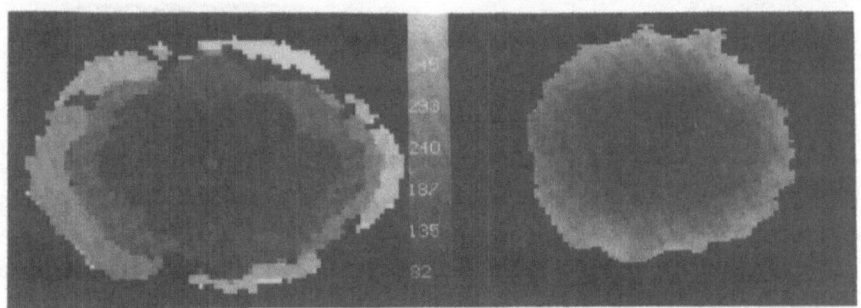

Ultrasonic
Depth Location Image

Infrared Thermographic
Depth Location Image

Figure 16
TIS / US comparison of the damage produced by impact.

The same sample was also immersed in water and C-scanned by the conventional US sensor described in the previous section. The measurements of the time of flight is used to produce the in-depth image once the sound velocity of the longitudinal wave is measured.

The comparison between the TIS and US in-depth images is given in figure 16. Though the rough features and many details are very close in both the images, they are nevertheless not totally identical. The object is so complex that it is difficult to say what it is the best information, but the important result is given by the comparison between the total extensions deduced from these two measurements : the TIS evaluation is only 15% larger than the US one. Since the US resolution is of the order of 1 mm^2, the extension of the defect deduced from this method (which can be, nevertheless perturbed by diffusion processes in such a material) is probably very close to the reality. Thus the TIS method appears to be able to restitute quantitatively the damage of an impact in a composite within an incertainty (by excess) of 15%.

4.2. Conclusion

The main NDT methods able to provide fast, contactless and quantitative images of subsurface defects have been tested on a unique artificial sample.

The TIS method gives the depth and the thermal resistance. The method is very fast and accurate, provide a good inversion procedure is chosen. The lateral heat diffusion is the main problem encountered. It destroys the resolution and limits practically the detectivity to half the sample thickness.

The global optical methods are rather fast but need mechanical excitations able to produce surface deformations of the order of a few microns. The problem of inversion seems difficult to solve and, for the moment, these methods cannot be considered as quantitative.

Laser ultrasonics are highly promising techniques but, at least with the system used in this study, are difficult to use when the surface state is not perfect. If this problem can be solved, this method is able to give remarkable results : excellent resolution, quantitative in-depth measurements, easy inversion.

Finally, in the special case of the quantitative detection of impact damages in composites, the TIS method appears very competitive with US since it provides the same damage extension within 15 %.

REFERENCES

/1/ S. G. Sampath, "National Aging Aircraft Research Program", International Conference on Nondestructive Inspection, AAAF Proceedings, Paris 1991

/2/ A. Deom, D. Bosher and D. Balageas, Rev. Progress in QNDE 9A, 525, 1990.

/3/ P. Delpech, D. Bosher, F. Lepoutre, A. Deom and D. Balageas, Proceedings of QIRT 92, Eurotherm Seminar Series 27, Editions Européennes Thermique et Industrie, Paris 1992.

/4/ D. Balageas, A. Déom and D. Bosher, Materials Evaluation, 45, 461, 1987.

/5/ J.C. Krapez, D. Boscher, P. Delpech, A. Déom, G. Gardette and D. Balageas, Proceedings of QIRT 92, Eurotherm Seminar Series 27, Editions Européennes Thermique et Industrie, Paris 1992.

/6/ J.C. Krapez, Ph. D. Thesis, "Contribution à la caractérisation des défauts de type délaminage ou cavité par thermographie stimulée", Ecole Centrale de Paris, 1991.

/7/ L. D. Favro, D. J. Crowther, P.K. Kuo and R.L. Thomas, "Inverse Scattering of Pulsed Thermal Waves", in this NATO Advanced Research Workshop Proceedings, Quebec 1992.

/8/ D. Royer et E. Dieulesaint, Revue de Physique Appliquée, 24, 833, 1989.

/9/ P. Cielo, "Optical Techniques for Industrial Inspection", Academic Press, New York, 1992.

/10/ E. Dieulsaint et D. Royer, "Ondes élastiques dans les solides", Masson Paris 1974.

/11/ R. K. Erf, "Holographic Nondestructive Testing", Academic Press, 1965
J.C. Vienot, P. Smigielski et H. Royer, 'Holographie optique, développements et applications", Dunod, Paris, 1971.

/12/ Y. Y. Hung, J. Non Destructive Evaluation, 8, 2, 1989.

/13/ G. Tober and R. Henrich, "In-Service Inspection concept for impact damage on CFRP structures", International Conference on Nondestructive Inspection, AAAF Proceedings, Paris 1991

/14/ J. C. Krapez, X. Maldague and P. Cielo, Research in NDE 3, 81, 1991

Combination of FEM and HI for NDE purposes

W. Jüptner, Th. Bischof
Bremen Institute for Applied Beam Technology
Klagenfurter-Straße 2, 28359 Bremen
Germany

Abstract. The holographic interferometry is a tool, to detect the displacement on the surface of a specimen, with an accurancy of less then the wavelength of the used laser. The Finite-Element-Method can be used, for instance to model problems of the structural mechanics. To verify the model with the behaviour of the real specimen, there must be an interface.

Two investigations were done on overlap-adhesive-bonds. First the material proporties of the adhesive-layer could be determine and second the inhomogeneous surface displacement over an adhesive defect under thermal load could be explained, both with the surface deformation as interface between modell and experiment.

1. Introduction

Adhesive bonds are used increasingly to join materials in modern industries like aircraft or automobile production. However, one of the main problems of this technology is still the quality control, due to the fact that standard methods of non-destructive testing (NDT) are very often not sufficient to the safety requirements. Therefore new methods are needed.

Holographic interferometry combined with the corresponding FEM-calculation is a NDT technique capable to solve the quality control problems, especially with compound materials [1, 2]. It is a very sensitive measuring methode capable of determining surface displacement in a range, which goes below 0.5 μm with a deviation of less than 5%.

Using this measurement technique in BIAS two investigations were done on adhesive bonds, using a combined method of holographic interferometry and calculations with the finite-element-method.

- The displacement caused by tension and perpendicular to the surface on the upper and lower sides of a intact overlapping adhesive bond can be holographically recorded. If the distortion of both sides of the specimen are subtracted from each other, the result is the total change in thickness.

But with these measurements alone it is not possible to determine how the total change in form is devided up among the single components of the specimen. Therefore calculations facilitated by the method of Finite Elements must be carried out, with which a complete figure of the stress and strain conditions can be created.

This information makes it possible to determine to what degree each of the components has taken part in the total displacement.

X. P. V. Malague (ed.), Advances in Signal Processing for Nondestructive Evaluation of Materials, 323–335.
© *1994 Kluwer Academic Publishers.*

- The experience with non-destructive testing of adhesive bonds proves that a heat flux through the specimen is a reliable and easy to be applied load for the problem of flaw detection by an inhomogenity in the displacement field. However, the mechanism, which leads to this inhomogenity, is rarely known. In principle there are two main physical effects which may cause the changes in the displacement field:

  - The pressure of the gas inside the flaw can be rised

  and/or

  - the thermal inhomogenity in the boundary area between the adhesive layer and the metal sheets may induce mechanical stresses and change the surface strains.

In order to decide which of the both effects has the dominant influence under thermal load, the displacement characteristics of the specimen were calculated by means of FEM. In this first investigation a stationary heat flow was analysed as well theoretically and experimentally. With the achieved knowledge a more targeted application of holographic interferometry for NDT is expected.

2. Principle of holographic interferometry

Holographic interferometry is a coherent optical method for measuring surface displacements of opaque objects: The object is holographically recorded in a load state one. Then a load is applied to the object. Now the reconstructed object wave from the first load state is combined either with the recorded and reconstructed image or with the object wave directly from the second load state. The two object waves interfere and form an interference fringe pattern which seams to cover the object. The shape and position of these fringes depend strongly on the displacement caused by the load. The interference pattern represent the intensity distribution $I(P)$

$$I(P) = a(P) + b(P) \cos(\phi(P)) \tag{1}$$

in point P, in which case $a(P)$ is the variable for the background lighting, and $b(P)$ describes the local contrast and the speckles. $\phi(P)$ is the interference phase at point P on the surface being observed. The stress change moves the surface point P by the amount of the displacement-vector $\vec{d}(P)$, thereby changing the phase of the optical wave field by as much as the interference phase $\phi(P)$ to

$$\phi(P) = \frac{2\pi}{\lambda} \vec{d}(P) \left[\vec{b}(P) - \vec{s}(P) \right]. \tag{2}$$

λ is the wave-length of the laserbeam, $\vec{b}(P)$ is the unit vector for the observation direction, $\vec{s}(P)$ is the unit vector in illumination direction, figure 1.

In the quantitive evaluation of the interference pattern $I(P)$ a computer aided calculation of the interference-phase-distribution $\phi(P)$ was done after which, by inverting the equation 2, the displacement vector field could be determined. Evaluation processes such as the phase-shift method or the Fourier Transformation Evaluation yield a resolution of better than 1/20 of the light wavelength (in this case He-Ne-Laser with 632.8 nm). With the Heterodyne method it is even possible to reach resolutions of 1/1000 of the light wavelength [3, 4, 5].

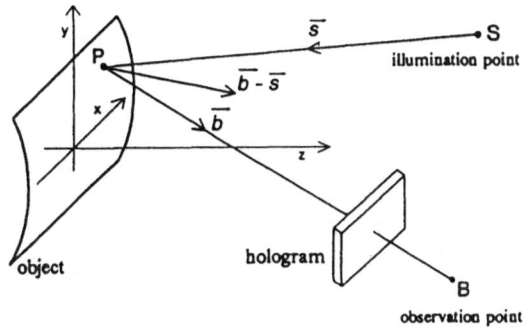

Figure 1: Geometry of the holographic set-up, schematically

In a lot of applications of HNDT the holographic interference patterns are analysed qualitatively for local and global deviations. Areas with higher fringe density than in the surrounding surface of the object indicate high stress concentrations. Those where the interference lines are interrupted contain surface cracks and others with a sudden change in the fringe pattern density are influenced by bends in the surface. The qualitative evaluation of the interference patterns is often sufficient for the quality control.

Different loads may be used to induce the displacement which should be measured. These loads can be of a variety of different types, eg. of the following:

- Direct mechanical strain, such as bending moment, tension, compression, torsion or gravity is possible.

- The object can be placed in a pressure or vacuum chamber and subjected to a pressure change; closed objects such as tanks can be put under internal pressure.

- Thermal load can be induced by means of radiation heat sources, hot air blowers, evaporation of liquids or by high power direct current.

- Vibrations within the object can be caused acoustically or with impact.

The following investigations were done under thermal and tensil load and correspond with the FEM-calculations.

3. Investigation on stress distribution in intact bonds

The recognition and valuation of stress on adhesive-layers is very important for the designing of stuctures incorperating overlapping adhesive bonds.

The displacement qualities of the components, which are very often difficult to define, influence the strength of the adhesive structure considerably. These qualities must be known in order to carry out strength and stability calculations.

3.1. HOLOGRAPHIC ANALYSIS

For the holographic examination and the calculations that followed, an overlapping adhesive bond was tested for tensile strength as shown in figure 2.

The specimen was subjected to a base load (Stress in the direction of the y-axis), so that settling effects could be ruled out. The base load, which was 12.5 N/mm^2 in the sheet-metal component of the specimen, was equal to medium thrust and shear-load in the adhesive layer of approximetly 2 N/mm^2. Then The specimen was holographically recorded. The holographic testing setup used during the experiments is shown schematically in figure 3.

After raising the load by 3%, the new image of the object was used to interfere with that of the object under base load, which was frozen in the hologram plate. This created the interference patterns on the upper and lower sides of the specimen, which were made visible by way of the mirrors.

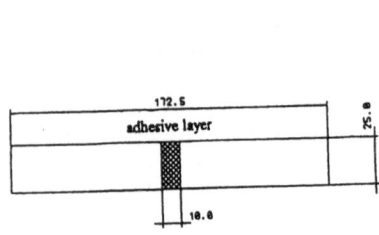

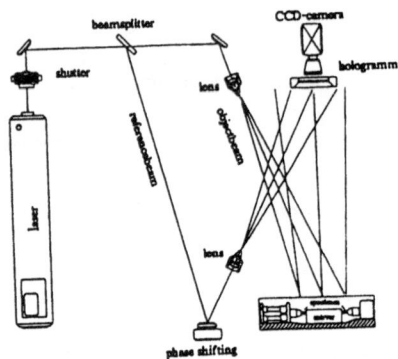

Figure 2: Schematically showing the specimen   Figure 3: Holographic Testing Setup

3.1.1. Results of the Holographic Tests Figure 4 shows the interference patterns on the upper and lower sides of the specimen.

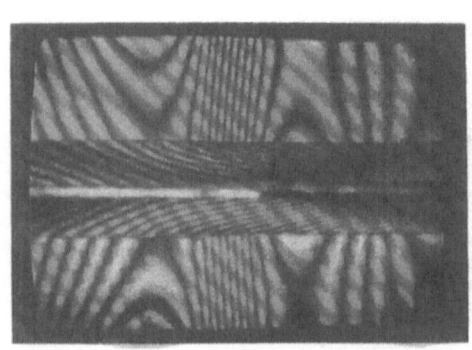

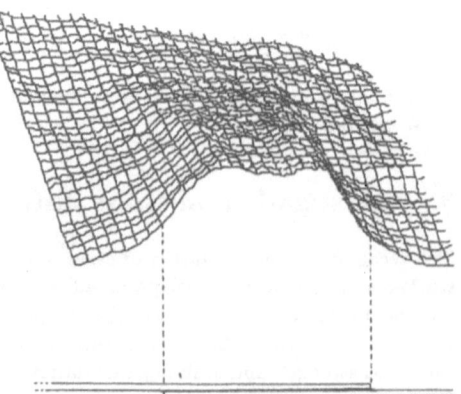

Figure 4: Interference patterns on upper and lower sides of the specimen

Figure 5: Pseudo-3-D appearence of the difference in displacement in the area of the overlap zone

The total change in thickness of the object along the z-axis can be obtained by subtracting the single surface displacement of the two sides from each other.

Displacement will also appear in the directions of the x- and y-axis. On the one hand, the chosen holographic setup is insensitive to these distortions. On the other hand, when the deformation of the upper and lower sides is subtracted from each other, these in-plane movements will be eliminated, so that only the displacement in the direction of the z-axis can be taken into consideration.

Figure 5 shows the total deformation during the surface evaluation of the interference pattern in the overlap zone.

The total displacement reaches a maximum at the edge of this zone, meaning that the adhesive layer in this area is being subjected to the maximum strain. The qualiative course of displacement along the adhesive layer corresponds with that of the theoretical course of an overlapping adhesive bond.

3.2. FEM-CALCULATIONS

3.2.1. Discretisation of the specimen geometry for FEM-calculations The measurements of the specimen along the x-axis are much longer than those of the z-axis (proportions: 25/1.6). In a cross-section view of the specimen in the y-z-plane (figure 2), a plane state of deformation can be assumed, so that only a 2D-mesh of the specimen had to be modeled. The distortion characteristics of the adhesive layer as well as these of the sheet-metal were ideally assumed to be linear.

In the experiment the specimen was held with two clamps on the right and left ends. For the calculations these conditions were realized with four supports as show in figure 2. Because the materials being tested showed linear elastical characteristics when subjected to smaller loads, and due to the fact that the deformation caused by nonlinear geometrical behaviour in comparison to the measurements of the specimen was very small and could therefore be neglected, a linear elastical calculation was considered sufficient.

Figure 6 shows the cross-section of the specimen with corrosponding two-dimensional elements.

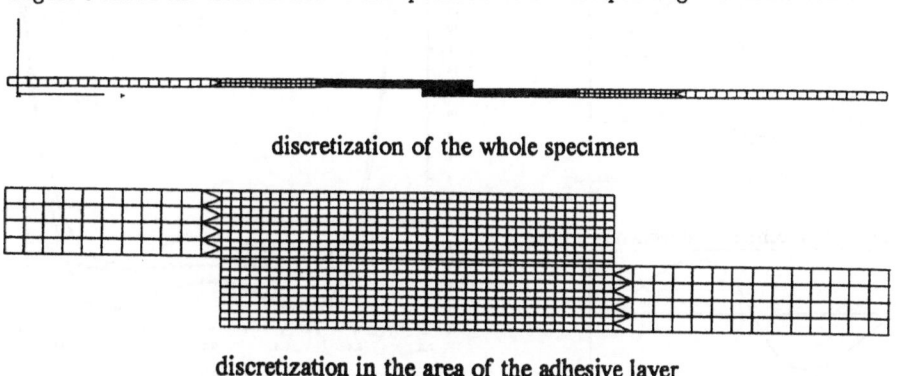

discretization of the whole specimen

discretization in the area of the adhesive layer

Figure 6: Meshing of the specimen

3.2.2. Results of the FEM-Calculations and comparison with the experiment The calculations were done for a base load of 12.5 N/mm^2. It was verified that the resulting deformation in comparison to the measurements of the specimen is very small.

In figure 7 the calculated specimen displacement in the direction of the z-axis is compared with results of the experiment itself.

The maximum calculated displacement along the z-axis amounts to 1.73 μm. The maximum which was measured during the experiment was 1.75 μm. The qualitive course of both graphs also correspond to each other.

With the help of the FEM-calculations it is possible to create a complete figure of the stress and strain conditions in the specimen, which in turn makes it possible to determine the distribution of the total deformation in the area of the adhesive layer and to examine its effect on each of the components.

328

Figure 8 graphically shows each of the calculated portions of the total displacement of the specimen in the direction of the z-axis and the amounts to which the adhesive and the sheet-metal layers took part in the total displacement.

The middle of the adhesive layer is due to its low load level (only about a third of that in the margin areas) not relevant for the stress-analysis of the whole adhesive layer. The adhesive layer is compressed in this area. The change in thickness of the glue in this area amounts only to about 28% of the total displacement.

On both the margin areas of the adhesive layer it is appearent, that there the distortion of the adhesive layer is even higher than the total displacement. The glue thickness change in this area is about 88% of the total displacement.

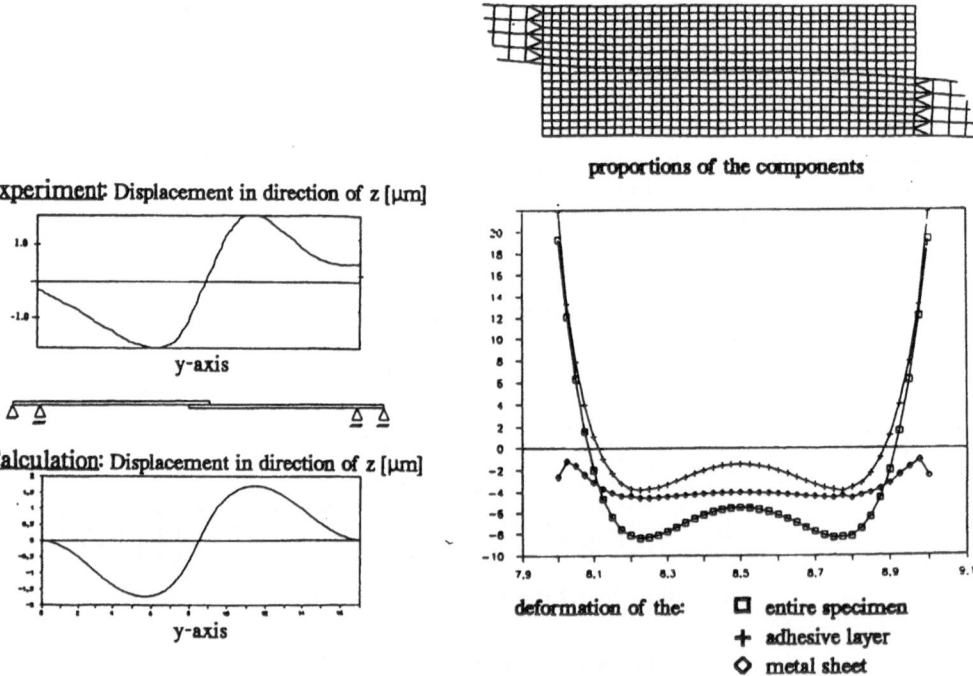

Figure 7: Comparison between the holograph- ically measured (top) and the calculated (bot- tom) displacement along a line

Figure 8: Proportions of the components (glue, steel sheeting) on the total displacement

The portion of displacement, which takes place in the sheet-metal, is very small and can therefore be neglected. This distortion of the sheet-metal component is negative over the whole surface of the specimen, due to the fact, that it is being subjected to tension. In the center of adhesive layer zone the deformation is negative, the glue is therefore being compressed.

The portion of adhesive layer deformation in comparison to that of the total dispalcement amounts to 28% in this area.

Further calculations proved, that even with the use of glues with higher moduls of elasticity (the modulus was gradually increased from 2500 to 10000 N/mm^2) the total displacement in the margin area took mainly place in the adhesive layer itself.

4. Investigation in mechanisms for the detectability of adhesive defects under thermal load

The specimen chosen for this investigation is shown in figure 9, the corresponding geometric data are sumerized in table 1.

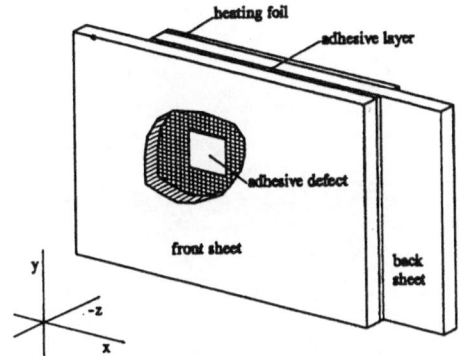

Figure 9: Overlap adhesive bonded joint with a defect in adhesive layer

Width of the plates (x-axis)	225.0 mm
Height of the plates (y-axis)	130.0 mm
Length of the overlapping zone	170.0 mm
Pos. adh. defect center (x-axis)	65.0 mm
Pos. adh. defect center (y-axis)	85.0 mm

Table 1: Geometry of the standard specimen

4.1. HOLOGRAPHIC TESTING SETUP

The setup is shown schematically in figure 1. The specimen under investigation was clamped in such a way that the specimen was able to expand unrestricted in all directions. The beam from argon-ion-laser, with an optical power output of 1 W, is directed to the beam splitter, which divides the beam into the object beam and the reference beam. The object beam illuminates the diffuse reflecting surface of the specimen in the direction of the load.

The object-beam reflected from the specimen is then overlapped with the reference beam on the hologram plate and recorded.

In these investigations the thermal load was achieved by means of a heating foil positioned on the backside of the specimen. The adhesive defect was caused by placing a piece of teflon tape ($20x20$ mm^2) in the adhesive layer.

The holographic real-time process was used. The recorded state of the specimen before the loading and the object under the load were viewed simultaneously. The resulting interferogram was recorded with a CCD-camera and stored in an imageprocessing system developed and built at BIAS [6]. With this system the interference fringes could be processed digitally.

4.1.1. Results of the holographic testing A standard bonding specimen was investigated with the parameters according to table 1 and figure 9.

First a specimen without any bond defects was tested. The displacement characteristics corresponded with those of the calculated deformation, figure 10.

After that, a specimen with adhesive defects was tested whereby size, position and the thickness of the adhesive layer as well as the heat flow were varied. However, in all cases the measurements were performed with a steady state heatflow. After a stationary state was reached the holographic interference pattern of the surface was recorded, figure 11, and processed.

The quantitative evaluation of this interference pattern yielded the displacement fields is shown in the figure 11. These drawings show the pseudo-3D-display of displacement component normal to the specimen surface.

The non-uniform distortion in the area of the defect is to be recognized easily, although the added displacement in this area caused by the defect is less than 0.4 mikrometers.

The experiment proved, that adhesive defects in overlapping bonded joints may be detected by means of holographic interferometry. However, theoretical calculations have to be performed by means of the Finite Element Method, in order to answer the questions concerning to influence of the parameters on the detectablity of defects.

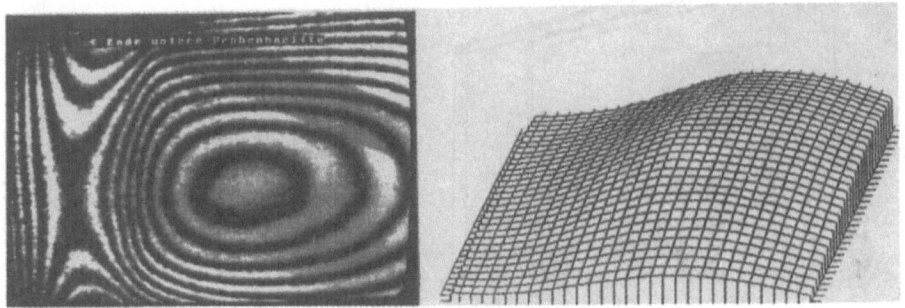

Figure 10: Surface displacement of a specimen without defect, holographically recorded

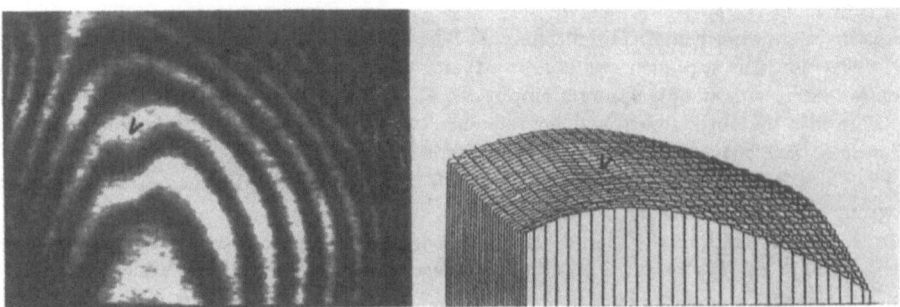

Figure 11: Surface displacement of a specimen with defect, holographically recorded

4.2. FEM-CALCULATIONS

4.2.1. Modelling of the specimen

Discretisation The bonded specimen consisted of three parts: the front plate, the adhesive layer and the back plate. Figure 12 shows the discretisation of the different parts of the specimen.

The front and back plates of the specimen were subdivided into three layers each. The thickness of the subdivisions becomes less when moving in the direction of the adhesive layer so that the nonlinear effect in the area of the built-in defect becomes more visible. For the same reason this area also received the finest discretisation in reference to the height and width of the elements.

Figure 12: Discretisation of the specimen

Geometric Data The thickness and width of the plates, the amount of overlapping and the position of the defect itself were the same through out all the calculations, table 1.

Choice of Elements For the calculation of the temperature distribution as well as for the displacement of the surface deformation, 8-node isoparametric elements were used [7, 8, 9].

4.2.2. Boundary Conditions

Thermal Analysis There are two different types of boundary conditions. On all surfaces of the object the surface temperature or the heat flow through the surface must be known.

The energy, which the specimen receives from the heating foil, radiates to the surrounding air by free convection. The equation for this boundary condition is:

$$q^s = h(T_e - T_s) \tag{3}$$

with: h :   convection rate (heatflow/(area*temperature diff.))
q^s :   heatflow, radiated to the surrounding air
T_e :   Temperature on the surface
T_e :   surrounding temperature

During all calculations the surrounding temperature was held at a constant 293.15 K.

The heat flow is achieved via the heating foil on the backside of the specimen and distributed homogeneously. This heat source is represented by taking every element of the back plate, which is covered by the heat foil and specifying it with a heat source, which corresponded with the heat flow. The backside of the elements were considered adiabatically insulated.

The influence of an adhesive defect on the temperature distribution was represented by a local change of the heat conduction coefficient to lower values in the adhesive layer. The specimen then was subjected to the same heat power as in the case of the flaw-free calculation.

Deformation Analysis The specimen expanded in dependency only to the warming of the single elements: The specimen was clamped in such a way, that it was able to expand freely, so that no other distortion could happen, than that caused by thermal expansion.

4.2.3. Results of the FEM-Calculations

Temperature distribution First the temperature distributions, caused by the thermal load for a specimen, were calculated for both a specimen with and without artificial defect. Figure 13 shows the temperature distribution on the front side of the specimen in a false color represent.

The temperature difference between two colors amounts to about 2/1000 K. These temperature differences are too small be displayed with a thermography camera.

In the case of a flaw free specimen the temperature distribution is uniform, as expected. But in the temperature distribution on the specimen with the adhesive defect a nonuniformity in the temperature distribution is noticeable.

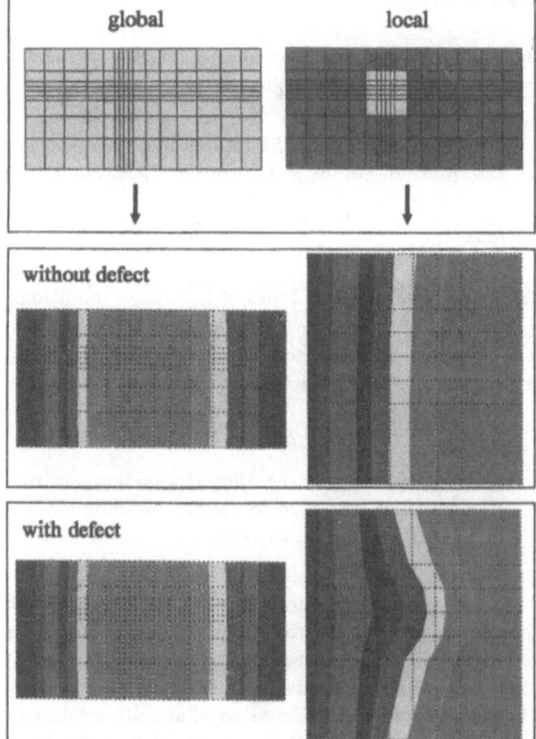

Figure 13: Temperature distribution on the surface of the specimen

Calculation of the surface deformation For the estimation of the deformation caused by the thermal load, the results of the temperature field calculations were used, so that the expansion of the specimen during the rise in temperature could be determined, figure 14. This also included the thermally induced rise in pressure in side the flaw.

The deformation of the specimen without a defect was calculated for comparison. The specimen is deformed spherically comperable to a bi-metallic bar: the back side, on which the heating foil is attached, expands more than the front side.

A nonuniformity in the deformation of the specimen with the adhesive defect is clearly detectable.

The FEM-model can therefore be used to show the reasons why the defect is visible on the surface, and to determine the influence of geometric factors such as plate thickness, adhesive layer thickness.

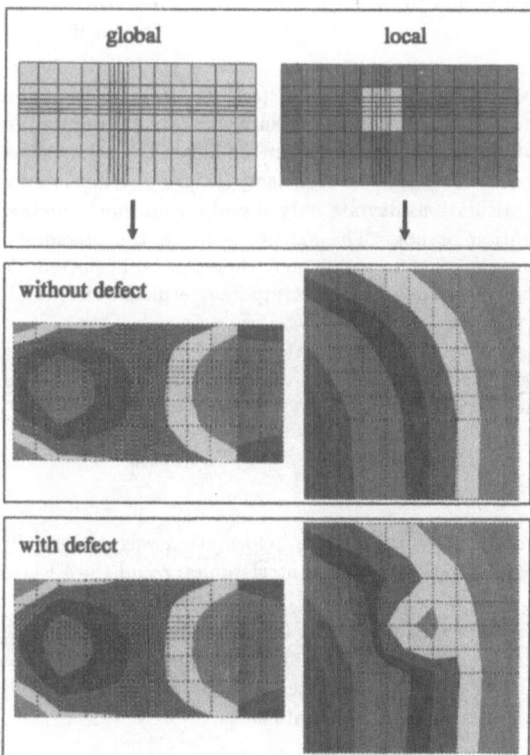

Figure 14: Deformation of the surface of the specimen

4.3. RECORDING THE CAUSES OF NONUNIFORM DEFORMATION

4.3.1. Inhomogeneous temperature distribution On the back side of the specimen the area of the defect was heated more than the surrounding area because of the worse heat conduction through the adhesive layer. Due to the higher temperature and the therefore higher thermal expansion, a larger amount of compression is present in this area.

The front side in the region of the defect is not heated as well because of the disturbance in the heat conduction. Consequently the specimen will not expand as much locally and therefore induce higher tension in that area.

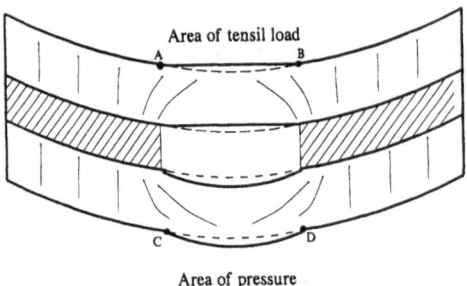

Figure 15: Heat conduction and displacement in the area of the defect

Figure 15 shows schematically the effects of the induced tension and pressure. The stress distribution on the backside with the added compression due to the higher temperature between the points C and D cause an expansion in this area, which increases the bi-metallic effect in the surrounding region.

The tension on the front side of the specimen works against the vault. The specimen attempts to realize the shortest distance between the points A and B.

4.3.2. Increased pressure inside the adhesive defect In order to decide up to which degree the total deformation is caused by the increase in pressure, the calculations for the deformation were also carried out with and without internal pressure in the flaw. The internal pressure was determined from the laws of gas expansion with the known temperature distribution around the flaw.

The results of these calculations deviate only slightly from those whether the pressure increase was taken into consideration or not. The extent to which the pressure increase influences the fluctuations in the deformation are so small, that they can be neglected. The largest part of the deformation is caused by the inhomogeneous temperature distribution and the resulting thermal expansion.

The theoretical evaluation of the case of stationary heat conduction proves, that the indication of a defect is caused largely by the nonuniform thermal expansion and not, as is often assumed, by the expansion of possible gas inclusions.

5. Conclusion

Based on these two investigations, the surface deformation was successfully used as an interface between the experiment with the holographic interferometry and the FEM-calculation.

- The results of the first investigations allow to explain the effects which cause the inhomogenity: Mainly the local change of the heat flow influences the deformation inhomogenity and not the internal pressure in the flaw. A parameter study allows to calculate the different influences of the material properties and geometrical effects.

- In the second part, the total change in thickness of a adhesive layer under tensil load was holographically recorded. By way of calculations, facilitated by the use of the Finite-Element-Method, a complete figure of the stress and strain conditions within the specimen was able to be made. This in turn made it possible to determine to what proportions the single components of the bonded joint (adhesive and steel-sheet layers) took part in the total deformation of the specimen.

6. A word of thanks

The authors of this report would like to thank the Federal Ministry for Research and Technology for funding the work, which was done in connection with the project No. 02 FT 47330, and the German Research Association DFG for sponsoring this research work (No.: Ju 142/12). The authors want to thank J. Geldmacher for his experimental work and the helpful discussion during the research work.

REFERENCES

[1] **Th. Bischof, W. Jüptner:** *Holografische Interferometrie an Klebverbindungen*, in Handbuch Fertigungstechnologie Kleben, O.-D. Hennemann, W. Brockmann, H. Kollek (Hrsg.), Carl Hanser Verlag, München 1992

[2] **Jüptner, W., Bischof, Th.:** *Investigations in Bond Defects by a Combined Method of Holographic Interferometry an FEM Calculations*, Proceedings of the 10th International Invitational UFEM Symposium, Worcester July 1991

[3] **Jüptner, W.:** *Automatisierte Auswertung holografischer Interferogramme mit dem Zeilen-Scan-Verfahren*, Proc. Frühjahrsschule 78, Holografische Interferometrie in Technik und Medizin (1978)

[4] **Jüptner, W., Kreis, Th., Kreitlow, H.:** *Automatic evaluation of holographic interferograms by reference beam phase shifting*, Proc. SPIE, Vol. 398, (1983)

[5] **Kreis, Th.:** *Methoden zur Auswertung holografischer Interferenzmuster, ein Vergleich*, Laser und Optoelektronik 21(2),54-61 (1989)

[6] **Telepski,J.:** *BIAS Bildverarbeitung*, Vision and Voice, 2/90

[7] **Barthe, K. J.:** *Finite Element Procedures in Engineering Analysis*, Prentice-Hall, 1982,

[8] **DeSalvo, G. J., Gorman, R. W.:** *ANSYS Engineering Analysis System User's Manual*, Swanson Analysis Systems, 1989

[9] **Zienkiewicz, O. C.:** *The Finite Element Method*, McGraw-Hill Book Company, 1977

PROCESSING SUITED TO IN-THE-FIELD APPLICATIONS OF NDT

C.P. Hobbs
AEA Technology,
National NDT Centre,
447 Harwell, Oxfordshire.
OX11 0RA. United Kingdom.

ABSTRACT. Whilst there are several key NDT techniques, and others presently developing, it is clear that in almost every area, image processing can significantly enhance the information that the NDT technician can provide. However, as these developments proceed, we should not loose sight of the user's requirements, namely, that the NDT results are produced at the inspection site, rather than requiring post-processing or careful analysis at a remote base, and that any on-site processing should be fast. This need is reflected in the general move to PC-based systems, where a degree of data enhancement in real-time is available to permit on-site analysis. In this presentation, the work at AEA Technology, within the National NDT Centre (NNDTC), is used to demonstrate such processing and their benefits in the fields of thermography, ultrasonics, X-ray computerised tomography and radar.

It is concluded that the NDT community benefits a great deal from the advances in signal processing presently being made. It is shown that these advances are not restricted to laboratory applications, and that the future should see a number of them in common usage at the inspection site.

1. Introduction

A picture is worth a thousand words. This common phrase is so true in the field of non-destructive testing and evaluation that it would be difficult to find anyone to disagree. The image, whether it be for two or even three-dimensions, can be far more instructive than a parameter's value at a particular point, such as ultrasonic velocity/amplitude or X-ray intensity, largely because it is the value, in the context of the surrounds, that will determine if the component is defect-free at this point. The component itself can only be reported as fit-for-purpose once all points have been given the all-clear. Thus, it is obvious that an image will, in most cases, be far more useful to this end.

The problem of the image is the shear volume of data that it contains. This fact can become overlooked, because a trained inspector can perform such rapid assimilation of the data presented to him. However, it is usual for the inspection to result in a report to a colleague or superior, from which a judgement will be made regarding the acceptability of the component. The need here is to summarise all of the available data

337

X. P. V. Malague (ed.), Advances in Signal Processing for Nondestructive Evaluation of Materials, 337–352.
© *1994 Kluwer Academic Publishers.*

as concisely as possible, but without corrupting or overlooking any fact that will have an effect on the overall assessment. The analysis of fatigue cracks is one good example of this situation. The simplest summary for a component would be a statement that the component is free of defects, or contains a number of defects. In the former case, we do not require to go any further, however, in the latter case, it is the amount of information that we can provide (such as defect size, position and character) that will determine the outcome of the final assessment. It could be a tedious task to perform all of the actions necessary to determine these quantities. Given that the data contains this information, it is therefore very necessary to apply image processing in an endeavour to allow such details to be obtained rapidly, and presented to the operator in the most suitable way.

Much work has been done throughout the world with the above in mind, with impressive results. However, the need for processing usually results in a system whereby the operations are restricted to a specialist facility, quite often of considerable value.

The National NDT Centre (NNDTC) of AEA Technology (AEA) provides state-of-the-art NDT to many companies throughout the world. A considerable amount of this work has to be performed on the customer's site. Therefore, there is a need to be able to apply these enhancements remote from base. The rapid growth of the PC market and their accessories has made this a viable proposition. This aim of this paper is to describe some of the sorts of processing that are both beneficial and practical in-the-field. It will also be shown that the expertise from more established NDT techniques can be adapted into developing ones, in order to help them obtain the acceptance they rightfully deserve. To do this, the paper will draw on data obtained by the National NDT Centre to demonstrate such processing and their benefits in the fields of thermography, ultrasonics, X-ray computerised tomography and radar.

2. Thermography

Thermal imaging, otherwise known as thermography, is a key technique for the assessment of quality assurance. This is because it offers several advantages, namely, it is totally non-contacting and non-invasive, the image is presented instantly in pictorial form, and the data is easy to record and archive for quality assurance purposes.

These attributes can be employed provided that the component to be inspected is in an environment where potential defects interrupt the flow of heat to the inspected surface, as this will affect the rate of thermal emission, which enables the defective area to be differentiated from the surrounds with a thermal (IR) imager. If there is no suitable flux of heat, the defects presence may be established by using a rapid thermal transient. This technique, known as Transient Thermography, has all of the above advantages, combined with the fact that it can be used to inspect relatively large areas rapidly, because a considerable area (typically several square feet) can be inspected simultaneously using an event that lasts only a few seconds.

In order to do this, a heat source, such as a Xenon flash tube, is used to generate a moderate but rapid rise of surface temperature (usually only $10^{\circ}C$). Using a thermal imager, defects and structure can be observed because they inhibit the passage of heat from the surface into the bulk material, and thereby influence the pattern of surface temperature directly after this heating.

Transient Thermography offers two complementary modes of operation - single-sided and transmission. Single-sided testing is typical of in-service inspection where access to the component is limited; in this instance, the heat is applied to the surface being viewed by the thermal imager. Where access can be gained to each side of the component, a transmission operation, where the heat is applied to the surface opposite the thermal imager, offers even greater detail for qualification and defect detection. The NNDTC has demonstrated the use of this technique in a number of key areas over the years, for example, in the quality control of coatings, both metallic and ceramic (refs 1-3); checking the integrity of composites (GRP and CFRP) for unit-cell dimensions, fibre lay-up and the detection of delaminations, voids, impact damage and moisture ingress (ref 1); the detection of near-surface delaminations, disbonds and pores in metallic and ceramic materials (ref 3); and the qualification of bond uniformity and thickness. The technique, equipment and examples are documented in much greater details in other publications (refs 4-6).

The beauty of the technique is the very visual nature of the data, often provided in a pseudo-colour scale than the raw grey-scale images more prevalent years ago. However, one problem with the transient approach is that the image develops over a time interval, resulting in a series of images which are relevant to the analysis. For thin coatings, only one or two images may result, in which case assessment is fairly straight-forward, however, for materials and systems where the transient develops over several seconds, there can be a considerable amount of data. If the event is exceedingly slow, then the image can be reviewed in real time, with the operator mapping out the positions of defect indications as he observes them. However, for the faster events, there is a need to review the data in slow-motion. This is simple using a frame-grabber, but is still a somewhat cumbersome way of dealing with the data, because of the way thermal contrasts vary in the time domain, dependent on defect depth, type and size.

With this problem in mind, the NNDTC has demonstrated that the event from a thermal transient can be compressed into a single picture, retaining and enhancing all of the available information (ref 4). This is done by forming a map of a suitable thermal parameter, based on all of the information within the thermal event, processed by relating the observed temperature differences to the well-established mathematical descriptions of these two operational modes (refs 7,8). To do this accurately, it is necessary to use a system fit for the purpose, and those with the ability to acquire a sequence of calibrated thermal images under total control of the operator cannot be over-emphasized. There are now several such systems available, making this form of presentation a practical proposition.

In the transmissive mode, it is the thermal diffusivity parameter that can be mapped. This measures the rate at which heat from the rear (excited) surface reaches the observed surface, and as it is usual for defects to inhibit the heat-flow, they are revealed as areas where the diffusivity is lower than the norm. This effect is shown in figure 1, where the sequence of thermal images from a 2.5D carbon-carbon composite plate, 200 mm square and 3 mm thick, is shown as a map of diffusivity. The component presents a relatively constant value of diffusivity, of 7×10^{-6} m^2s^{-1}, over the majority of the surface; this value is close to the value to be expected of good material. However, there are also four very distinct areas where the diffusivity is substantially lower. The beauty of this form of presentation is two fold. Firstly, any defects in the field-of-view

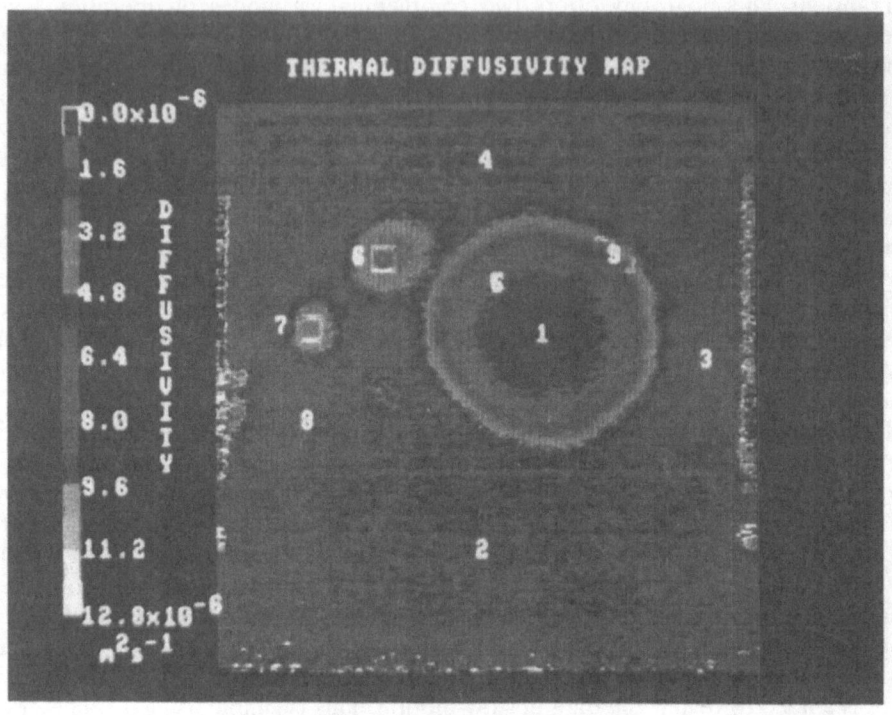

Figure 1 : A map of thermal diffusivity on a CFRP sample, as derived from a Transient Thermography event.

are enhanced, with more distinct edges, and a very tangible measure of their resistance to heat-flow across them giving some indication of defect-type. Secondly, with the image as a two-dimensional map, it would now be straight-forward for further image-analysis to provide a statistical parameter that could be used as an accept/reject criteria for this particular component. It can be seen that, in this way, the tolerance of good components could be quoted as an additional factor for quality control. Other additional spin-offs would be the ability to archive such details with a minimal storage requirement, and the ability to create acceptance templates for routine on-line inspection, since the processing is not-interactive and so can be wired in directly to a production-line.

The NNDTC has also shown the idea of mapping to be viable for the single-sided inspection (ref 4). In this case, it is the thermal effusivity parameter that is mapped. The algorithms are somewhat more complex, because in this situation, it is the whole time-dependence of the event, combined with the temperature differences and the amount of deposited energy that must be used to determine the effusivity at every point in the field-of-view. This is just as feasible, as demonstrated by the map of thermal effusivity obtained from the thermal event taken for five adjacently mounted alumina discs of differing porosity (figure 2). The defects in two of the samples are clearly evident as areas of considerably lower thermal effusivity (that is, an increased resistance to cooling by the bulk material) ; these defects are delaminations which are shown up as hot-spots in the actual thermal event. Further work demonstrated that, for the non-defective samples, the value of effusivity correlated with the amount of porosity within the sample. It is thus obvious that a suitable calibration experiment, with the results written into a further processing module, could result in a map of porosity itself. Other quantities, such as film-thickness, could also be measured in this way.

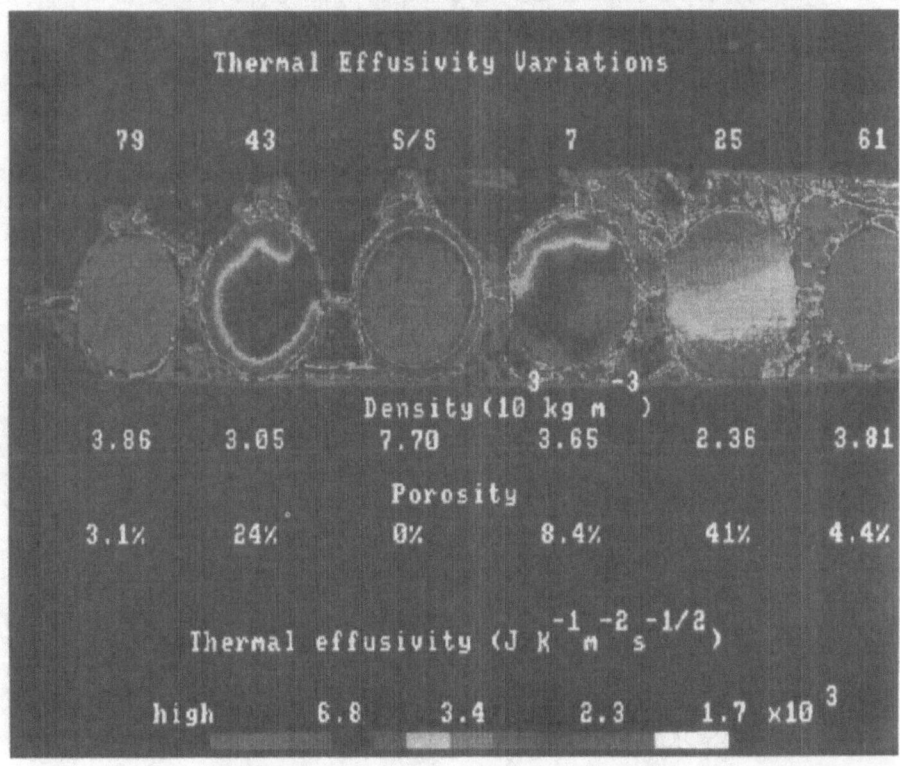

Figure 2 : A map of thermal effusivity over the surfaces of five alumina samples (numbered), as determined by single-sided Transient Thermography.

At the time these demonstrations were performed, the imagers did not have the capacity for data storage or acquisition. However, recent developments, largely in the PC field, have fed into the new breed of imagers available, and the processing described above could now be supported by some of these stand-alone, portable systems. It is therefore a very real possibility to provide this form of processed data at the inspection site.

Many other workers have recognised the benefit of summarising the thermal event. The majority of these use an alternative to the thermal parameter mapping, called inversion. In this situation, the data from the thermal event, in terms of the time of maximum thermal contrast, is used in conjunction with a model or mathematical description to extract values of depth, size and thermal resistance of defects. Maldague (ref 9) demonstrates how the defect's depth affects both the value of maximum contrast and the time at which this occurs. This shows the very basis of inversion processing, namely relating time to depth. Whilst it has been demonstrated as a powerful way of improving the thermal images, a word of caution must be raised. In this form of processing, it is assumed that contrast for a defect at a particular depth is not affected by it's size. This is clearly not the case as the size of the defect decreases (ref 10). Data obtained by the NNDTC emphasizes this point, as figure 3 shows several defects at the same depth (along each horizontal row), and at the same instant of time, with considerably different temperature values and hence contrast (against the general background). The effect is very easy to understand. For any particular depth, large sizes will present the same thermal impass to the heat-front, resulting in the same amount of thermal energy returned to the inspected surface. However, as it's size reduces, there comes a point where the defect's depth and size become comparable. Consequently, a reduced amount of energy is returned to the surface as the majority of thermal energy is able to pass around the defect as opposed to being reflected back to the surface.

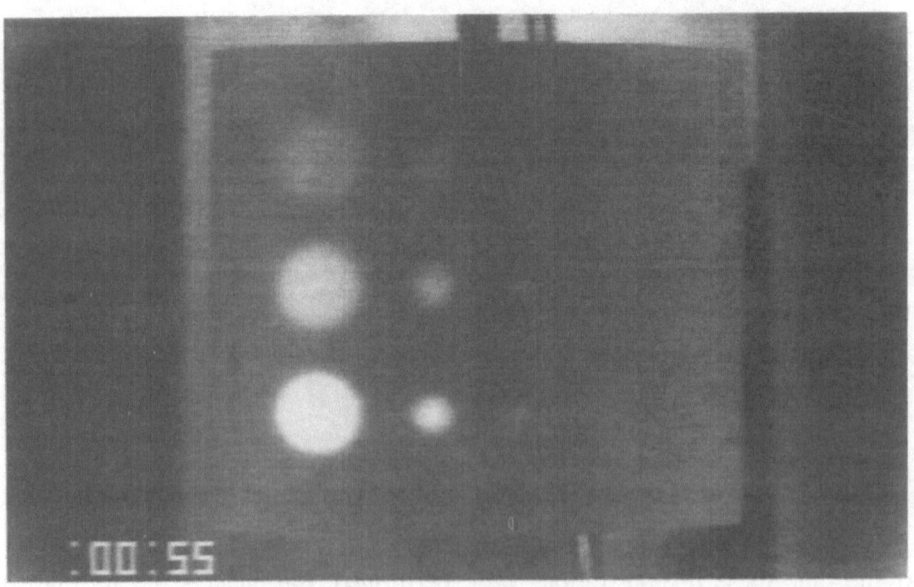

Figure 3 : A thermal image (taken at 0.55 seconds) of rear-drilled flat-bottomed holes in a mild steel plate.The diameters decrease left-to-right along a row, while the depth increases upwards along the verticals. The influence of diameter on contrast is obvious along each individual row. The plate is 180 mm square.

The very controlled manner of these data ensures that this is a true effect of the defects and not one due to inconsistencies across the field-of-view (such as heating). This data, shortly to be published in greater detail, is important to those involved with inversion techniques, as it shows that the algorithms will produce false readings for defects below particular limits in that material. This is not all bad news, because it means there are now established limits above which the inversion algorithms will perform correctly. It also means that we are aware of the possible failings and can build-in appropriate safeguards to the processing outputs.

3. Computerised X-ray Tomography (TOMOHAWK)

X-ray computerised tomography (CT) is another area where image processing is vital to interface the raw data to the end-user. CT is an inspection method that provides true cross-sectional images of the internal details of a component, with full through-thickness information that is absent from conventional radiography. Quantitative estimates of the densities over the entire section can be reported. This additional information is invaluable for many applications, including the measurement of wall-thickness of complex components (such as turbine blades and automotive castings), visualisation of the internal details of complex components to verify correct part assembly, measurement of size, shape and location of internal flaws such as cracks, voids, inclusions and porosity, and the study of density variations across advanced materials.

To date, CT scanners for industrial applications have been high-cost dedicated pieces of equipment. This has limited their application to the inspection of high-value items, typically in the aerospace and defence industries. This cost arises from the use of specialist detector arrays and the computer power needed to deal with the considerable amount of data processing required.

Over the past few years, the NNDTC has been evaluating an alternative approach to CT for industrial applications, in order to provide a high-performance CT capability for a fraction of the cost of traditional dedicated facilities (ref 11). This exploits the recent developments in IBM-compatible PC technology and X-ray sensitive detectors. In this approach, the discrete detectors used in conventional CT scanners are replaced by a real-time imaging system, such as an image-intensifier or low-light level TV camera with separate fluorescent screen. The data needed for CT are then acquired by digitising the analogue video signal of the real-time system using an IBM-compatible PC linked to an encoded turntable on which the specimen is rotated. By using the enhanced processing capabilities now made possible using accelerator cards, it is possible for the PC to process the data into the CT image at a speed comparable with more expensive computing facilities.

This adaptation package, known as TOMOHAWK, allows any existing real-time imaging system to be upgraded to a CT system (ref 11). The specific features of interest are shown in table 1. The acquisition of images 512 x 512 elements in size is performed at standard video rates. These images have a relatively low signal-to-noise ratio, and therefore, it is usual for the hardware to provide summation of a number of successive frames at each point in the rotation. This number can be any value up to 256. This type of noise reduction is exactly the same as used for A-scans in Time-of-Flight Diffraction (TOFD) of ultrasound, and consequently, the user makes the same trade-off between noise reduction and acquisition time. Typically, we have found that integration of 16 frames give significantly improved data. Each batch of sixteen will require about 0.6 seconds, which means that typical acquisition times for a 180 degree rotation in steps of 2 degrees would be about 60 seconds.

344

The reconstruction performed to produce the cross-sectional CT images is a filtered back-projection algorithm. As this is common to many standard CT systems, it will not be elaborated here (ref 12). The important feature is that this processing is a highly computer intensive operation. On TOMOHAWK, all of this is performed on a PC with an accelerator card, and nothing more. The images are corrected for effects that would degrade them, and presently this includes corrections for spatial non-linearity in the real-time imaging system, compensation for the unevenness of the X-ray beam intensity across the field-of-view, and a correction for the reduced sensitivity towards the edges of the system (a common effect in most real-time systems). On a standard 80386 based PC running at 20 MHz, this reconstruction takes about 15 minutes per image. One of the real benefits within TOMOHAWK is that, with the addition of the accelerator card, this time is reduced to a mere 20 seconds, a reduction (in time) by a factor of 45. This makes the whole process of CT much more useful for routine examination.

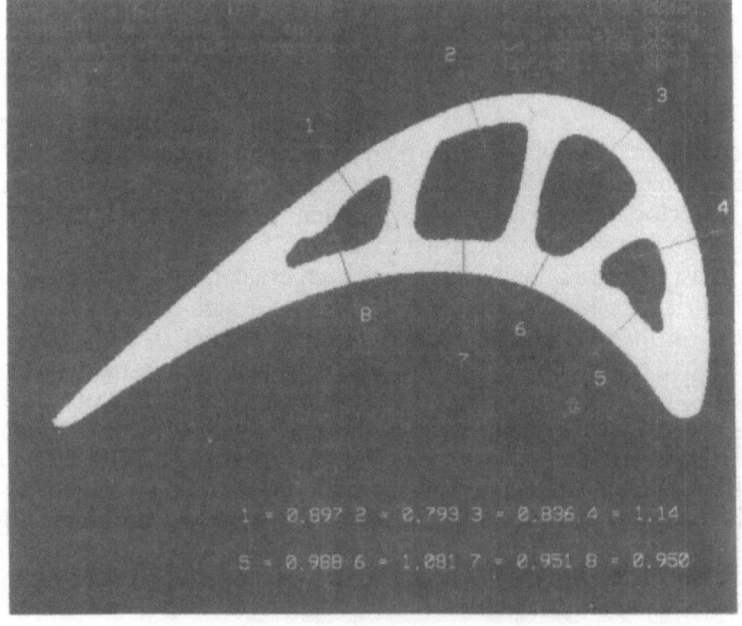

Figure 4 : The CT image of a turbine blade, showing the on-screen measurement of wall-thickness (in mm) in eight places.

The bulk of the image processing suited to CT is in the formation of the image itself, where the physics of the data is used to deconvolve the various projections of the X-ray beams taken as the component is turned. However, as we have already noted, it is the way in which the user can determine the important facts that makes for the most cost-effective inspections. To this end, the software for image interpretation has been written to provide the user with rapid image analysis tools that output parameters which can immediately quantify the component under interrogation. One of the most useful features permits rapid measurement of dimensional features. For example, the user is able to mark a line across a feature, using the mouse; the software then outputs the distance between these points. This simple data extraction has been taken the next logical step, to remove the need to actually mark the edges of the feature itself. In the newer algorithm, the density-profile along the user-defined line is extracted. This profile is displayed, and further processing allows the user to gain very accurate dimensional values. This is done by the user marking the upper and lower density values that form an edge. The algorithm then uses this template to find the position of the edge to a fraction of a pixel, by interpolation. This method has been extended to allow rapid measurements of wall-thickness. An example of the output is shown in figure 4. One density calibration can be used to make a multitude of wall-thickness measurements, as demonstrated in the figure. Alternatively, for production line situations, the image could rapidly be compared to a template for a check of tolerance. It is clear that image processing is a very successful means of reducing the data to a level that retains all of the useful information but also permits rapid assessment.

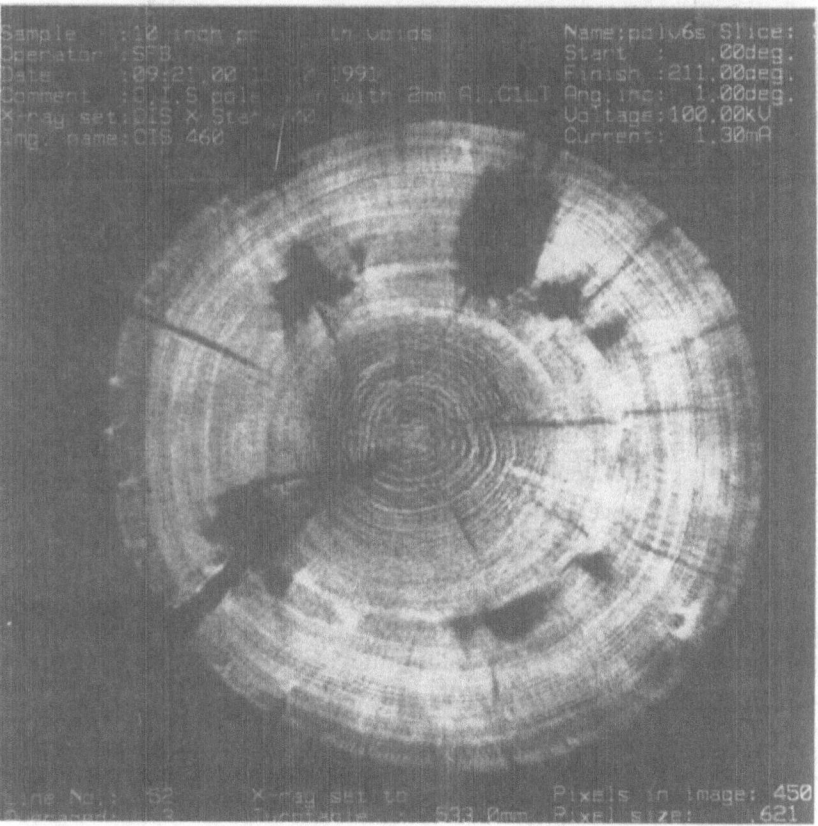

Figure 5 : A CT image of a wooden pole, as obtained at the inspection site, showing the presence of large voids and significant cracking along the radial directions.

One other feature of the system is that of portability. This is something not usually associated with CT imaging, since a dedicated facility is usually the norm. However, since the processing can be performed on a PC, it is simple to move the processing units around. The NNDTC has actually performed CT literally in-the-field. The work was to inspect a number of wooden poles for a utility company on agricultural land, with a view to assessing their structural fitness, in particular, to detect any large cracks or voids. A special rotational head (including X-ray source and detector) was manufactured that could be attached to the poles in question, and the data from this imaging head, and the position sensors, downloaded onto the TOMOHAWK PC system at the site. The data was then reconstructed as normal to give cross-sectional images of the poles. A typical image is shown in figure 5, and demonstrates the very dramatic failures that can develop within these poles. The advantage was that, with all of the equipment operating from a standard generator, on the customer's site, a series of rapid qualifications could be performed that were of immediate benefit. The ability to perform such work in-the-field is clear in this example.

Table 1 : Technical Details of the TOMOHAWK system

Typical data acquisition time	1 - 3 mins
Typical angular increment used	1 - 2 degrees
Number of tomographic images collected	50 (max)
Pixels in tomographic image	512 x 512
Processing time per image (basic PC)	c. 15 mins
Processing time per image (with accelerator card)	c. 20 secs
Video display resolution	512 x 512
Display	256 grey/16 colours

4. Ultrasonics

The area of ultrasonics has received a great deal of attention over the years, and whilst the pulse-echo techniques used for routine testing tend to remain dominated by simple operations with no improvement on the image, there has been an increasing move to dedicated systems with significant imaging capabilities. For example, immersion tests provide C-scans that are now regularly collected in a digitised format that permits the images to be processed in terms of amplitude or depth. The advent of Time-of-Flight Diffraction (TOFD) also brought about the need to provide robust image processing that could deal with the increased amount of data being provided. The B-scan, which plots a series of A-scans, is of great use, but once again, we really want to reduce this image to a decision, through a series of facts drawn from that image. Large processing libraries have resulted. Parts of these deal with the practical problems that can occur within an image (such as changes in couplant which result in time-shifts of the signals). Others specifically allow the image to be calibrated and analysed, so that a report can be issued and a decision made. Most of these routines have been written into the standard software of the NNDTC ZIPSCAN units, which have been the backbone of the NNDTC TOFD service around the world. These units are totally portable data-acquisition and analysis systems, capable of being transported around the world and man-handled into the inspection points on typical petro-chemical plants and factory sites. These units permit an array of processing to be performed, including techniques such as synthetic aperture focusing, to reduce the diffraction arcs to the singular defects from which they result. This is very useful, as it permits accurate measurements of defect shape to be performed. More computationally intensive processes, such as split-spectrum processing (SSP), have been restricted to special facilities until recently, but as with the tomography, the recent advances in PC technology are now enabling these specialised techniques to be made available at the inspection site. Consequently, the NNDTC has been building the next generation of units around PC technology, with a view to these advanced features being standard options upon which the operator can draw.

The reason why such processing would be beneficial in-the-field is clear from the next example. In figure 6, the B-scan from an ultrasonically-noisy material is shown compared to the same image after split-spectrum processing. The positions of the defects within the structure are as labelled, but it is clearly evident that it would be impossible for the operator to report anything from the raw datafile. With the processed data in front of him, it is likely that the defects as small as 1 mm would be reported. The important point is that, by having this information on site, further tests can be carried out, should they be required.

The idea of split-spectrum processing (SSP) is not new, and is a well documented form of frequency averaging (refs 13,14). In summary, it works because the echo from the defect is observable at a relatively constant level over a specific range of frequencies, whereas, over this same range, the noise from scattering is totally random or very variable. In this situation, an average of the actual ultrasonic signal at any point, taken over the above frequency range, will tend to gradually enhance the coherent defect echo at the expense of the noise. One of the specific features of the processing is the need to split the spectrum up in the frequency domain using a series of bands, between a minimum and maximum value defined from the frequency spectrum of the original signal. This is done using a series of filters of specific width and spacing in the frequency domain. For each filter, the resultant frequency signal is reconstituted into the A-scan form. It is the way that these inverse transforms are then added together that can significantly improve or degrade the output of the processing when compared to the original A-scan.

The number of filters, their width and spacing are all features that must be determined by careful experimentation on the materials in question. The NNDTC has been doing this to great effect for a number of years (ref 15). However, the work must also consider the best means of reconstruction, that is, whether to use minimisation, polarity thresholding or some adaptation of these or other criteria. One of the general points that has arisen from this work is the great improvement that is seen using SSP on B-scans, rather than A-scans. Ofcourse, it is true that this puts additional burdens on the computational time, but our work has shown that the improvement on an image is far in excess of the improvement in any single A-scan within that image. One other important finding was that it is relatively easy to optimise the SSP for the B-scan, whereas it is very difficult to do this for A-scans. This is perhaps an effect of the way that the B-scan presentation itself can help to reduce the importance of the noise within the image. Interestingly, it has been said that polarity thresholding works well for A-scans, but it was our experience that, however good this was, it is much better, and certainly more robust, when applied to a B-scan.

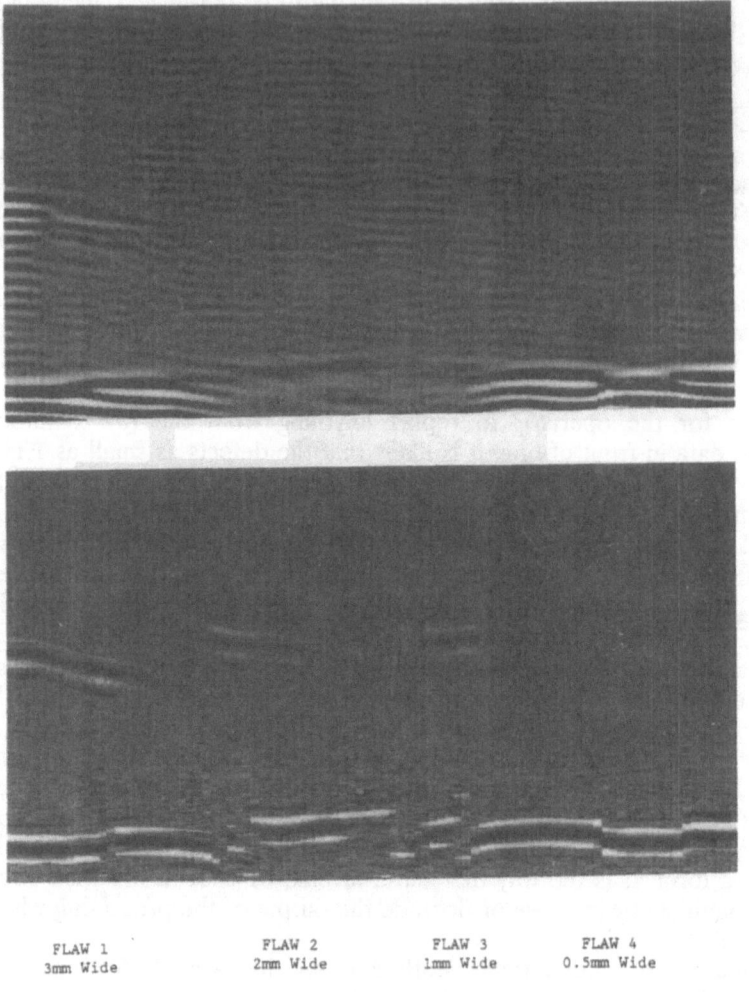

FLAW 1   FLAW 2   FLAW 3   FLAW 4
3mm Wide   2mm Wide   1mm Wide   0.5mm Wide

Figure 6 : An ultrasonic B-scan before (top) and after (bottom) split-spectrum processing. The positions of flaws 3.0 - 0.5 mm wide are as indicated.

So what of processing suited to use in-the-field ? As mentioned earlier in this section, the NNDTC is now producing the successor to its ZIPSCAN systems based around PC technology. Recent work has shown that, using the same principles as with TOMOHAWK, accelerator cards can be used to allow computationally intensive image processing to be made a viable proposition for use at the site of data acquisition. It has been demonstrated that the SSP algorithms used on older mainframe computers can be employed on such PC systems, and we are close to our goal of providing SSP as a standard option that will take no longer than one minute on a typical B-scan image.

5. Ground Penetrating Radar

In the field of inspection in civil engineering, the choice of NDT can be somewhat limited, particularly where the inspection of concrete from a single-side is concerned. Ground penetrating radar (GPR) has much to offer in this area, as it permits defects in concrete to be imaged in much the same way that we use ultrasonics in metals. However, industry has been slow to adopt the technology, partly because some claims have been overstated, and partly because the data can be difficult to understand. There are some very distinct differences between radar and ultrasonics, such as the bulk of the equipment, and the physical mechanism by which signals arise (ref 16), but there is enough similarity for a significant amount of our ultrasonic background to be used beneficially in radar inspection using pulsed systems.

One of the main similarities is the mode of operation. In radar, just as with ultrasonics, the source is scanned along a line across the surface of the inspection site. An image of radar reflections is thus obtained using a transmitter/receiver (and others as required) in exactly the same way as one would go about this for ultrasonics. As a result, the images from a radar scan have a distinct familiarity about them. This does not mean that they are as easy to understand. Indeed, the reason that GPR is not one of the 'big five' is precisely that the images are much less easy to interpret. For example, signal velocity for radar varies between typical engineering materials considerably more than ultrasonic velocity in metals, and consequently a linear depth-scale is very unlikely to be present. These difficulties, combined with the fact that there are no codes of practice unlike other inspection techniques means that to date, radar has not found its rightful place in the inspection market. It is clear that the potential is there, since the subjectivity presently involved in radar analysis could be significantly reduced by suitable signal processing and data presentation. Work performed within the NNDTC has shown that the thorough background in processing of ultrasonic images can be put to good use on radar images. In figure 7, the raw radar image from a 1 GHz probe, enhanced purely by simple presentational means (upper image), is significantly improved by signal processing not too dissimilar from that used in ultrasonics. The individual reinforcing bars in the specimen are much more evident in this processed image. More importantly in this field than ultrasonics is the fact that the results of the processing put the data into a form more akin to an engineering drawing. This makes it considerably easier for the decision makers, namely engineers, to act upon the data that radar can supply, by processing the data into the form most suited to the end-user.

With this early work showing what can be done, AEA Technology, through the NNDTC, is setting up a radar inspection for civil engineering club, in which industrial partners will direct the work so that radar inspection is improved in the ways most suited to likely users. It is expected that, drawing on our experience from other fields, image processing, combined with improved inspection procedures and predictive tools will permit radar to become an increasingly beneficial tool for inspection of civil sites to locate voids, buried services, leaks from pipes, reinforcements and other information on construction and state of repair. One very real objective that will result from an application of image processing allied with procedures akin to those used in other techniques, will be the provision of the first true depth scale.

In all of this work, it is necessary to provide the images on-site, because of the way the civil engineering community need to act on such data. Consequently, all of the developments build around the digital nature of the images from the radar unit and what can then be done on a suitable PC. Therefore, this field is an ideal example of the theme we have been presenting, namely the need to supply appropriate processing that can be provided on-site.

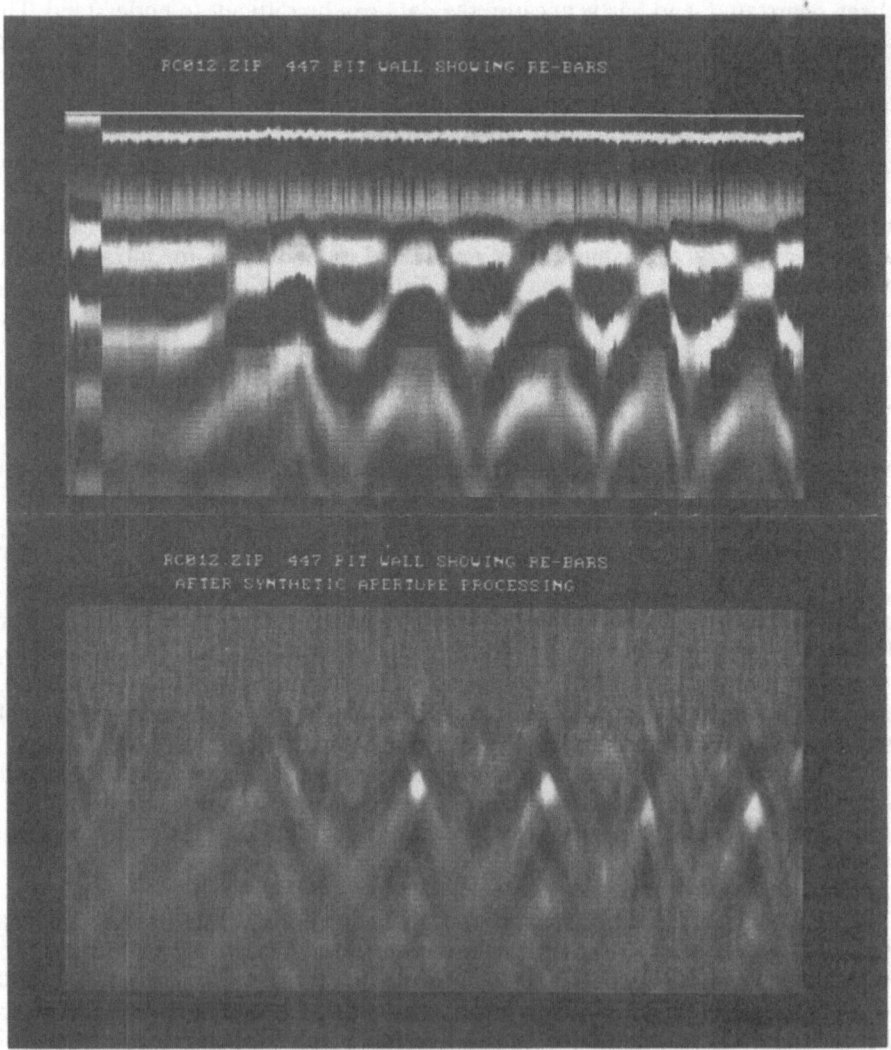

Figure 7 : Radar image of reinforcing bars (12" separation) in concrete, and below, the result of a first attempt at image processing to reveal the rebars themselves.

6. Conclusions

The paper has aimed to show that image processing can be provided as a powerful aid to NDT at the point where the inspections are performed. This has been illustrated by a number of specific examples drawn from several technical areas.

In the field of thermography, it has been demonstrated that the event from a thermal transient can be compressed into a single picture, retaining all of the available information, in the form of a plot of thermal diffusivity or thermal effusivity. It is now becoming routine for thermographers to try and solve the inversion problem of this event, and such processing will have similar gains. The work within the NNDTC has recently quantified defect detection limits in a number of key materials, and this information will be useful to those involved in the inversion problem, because the work establishes the working limits of such algorithms, and also indicates the true levels of signal-to-noise, which has some bearing on the robustness of such processing.

In the area of ultrasonics, the work at the NNDTC has demonstrated that, by application of physical rules through suitable signal processing, it is possible to retrieve defect signals from raw datafiles where noise appears to completely obscure their presence. It has been found that such processing can be performed in the field quite routinely, using standard PC-based equipment. Thus, such improvements can be provided on-site at the time of inspection.

In the field of radiography, the NNDTC has launched TOMOHAWK, a computerised X-ray tomography system operating from a standard PC. Some examples of the rapid analysis available, requiring no more computer power than the PC provides, have be demonstrated, with the theme of rapid processing for routine, rather than one-off, usage.

In the field of inspection in civil engineering, AEA is now embarking on an activity to improve inspection using radar, through a careful application of image processing and data presentation. It has become clear that cross-fertilisation from other fields, including those above, will rapidly improve the reliability and usage of radar, provided the goal of processing on-site is always retained as paramount.

It is concluded that there are many examples where image processing is providing significant improvements in the detail and assessment that the inspector is able to provide on-site. It is hoped that, with careful consideration of what is required, the continuing developments in the PC and hardware fields will permit further advances to be imported into the inspection site for routine use.

7. Acknowledgements

The author acknowledges the efforts of many of the staff of the NNDTC who have contributed information for the presentation. He is particularly grateful to Maurice Silk and Andrew Temple for providing useful information for, and comments on, the contents of this paper. The financial support of ESA/ESTEC in the area of thermography, in particular, the provision of the diffusivity and effusivity mapping, is gratefully acknowledged. The work in the area of radar is supported by the Corporate Research Programme of AEA Technology.

352

8. References

1. J.M. Milne and W.N. Reynolds, Proceedings of the SPIE vol 590, 293-297, 1985.
2. W.N. Reynolds, Canadian Journal of Physics vol 64(9) 1150-1154, 1986.
3. R.L. Smith, VTT Symposium 93 Proceedings of Anglo-Finnish Joint Symposium on Advances in Nondestructive Testing, 26-32, 1988.
4. C. Hobbs, D. Kenway-Jackson and J.M. Milne, Proceedings of the SPIE vol 1467, 264-277, 1991.
5. C. Hobbs, Sensor Review vol 12(1), 8-13, 1992.
6. C.P. Hobbs and J.A.G.Temple, British Journal of Nondestructive Testing vol 35(4), 183-189, 1993.
7. H.S. Carslaw and J.C. Jaeger, Conduction of Heat in Solids, Oxford University Press, Oxford, 1947.
8. W.N. Reynolds, NDT International vol 20(3), 1987.
9. X.P. Maldague, Nondestructive Evaluation of Materials by Infrared Thermography, Springer-Verlag, London, 1993.
10. D.P. Almond and S.K. Lau, Proceedings of the Eurotherm Seminar number 27, 207-211, 1992.
11. S.F. Burch and P.F. Lawrence, British Journal of Nondestructive Testing vol 34(3), 129-133, 1992.
12. A.C. Kak, Proc. IEEE vol 67, 1245-1271, 1979.
13. N.M. Bilgutay, E.S. Fergason and V.L. Newhouse, IEEE Transactions on Sonics and Ultrasonics vol SU-23(5), 329-333, 1976.
14. V.L. Newhouse, N.M. Bilgutay, J. Saniie and E.S Fergason, Ultrasonics vol 20(2), 59-68, 1982.
15. M.G. Silk, Proceedings of the 4th European Conference on Non-destructive Testing vol 3, 1647-1660, 1988.
16. C.P. Hobbs, J.A.G. Temple, M.J. Hillier, M.G. Silk and H.G. Tattersall, Proceedings of the International Conference on Non-destructive Testing in Civil Engineering vol 1, 79-96, 1993.

IMAGING OF MAGNETIC PROPERTIES USING PHOTOTHERMAL TECHNIQUES IN MICROWAVE RESONANCE[1]

J. PELZL, O. VON GEISAU, TH. ORTH, F. SCHREIBER
AND K. FRIEDRICH*
Institut für Experimentalphysik III,
AG Festkörperspektroskopie, Ruhr Universität
Bochum, D-44780 Bochum, Germany
* *Institut für Optik und Quantenelektronik,*
Friedrich-Schiller-Universität, D-0-6900 Jena, Germany

The interaction of microwaves with magnetic materials in determined by the high frequency magnetic susceptibility tensor. To achieve a spatially resolved measurement of this property two different photothermal approaches have been developed and applied [1]. One technique makes fo the heat dissipated in the material in the course of the absorption of microwaves. Modulating the microwave input power thermal waves are generated which can be detected by the photoacoustic effect or by laser beam deflection. The photoacoustic mehtod was successfully applied to image the depth variation of the magnetization in layered magnetic tapes [2]. The laser beam deflection technique, in addition, provides a lateral resolution which has been used to visualize collective magnetic excitations in soft ferrites and in yttrium iron garnet slab [3,4]. In the latter material also the spatial evolution of magnetostatic modes in the linear and in the non-linear regime could be studied. To improve the spatial resolution and the sensitivity, a second technique has been developed that relies on the local modulation of the high frequency susceptibility by a thermal wave generated with an intensity modulated laser beam incident on the sample [5,6]. This technique has been used, for the first time, to obtain lateral resolved images of spherical magnetostatic modes excited at a microwave frequency of 9.2 GHz in an yttrium iron garnet sphere [7]. A further benefit of the photothermally modulated ferromagnetic resonance is the ability to visualize magnetic signals. The technique has been applied for the imaging of the magnetization distribution due to a magnetic signal recorded by a tape recorder on a particulate tape. As compared to the photoacoustic method and to the laser beam deflection technique, the photothermally modulated magnetic resonance provides a true three-dimensional tomographic method for magnetic properties with a superior sensitivity and spatial resolution. The signal strenth and shape, on the other hand, depends beside the magnitude of the high frequency susceptibility also on its temperature derivative, which complicates the quantitative description of the signal generation process. In most systems investigated untill now, the photothermally modulated ferromagnetic reso-

---

[1]this contribution: abstract only

X. P. V. Malague (ed.), Advances in Signal Processing for Nondestructive Evaluation of Materials, 353–354.
© 1994 *Kluwer Academic Publishers.*

354

nance signal is governed by the temperature coefficient of the spontaneous magnetization. Here the signal can be modelled straight forward on the basis of the magnetization behaviour. A dramatic intensity enhancement accompanied by a change of the signal shape is predicted on the basis of this mechanism when the magnetic transition temperature is approached. The predicted behaviour could be confirmed experimentally in the case of a yttrium iron garnet sample [8].

This work is supported by the Deutsche Forschungsgemeinschaft, Kennwort: Photothermik.

1.   J. Pelzl ans U. Netzelmann, "Locally Resolved Magnetic Resonance in Ferromagnetic Layers and Films", in TOPICS OF CURRENT PHYSICS Vol. 47: Photoacoustic, Photothermal and Photochemical Processes at Surfaces and in Thin Films, (P. Hess, ed.), Springer Verlag, Berlin, Heidelberg, 1989, p. 313-365.

2.   U. Netzelmann and J. Pelzl, "Ferromagnetic resonance dept profiles of magnetic tapes using frequency dependent photoacoustic detection", Appl. Phys. Lett. 44, 854-856 (1984).

3.   O. v. Geisau, U. Netzelmann and J. Pelzl, "Photothermal Characterisation of Inhomogeneous Ferromagnetic Resonance Absorption in Ferrites" J. Appl. Phys 63, 3347-49 (1988).

4.   O. von Geisau, U. Netzelmann, S.M. Rezende and J. Pelzl, Photothermal Investigation of Magnetostatic Modes in Yttrium-Iron-Garnet at High Microwave Power Levels, IEEE-Trans. 26, 1471-73 (1990).

5.   Th. Orth, U. Netzelmann and J. Pelzl, "Thermal wave imaging by photothermal modulated ferromagnetic resonance", Appl. Phys. Lett. 53, 1979-81 (1988).

6.   J. Pelzl, U. Netzelmann, Th. Orth and R. Kordecki,. "Photothermal Imaging Using Microwave Detection" In "Photoacoustic and Photothermal Phenomena", B. Royce and J. Murphy, eds. Springer Series in Optical Sciences, A.L. Schawlow, Series Ed., (Springer Verlag, Berlin, Heidelberg 1990), 2-12.

7.   J. Pelzl and O. von Geisau, "Spatially resolved detection of microwave absorption in ferrite materials", in: Progress in Photothermal and Photoacoustic Science and Technology, Vol. II, ed. A. Mandelis, Elsiver Science Publ., N.Y. 1993, in press.

8.   F. Schreiber, M. Hoffmann, O. von Geisau and J. Pelzl, "Investigation of Photothermally Modulated Ferromagnetic Resonance Signal of Magnetostatic Modes in Yttrium Iron Garnet", Appl. Phys. D. (Zeitschr. Physik C), 1993, in press.

LIMITED VIEW COMPUTED TOMOGRAPHY IN NUCLEAR INSPECTION[1]

RICHARD V. MURPHY
Engineering Technology Div.
AECL - Research, Chalk River Labs.
Chalk River, Ontario Canada K0J 1J0

This paper discusses the theoretical and experimental development of Limited View Computed Tomography (LVCT) methods based upon Algebraic Reconstruction Techniques (ART). The LVCT/ART process was sucessfully applied to image the distribution of fouling on steam generator tubes.

Some nuclear steam generators contain approximately 3,000 Inconel 600 U-bend tubes of 30 metre length and 13 mm diameter. These are closely spaced, with a tube-to-tube gap of 7 mm. Tube separators, termed support plates, are used to hold the tubes in place. The gap between a tube and support plate (termed broach hole) is only 3 mm. A build-up of fouling in these holes can cause abnormal flow surges and level fluctuations during operation. These is a need for and inspection tool to image the extent of broach hole blockage before deciding on corrective maintenance.

Chalk River Laboratories (CRL) investigated methods to measure the degree of blockage using penetrating radiation. Theoretical calculations showed that sufficient density resolution could not be obtained by simultaneous movement of a single radioactive source in one tube and a single detector in an adjacent tube. Computer modelling revealed that a useful image of broach hole fouling could be constructed using several source and detector position. CRL conducted theoretical and experimental studies to measure broach hole fouling by Limited View Computed Tomography.

An Iridium-192 source was placed inside one steam generator tube and an array of gamma-ray detectors was placed in an adjacent tube. The source was moved in 1 mm steps over a range of 300 mm. For each source step, the radiation intensity recorded by each element of the detector array was measured. This information was transmitted out of the steam generator tube by a fibre-optic cable, which was connected to a high resolution charge coupled device camera operating under computer control. As each of the 300 x-ray images was captured, the information was processed by Algebraic Reconstruction Techniques. ART uses iterative algorithms to approximate the image that otherwise could not be attained through analytical solution. The result was a Limited View Computed Tomographic image showing a cross section of the broach plate region. The severity of broach plate erosion and the degree of broach hole fouling were clearly visible.

---

[1]this contribution: abstract only

X. P. V. Malague (ed.), Advances in Signal Processing for Nondestructive Evaluation of Materials, 355.
© 1994 *Government of Canada.*

DEVELOPMENT OF VISION TOOLS FOR WELDING APPLICATIONS

B. V. HESS
E. A. FUCHS
A. R. ORTEGA
Center for Materials and Applied Mechanics
Sandia National Laboratories
P. O. Box 969
Livermore, CA 94551
USA

ABSTRACT In order to adequately support our weld modeling work, new ways of acquiring and analyzing data were required. This involved not only instrumentation of the welding process with a variety of vision tools to provide the data required, but also developing image analysis tools for reducing the data and delivering the data to the modeler in a timely and accurate manner. Temperature and displacement diagnostics have been used most frequently and will be discussed in this paper. The availability and increased speed of workstations have enabled us to extend these tools from use for modeling to weld process control.

1. Introduction

Modeling of GMA (gas metal arc) welding and GTA (gas tungsten arc welding) requires experimentally measured input parameters as well as code validation data. Typical input to each process is the integrated power curve. GMA welding also requires wire feed and metal droplet temperature. Most of the full field diagnostics   have been developed for welding apply to the code validation aspects of modeling [1]. A number of parameters must be measured in order to validate the thermal and mechanical aspects of the code: spatial and temporal temperature distributions, plate distortions that occur during welding as well as final deformed shape, and residual stresses produced by the welding process. Although most of these diagnostics have been used on welding or other processes for many years, the use of the workstation and image analysis tools to improve the data

357

X. P. V. Malague (ed.), Advances in Signal Processing for Nondestructive Evaluation of Materials, 357–370.
© 1994 *Kluwer Academic Publishers.*

reduction process has made use of vision tools feasible for real time process control as well as decreasing the time required to reduce and deliver the data to the analyst.

The vision tools that have been used to provide validation can be extended to provide input data or process control information. Displacement tools can be used to track arc gap, which provides a measure of input voltage drop, seam tracking, and real time planar distortions. Full field temperature sensors provide thermal gradient information. These thermal sensors can also be used to track metal deposition rate, seam location and arc gap.

2. Thermal Diagnostics

The automation of the thermal diagnostic calibration and data reduction has greatly reduced the amount of time required to process the full field data and has made "real time" temperature acquisition possible. In all of the cases cited below, the examples serve to illustrate the capabilities of the diagnostics developed in the workstation environment.

2.1 "CALIBRATION" OF THE SYSTEM

In order to calculate temperature using a thermal imaging system, it is necessary to map radiance, output in gray levels, to temperature. This can only be accomplished by measuring how the emissivity changes as a function of temperature and surface condition. By monitoring the temperature on a specimen fabricated out of the welded material and comparing the temperature recorded by an independent means (e.g., thermocouple or resistance thermometer) with the temperature calculated using the output from the thermal imaging system during a controlled heating procedure, a lookup table or calibration curve can then be developed that relates the output radiance to the temperature of the specimen. If the temperature is experimentally monitored in areas that reflect different surface conditions, e.g., virgin, oxidized, welded, etc., and the temperature excursion reflects the temperatures seen during the actual welding process, then the surface of the material can be characterized and the temperature calculated from the radiance. This technique has been used successfully on a number of previous applications.[2,3] The agreement between data collected by thermocouple and thermal imaging system using this calibration technique produces good agreement, Figure 1.

When the above experimental procedure was carried out on a PC or microVAX II, The process was completely inflexible and required an extensive menu system or a complicated command line code. Some of the problems were that in order to insure that the data from the thermal imaging system was only collected during periods when the temperature was stable (changed less than 1%- 5% between thermocouple readings) for a thirty second period, the data acquisition system and the heating system were tied together. This stability requirement is required because of the lag between the collection of the thermocouple data and the infrared data. There was not enough computing power in the image analysis system to calculate whether the system was stable based on the radiance output. Even with minimal calculations being done in the image analysis system, the maximum data rates were around 2 Hz. In order to change any of the parameters, a time consuming procedure was required. The creation of the look-up table was done after all of the data had been recorded and stored on disk. It was not possible to monitor the process that developed the look-up table [4].

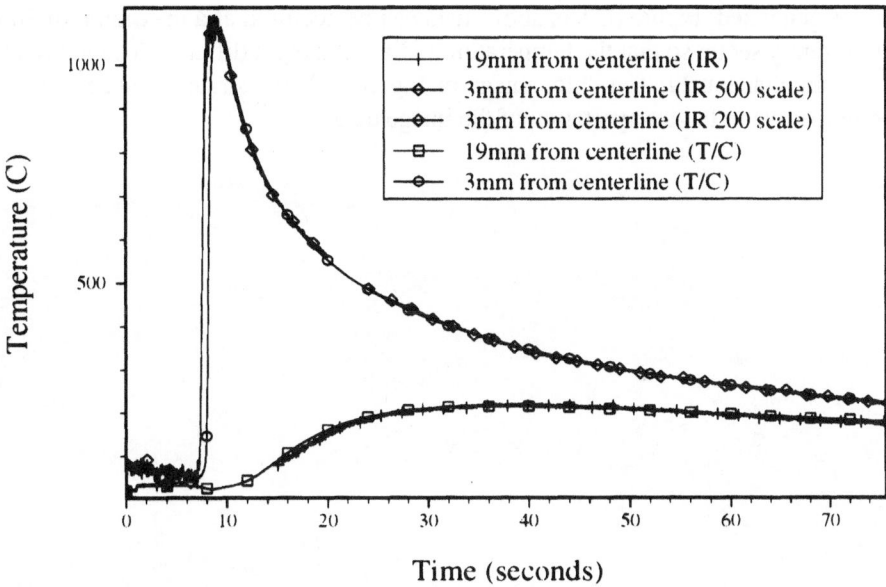

Figure 1. Comparison of thermocouple and infrared camera data for 304L, stainless steel. One thermocouple was located 3 mm from the center of the weld pool (1 mm from the edge of the pool). A second thermocouple was located 19 mm from the centerline. The plate was nominally 1.27 mm (050 inches) thick, 152.4 mm (6 inches) long, 101.6 cm (4 inches) wide, and welded using GTA. The weld was run at 12 mm/sec in a relatively inert, argon environment.

By using a workstation with much faster computing speeds and a system architecture that facilitates the handling of large arrays, the image analysis routines can process the data acquired from the above experiment faster and provide greater programming flexibility. The program that controls the heating of the specimen and the image analysis routine that reduces the full field data are completely independent. The heater controller can be

360

changed or upgraded at any time without necessitating a change in the full field data reduction program. The stability of the heating process is determined within the image analysis program and can be done in a number of ways: the percent of temperature excursion can be varied, the amount of "stable temperature" time can be varied, the stability requirement can be shut off (useful in cases where the material will change phase and surface condition when resident at a high temperature for too long a period, e.g., titanium). The temperature vs. time, gray level vs. time, and the temperature vs. gray level curves are all displayed in real time during program operation, Figure 2. When the experiment is finished, the calibration data file is then written automatically. If a "calibration" curve is required, the data file can be "dropped" into a fitting routine where a curve is generated, Figure 3. The above data can be acquired at a maximum of 30 Hz and is currently set up so that the temperature/radiance in eight different locations can be acquired simultaneously. The color image of Figure 4 demonstrates the necessity of a calibration curve in the interpretation of the image data.

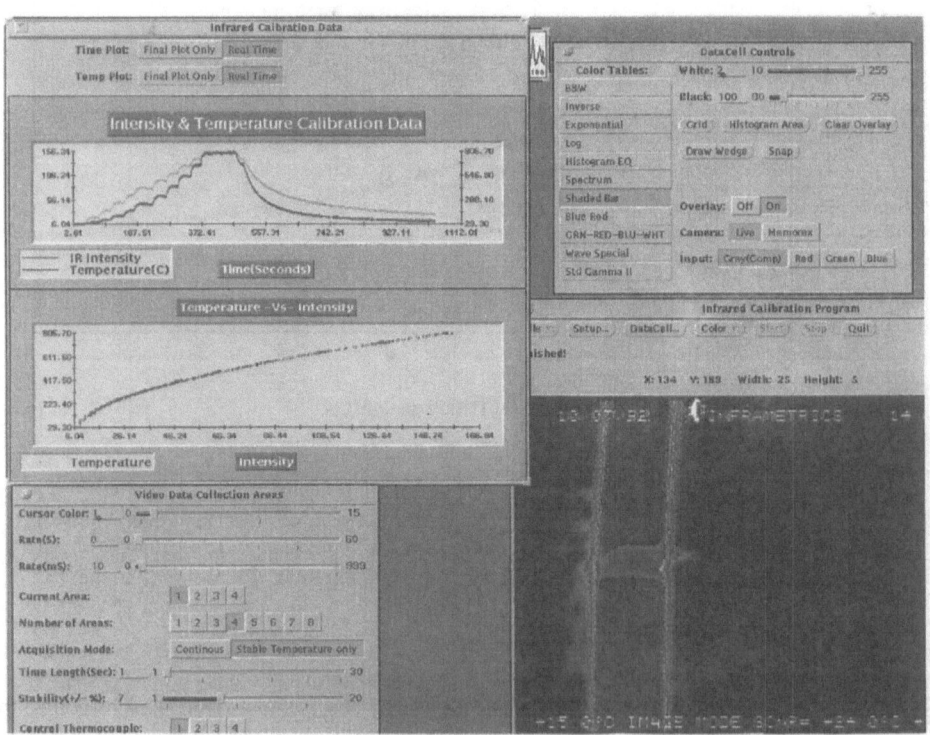

Figure 2 Sample screen from the "calibration" program. Shows typical windows available and pull down menus.

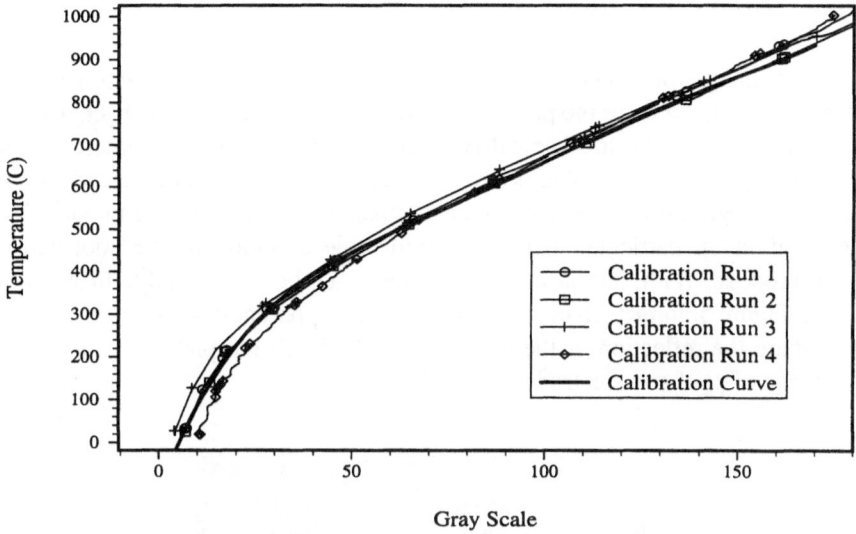

Figure 3. Data for 304L stainless steel. Comparison of four different test runs with a calibration curve fit to the data. The specimen was heated to approximately 1100° C in 100° C increments. The "calibration" runs were done in a vacuum to prevent further oxidation to the sample.

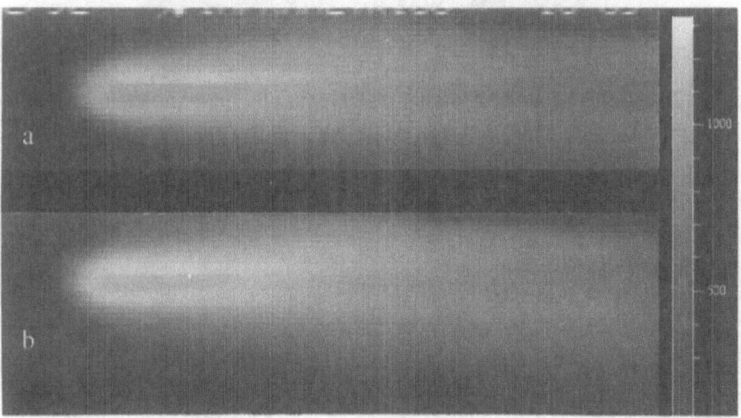

Figure 4. Comparison of the image of the raw output gray levels and the image processed using the calibration curve. Figure 4a is from a GTA welded plate with raw data. Figure 4b is the same image processed through the calibration curve.

2.2 DATA PROCESSING TOOLS

The thermal images produced by a scanning camera have a common feature when a very hot region (the electrode or the pool) is next to a very cool region (the background or the plate). The effect of the mechanical scanner in thermal scanning camera, produces a noticeable, ragged edge. Another issue for imaging cameras with interlaced video is that if there is enough motion, there can be a completely different image in each field. For GMA welding, in particular, there is considerable agitation in the pool and droplet transfer process. With interlaced video, there can be two totally different images recorded, Figure 5. In fact, changes in the image are apparent from top to bottom in some fields, where the reflection of the pool and electrode on the plate is different than the direct image of pool and electrode, Figure 6b.

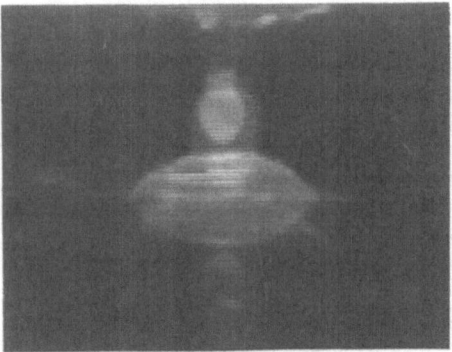

Figure 5. Two fields of a GMA weld on 304L stainless steel superimposed on a single frame.

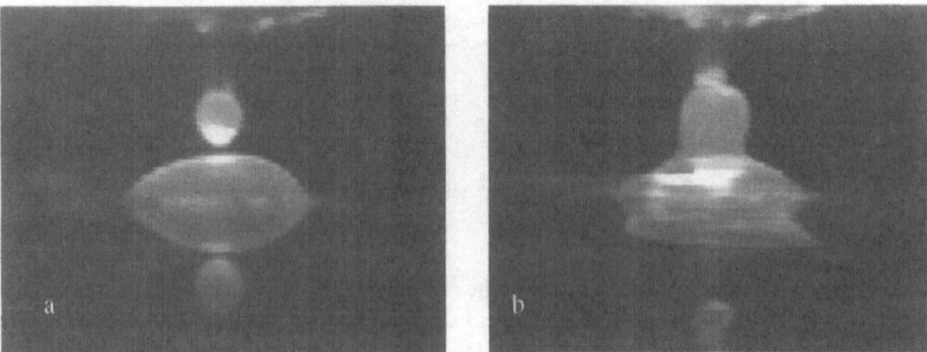

Figure 6. Two fields of a GMA weld. Figure 6a is from the even lines and 6b is the odd lines. The ragged edge produced by the scanning mechanism is readily apparent.

A clean-up routine that first separated the frame into two fields, then used filtering routines was developed in order to gain access to each field and to clean-up the ragged edges. About three minutes were required for each frame on the microVAX II. For a weld taking ten seconds, it took about 15 hours of constant computation time to clean-up the images and get them ready for data reduction. On the workstation the processing time for each frame is about 1.1 seconds. The complete ten second weld now only takes 5.5 minutes to process, Figure 7. While this procedure would not be used during feedback control, it is a useful tool for post processing the data. One of the reasons that the processing is faster is that the workstation can operate on much larger arrays in program memory rather than swapping data between the disk and program.

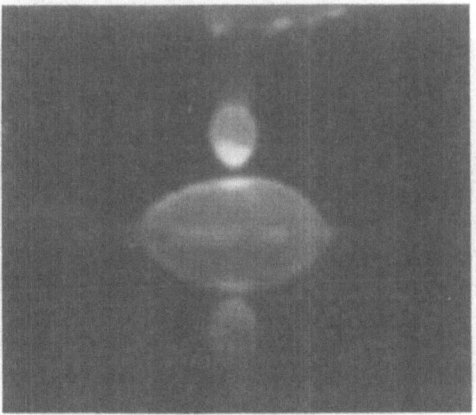

Figure 7. Figure 5a after cleanup.

2.3 "REAL TIME" DATA REDUCTION

With the help of modeling tools, feedback control parameters can be determined based on stresses or temperatures generated by the welding process. For example, if a weld is one of the final assembly steps, it might be necessary to protect internal components. Some components might crack in the presence of process high temperatures or in the presence of stresses induced by thermal expansion. It might be determined that the residual stress field can be affected by controlling the thermal gradients. Then the input power needs to be controlled by feeding in thermal gradient information, which can readily be obtained using a thermal imaging system.

Thermal imaging has been used to provide a full field view and to provide a diagnostic of inconsistencies seen in an infrared fiber optic sensor used for feedback control, but we have not used thermal imaging directly for feedback control [5]. In order to use data generated by full field diagnostics in a feedback control scheme, real time data processing is essential. *Real time* can have very different meanings depending on the process trying to be controlled. Presently for arc welding processes, the welding machine cannot be changed any faster than about 10 Hz. Most of the standard full field diagnostics operate at 30 Hz. With a fast enough CPU, the data can be processed and output to the machine in a timely fashion. One of the tools that has already been developed is the temperature line scan. While it may seem like the familiar line scan mode, this tool processes the data and produces the temperature. The lines are written to a window during processing, Figure 8, and is stored as an image. Individual lines can be interrogated post-test to determine the temperature history of the line or a point on the line. Because the data is stored as unique lines, these lines of data can be interpreted by a workstation plotting pac

Figure 8. An image of the temperature line scan feature. Each line of this image represents a line across the plate and through the weld pool.

3. Displacement Diagnostics

The displacement diagnostics have been developed out of the need to validate the mechanical part of the finite element analysis. The final distorted shape or the residual stresses can be used to determined whether the end result is correct. However, it is important to understand whether or not the intermediate results are correct. By tracking the distortion on the plate during welding, the analyst has the opportunity to verify that the code is following the correct path to correct result.

3.1 OUT OF PLANE DISTORTION: SHADOW MOIRE AND LASER STRIPE

Shadow moire is one of the simplest tools available for determination, post-test, of the out-of-plane distortion. It provides a flexible technique to measure distortion: the resolution can be altered by changing the angle of incidence of the light source;

displacements as small as about 25 μm (.98 μinches) can be accurately determined; the output is a fringe pattern that depicts the distortion field as a set of fringes, Figure 9. This fringe pattern can be qualitatively compared directly to a similar plot produced by a finite element analysis. If the values of the displacement are required at various locations on the specimen, then a tedious fringe counting procedure must be used to determine the loci of the center of the fringes. Another procedure would be to pick a particular line scan and determine the fringe count as each fringe is crossed. If more than one line of data is required, this can become very tedious.

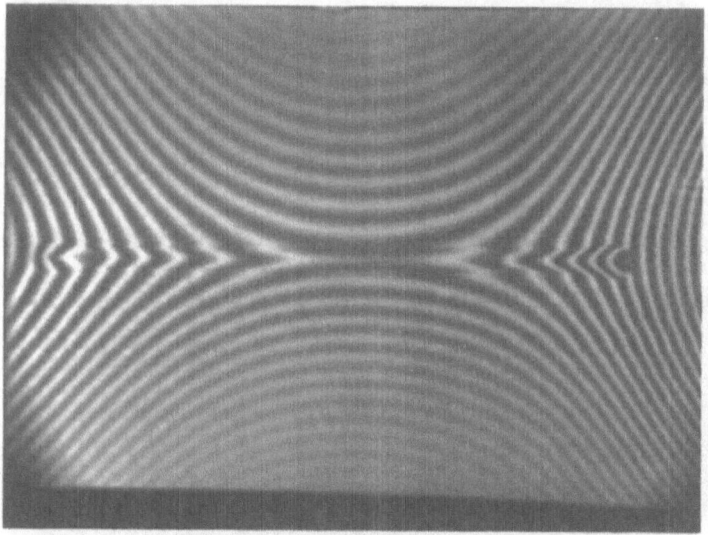

Figure 9. Shadow moire pattern for Ti 6-4 plate distorted into a saddle shape. The plate was welded using GTA, is 1.27 mm thick, 152.4 mm by 101.6 mm in area. Each fringe represents a displacement of approximately .3 mm (.012 inches). The pitch of the master grating is .123 mm/line (.005 inches/line), the angle of illumination is about 23 degrees.

Another way to solve the problem is to design a filter that finds the center of each half fringe, Figure 10 and 11. Now the tedious work of locating the true center is done, the fringes can be easily counted and displacements calculated automatically. The qualitative comparison with the analysis is easier to visualize, as well. By designing the appropriate filter, the tedious job of quantitative data reduction becomes automated. Although we are currently using this technique only as a post-test processing tool, it can be used, with modification, as a tool for distortion measurement during welding.

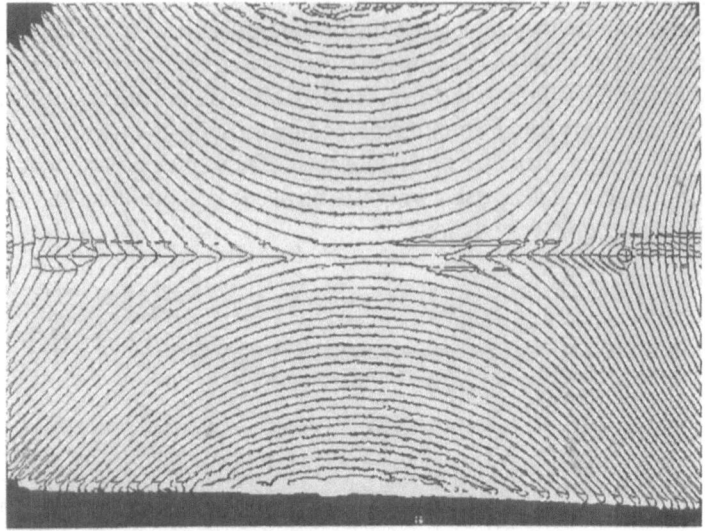

Figure 10. The peaks and valleys of the fringe pattern shown in Figure 9.

Figure 11. The superposition of Figure 9 and 10.

A coarser but similar technique to shadow moire is the laser stripe technique. In this case structured light produced with a laser illuminates a test object. By knowing the angle of incidence and angle of observation, the distortion at any point of illumination can be determined, Figure 12. The distortion that can be measured using this technique is about .013 mm (.0005 inches). This technique is relatively easy to use. Commercial diagnostic systems are available. However, they are in general very expensive and inflexible. This technique has applications in seam tracking and arc gap determination for process control. It can also be used to provide real time distortion data on parts that see displacements. By comparing the distortion resolution in the weld pool, the figures 9 and 11 illustrate the difference in resolution of the two techniques.

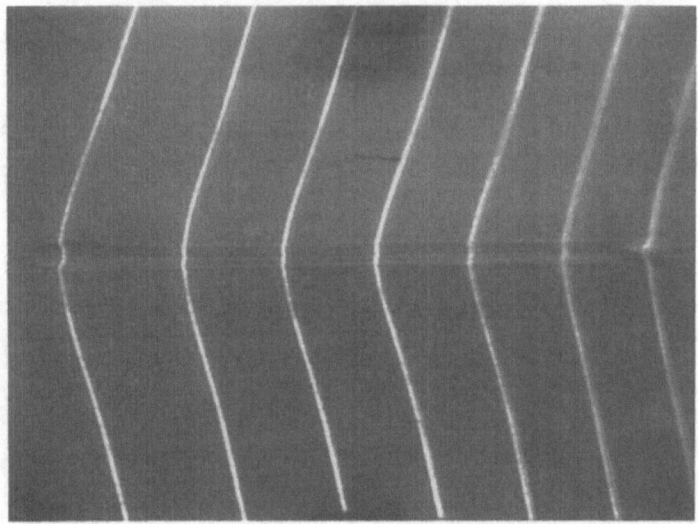

Figure 12. Image of a plate with structured light illumination.

3.2 EDGE DISPLACEMENT TRACKING

In order to validate analytical distortion predictions on a welding specimen with large displacements, approximately 6.9 mm (.27 inches), cameras were placed to record the motion of the various points on the edge of the plate. A corner, the mid point of one side, and midpoint of the end where the weld finishes were monitored, Figure 13. The displacement data was linked with time code to the thermal data described in section 2 above. This linkage allows the interactions between the thermal and mechanical aspects of this problem to be interpreted. This can be important in the case of a material, such as titanium, that undergoes a phase transformation at specific temperatures.

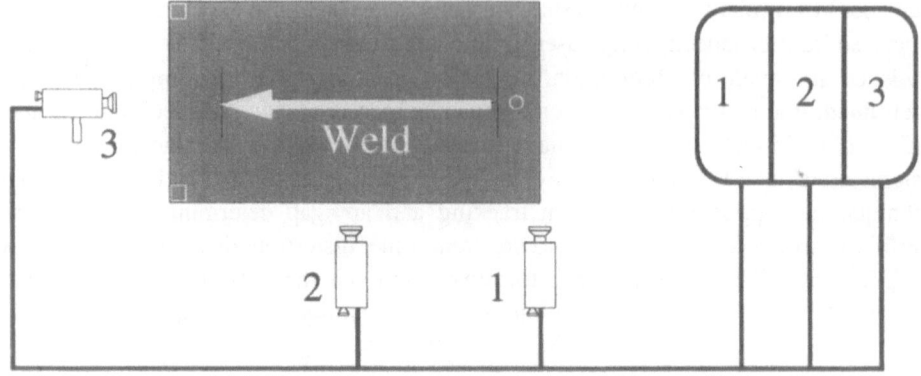

Figure 13. Figure 13 is a schematic of the plate and points where the displacement was monitored. (Same parameters as Figure 9.)

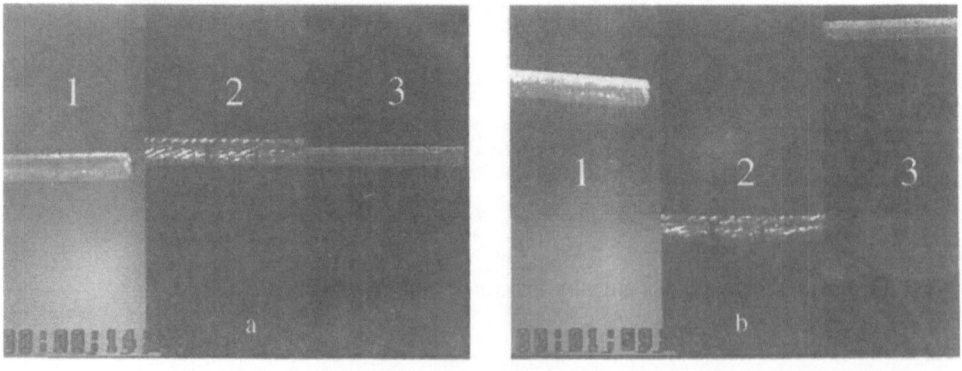

Figure 14. Figure 14a is the image of the initial position of the edges; Figure 14b is the image of the final position of the edges.

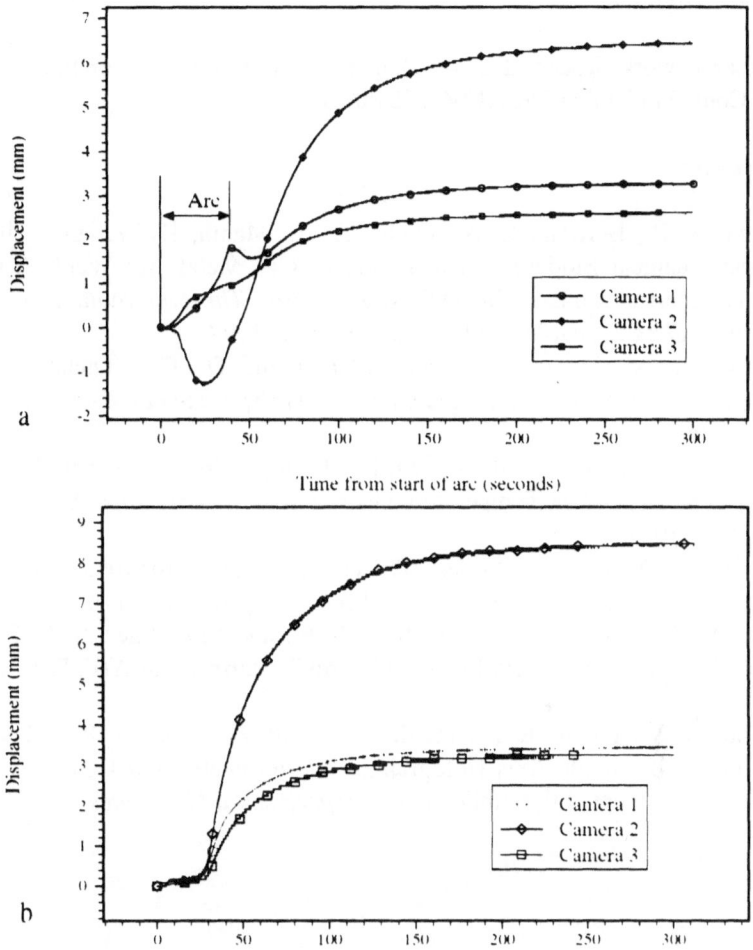

Figure 15. Figure 15 a is the plot of the displacement of the plates shown in Figure 14 as a function of time. For comparison, Figure 15b is a plot of a 304L stainless steel plate that had roughly the same dimensions and welding configuration.

4. Conclusions

Temperature and distortion vision sensors are currently being used to obtain data for weld modeling applications. The same sensors are being extended for use in process control. There are many other sensors that are being used for verification of the weld modeling analysis that have not been discussed here, e.g., holographic hole drilling for the measurement of residual stress [6].

5. Acknowledgment

The research work described in this paper was performed under the Department of Energy Contract Number DE-AC04-76DP00789

6. References

1. Ortega, A. R., Bertram, L. A., Fuchs, E. A., Mahin, K. W., and Nelson, D. V., "Thermomechanical Modeling of a Stationary Gas Metal Arc Weld: A Comparison Between Numerical and Experimental Results," *Proc. Third International Conference on Trends in Welding Research*, Gatlinburg, TN, June, 1992.
2. Fuchs, E. A., Hughett, P., and Macmillan, D. C., "Quantitative Infrared Thermography," *Proc. 1987 Spring Conference on Experimental Mechanics*, June 15-19, 1987, Houston, TX
3. Fuchs, E. A., Lipkin, J., and Hughett, P., "Using an Imaging Infrared Radiometer to Measure Time Dependent Temperature Distributions, *Thermosense X*, Proc. SPIE, 371, pg, 230, May 1987.
4. Hughett, P., Fuchs, E. A., "Image Processing Software for Real Time Quantitative Infrared Thermography," *Thermosense X*, Proc. SPIE 780, page 176, May, 1987
5. Fuchs, E. A., Bertram, L. A., Bentley, A. E., and Marburger, S. J., Evaluation of Feedback Parameters for Weld Process Control," *Thermosense XIV*, Proc. SPIE, April, 1992.
6. Nelson, D. V., Fuchs, E. A., Makino, A., and Williams, D. R., "Residual Stress Determination by Single-Axis Holographic Interferometry and Hole Drilling, Part II: Experiments," Accepted for publication in *Experimental Mechanics*.

A NOVEL TECHNIQUE FOR DEFECTS CLASSIFICATION FROM THEIR ULTRASONIC PULSE ECHOES

AHMED YAMANI and S.Z. AL AKHDHAR
Department of Electrical Engineering
King Fahd University of Petroleum and Minerals
Dhahran 31261
SAUDI ARABIA.

ABSTRACT. A new technique for defects classification from their ultrasonic pulse echo signals is proposed. Conformal mapping of the input-output defect model is used to enable the formulation of the defect characteristic function. Features from this function are extracted for use in a hidden flaw discrimination.

1. Introduction

Identification of defects from their pulse echo signals is equivalent to finding the correlation between the signal seen on the CRT and the nature of the defect that causes it. In recent years, considerable efforts have been devoted to this problem [1-3]. Chen [3], for instance, has suggested the use of signal processing and pattern recognition techniques. By suitable modeling of the ultrasonic pulse echo, features from the impulse response (IR) domain can be extracted for use in a hidden flaw discrimination. However, the highly complex interaction between the defect geometry and the backscattered ultrasonic wave inside the material cannot; in general; be assumed as a linear process and therefore, the impulse response signature is limited to a very special class of defects.

In this paper, the ultrasonic pulse echo model [3] is used to generate the reference signal (input to the defect model) such that the measured pulse echo can be considered as the output of the defect model. The input and output of this defect system are mapped into a complex trajectory plane described by the characteristic function of the defect. Features of this function are extracted to represent a generalized defect signature. Simulation and experimental results are presented to support this finding.

2. System Identification Problem

2.1 THE CONVOLUTIONAL MODEL

In the pulse-echo mode of ultrasonic testing, if the interrogating signal is denoted by $x(t)$,

X. P. V. Malague (ed.), Advances in Signal Processing for Nondestructive Evaluation of Materials, 371–384.
© 1994 *Kluwer Academic Publishers.*

then under various simplifying assumptions, the reflected signal $y(t)$ can be expressed as a convolution integral:

$$y(t) = \int_{-\infty}^{\infty} h(\tau)x(t-\tau)d\tau \tag{1}$$

where $h(t)$, called the impulse response, is a function of time related to the reflections that occur as the sound propagates through the medium in the direction along the beam. The problem of calculating $h(t)$ given the measurements $x(t)$ and $y(t)$ is called *system identification*. In practice, the system identification problem can be very difficult, and a great deal of the literature has been dedicated to it [4-8]. In ultrasonics, measurements are always band-limited, and often noisy. In addition, if the impulse response of a flaw is to be estimated, one must find a meaningful way to generate (or measure) an input (or a reference) signal $x(t)$.

2.2 GENERATION OF A REFERENCE SIGNAL

For known artificial defects in a homogeneous medium, we can assume that the transducer, the wave propagation paths and the defect as linear, time invariant systems, and thus we can write the received signal $y(t)$ as:

$$y(t) = u(t) \otimes T_1(t) \otimes P_1(t) \otimes h(t) \otimes P_2(t) \otimes T_2(t) \tag{2}$$

where $u(t)$ is the electrical impulse driving the transducer, $T_1(t)$ is the forward transducer impulse response, P_1 is the forward propagation path impulse response, $h(t)$ is the impulse response of the defect, P_2 is the return propagation path impulse response, T_2 is the backward transducer impulse response, t denotes time and $\otimes$ denotes the convolution operation.

The reference signal can be generated (measured) by using a scatterer with an impulse response that approximates the Dirac delta function. This can be done by measuring the reflection of an ultrasonic pulse from a flat back surface of the same medium. For our problem, we can write $x(t)$ as:

$$x(t) = u(t) \otimes T_1(t) \otimes P_1(t) \otimes \delta(t) \otimes P_2(t) \otimes T_2(t) \tag{4}$$

and thus

$$y(t) = x(t) \otimes h(t) \tag{5}$$

However, the highly complex interaction between the defect geometry and the backscattered ultrasonic wave inside the medium cannot in general, be assumed as a linear process and thus the defect impulse response is limited to a very special class of defects.

2.3 GENERALIZED SYSTEM IDENTIFICATION PROBLEM

For a given defect, equation (5) can be rewritten as:

$$y(t) = x(t) \odot s(t) \qquad (6)$$

where $s(t)$ is the defect signature, and $\odot$ is an operator that defines the relation between the input and output of a given system signature. If the defect system is linear then the operator becomes a convolution and the defect signature becomes the defect impulse response. Equation (6) requires, in addition to the input-output defect signals, the knowledge of the operator or the defect signature to be solved. In this paper we take the operator as the *circle* operator used in function decomposition, such that equation (6) becomes:

$$y(t) = x(t) \text{o} f(t)$$

$$= f(x(t)) \qquad (7)$$

where f is the characteristic "*function*". Features of the defect characteristic *function* can be considered as the defect signature.

3. Fourier Descriptor Approach

One way of extracting the input-output relationship of the defect model is to find the characteristic function of this model. This function can be thereafter, qualitatively mapped into a domain that makes it amenable to mathematical formulation. The defect model is shown in figure 1.

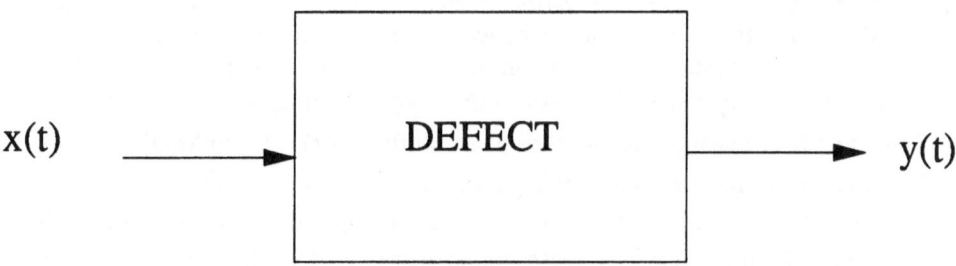

Figure 1: Defect model.

Plots of the output $y(t)$ versus the input $x(t)$ is by definition called the characteristic function. In this paper, we define the functional mapping as:

$$v(t) = x(t) + jy(t) \qquad (8)$$

where $j = \sqrt{-1}$.
The tip of the phasor $v(t)$ describes a curve γ in the complex trajectory plane for every input-output pair. This curve, being independent of time, shows the variation

$y = f(x)$ where f is the characteristic function of the defect model. For periodic signals, as is the case in ultrasonic pulse echo signals, the curve γ is closed and thus $v(l)$ is periodic that is:

$$v(l+L) = v(l) \tag{9}$$

where l is an arc length along the curve γ and L is the perimeter of the curve γ. Fourier Descriptors [4] of the curve γ (FD_γ) can be obtained.

Although FD_γ are unique to the curve γ, they do not represent the ensemble $\{\gamma\}$ that carries the same information (i.e. defect signature) regardless of the input-output signal type.

Alternative quantities that are insensitive to many transformations of the curve γ, but dependent solely on the "shape" of γ can be derived [4]. Amongst these quantities, we define

$$b_n = \frac{C_n C_{-n}}{C_0^2} \tag{10}$$

where C_n is the nth Fourier Descriptor of the curve γ.

4.   Validation of the Results

4.1   SIMULATION RESULTS

To validate the proposed approach, various systems with different characteristic functions are considered. The input is taken as a sine wave over one period for all systems.
*   * System A: linear system with no phase shift; $y(t) = x(t) = \cos(\omega_0 t)$
*   * System B: linear system with a phase shift; $y(t) = \cos(\omega_0 t + \theta)$
*   * System C: Non-linear system with no phase shift; $y(t) = 1 - \sin^2(\omega_0 t)$
*   * System D: Non-linear system with a phase shift; $y(t) = 1 - \sin^2(\omega_0 t + \theta)$

Figures 2 and 3 show the real and imaginary parts of C_n and b_n for systems (A,B) and (C,D) respectively. It is clear from these figures that C_n and b_n are very much sensitive to the "shape" of the characteristic function (or system). To demonstrate the insensitivity of b_n to the starting point, system A is considered with the input taken as $x(t) = \cos(\omega_0 t + 20°)$. It can be seen from figures 2 and 4 that whereas C_n is sensitive to the starting point, b_n is invariant and thus, it can be taken as the defect signature. The characteristic functions of system A,B,C, and D are shown in figure 5.

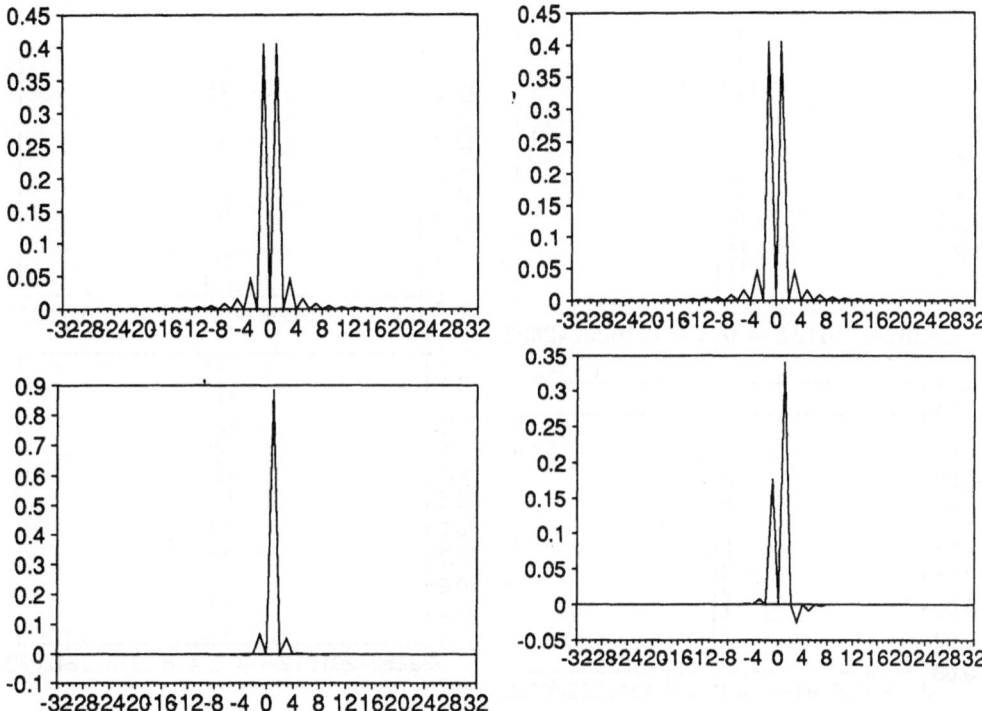

Fig.2: a) Real and imaginary parts of C_n [A,B]

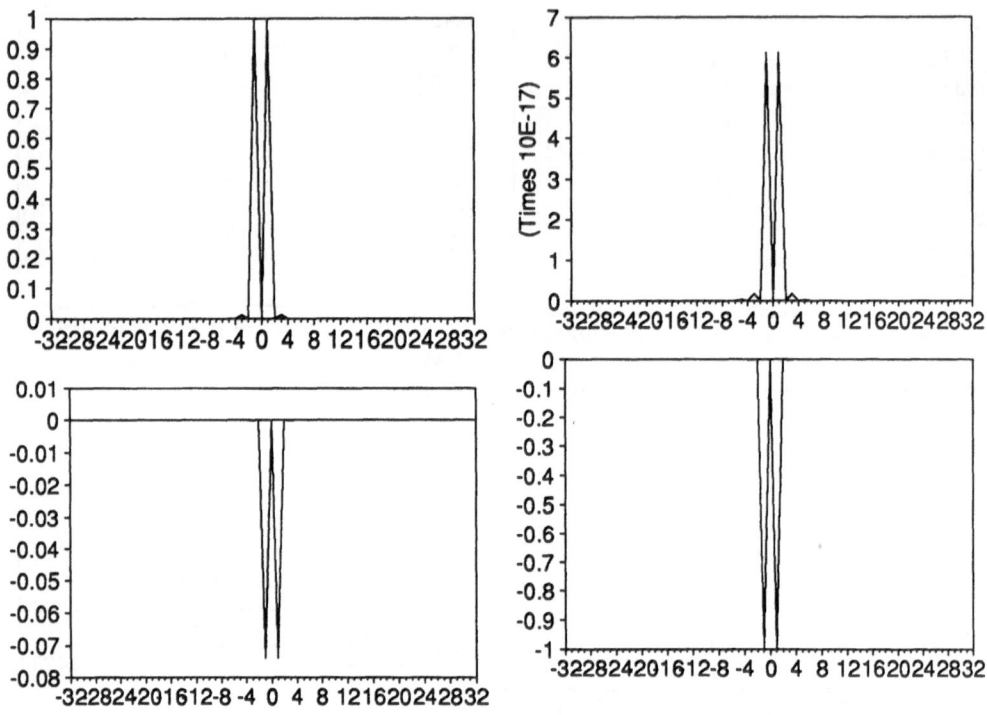

Fig.2 b) Real and imaginary parts of b_n [A,B]

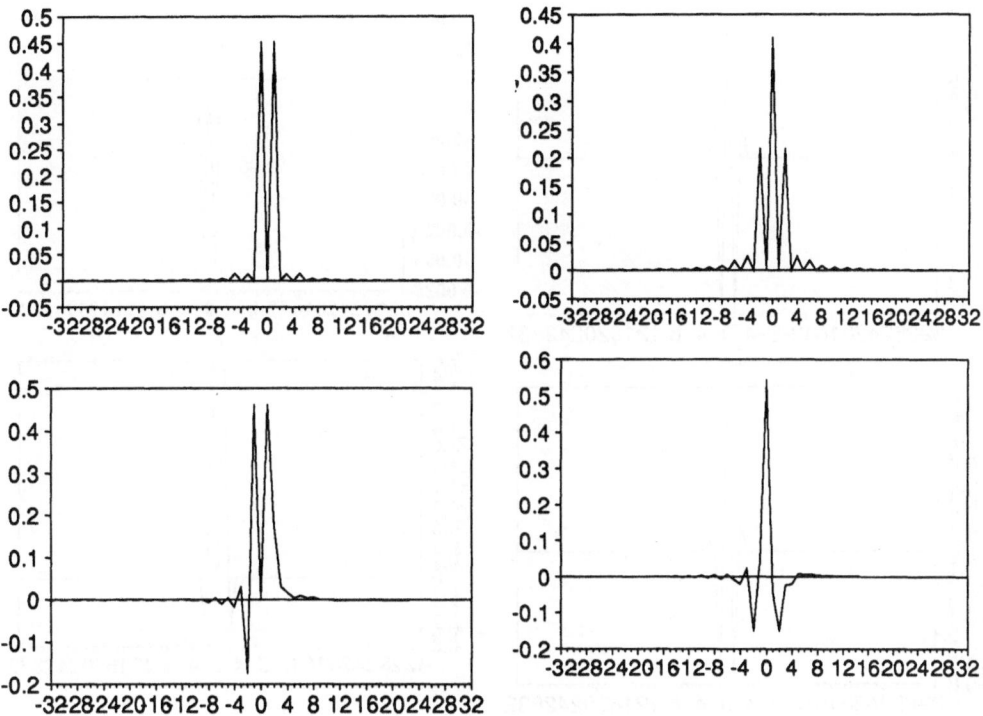

Fig.3: a) Real and imaginary parts of C_n [C,D]

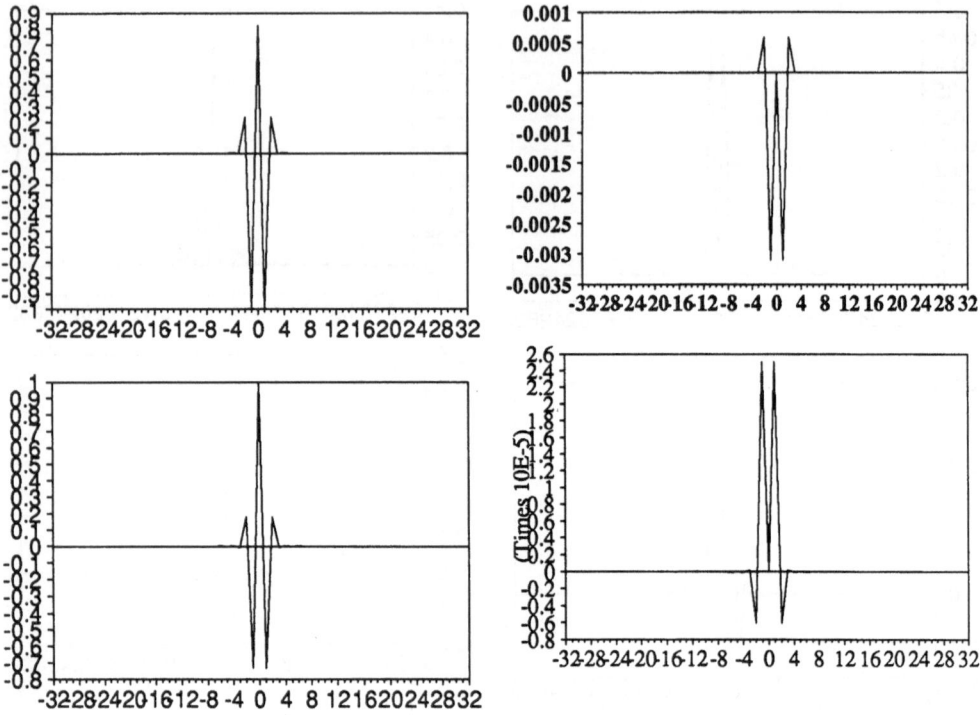

Fig.3: b) Real and imaginary of b_n [C,D]

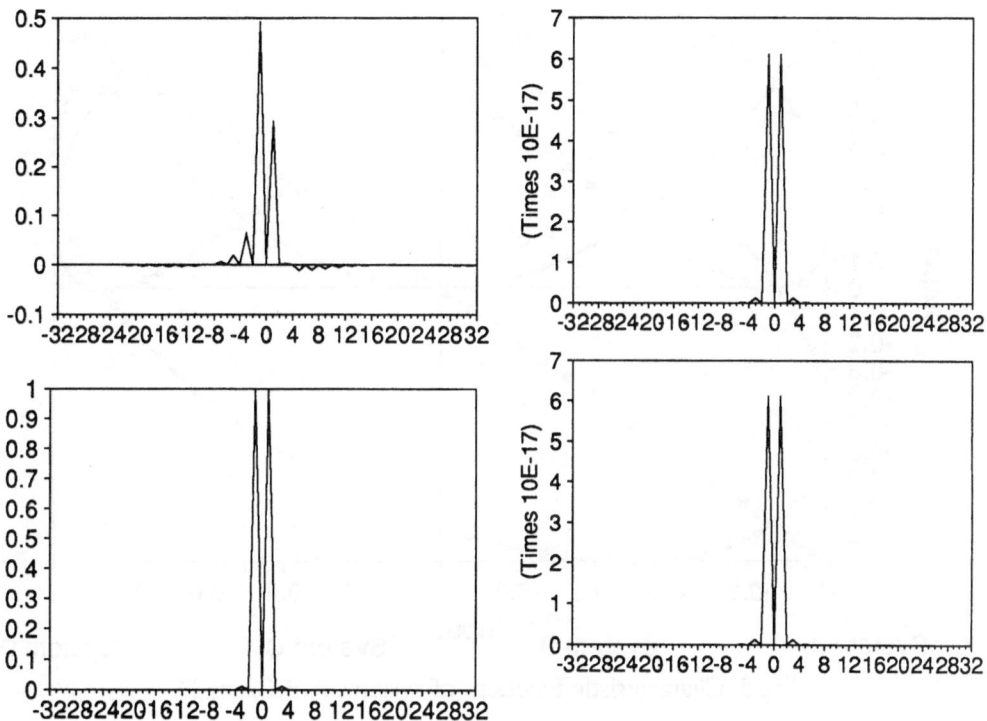

Fig.4: Real and imaginary parts of C_n and b_n of system A

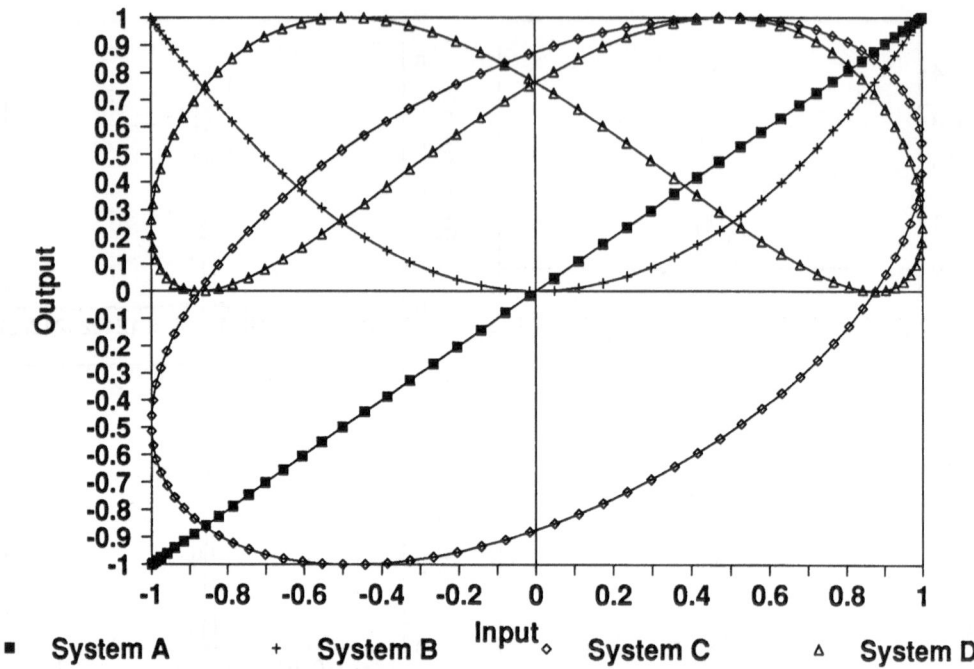

Fig.5: Characteristic functions of systems A,B,C, and D.

4.2 EXPERIMENTAL RESULTS

Two different defect geometries are used; namely, a *flat cut,* and a *circular cut hole.* A test block of each is sketched in figure 6. The ultrasonic returns used in the experiment are shown in figure 7, where T15A0 is taken as the reference signal, while T15A1, and T15A3 are, respectively, pulse echoes from a flat cut, and an circular cut hole inside a cylindrical aluminium test block. The center frequency of the driving ultrasonic pulse is 15 MHz and the returns are all sampled at the rate of 100 MHz. The same set of ultrasonic returns at 5 MHz is used in the experiment.

First, the characteristic "functions" of defect A1 and A3 are drawn, where the returns T15A1 and T15A3 are ploted versus T15A0 respectively. This is shown in figure 8. For the same defect, the computed b_n using 5 MHz transducer frequency data is found to be remarkably similar to that computed from the 15 MHz data set. This result is shown in figures 9 and 10 for defect (T5A1,T15A1) and (T5A3,T15A3) respectively.

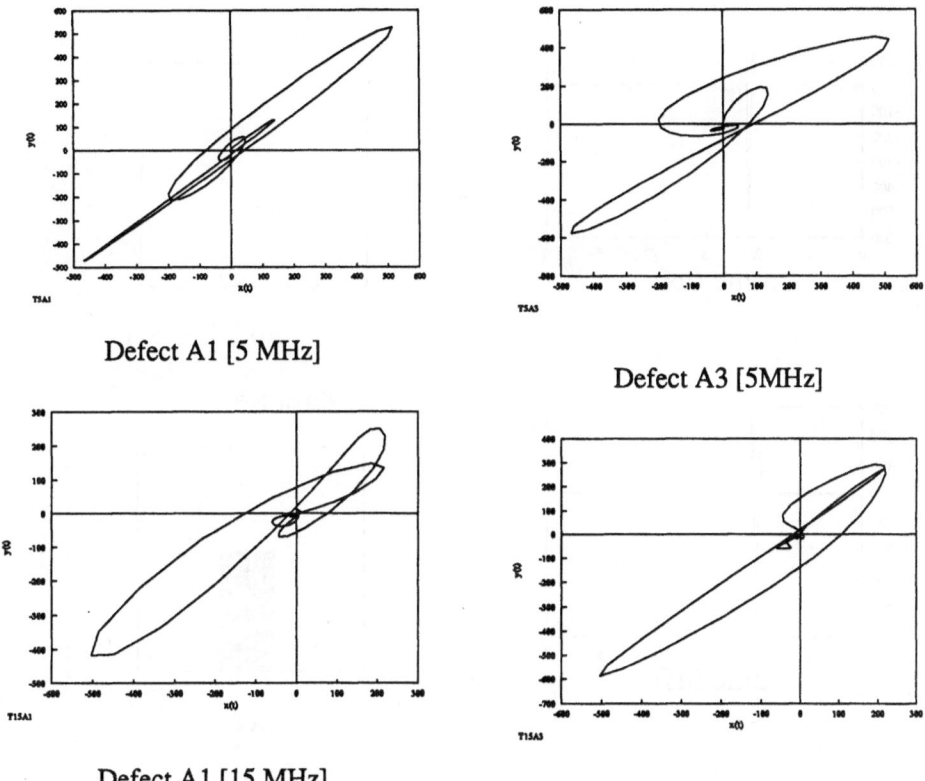

Defect A1 [5 MHz]

Defect A3 [5MHz]

Defect A1 [15 MHz]

Defect A3 [15 MHz]

Fig.8: Characteristic functions of defects A1 and A3.

382

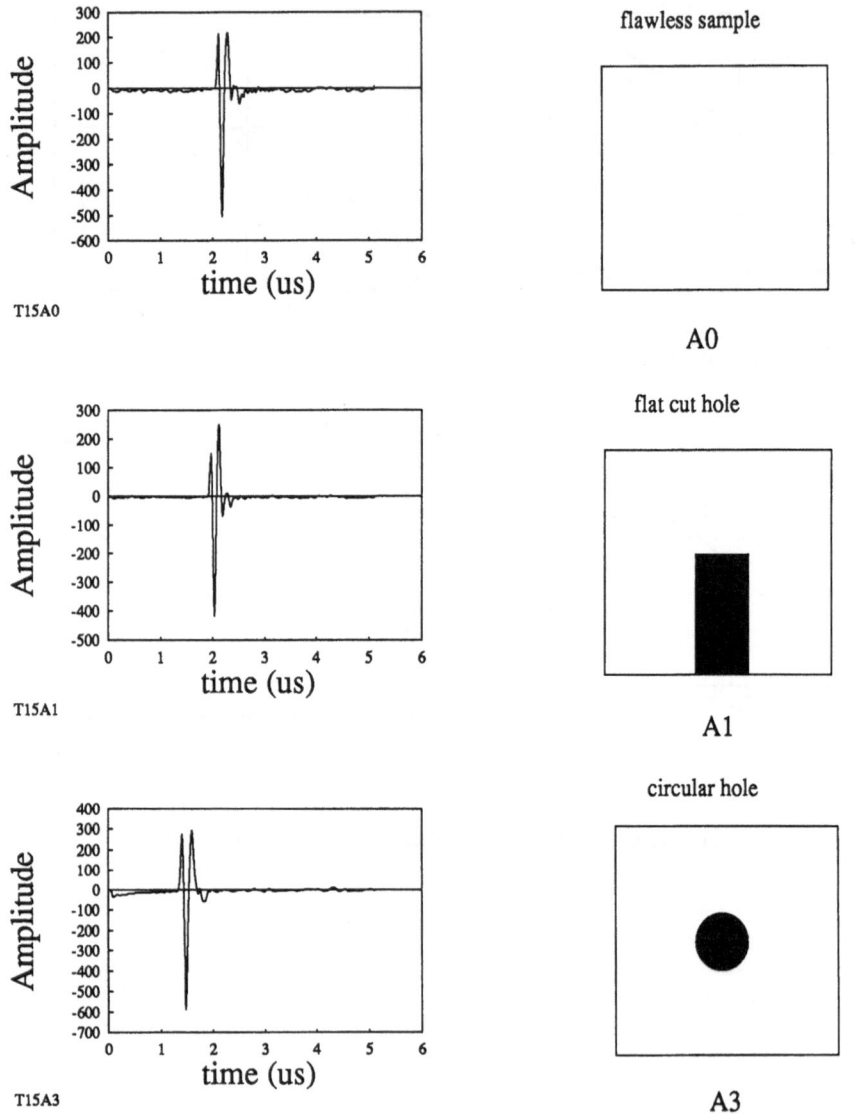

T15A0

T15A1

T15A3

flawless sample

A0

flat cut hole

A1

circular hole

A3

Fig.7: Pulse echoes of the defects [3].

Fig.6: Test specimen with different geometries. (A0) flawless, (A1) flat cut, (A3) circular cut hole [3].

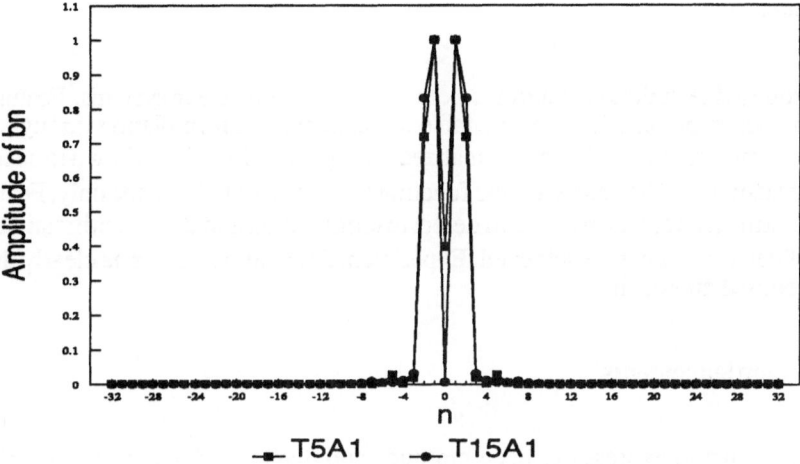

Fig.9 : Amplitude of b_n for defect A1.

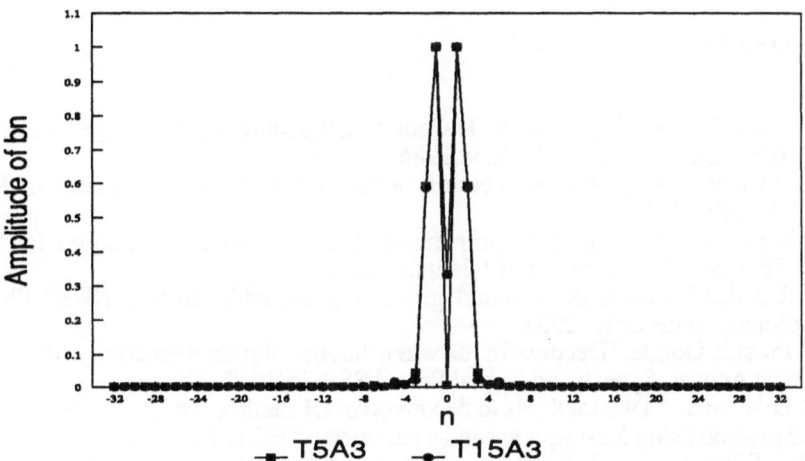

Fig.10: Amplitude of b_n for defect A3.

384

5. Conclusion

The proposed technique is valid for both linear and non-linear systems. Features of the defect characteristic function are amenable to mathematical formulation through the use of Fourier descriptors approach. The normalized Fourier descriptor b_n is shown to be insensitive to transformations of the characteristic function but sensitive to its shape only. Features from b_n can be extracted for use in a hidden defect discrimination and thus identification of a large variety of defect class can be achieved. Experimental results demonstrate clearly the validity of the proposed approach.

6. Acknowledgements

We would like to express our deep gratitude and thanks to the King Fahd University of Petroleum and Minerals for the research facilities it provides and the help we received throughout this research.
We also would like to thank Prof.C.H. Chen (University of Massachusetts, North Dartmouth) for the experimental data used in this paper.

7. References

[1]   G.A.Clark,D.M.Tilly and W.D.Cook," Ultrasonic signal/image restoration for QNDE", NDT Inter. vol.19,June 1986.
[2]   C.H.Chen," A signal processing study of US NDE", tech.rep. MTL TR 87-11.Feb.1987.
[3]   S.K.Sin and C.H.Chen,"A comparison of deconvolution techniques for USNDE", IEEE Trans. Image Proc. vol.1,January 1992.
[4]   S.R.Satish,"A parametric signal processing for eddy current NDT" Ph.D thesis, Colorado State Univ. 1983.
[5]   R.Prost,R.Goutte,"Deconvolution when the convolution Kernal has no inverse."IEEE Trans.Acoust.,Speech and Signal Proc.ASSP-25 1977
[6]   B.Drachman, "Two methods to deconvolve: L1 method using simplex algorithm and L2 method using least squares and a parameter." IEEE Trans. Ant. Prop. AP-32 1984
      H.F. Silverman, A.E. Pearson," On deconvolution using the discrete Fourier trans-
[7]   form.",IEEE Trans.Audioelectroacoustics AU-21 1973
      G.A. Clark, B.D. Cook, R.L. McKinney,G.D. Poe,"Impulse response of a crack: case
[8]   study of signal processing of ultrasonic signals.", Symp. On review of Progress in Quantitative NDE, July 1984.

CHARACTERIZATION OF ADVANCED COMPOSITE MATERIALS BY X-RAY COMPTON SCATTERING

Daniel BABOT, Gilles PEIX
Contrôle Non Destructif par Rayonnements Ionisants
INSA, Bât.303,
69621 VILLEURBANNE Cedex
France

ABSTRACT.Non destructive evaluation by Compton scattering uses an industriel X-ray tube and allows three-dimensional (3D) imaging of materials. The X-ray tube and detector are set on the same side of the object. By that way, non destructive evaluation of the wall of a tank, even full, is possible without the use of very high energy, because X-ray do not cross the whole object. Beside 3D imaging, this method allows point by point density measurements in the near-surface zone of any component (even dense and bulky). An accuracy of 1% was achieved not only on light composite materials, but also on dense ($\rho \approx 6.5$) components provided by powder metallurgy. A method allowing thickness measurement of a wall was also developed, especially suitable for multilayers compounds. The accuracy is $\pm$ 0.01 mm.

INTRODUCTION

Technology of spacecrafts requires the use of advanced composite materials. On such components, non destructive evaluation plays a major role for two reasons. Firstly, integrity and absence of voids must be proved. Secondly, the perfect knowledge of thickness and density may be essential for some applications.

On such components, an access to the inner part of the structure is not always available. Compton scattering tomography (CST) allows the opportunity of performing 3D imaging, as well as accurate thickness and density measurements, from a single side of the part. In the field of industrial nondestructive testing, CST is a relatively new technique[1 to 5].

1. Physical basis

Figure 1a shows the source (or x-ray tube focus) emitting a finely collimated beam into the material. This beam is named "incident beam" or "primary beam". The detector is also equiped with a collimator. The intersection of the two collimated beams defines the

X. P. V. Malague (ed.), Advances in Signal Processing for Nondestructive Evaluation of Materials, 385–396.
© *1994 Kluwer Academic Publishers.*

"measurement volume" also called "volume V". Detected photons, are necessarily scattered inside the measurement volume, and their number (or dose rate) is correlated to the density of the material at that point[6]. A precise description of the equipment dedicated to those measurements was given elsewhere[7].

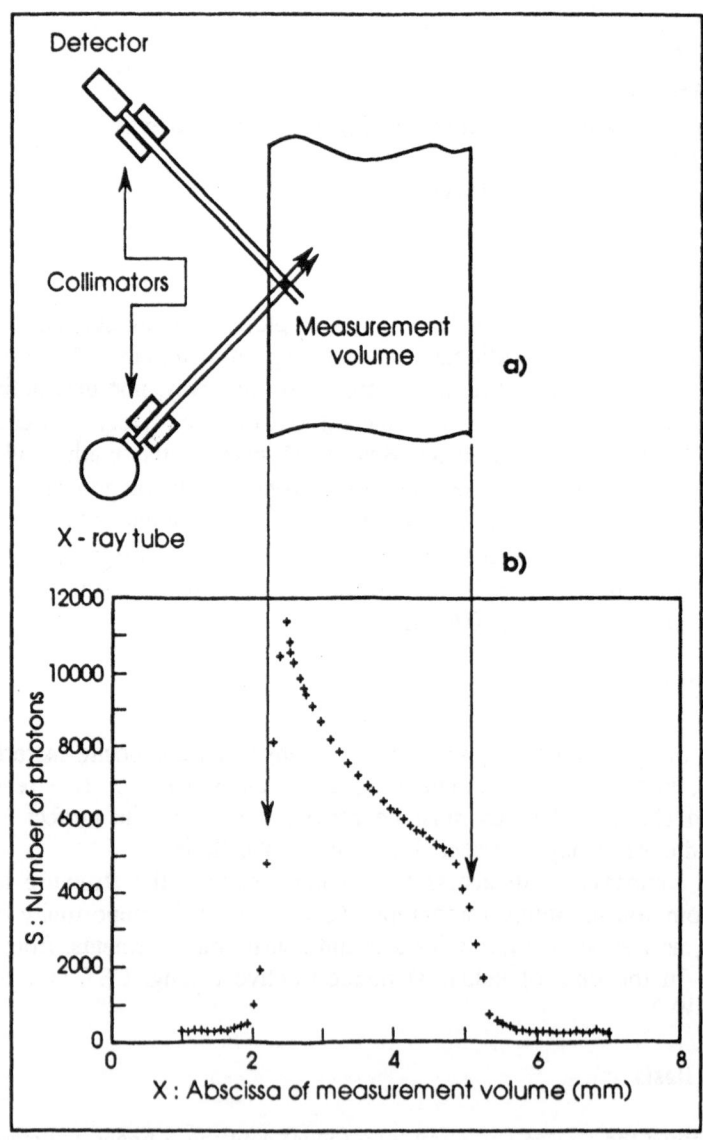

Fig. 1 *Principle of Compton scattering tomography. a) equipment b) result of a scan through a sample.*

Figure 1b shows the result of a scanning of volume V throughout the tested component. The number of counts increases sharply when measurement volume enters the wall. The exponentially decreasing part of the curve corresponds to the increase in photon attenuation along both primary and scattered beams, when volume V is located deeper and deeper. The sharp fall then corresponds to the exit through the rear face.

2. Thickness measurements

2.1. COLLIMATORS

The shape of the sharp increase in count rate (fig.1b) is the result of the convolution of some rectangular edge function with the response of our apparatus to a thin sheet (a kind of impulse response along a line, also called "line spread function" or LSF). For the purpose of thickness measurement, we chose very thin widths for both collimated beams and a small value of Compton angle θ (figure 2). By that way, the measurement volume is elongated along a direction which is parallel to the face of the tested component, reducing the width of the LSF curve.

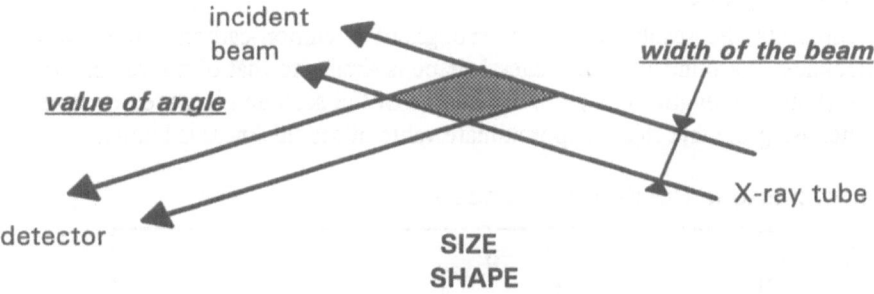

Fig 2. Influence of beam width and Compton angle on the shape of volume V

For practical reasons, the collimator attached to the detector is constituted of 16 converging slits, milled in a lead block (figure 3). The purpose here is to increase the photon flux reaching the detector. It can be seen on figure 3 that the detector, an NaI crystal with 2" x 2" dimensions, covers all the slits. The shape and size of measurement volume is not severely changed, provided that convergence of the 16 slits with the single "primary beam" slit is good. In such a way to ensure a precise convergence of the slits and a high stability along time, the 16 + 1 slits are milled in the same bulky lead block. Localization and dimensions of volume V are therefore stable. The full width at half maximum (FWHM) of the LSF of our collimator is 0.5 mm.

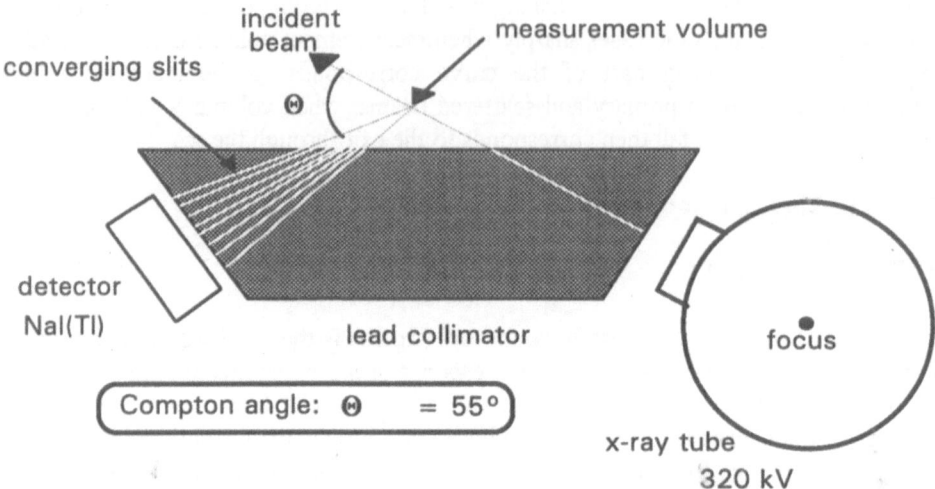

Fig.3. General arrangement of the lead collimators; sixteen slits are milled (instead of height reported here for clarity).

2.2. MEASUREMENTS

Figure 4 represents the result of a scan throughout a carbon-carbon sample whose expected thickness is around 11 mm. General shape is similar to that of figure 1b, except for the jump (8 mm in length) located in the middle of the scan and intended to save up counting time, using the fact that an approximate value of the thickness is known.

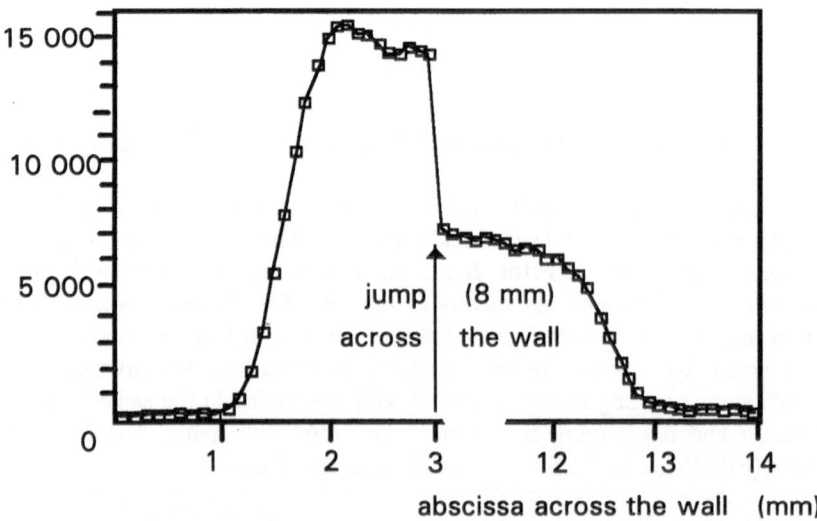

Fig.4. Result of a scan through a carbon-carbon sample. A jump of 8mm was operated in the middle of the scan

The voltage applied here to the X-ray tube is 160 kV, with a 20 mA current. Each measurement point in figure 4 corresponds to a counting time of 5 seconds, the whole scan lasting around 5 minutes, for 60 values named N_i (number of counts at the abscissa with index i). The sampling length is 0.1 mm.

The discrete N_i values are then used to compute the differences

$$D_i = N_{i+2} - N_{i-2} \qquad (1)$$

which represents a good approximation of the derivative for each abscissa i. We use indexes i+2 and i-2 (instead of i+1 and i-1) in such a way to increase the absolute value of D_i, thus reducing the influence of the uncertainty that exists on N_i. Figure 5 shows the corresponding D_i values. A sharp positive peak corresponds to the entrance and a smoother negative one to the exit. After a thresholding, dedicated to eliminate values around zero, a computation of the abscissa x_G of the center of gravity of each peak is performed, using the formula :

$$x_G = \frac{\sum\limits_i N_i \, x_i}{\sum\limits_i N_i} \qquad (2)$$

where x_i represents the abscissa of the counting with index i, the summation being done for all the i indexes of the corresponding peak.

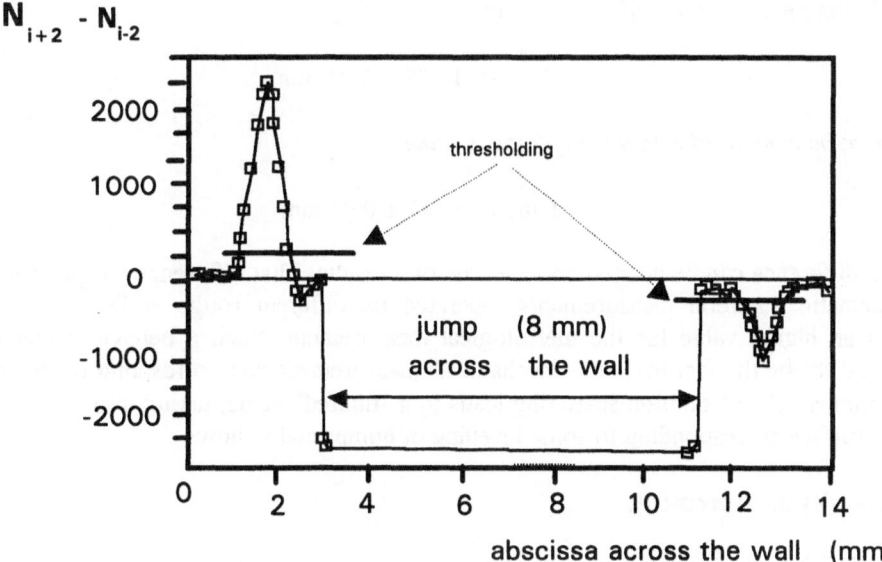

Fig.5. Value of differences D_i along the scan of figure 4.

It must be underlined that the use of the second neighbours for the computation of D_i (formula 1) reduces the influence of uncertainty mainly on the low differences corresponding to the feet of the peaks. It is of main consequence to reduce uncertainty on those values in such a way to cancel mistakes in the thresholding process. One single value added (or removed) at that level have a great influence on the computed value of x_G because those points are distant from the center of gravity.

The thickness T is then computed as the difference between abscissae of the two peaks :

$$T = x_{G\,'out} - x_{G\,'in} \qquad (3)$$

where "in" and "out" stand for the entrance and exit peak.

2.3. RESULTS

Within measurement conditions described in the preceeding section, a cylindrical tank made of carbon-carbon (woven along the three directions), 11 mm in thickness, was tested. Seventeen measurements were operated at the same place. The results are :

mean value : $\overline{T}$ = 10.98 mm

standard deviation σ = 0.005 mm

Considering an uncertainty of $\pm 2\sigma$ we get :

$$T \quad = \quad 10.98 \pm 0.01 \text{ mm}$$

which can be compared with a metrological value

$$T \text{ metr.} \quad = \quad 11 \pm 0.01 \text{ mm}$$

A slight difference can be noticed between the two results. That difference has proven to be systematic : several measurements, operated on different rough surfaces, always induced an higher value for the metrological measurement. Such a behaviour can be accounted for by the fact that any "mechanical" measurement will correspond to the top of the humps, while Compton scattering leads to a "filtered" value, taking into account a "mean" surface corresponding to some levelling of humps and hollows.

3. Density measurements

3.1. GENERAL FEATURES

Density was measured in the near-surface region, on several kinds of materials. The result of just one counting is sufficient to determine density in one point of a sample. We chose to measure density at the surface of the material, at this abscissa where the photon flux is maximum (figure 1). Such a choice has two advantages. Firstly it allows a short counting time with a good accuracy. Secondly, an uncertainty in the position of the measurement volume (in the direction of depth), will result in a minimum error on photon counting, because of the near zero derivative.

Measurements can be made inside the sample, at any depth, but the accuracy will be poorer.

Though measurements were operated superficially, the reproductibility in positioning of volume V remains the key problem. It exists three possibilities.

1. the use of a mechanical length gauge : such a solution needs a contact with the sample and is not allowed on fragile components, because it may generate a bump, or a streack ;
2. the use of an optical length gauge : does not work on a black or (and) woven composite ;
3. through an experimental determination of the abscissa of the maximum : this is the most time-consuming solution, but it is always available and will be described here.

3.2. RESEARCH OF THE ABSCISSA OF THE MAXIMUM

FIRST STAGE : determination of entrance abscissa

Starting from a position for which volume V is located outside the component, a scanning is started with :

coarse displacement	$\approx 1mm$
short counting time	$< 1s$

The scanning is stopped when the number of counts exceeds some threshold (fig.6).

SECOND STAGE : rough research of maximum

Scanning is carried on with :

shorter displacement	≈ 0.5 mm
longer counting time	a few seconds.

The scanning is stopped when the number of counts decreases markedly. A lowpass filter (size 3) is then applied to the results.

The abscissa of rough maximum is then claimed to be the abscissa for which the number of counts (after application of the lowpass filter) has the highest value.

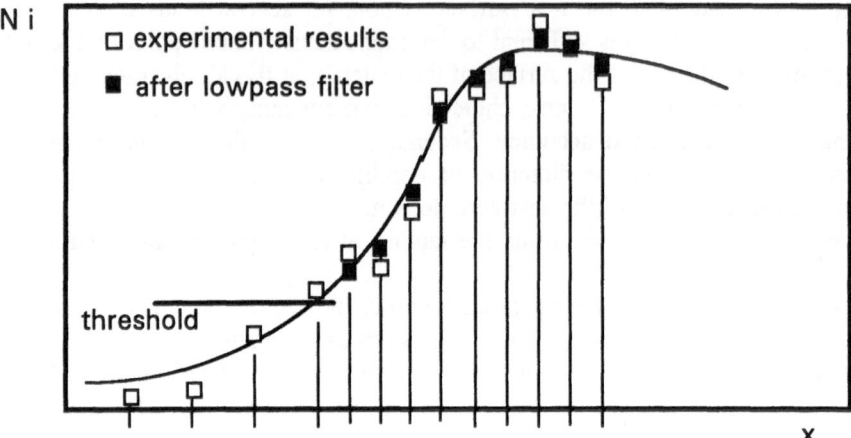

Fig.6. Diagram illustrating the first and second stage in the determination of the maximum

THIRD STAGE : fine research of maximum

Counting are carried out at eight abscissae equally distributed on both sides of rough maximum with :

        fine displacement   0.25 mm
        long counting time   10 s

A polynomial interpolation is computed :

$$N(x) = a_0 + a_1 x + a_2 x^2 + a_3 x^3 \qquad (3)$$

Abscissa of maximum is then easily defined from equation (3).

FOURTH STAGE : final counting

An unique counting at that abscissa is representative of the density of the component at that point. Duration of this last counting depends on the photon flux (and correlatively on the sample density) and on the expected accuracy (quantum noise in photon counting).

3.3.ACCURACY OF THE METHOD

α) Determination of the abscissa of maximum

Stages 1 to 3 were applied 100 times at the same point of some sample, and a statistical analysis of the abscissae determinated by that method was undertaken. Figure 7 shows that the position of maximum can be asserted with 0.01 mm accuracy. Such a value is sufficient to induce a negligible mistake on the result of the final counting.

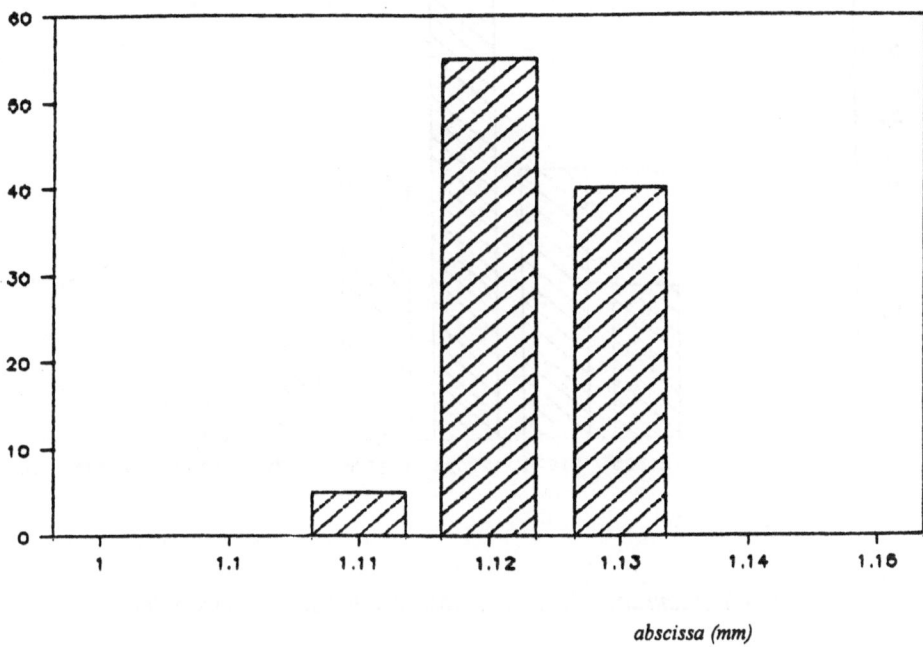

abscissa (mm)

Fig. 7. Determination of the abscissa of maximum: histogram corresponding to 100 measurements

β) overall accuracy

The preceeding test was applied from stages 1 to 4. Thus, beside uncertainty corresponding to the determination of maximum, occur two new sources of uncertainties
- quantum noise on final photon counting, which can be determinated according to Poisson's distribution law ;
- fluctuations of X-ray tube over a long period (several hours).

Figure 8 summarises the results, obtained for a counting time of 20 seconds and over 100 measurements operated on the same sample. A statistical analysis leads to the mean number of count $\overline{N}$ for the given point of that sample :

$$\overline{N} = 121\ 998$$
$$\text{and } \sigma = 415$$

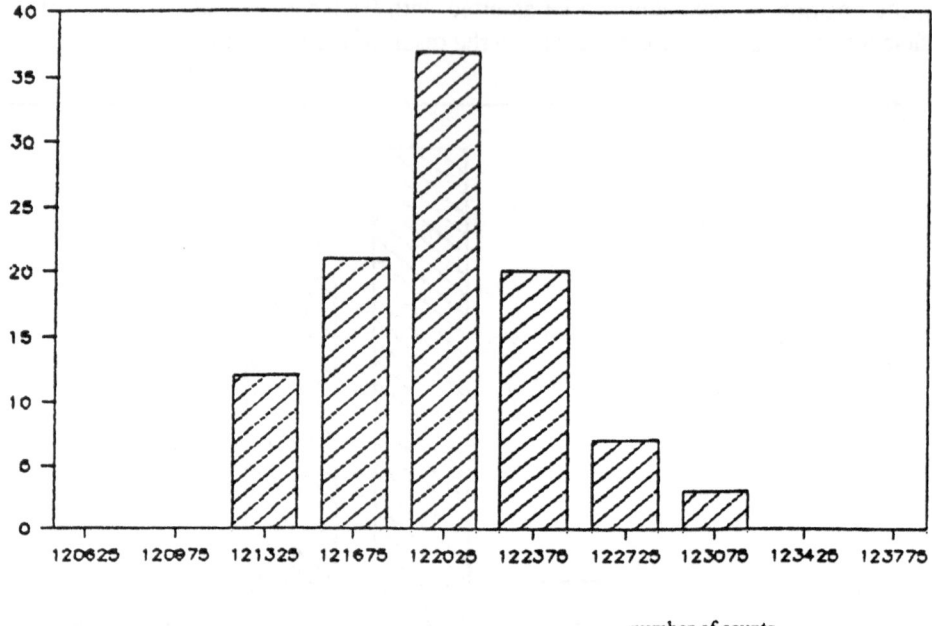

number of counts

Fig.8. Accuracy of density measurements on the same sample

From such results, it can be observed that uncertainties corresponding to the determination of maximum and to the fluctuations of X-ray tube are of minor influence in front of quantum noise. Quantum noise, whose value can be approximated by the standard deviation $\sqrt{N}$, represents, effectively, an important part of the experimental (total) noise. Let us compare

$$\sigma_{quantum} = \sqrt{\overline{N}} = 349 \qquad \text{with } \sigma = 415$$

d) results on industrial components

Figures 9 and 10 show that an accurate calibration can be carried on, as well on light "carbon-carbon" composites as on powder metallurgy components containing iron and exhibiting a density around 6.5. In both figures, the number of counts delivered by the Compton scattering method is compared with the "expected" density (measured by hydrostatic method).

The overall accuracy is estimated to be within ± 1%.

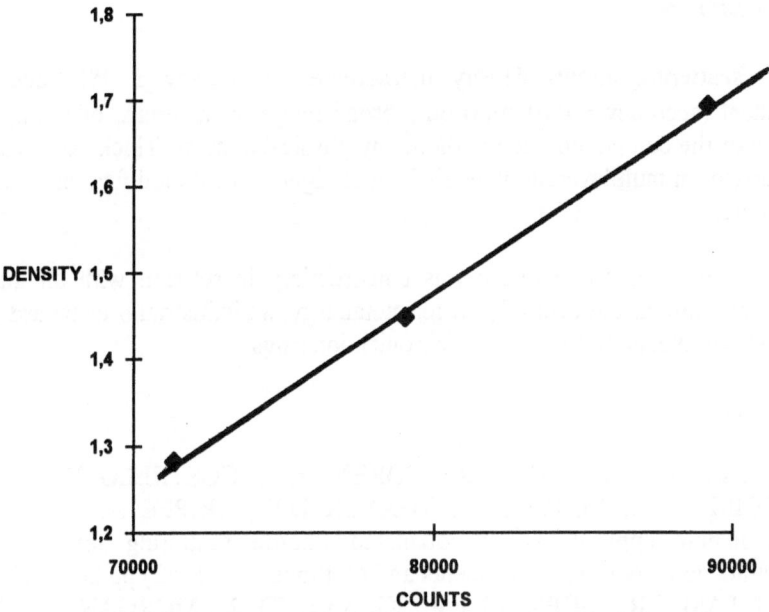

Fig.9. Results of measurements on three carbon-carbon samples.

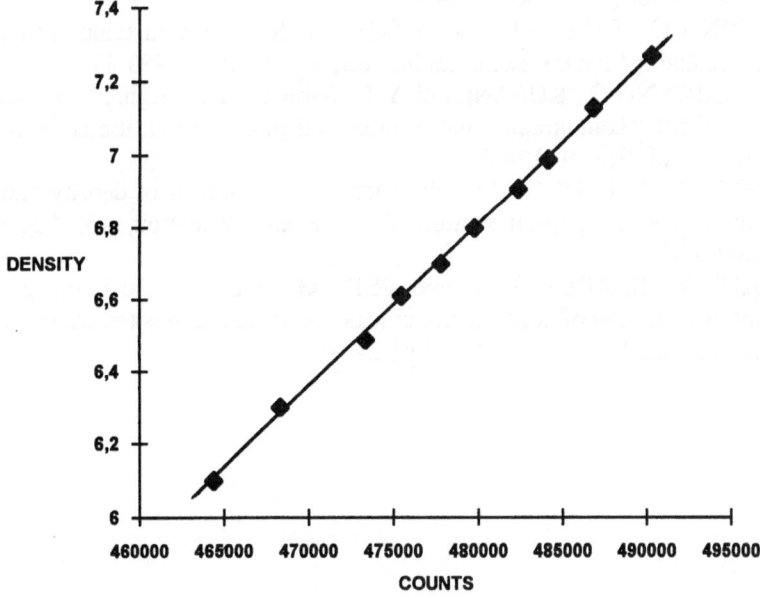

Fig. 10. Results on ten components elaborated by powder metallurgy.

CONCLUSIONS

Compton scattering allows density measurement (accuracy $\pm$ 1%) and thickness measurement (accuracy $\pm$ 0,01 mm) on a broad range of materials, operating from one single side of the component, and without any physical contact. Thickness measurements can be carried on multilayered materials by a straightforward modification of the technic exposed here.

At the present time, our laboratory is constructing, in relation with an international Company working in the field of powder metallurgy, an industrial breadboard device for density measurements on low cost synchronization rings.

REFERENCES

[1] STOKES J.A., ALVAR K.R., COREY R.L., COSTELLO D.G., JOHN J., KOCIMSKI S., LURIE N.A., THAYER D.D., TRIPPE A.P., YOUNG J.C., "Some new applications of collimated photon scattering for nondestructive examination", Nuclear Instruments and Methods 193 (1982) pp.261-267.

[2] GAUTAM S.R., HOPKINS F.F., KLINKSIED R., MORGAN I.L., "Compton Interaction Tomography. I. Feasibility studies for applications in earthquake engineering", IEEE Transactions on Nuclear Science, vol.30 (1983) pp1680-1684.

[3] BRIDGE B. "A theoretical feasibility study of the use of Compton backscatter gamma-ray tomography (CBGT) for underwater offshore NDT" British Journal of NDT November (1985) pp.357-363.

[4] HARDING G., STRECKER H., TISCHLER R. "X-ray imaging with Compton-scatter radiation" Philips Technical Review, Vol.41, n°2 (1983/84) pp.46-59.
[5] HARDING G., KOSANETZKY J. "X-ray scatter imaging in non-destructive testing" from "Tomography and Scatter Imaging ; Applications in NDT" IOP Publishing Ltd, Bristol (1988).

[6] BERODIAS G., PEIX G., "Nondestructive measurement of density and effective atomic number by photon scattering" Materials Evaluation, 46, August (1988) pp.1209-1213.

[7] BABOT D., BERODIAS G. and PEIX G. "Detection and sizing by X-ray Compton scattering of near-surface cracks under weld deposited cladding" NDT & E International, Vol.24, n°5 (1991) pp.247-251.

STEEL SHEET SURFACE QUALITY MONITORING BY MULTIPLE SPECTRAL REFLECTANCE SIGNAL PROCESSING

J.HENRIETTE (*), R.SIRJACOB (*), P.HERVE (+)
(*) Faculté Polytechnique de Mons (Belgium)
(+) IUT Paris X (France)

ABSTRACT. This paper summarizes the work carried out to date in order to design a device providing - permanently and on-line - a reliable index of oxidation presence and importance on steel sheets leaving a continuous annealing furnace. Reliability namely means that the effects of roughness should not affect this index.

1. Problem Analysis

1.1. SELECTION OF THE MONITORED VARIABLE

The selection of a radiative property is obvious, as contact is impossible. Logically enough, the first one considered is the emissivity, as the inspected sheet is, in the reference application, at high temperature. However, it is significantly affected by several uncontrolled effects such as temperature variation, and outer radiation reflection.

It is thus more convenient to select a "nonactive" property - such as reflectance - providing the user with the opportunity of choosing the spectrum of the radiation source as well as modulating this one in order to eliminate the "parasitic" direct and reflected radiations.

In our reference application, the property the most available to direct observation is the *bidirectional reflectance*.

It is useful to observe that these two characteristics - emissivity and reflectance - are intimately related, so that any information - either theoretical or experimental - concerning one of them may easily be transposed to the other. To that purpose, we recall, in Appendix 1, the fundamental relationships between them.

1.2. DISCRIMINATION BETWEEN ROUGHNESS AND OXIDATION

A priori, there is a choice between several criteria e.g. : spectral selectivity, spatial selectivity and polarization. In this work, the spectral selectivity was retained because it is very simple to

X. P. V. Malague (ed.), Advances in Signal Processing for Nondestructive Evaluation of Materials, 397–415.
© 1994 *Kluwer Academic Publishers.*

implement; it will be shown, in the next section, that there exists a range of wavelengths where the effect of oxidation is selective while the roughness is not. Furthermore, the effect of oxidation depends on the oxide layer thickness.

2. Radiation Reflection Modelling

The modelling is based on the electromagnetic theory of radiation, compounded with a statistical representation of roughness. We have made use of the work published by Sacadura [1] on one hand, for what concerns roughness, and by Huetz-Aubert [2] on the other hand, for what regards oxidation : their contributions are consistent and complementary.

2.1. EFFECT OF ROUGHNESS ON REFLECTANCE

If the surface roughness is assumed isotropic, it may be described by 2 statistical functions (Beckman) :
- the function of peak heigth distribution,
- the autocorrelation function of the spatial distribution.

In the GG version, presumably the most appropriate to our application, both of these functions are given the form of a gaussian (G) law. The essential parameters are thus :
- σ the standard deviation of the first law, also called the RMS amplitude of the roughness and
- T the "correlation distance" of the second law.
From a practical point of view, one may consider that σ characterizes the depth of the roughness, while T characterizes its "fineness". Accordingly, the group $m = \sqrt{2} . \sigma / T$ is called the "quadratic mean slope" of the roughness (Fig. 1).

Now, on the basis of the electromagnetic theory of radiation and of this statistical description of roughness, it is possible to work out the spatial distribution of the reflected radiation and to derive an expression for the bidirectional reflectance. More precisely, two different expressions are obtained according to the value of λ / σ :

- for $(\lambda / \sigma) > 1$, the expression of the reflectance depends on λ / σ, which means a selective effect of the roughness on the different wavelengths. This expression will not be given here as we shall not use it. Suffice it to say that σ is mandatorily an upper bond for the wavelengths of our device.

- for $(\lambda / \sigma) < 1$, the expression is as follows :

$$\rho''_\lambda(\Delta ; \Delta_r) = \sum | R_{ij} |^2 . \frac{f_1}{m^2} . \exp\left(-\frac{f_2}{m^2} \right)$$

where f_1 and f_2 depend only on Δ and Δ_r i.e. on θ , θ_r and ϕ_r (fig.2),

m^2 is the roughness parameter (mean slope),

R_{ij} , coefficients of the polarization matrix, characterize the nature of the material(s) of the outer layer(s) : they are functions of the Fresnel reflection coefficients R_1 and R_2 as well as of θ , θ_r and ϕ_r . We shall write out their expression later on, for the case of interest.

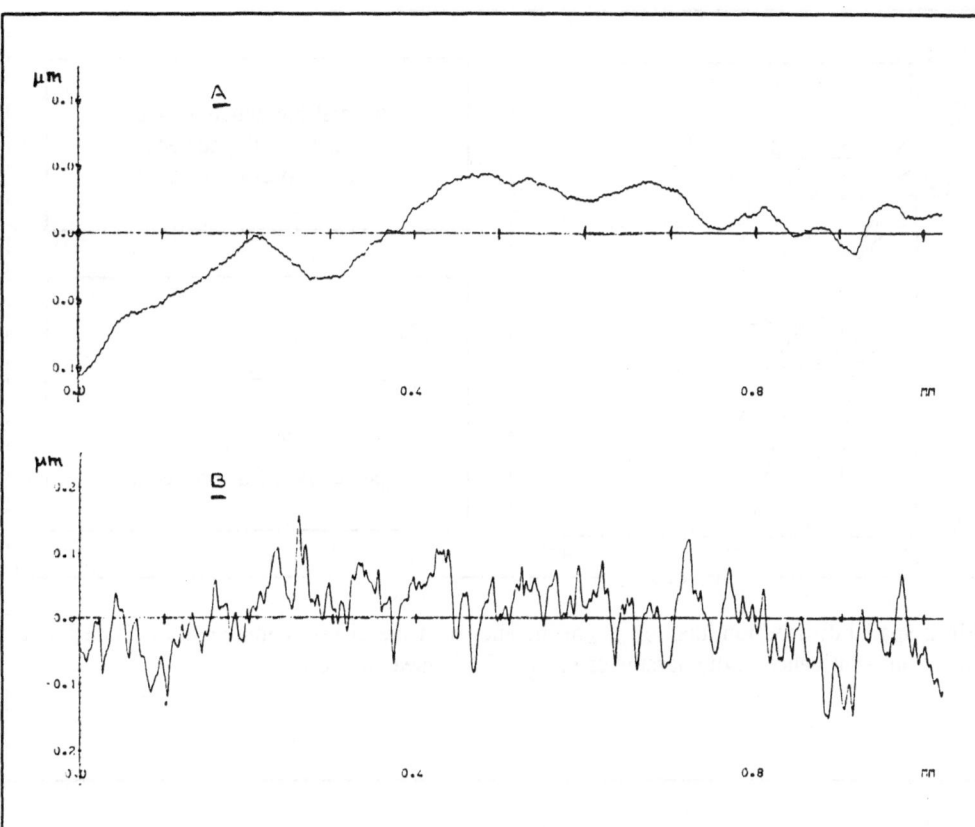

Sample	Surface finish	Ra [μm]	σ [μm]	m [SD]
A	polished	0.03	0.034	8.54E-4
B	roughened with abbrasive #1000	0.04	0.054	9.72E-3

Fig.1 Roughness profile and characteristics of two stainless steel (304 L) samples

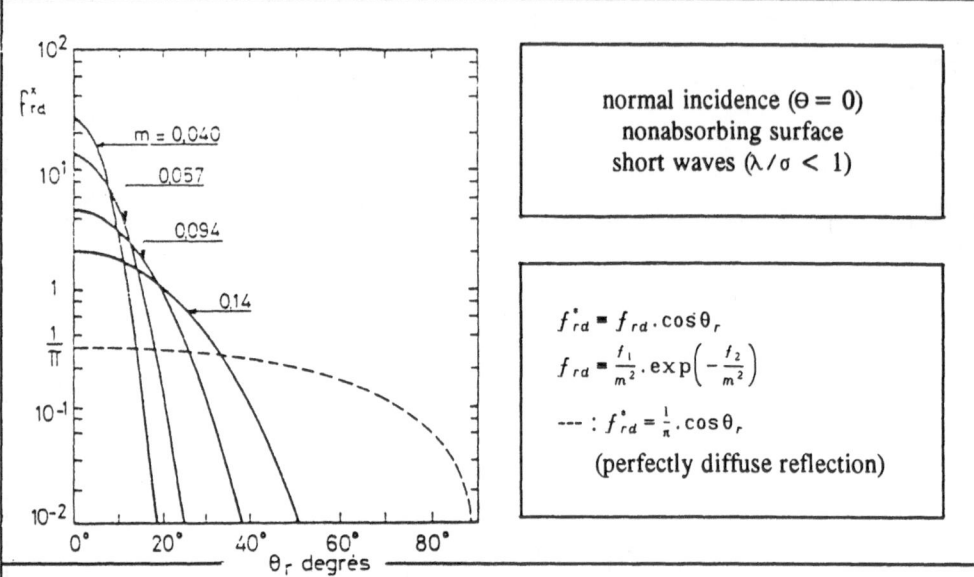

Fig.2 Spatial distribution function f_{rd}^* of the energy of the diffuse component of the reflected radiation. - Influence of the parameter m (quadratic mean slope)

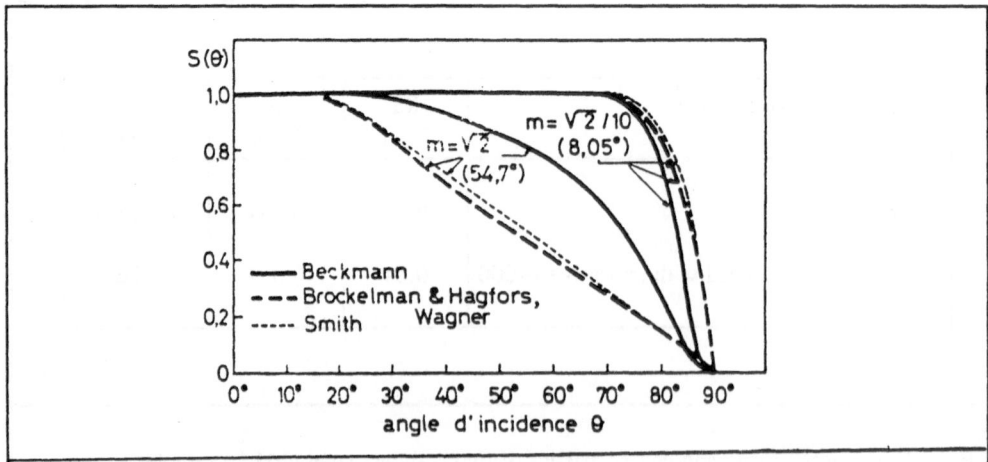

Fig.3 Mask function (accounting for the shadowing effect). - Different versions.

The validity of this expression is restricted to lower values of Θ and m. At higher values, it is necessary to multiply by a "mask function", but this does not add any new variables, since the latter function depends only on Θ and m (fig. 3).

Thus, for the range of λ satisfying the criterion ($\sigma/\lambda > 1$), the roughness affects all the wavelengths uniformly. It is the sought situation. It remains now to show that oxidation, on the contrary, induces selective effects.

2.2. EFFECT OF OXIDATION ON REFLECTANCE

The presence of oxide changes the nature of the reflecting surface and modifies the value of the coefficients R_{ij}. It must be kept in mind, on the other hand, that what we are trying to detect in this application is a "spread" oxidation resulting from a poor process control. According to the experts, this oxidation consists of a very thin layer (a few hundreds of A°) of Fe_2O_3, or possibly of Fe_2O_3/Fe_3O_4, if the malfunctionment is more severe. In those conditions, one can expect a semi-transparent behaviour of the oxide layers, with interference effects.

In order to get an insight into this behaviour, we shall use the results of Ane and Huetz Aubert although they are established for surfaces perfectly smooth and plane. Their application to our case is justified by the fact that - in the conditions of operation $(\lambda/\sigma) < 1$ - the rough surface behaves like a set of perfectly plane facets.

Now, after [2], the polarized reflection factors R_1 and R_2 of a "system" of multiple thin semi-transparent layers may be expressed as follows :

$$R_1 = \frac{Det\left(\begin{matrix} M_1 \cdot \underline{k}_s^x & I \cdot \underline{k}_0^x \end{matrix} \right)}{Det\left(\begin{matrix} I \cdot \overline{\underline{k}}_0^x & M_1 \cdot \underline{k}_s^x \end{matrix} \right)}$$

$$R_2 = \frac{Det\left(\begin{matrix} M_2 \cdot \overline{\underline{k}}_s^{xx} & I \overline{\underline{k}}_0^{xx} \end{matrix} \right)}{Det\left(\begin{matrix} I \cdot \underline{k}_0^{xx} & M_2 \cdot \overline{\underline{k}}_s^{xx} \end{matrix} \right)}$$

In these expressions,

the vectors k^x and k^{xx}, $\overline{k}^x$ and $\overline{k}^{xx}$ are related to the propagation vector $\vec{k}$ - more precisely

to the component normal to the reflecting surface - as well as to the wavelength and the electromagnetic properties μ and ϵ (see Appendix 2),

subscripts o and S correspond respectively to ambiance and to substratum,

I is the unit matrix,

M_1 and M_2 are the global transmission matrices, products of the individual transmission

matrices of the different layers (see Appendix 2). For each layer, the matrix coefficients are functions of k_z^x or k_z^{xx}, and of $k_z h$, where h is the layer thickness.

It means that the coefficients $M_{1,ij}$ *and* $M_{2,ij}$ *depend on the nature of the intermediate layers, of their thickness, of the incident radiation wavelength and direction.* Similarly for R_1 and R_2 which further depend on ambiance and substratum.

Interference - mathematically expressed by the presence, in the transmission matrices coefficients, of trigonometric functions of $(k_z h)$ - manifests itself by the fluctuation of the value of R_1 and R_2 and consequently ρ_λ, versus λ on one hand, and h on the other hand. This effect may be observed on figure 4 for λ and figure 5 for the thickness.

It is thus well established that the effect of oxidation is selective. Nevertheless, in order that this selectivity is observable within a small sample of wavelengths, the latter have to be purposely selected, with due regard to the nature of the expected oxides and to their thickness. Figure 5 suggests that for thicknesses of the order of 10^{-2} to $10^{-1} \mu m$, the appropriate range is $0,5 < \lambda < 1,5$ μm.

3. Apparatus

3.1. SELECTED WAVELENGTHS

On the basis of the above analysis, we have selected 0.5, 0.6 and $1.0 \mu m$. This set of wavelengths, as a matter of fact, satisfies the double suitability criterion :

- nonselectivity of the roughness effects;

the RMS amplitude of the pickled sheet roughness being of the order of $1.5 \mu m$, the selected range of λ satisfies the condition $\sigma/\lambda > 1$, required for nonselectivity .

- selectivity of the oxidation effects;

the thickness of the Fe_2O_3 layer considered as critical is 100 to 200 A°. Figure 4 shows that, in this range, at a given thickness, the reflectance increases significantly when λ increases in the range $0.5 - 1\mu m$. On the other hand, see figure 5, ρ decreases sharply when the thickness increases.

3.2. CONSTITUTION OF THE APPARATUS

Let us consider, for the sake of this description that the apparatus is made of three main parts : the modulated light source, the measuring unit, and the transmitting unit.

3.2.1. Modulated Light Source. The source itself is a Quartz Tungsten Halogen lamp, powered with a stabilized dc voltage. It emits with a quasi white spectrum, centered on $1 \mu m$ (3000 K). It is completed by a reflector and a condensing lens set. The resulting source emission is thus a beam, 33 mm diameter, with a continuous spectrum limited to the range $0.36 - 2.5 \mu m$. This beam is chopped by a disk rotating at a controlled speed, making the chopping frequency adjustable up to 500 Hz.

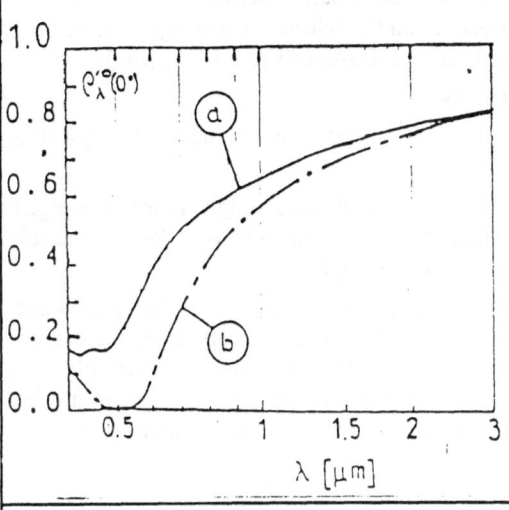

Fig.4 Spectral Directional Hemispherical (SDH) reflectance ρ'_λ
of an oxidized stainless steel surface , plotted versus the radiation wavelength λ.

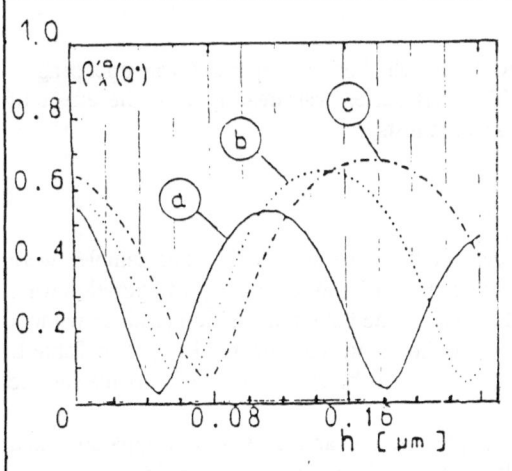

Fig.5 Spectral Directional Hemispherical (SDH) reflectance ρ'_λ
of an oxidized stainless steel surface , plotted versus the oxide layer thickness.

3.2.2. *Measuring Unit.* The essential part is a set of three radiation detectors, each associated with a "monochromatic" interference filter. Depending on the selected wavelength, the detector is either a Si photodiode (for $\lambda = 0.5$ and $0.6\,\mu m$), or an InGaAs cell (for $\lambda = 1\,\mu m$).

It is completed by signal handling electronics, namely :

- the detector output selector - i.e. wavelength selector -, actuated by the monitoring system at the shifting frequency;

- a lock-in amplifier, tuned to the chopper frequency, which accordingly rejects the signal component not correlated with the source emission. Its output will be taken as suitably representing the spectral reflectance, at the selected wavelength.

- the logging and monitoring system which displays the three above mentioned reflectance indices or, actually, any combination or even function of these indices. In any case, the original reflectance indices are disk stored for possible further analysis. Although it is presently - i.e. under the development phase - totally independent of the process monitoring system, it could be easily integrated in the hierarchical distributed control system. An alarm should be provided when excessive oxidation is detected.

3.2.3. *Transmitting Unit.* In order to keep the two previous delicate devices out of the furnace heat, use is made of a transmitting unit consisting basically of two 1.5 m long bundles of optical fibers - one "forth" and one "back" - as well as the required injection and output optics. For the relevant wavelengths, standard quality glass is acceptable. The "back" bundle is subdivided, at its outlet end, into three sub-bundles, corresponding to the three different detectors.

4. Validation Tests

Two series of tests were carried out in order to check whether the apparatus was operating in accordance with the theoretical predictions. The first series was dealing with the effect of roughness, while the second one was made of oxidaton tests.

4.1. EFFECT OF ROUGHNESS.

For this test, 17 sheet samples were used, all cut out from the same coil. The samples were submitted to industrial-like acid pickling. Five somewhat different conditions of operation were realized by varying either the contact time (1, 2 or 6 s) or the bath composition (with or without inhibitor). Accordingly, five batches of samples may be identified, denoted I, .., V in Table I. It must be pointed out that, according to the experts, from one batch to the other, only the surface "mechanical finish" will be different.

  Immediately after the pickling treatment, the samples were examined with our apparatus, and their roughness (Ra) was measured mechanically. The results are presented on figure 6, where, in order to verify the theotetically predicted correlation ($\rho''_\lambda \sim T^2 / Ra^2$), the different signals (s_i) are plotted against $1/Ra^2$.

  One observes that :

- the points corresponding to a same bath composition are pretty well aligned, for each wavelength, with only a minute influence of the contact time ;

Table I : Test relative to the effect of roughness

a) Pickling Operating Conditions

Batch No	Number of samples	Contact Time [s]	Inhibitor
I	5	1	NO
II	5	2	NO
III	1	6	NO
IV	5	2	YES
V	1	6	YES

b) Results of the test

Sample No	Ra μm	$1/Ra^2$	s_1	s_2	s_3	s_1/s_2	s_1/s_3
I.1	1.033	0.937	439	546	309	0.804	1.421
I.2	0.800	1.536	520	697	367	0.746	1.417
I.3	0.950	1.108	530	687	360	0.771	1.472
I.4	1.000	1.000	476	637	361	0.747	1.319
I.5	1.033	0.937	(337)	576	337	(0.585)	(1.000)
II.1	1.100	0.826	425	587	313	0.724	1.358
II.2	1.000	1.000	511	658	342	0.777	1.494
II.3	0.983	1.035	417	546	291	0.714	1.433
II.4	0.917	1.189	467	617	338	0.757	1.382
II.5	1.100	0.826	347	462	272	0.751	1.276
III.1	1.316	0.577	400	512	301	0.781	1.329
IV.1	1.017	0.967	990	1220	669	0.811	1.480
IV.2	0.883	1.283	1246	1545	920	0.806	1.354
IV.3	1.000	1.000	1097	1382	785	0.794	1.397
IV.4	0.983	1.035	1053	1352	769	0.779	1.369
IV.5	1.117	0.802	902	1135	634	0.795	1.423
V.1	0.900	1.236	1010	1225	680	0.824	1.485
Mean V.						0.774	1.401

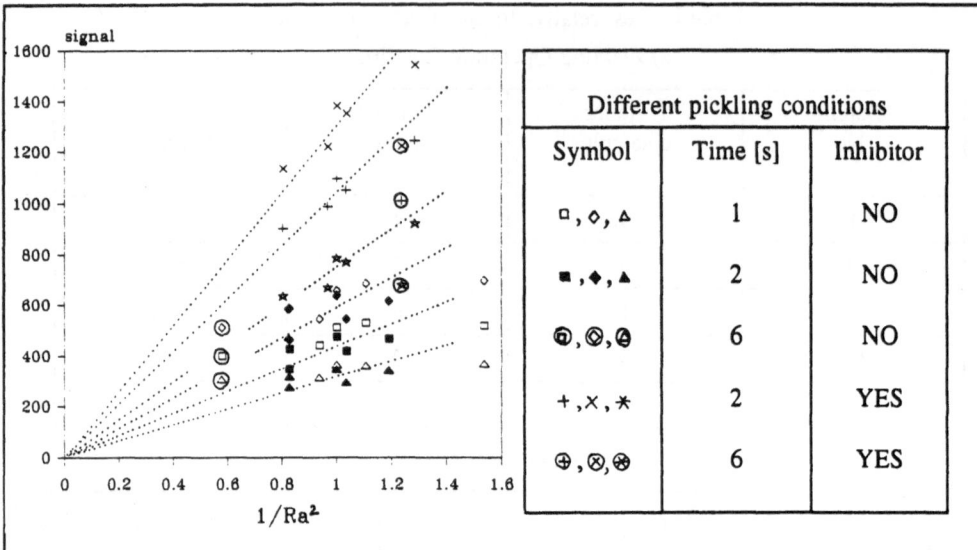

Fig.6 Spectral Bidirectional (SB) reflectance signals s_1, s_2 and s_3 versus $1/Ra^2$ for 17 samples of pickled steel sheet.

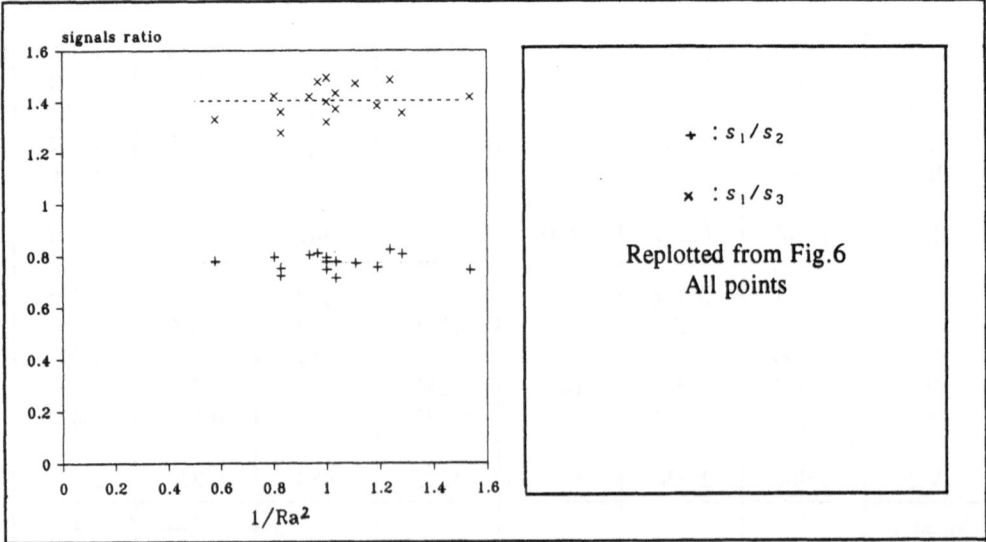

Fig.7 Spectral Bidirectional (SB) reflectance signals ratios s_1/s_2 and s_1/s_3 versus $1/Ra^2$ for 17 samples of pickled steel sheet.

- the different (straight) lines corresponding to the different wavelengths are convergent to zero ; in other words, for each batch, the ratios s_i / s_j are independant of the roughness.

It is further shown on figure 7, that, for each pair of wavelengths, the ratio takes the same value for all the batches, i.e. is independant of the surface mechanical finish. We may thus conclude that, on that point, the theoretical predictions are fully validated.

4.2. OXIDATION TEST.

The samples are small pieces of steel sheet, 100 mm x 100 mm. The surface preparation - taking place just before the test itself - is, like previously, acid pickling followed by rinsing and wiping.

For the test, the sample, standing vertically in front of the apparatus, is heated up in open air, by a flame of natural gas, sweaping its rear face. This lasts for 15 to 30 minutes ; the sample temperature reaches some 450 °C.

The results of a test are shown on figure 8 a) and b). Fig. 8 a) presents the timewise evolution of the three reflectance signals : the three of them exhibit a sharp decrease during the first few minutes after the heating is started, but there are some noticeable differences from one wavelength to the other. On fig. 8 b), the same results are plotted in terms of signal ratios, s_1 / s_2 and s_1 / s_3. This time, the evolution looks more complex : first a slight decrease, followed by a high peak and finally a stabilization at a level comparable to the original one.

In order to appreciate the significance of this evolution, it is necessary to compare it with the one predicted by theory. To that purpose, on figure 9, replotted from figure 5, is shown the evolution of the reflectance ratios - $\rho'_{0.71} / \rho'_{0.85}$ and $\rho'_{0.85} / \rho'_{1.00}$ - with the thickness of the oxide layer. The comparison between figure 8 b) and figure 9 makes sense since, in the conditions of the test, it may be expected that the oxidation penetration monotonically increases with time and, on the other hand, the wavelengths of our apparatus are sufficiently close to the triplet ($0.71\,\mu m$, $0.85\,\mu m$, $1.00\,\mu m$).

Now, as a matter of fact, there exists an essential similarity between the two figures : not only the general shape is the same, but also the relative amplitudes of the two ratios as well as the time lag between them. The principle of our apparatus is thus validated on that point too.

5. Defect Detection Capability Test

On the basis of the validation tests, it appears that - as far as we are dealing with current steel sheets - a variation of the signals ratio reveals a surface contamination. This was further checked by way of an other test where the samples were, this time, cut out directly from sheets issuing from an industrial pickling installation. The operations were driven so as to obtain sheets exhibiting the different most typical defects.

Eleven samples were selected; they are presented in Table II, with the comments of the expert in charge of the operations. As announced, this time, we are dealing with problems of surface contamination : either oxidation (although not of the spread type) or salt fouling or oil traces.

A second basis of appreciation was the "quality index " (QP) provided by a colorimeter, HUNTERLAB Qualprobe, calibrated to that purpose. This index is also shown in Table II. For our apparatus, as previously, both the three reflectance signals and their ratios are considered.

The straightest way to make a comparison between the three methods is on the basis of the samples ranking. This is made in figure 10, where the expert's ranking, considered as the reference,

Table II : Defect Detection Capability Test
a) Expert's Comments and Ranking

Sample No	Comments	Ranking
1	corrosion pits, but very bright	8
2	corrosion pits, darker	9
3	similar to 2	10
4	Laser artificial roughness, dark coloured	11
5	correct shade, but oily	4
6	bad shade : yellow	7
7	reasonably good, but yellowish	3
8	quite good, reference	1
9	not too bad, but very oily	6
10	good	2
11	very oily	5

b) Apparatuses Results

Sample No	QP	Ranking acc. QP	s_1	s_2	s_3	s_2/s_3	Ranking acc. s_2/s_3
1	48	8	430	270	100	2.70	8
2	48	9	252	175	67	2.61	9
3	48		342	290	72	4.02	
4	42	10	112	68	34	2.00	10
5	59	4	400	204	74	2.76	5
6	50	7	285	153	56	2.73	7
7	58	5	403	200	68	2.95	3
8	64	1	399	221	60	3.69	2
9	58	6	280	135	47	2.87	4
10	61	2	414	200	54	3.70	1
11	60	3	340	174	63	2.76	6

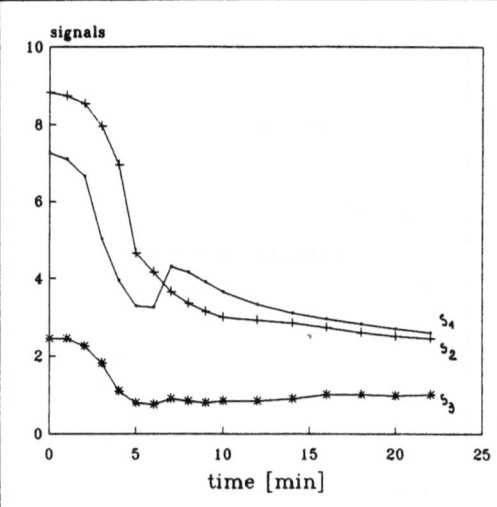

Operating conditions :

sample heated up in open air
starting at time 0
temperature max. ca 450 °C

Fig.8a Time evolution of the SB reflectance signals s_1, s_2 and s_3 of a steel sheet sample during oxidation test.

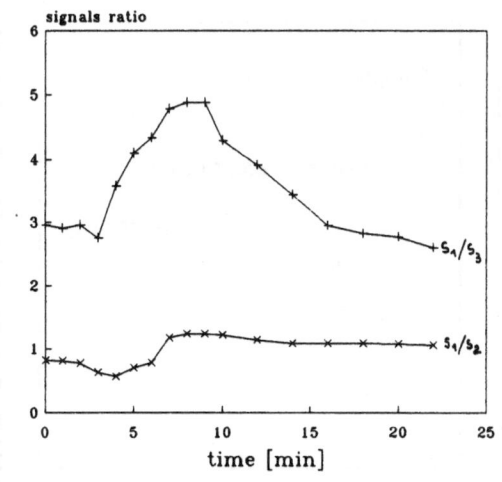

replotted from Fig.8a

Fig.8b Time evolution of the SB reflectance signals ratios s_1/s_2 and s_1/s_3 of a steel sheet sample during oxidation test.

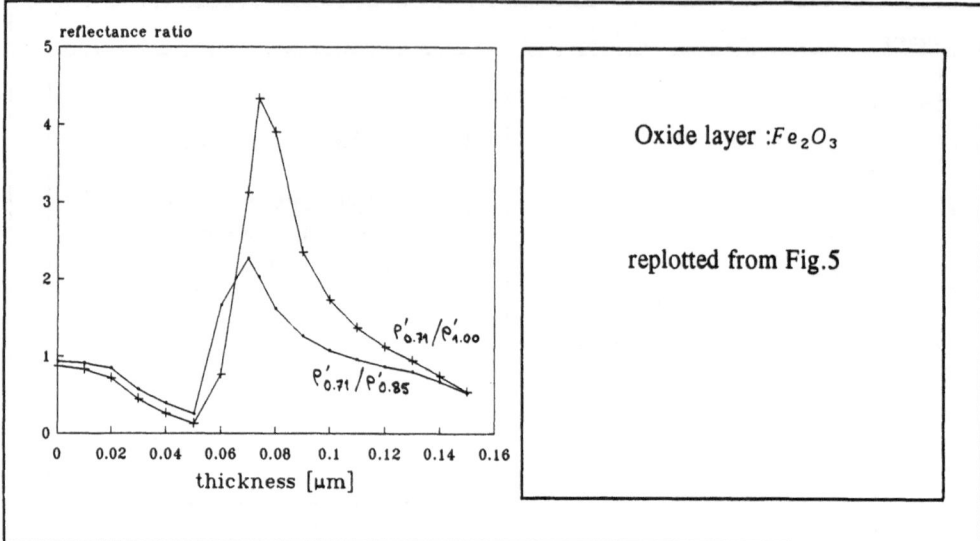

Fig.9 SDH reflectance ratios $\rho'_{0.71}/\rho'_{0.85}$ and $\rho'_{0.71}/\rho'_{1.00}$ of an oxidized stainless steel surface plotted versus the oxide layer thickness.

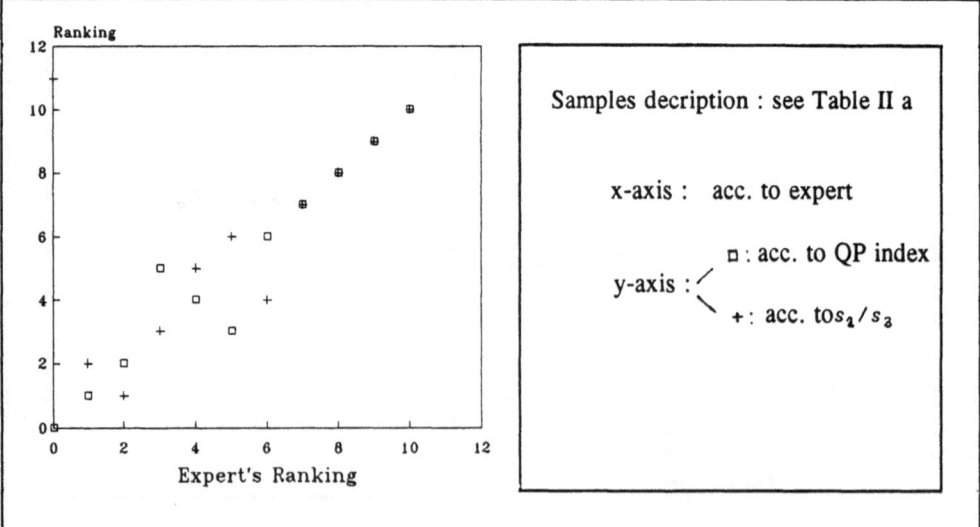

Fig.10 Comparison of three different rankings of 11 samples of pickled steel sheet

is plotted on the x-axis, while the rankings established by the two apparatuses are plotted on the y-axis.

We note that both the QP index and the reflectance signals ratio s_2/s_3 are well correlated with the expert's opinion. Despite this good result, it is only a partial success because the ranking makes no difference between the different contaminations, what the expert was able to do. Several tentatives were made to discriminate the nature of the contaminant, by using more than just one signals ratio : for instance, the two signals ratios or a combination of one signals ratio and of the signals amplitudes. But, although not completely fruitless, these tentatives were not convincing enouhg to be recommended. Further investigation is required.

6. "On-Line" Testing

Finally, "on-line" testings were also carried out. The complete analysis of these tests would deserve a full paper, but we think interesting to mention here three of the most significant results.

As can be seen from figure 11, the recorded signals are moderately noisy. Part of this noise is unidentified (say "electronic") while part of it is due to the random variations of (θ, θ_r, ϕ_r) corresponding to the vibrations of the moving sheet. These vibrations depend significantly on the speed.

In regular operation (namely at constant speed), the reflectance signals exhibit a characteristic profile (see figure 12), that is properly a "signature" of the coil : it may be very different depending on the origin of the coil. It is an effect of the surface finish, as the signals ratios remain constant.

The operation anomalies are well detected. For instance, on figure 13, one can see a record taken at the outlet from the pickling section of an annealing furnace, showing the transition from the starting phase to the production phase : the left part of the record corresponds to the end of the "return coil", used for starting, while the right side corresponds to a standard coil. One easily identifies on the latter, successively :
- some "defect", revealed by the selective decrease of the different reflectance signals, and
- the beneficial effect of the finishing rolls ("skin pass"), corresponding to the sudden increase of the signals.

But, according to this test, the choice of the wavelengths may not be optimal to "diagnose" the presence and the importance of a starting oxidation. Here again, further investigation is required.

7. Conclusion

This work has shown that the processing of multiple well selected spectral reflectance signals is a potential tool for monitoring the surface quality of steel sheets. In particular, it is well established that the signals ratios may be valuable indicators of surface contamination. However, important developments are still necessary, namely in order to determine :
- the number and optimal values of the wavelengths,
- the most appropriate combination of the signals and their ratios providing a reliable "diagnosis" of the oxidation status.

412

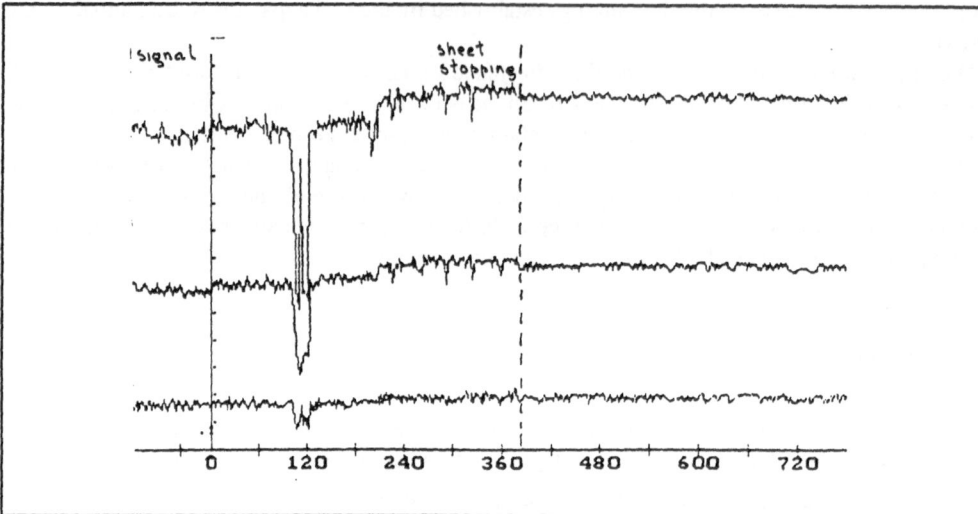

Fig.11 Time evolution of the SB reflectance signals s_1, s_2 and s_3 during on-line testing at the outlet from a continuous annealing furnace. - Influence of sheet movement on the signals noise

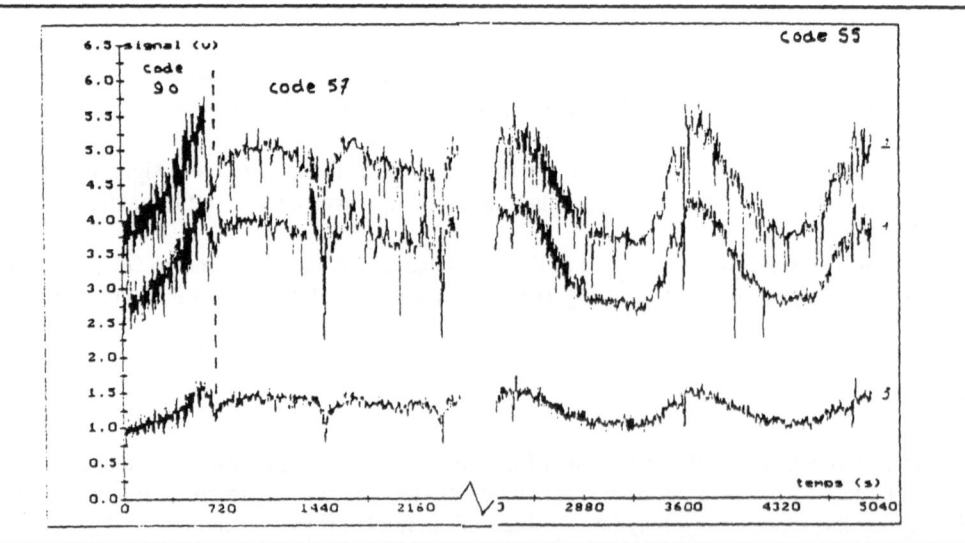

Fig.12 Time evolution of the SB reflectance signals s_1, s_2 and s_3 during on-line testing at the outlet from the pickling section of a continuous gavanizing furnace. - Signature of different coils.

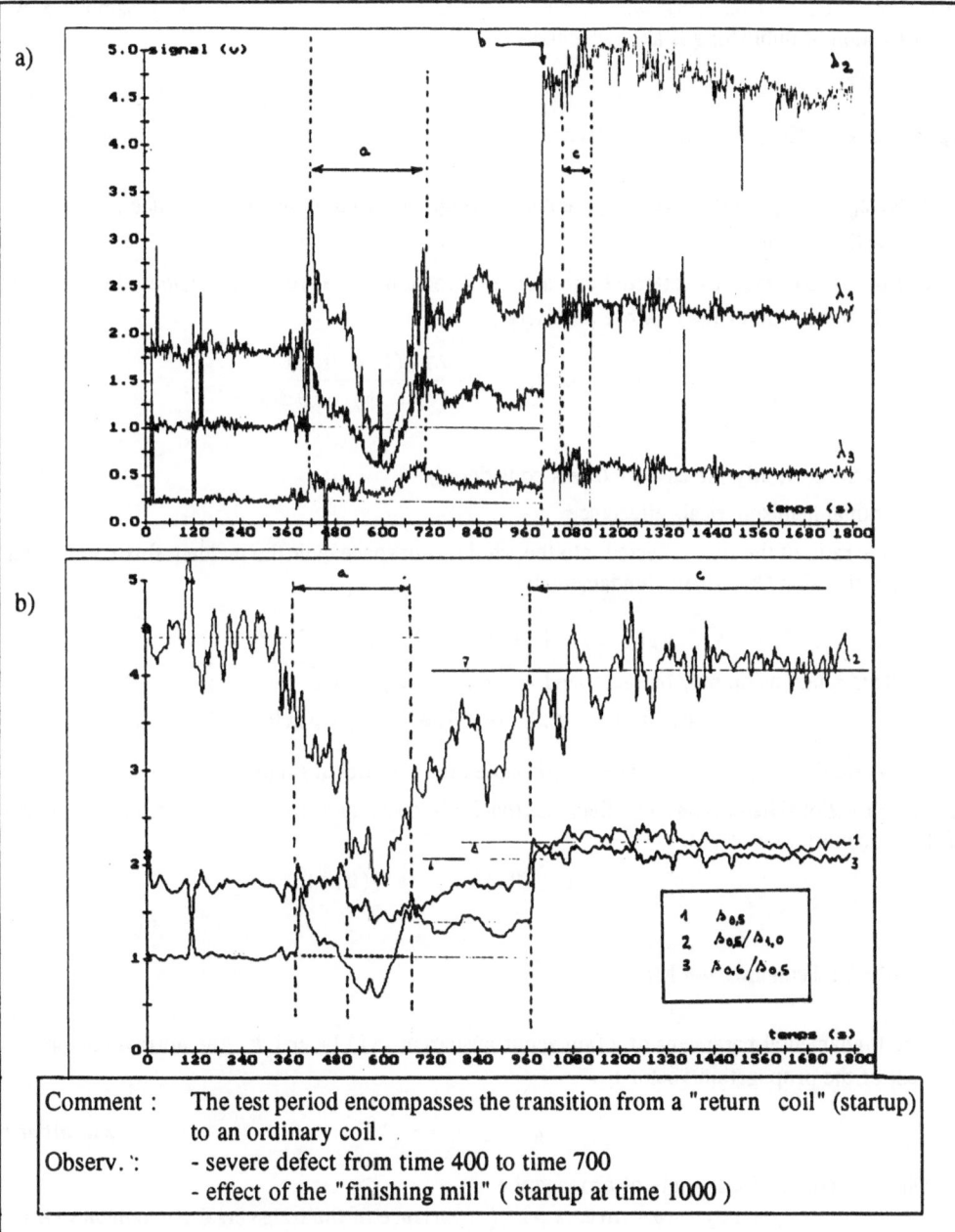

Comment :   The test period encompasses the transition from a "return coil" (startup) to an ordinary coil.

Observ. :   - severe defect from time 400 to time 700
           - effect of the "finishing mill" (startup at time 1000)

Fig.13a Time evolution of the SB reflectance signals s_1, s_2 and s_3 during on-line testing at the outlet from a continuous annealing furnace . - Anomaly detection capability.
Fig.13b idem for signals ratios s_1 / s_2 and s_1 / s_3

Aknowledgement. This work was carried out thanks to a grant of Cockerill Sambre S.A. Authorization of publishing is aknowledged.

Appendix 1. Reflectance and Emissivity

Considering only spectral properties, we may distinguish two notions of reflectance :
-   bidirectional reflectance

    It is the ratio of the flux reflected into a given direction (θ_r, ϕ_r) over the incident flux coming out of the direction (θ, ϕ). More precisely :

$$\rho_\lambda^{\cdot}(\theta, \phi; \theta_r, \phi_r) = \frac{L_{\lambda, r}^{\cdot}(\theta, \phi; \theta_r, \phi_r)}{L_{\lambda i}^{\cdot}(\theta, \phi) \cos\theta\, dw}$$

(In the text, only the case $\phi = 0$ was considered)
-   directional hemispherical reflectance

    It is the ratio of the flux reflected into the whole front space over the incident flux coming out of the direction (θ, ϕ). So, fundamentally :

$$\rho_\lambda^{\cdot}(\theta, \phi) = \int_{2\pi} \rho_\lambda^{\cdot}(\theta, \phi; \theta_r, \phi_r) \cos\theta_r . dw_r$$

This relationship may be factorized in the following way :
$$\rho_\lambda^{\cdot}(\theta, \phi; \theta_r, \phi_r) = G_\lambda(\theta, \phi; \theta_r, \phi_r) . \rho_\lambda^{\cdot}(\theta_r, \phi_r)$$

where the function $G_\lambda(\theta, \phi; \theta_r, \phi_r)$ is called the " Reflexion Indicatrix ".

It is the directional hemispherical reflectance that is directly related to emissivity, as a consequence of the Kirchhoff's law :

$$\rho'_\lambda(\theta, \phi) = 1 - \epsilon'_\lambda(\theta, \phi)$$

Appendix 2 : Transmission Matrix

The fundamental parameter of a thin semitransparent (ST) layer is k_z, the normal component of the propagation vector :

$$k_z \stackrel{\Delta}{=} k' \cos\theta + j k'' \qquad (complex\, number)$$

where θ is the local angle of propagation

given by $\sin\theta = \sin\theta_0 . k_0 / k'$ (invariance of the transversal component of $\vec{k}$)

$$k' \stackrel{\Delta}{=} \frac{2\pi}{\lambda_0} . n \qquad k'' \stackrel{\Delta}{=} \frac{2\pi}{\lambda_0} . \kappa$$

n being the refractive index and κ the extinction index .

Let us then denote :

$$k_z^x \overset{\Delta}{=} \frac{1}{\mu_0} \cdot \frac{\lambda_0}{2\pi c_0} \cdot k_z \qquad k_z^{xx} \overset{\Delta}{=} \frac{1}{\epsilon_0} \cdot \frac{\lambda_0}{2\pi c_0} \cdot \frac{k_z}{n^2 + \kappa^2}$$

ϵ_0 and μ_0 being respectively the dielectric and the magnetic permissivity of vacuum, and c_0 the speed of radiation in vacuum.

and :

$$\underline{k}^x \overset{\Delta}{=} (1 \quad k_z^x)^T \qquad \underline{k}^{xx} \overset{\Delta}{=} (1 \quad k_z^{xx})^T$$

$$\underline{\overline{k}}^x \overset{\Delta}{=} (1 \quad -k_z^x)^T \qquad \underline{\overline{k}}^{xx} \overset{\Delta}{=} (1 \quad -k_z^{xx})^T$$

Then, the individual layer transmission matrices M_1 and M_2, corresponding respectively to the two components of polarization, are expressed :

$$M_1 = \begin{pmatrix} \cos h^x & -j\sin h^x / k_z^x \\ -jk_z^x \sin h^x & \cos h^x \end{pmatrix}$$

$$M_2 = \begin{pmatrix} \cos h^x & j\sin h^x / k_z^{xx} \\ jk_z^{xx} \sin h^x & \cos h^x \end{pmatrix}$$

$$h^x \overset{\Delta}{=} k_z \cdot h$$

References.

1 SACADURA J.F. - Modélisation et étude expérimentale du rayonnement thermique des surfaces métalliques microrugueuses. Thèse de Doctorat d'Etat ès Sciences, Lyon, 1980.

2 ANE J.M. et HUETZ-AUBERT M. - Stratified media theory interpretation of measurements of the spectral polarized directional emissivity of some oxidized metals. Int. J. Thermophysics, Vol. 7 (1986),N° 6, pp. 1191-1208.

3 NEUER G., GUENTERT P. and WOERNER R. - Radiation thermometry as an appropriate method to observe the oxide growth on metal-surface. Thermochimica Acta 85(1985) pp.295-298.

4 HERVE Ph. - Influence de l'état de surface sur le rayonnement thermique des matériaux solides. Thèse de Doctorat d'Etat ès Sciences Physiques, Paris, 1977.

5 ABDUKADIR A. - Spectral directional emittance of roughened metal surfaces. Ph.D. thesis, Univ. Kentucky(USA), 1973.

INTEGRATED HARDWARE IMPLEMENTATIONS FOR DIGITAL SYSTEMS[1]

ROBERT BERGEVIN
Computer Vision and Systems Laboratory
Department of Electrical Engineering
Laval University, Québec, Canada G1K 7P4

Advanced digital systems interact with complex dynamic environments. They sense and process information and their actions modify the state of the environment. Such systems must be robust, reliable, and flexible. They must also satisfy performance and timeliness criteria. In many complex real-world applications, there is a clear need for distributed, parallel, and heterogeneous architectures with real-time and fault-tolerance capabilities.

We review modern parallel processing technology: taxonomy of systems and interconnection networks. Two parallel building-block processors are introduced and compared: the Transputer and the TMS320C40. Real-time and fault-tolerance concepts are also exposed. Finally, we conclude the presentation with a look at a few case studies, from current projects at the Computer Vision and Systems Laboratory.

George S. Almasi and Allan Gottlieb, "Highly Parallel Computing", The Benjamin/ Cummings Publishing Company, Inc., 1989.

Yann Hang Lee and C.M. Krishna, Eds., "Readings in Real-Time Systems", IEEE Computer Society Press, Los Alamitos, CA, 1993.

Hoang Pham, Ed., "Fault-Tolerant Software Systems: Techniques and Applications", IEEE Computer Society Press, Los Alamitos, CA, 1992.

Ian Graham and Tim King , "The Transputer Handbook", Prentice-Hall, 1990.

Texas Instruments, "TMS320C4x User's Guide", 1991.

Côté, J., Gosselin, C.M., and Laurendeau, D., "Tracking a Moving Object With a 6-DOF Manipulator", Proc. of SPIE Conference on Vision and Robotics, Orlando, FL, April 1993.

---

[1]this contribution: abstract only

X. P. V. Malague (ed.), Advances in Signal Processing for Nondestructive Evaluation of Materials, 417.
© 1994 *Kluwer Academic Publishers.*

INTEGRATED HARDWARE IMPLEMENTATIONS FOR DIGITAL SYSTEMS

ROBERT DE CLER

Computation and Systems Laboratory
Department of Electrical Engineering
Laval University, Québec, Canada G1K 7P4

PARALLELIZATION OF THE SYNTHETIC APERTURE FOCUSING TECHNIQUE FOR ULTRASONIC TESTING

Keith S. Pickens
Southwest Research Institute
P. O. Box 28510
San Antonio, Texas 78228-0510

ABSTRACT. A new parallel implementation of the synthetic aperture focusing technique for ultrasonic testing (SAFT-UT) has been developed. This algorithm runs on small, inexpensive workstations, which are easily transported to a testing site. The performance is achieved in *realtime*, in that data can be processed as fast as it is acquired. The SAFT-UT technique is reviewed, and details of the new implementation are discussed. Performance data are presented for two platforms, a shared-memory parallel processor and a distributed-memory parallel processor.

1. Introduction

The synthetic aperture focusing technique for ultrasonic testing (SAFT-UT) has been the subject of extensive laboratory evaluation and field-site validation testing for over 16 years. The methodology can accurately determine the spatial location and extent of a defect; but because of the computationally intense requirements of the SAFT algorithm, SAFT has not yet seen wide application. To be practical, SAFT-UT signal processing must be accomplished on site where data are collected; but the necessary super computers have been too expensive and immobile for field use. Through the use of a powerful, small microprocessor-based parallel computer, these barriers can be overcome to make SAFT-UT a practical technique.

Parallel processing can vastly increase performance over serial processing while reducing a system's cost by a similar magnitude. These changes in cost and performance are so large that computational solutions that were impractical with serial computing become attractive approaches.

A new parallel SAFT-UT algorithm has been developed. The implementation of this algorithm has processed data at *realtime* rates; that is, the data can be processed as fast as it is acquired. Processing at these speeds has never before been demonstrated in general SAFT-UT.

X. P. V. Malague (ed.), Advances in Signal Processing for Nondestructive Evaluation of Materials, 419–433.
© 1994 *Kluwer Academic Publishers.*

2. Background

2.1 WORK TO DATE

SAFT-UT has been the subject of extensive laboratory evaluation and field-site validation. The research and development programs in the United States have been sponsored primarily by the Nuclear Regulatory Commission (NRC). Work began in 1974 at the University of Michigan in Ann Arbor (1) and continued through 1983. This work successfully demonstrated that (1) expected improvements in axial and lateral resolution of defects during nondestructive evaluation (NDE) using SAFT-UT were achievable and (2) image-based color-display formats significantly improved ease of data interpretation. Early efforts included evaluating the influence of selected parameters on factors such as image quality and processing speeds and attempting to develop a special-purpose processor to perform realtime imaging.

In 1977, Southwest Research Institute (SwRI) began a 7-year project (2,3) designed to demonstrate that SAFT-UT technology could be applied to meet inspection requirements at field locations. A follow-on SwRI internal research program (4) in 1983 demonstrated the generality of the technique using unfocused angle-beam transducers in conjunction with contact inspection techniques typical of inservice inspection (ISI) procedures. In 1982, the NRC contracted Pacific Northwest Laboratories (PNL) for design and development of a realtime SAFT-UT imaging system (5) to use in performing ISI of nuclear reactor system components and to generate an engineering database (6) to support American Society of Mechanical Engineers (ASME) Code acceptance of the technique. PNL recently completed a SAFT-UT imaging system with limited success because of its slow speeds; however, they are actively promoting activities intended to gain Code acceptance.

SAFT-UT has also received extensive international attention and laboratory research; field validation has been performed in Germany (7,8), Japan (9,10), and England (11). The results of these programs successfully demonstrated the general applicability of SAFT-UT in providing improved inspection reliability (more accurate flaw characterization and sizing).

2.2 CURRENT FOCUS

The focus of current SAFT-UT projects is toward (1) supporting the process required to incorporate SAFT-UT formally into appropriate sections of the ASME Code and (2) improving image-generation efficiency and speed for heavy-section steel applications. The Code effort has centered on placing SAFT-UT in a new Appendix to Article 4 (Computer-Processed Imaging), Section V. Once this process is completed, SAFT-UT will be accepted by industry (utilities) and regulatory agencies as a practical field-validated inspection technique for use to improve inspection reliability. Work also continues to reduce image-generation processing times for volumetric SAFT-UT images. While important progress has been reported (12,13), SAFT-UT image generation in heavy-section steel components [i.e., reactor pressure vessels (RPV)] must be improved further by reducing time and cost to permit practical routine usage during NDE ISIs.

3. Technical Approach

3.1 SAFT IMAGING PHILOSOPHY

SAFT-UT provides the capability of more accurately determining the spatial location and extent of acoustically reflective surfaces (flaws) than is possible with conventional ultrasonic inspection systems. The improved resolution is attained by the image reconstruction algorithm that accomplishes focusing.

3.2 FUNDAMENTALS OF SAFT IMAGING

The diffraction pattern existing at the focal plane of a lens is produced by the constructive and destructive interference occurring among all rays contributing to the beam. With a physical lens, all of the rays occur simultaneously, and the resulting interference pattern is produced in realtime. SAFT processing is performed without a physical lens. The diffraction pattern results from the summation of various rays, but the rays do not have to coexist and be summed simultaneously. It is only necessary that all rays exist at some point in time and that, eventually, they are all summed (coherently). These two requirements are necessary and sufficient for image formation and are accomplished in a SAFT imaging system in a sequential manner, as described in the following paragraphs.

3.3 SAFT RECONSTRUCTION ALGORITHM CONCEPTS

The SAFT reconstruction algorithm to be described is predicated on two assumptions:

(1)   Surfaces (flaws) to be imaged are diffusely reflective; that is, echoes can be detected from many small regions about the face of the reflector from many different directions of incidence.

(2)   Velocity of acoustic propagation is reasonably isotropic between the surface scanned and the flaws to be imaged.

Given that these conditions are satisfied, a reconstructive algorithm can be designed by making use of three concepts:

(1)   If there is a diffusely reflective irregularity within the component, that irregularity will reflect an echo directly back to the transducer for a wide variety of transducer locations.

(2)   If the shape of the component's surface is known or can be measured, it is possible to predict the arrival time of an echo reflected from any coordinate within the component as measured from any surface location.

(3)   The presence/absence of a diffuse reflector at a certain coordinate can be determined as follows: First, for each of many pulse-echo measurements taken about the surface, the

value of the signal at the exact time appropriate for an echo produced at that coordinate is recorded. The signal values are then summed together. If the resulting sum is a large number, evidence is strong that an irregularity at the coordinate being considered is present. If, on the other hand, the sum is a small number, one can conclude that the coordinate is not associated with a reflector.

This test can be repeated at closely spaced points throughout the volume to be imaged. Then, by noting where the summation process yields large amplitudes, one can determine the three-dimensional location of reflective surfaces (flaws). This process is central to the SAFT-UT reconstruction algorithm.

3.4 IMPLEMENTATION OF THE SwRI SAFT RECONSTRUCTION ALGORITHM

The SwRI reconstruction algorithm capitalizes on the data acquisition system's capability to accurately position and measure the location of the focused-beam entry point on the surface of the material being examined. This characteristic simplifies the accurate determination of the position of received signals because, for a given raw data set, all ray paths to the cells come from a single point. The requirement to accurately determine the x, y, and z sound-beam entry point is accomplished automatically by the scanning mechanism and associated electronic components. Therefore, no analytical determination is required during processing to compensate for irregularities, surface-roughness, and geometrically misshaped component surfaces.

3.4.1 General Description of the SwRI Algorithm. The SwRI reconstruction algorithm permits the operator to specify various parameters that define the volumetric region to be imaged. The operator controls the size of the surface aperture by specifying an included angle for the synthesized focusing. The reconstruction algorithm obtains a raw data set and determines, for every cell within the imaged volume, which waveform sample element contributes to the cell. Each cell has a local surface aperture (same included angle as defined for the volume), and only raw data sets recorded within this localized aperture contribute sample elements to the cell. The process is depicted in Figure 1. The operation of determining the contributing elements from every raw data set associated with a cell continues until all cells within the volume have been processed. As shown in Figure 1, the size of the localized surface aperture increases with increasing cell depth in the material. This characteristic provides for additional sample element contributions to deeper cells because a greater number of raw data sets will be contained in the surface aperture. In this way, the SwRI reconstruction algorithm helps compensate for signal attenuation within the material.

The final image is formed by plotting the computer signal amplitude as a function of cell location. The image is then interpreted by noting the location of high signal amplitude. The reconstruction algorithm does not attempt to directly determine the size and location of flaws. Rather, the algorithm is intended to test for the presence or the absence of a reflecting surface at the equivalent location of each cell. Flaw size and location are deduced by operators from the pattern formed by cells that show evidence of containing a reflector.

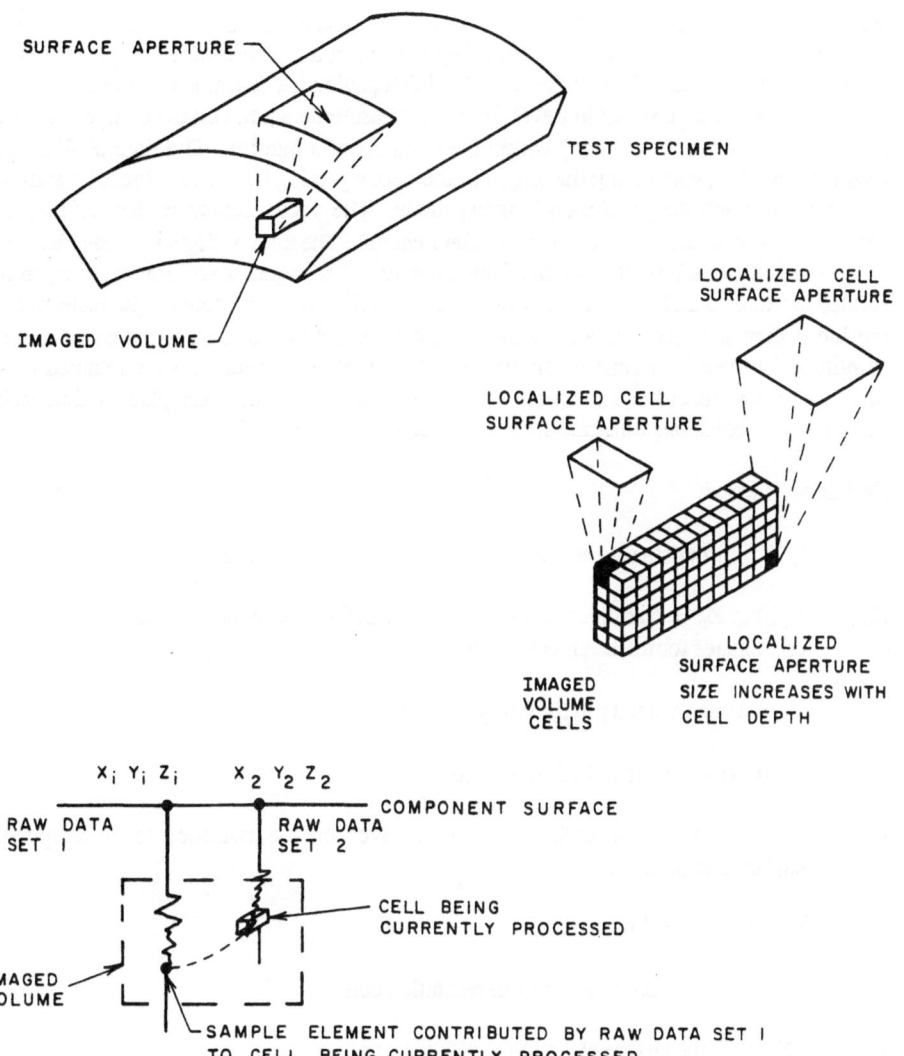

Figure 1. Illustration of essential elements of the SAFT
reconstruction algorithm

3.4.2 SwRI Reconstruction Algorithm Calculation. The basic calculation implemented in the reconstruction algorithm involves determining the contributing waveform sample element from a raw data set to a selected resolution cell. This calculation is determined by predicting the time of flight for an acoustic signal to travel from any measurement point on the component's surface to any other point (cell location) within the component's interior. The time of flight can be predicted by, first, calculating the length of the vector joining the surface location and the cell location for which the prediction is being made. The vector length is then divided by the acoustic velocity appropriate for the particular material. The time of flight thus calculated is then used to identify which small portion of the entire digitized signal waveform acquired from that surface location could have been reflected from the cell being considered. The identified signal portion is then summed into a computer memory location used to store the calculated signal amplitude for the cell. Because both phase and amplitude information were retained when the raw data set was recorded, constructive or destructive summation takes place within each cell. The following equations illustrate the generalized calculations used.

Definition of terms:

(1) DS_n = n'th data set contained within the synthetic aperture

(2) X_n, Y_n, Z_n = transducer coordinates for n'th data set (Note: X_n and Y_n are also the coordinates for the beam entry point)

(3) V_c = sound velocity in the coupling fluid

(4) V_s = sound velocity in the specimen

(5) T_{fs} = round-trip travel time of acoustic wave from the transducer to the component surface and back

(6) RC_i = i'th resolution cell

(7) X_i, Y_i, Z_i = coordinates of i'th resolution cell

(8) FS_n = front surface elevation at n'th data set

$$FS_n = Z_n + (T_{fs} \times V_c) + 2 \tag{1}$$

Let the sequence of data points (amplitude samples) within a data set start at the front surface and occur at an interval of I seconds.

$DP_{n,m}$ = amplitude of m'th data point contained within
        the n'th data set m = 1, 2, 3, ..., m

First, calculate:

$MP_{n,i}$ = metal path from the n'th beam entry point to the i'th resolution cell

$$MP_{n,i} = [(X_n - X_i)^2 + (Y_n - Y_i)^2 + (FS_n - Z_i)^2]^{1/2} \qquad (2)$$

Use this length, along with V_s and I, to calculate m for DS_n.

$$m = (2MP_{n,i}) + V_s I \text{ nearest integer.} \qquad (3)$$

Now use this m to address the proper sample that it may be summed into the intensity value of the i'th resolution cell.

4. Discussion and Results

4.1 OVERVIEW

The computational model selected for this work uses a central control (the SAFT "client"), which issues work requests to a group of parallel SAFT servers. The SAFT client handles the scheduling of work to each of the servers and collects the results. The client controls and sends the SAFT parameters to each of the servers, but it does not distribute the input data. The task of input data distribution is assigned to existing well-developed file-sharing techniques.

In SAFT-UT each output point depends on the input data set and the processing parameters. Thus, the minimum work unit is a single point, and the maximum, the entire input data set. In some cases, however, it is necessary to do post processing of the data in a per temporal data-set (A-scan) fashion. To allow this post processing to occur in the server, a single A-scan is the minimum work unit. An A-scan of data is typically of close order 10^3 points, and a typical data set is 10^3 to 10^4 A-scans. The A-scan was selected as the work unit issued to the SAFT servers.

At the start of a processing run, the client sends the servers the SAFT parameters and the input data-set file name. Each server loads the data set, builds its internal tables, and reports a ready status to the SAFT client. The client then issues requests to the SAFT servers to process A–scans until the entire input file has been completed. Since the A-scans are independent of each other, it is possible to achieve good load balancing by this type of simple dispatching algorithm.

4.2 IMPLEMENTATION

As many standard tools as possible were used to implement the parallel SAFT code. The implementation language was C++ with the Object Interface (OI) X Window System toolkit (14) and the U.S. Army Ballistic Research Laboratory "package" (PKG) protocol (15).

The client was written with an X-based graphical user interface. Figure 2 shows the top-level window. The parameter entry for the SAFT processing was handled by a popup window, shown in Figure 3.

The dispatching algorithm maintains a list of unprocessed A-scans. When a SAFT server returns a processed A-scan, a new request is issued to that server; and then the client stores the processed A-scan. There is some room for improvement here. The server could queue a request, avoiding the dead period between the time a result is processed and a new request is issued. Measurements indicate that this effect and the bandwidth limitations of the transport, if run over a network, have not been a significant factor in system performance.

The client communicates with the server via the PKG protocol layered on top of a TCP connection. The PKG protocol allows exchange of messages up to $2^{32}-1$ bytes and supports mixed synchronous and asynchronous message paradigms. The PKG protocol is also simple to integrate into event-driven, window-based code such as the SAFT client program.

The input data are distributed by "other means." In the case of shared memory machines, the data were loaded from the local file system. For distributed network operation, Sun NFS was used. The only requirement is that the data be available to the client and the severs under the same file name.

4.3 RESULTS

The prototype will run in a range of different environments. These can vary from a shared memory multiprocessor (MP) machine to a distributed configuration, which could range over much of North America. The servers must have access to the input data set, have a internal storage format compatible with the client (this could be relaxed), and be reachable with a low-latency, high-bandwidth TCP connection. The requirement for a low-latency, high-bandwidth network connection limits testing to a single local area net. This restriction can be relaxed as wide area networks become faster.

Two major configuration have been tested: a shared-memory MP machine (Sun 4/670MP) and a distributed configuration made up of Sun 4/75s and 4/670MP machines running in a single- processor mode. Figure 4 shows the typical MP configuration, and Figure 5 shows the distributed configuration. The test data set contained 2601 A-scans (each 1024 samples long) taken on 0.04-inch intervals in a square grid from a block.

Figure 6 shows the performance of the MP configuration. The sharp roll-off above two processors is attributed to main-memory bus saturation. The SAFT algorithm is very data intensive, and the small caches (typically 64Kbytes) on these machines are ineffective. Figure 7 shows the distributed case. Here the results are much better, since each processor has a local memory and bus and the saturation problem of the MP machine does not occur. Figure 8 shows the time to process the test data together with the measured overhead. The overhead includes delays in the network, dispatching algorithm, client, and servers.

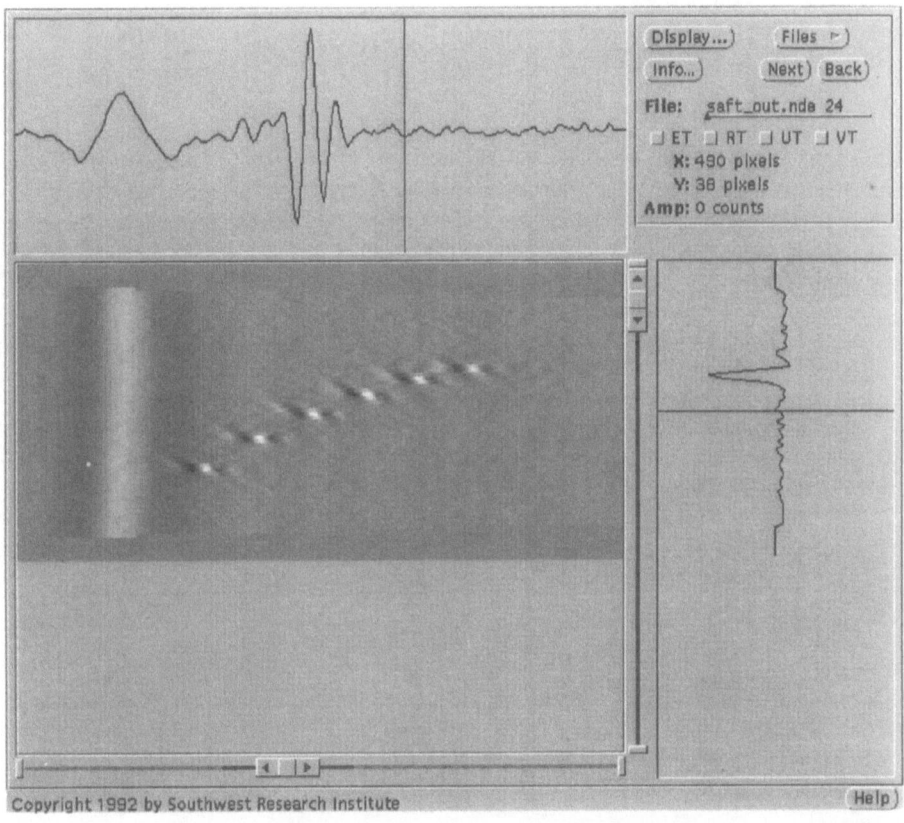

Figure 2. Top level window for the parallel SAFT prototype. It consists of an image display, two interactive cross-section displays, and command panel.

428

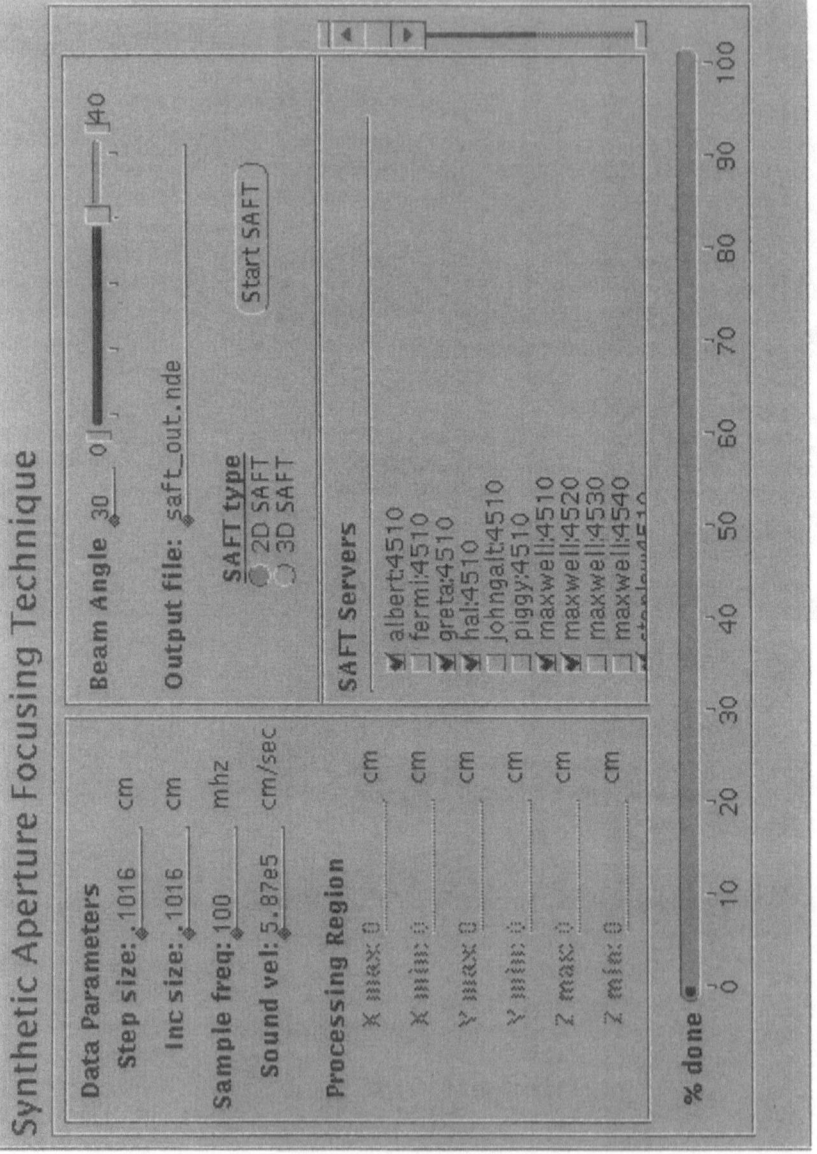

Figure 3.

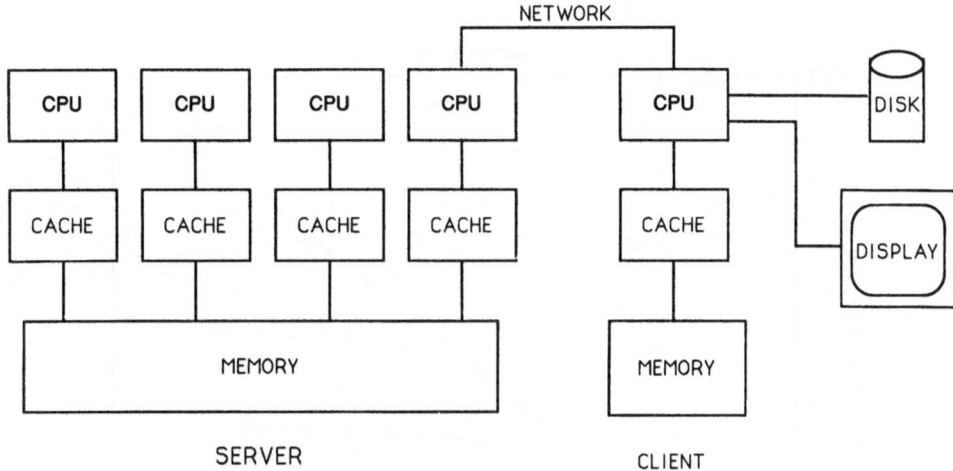

Figure 4. Shared memory, multi-processor, parallel SAFT configuration used for testing

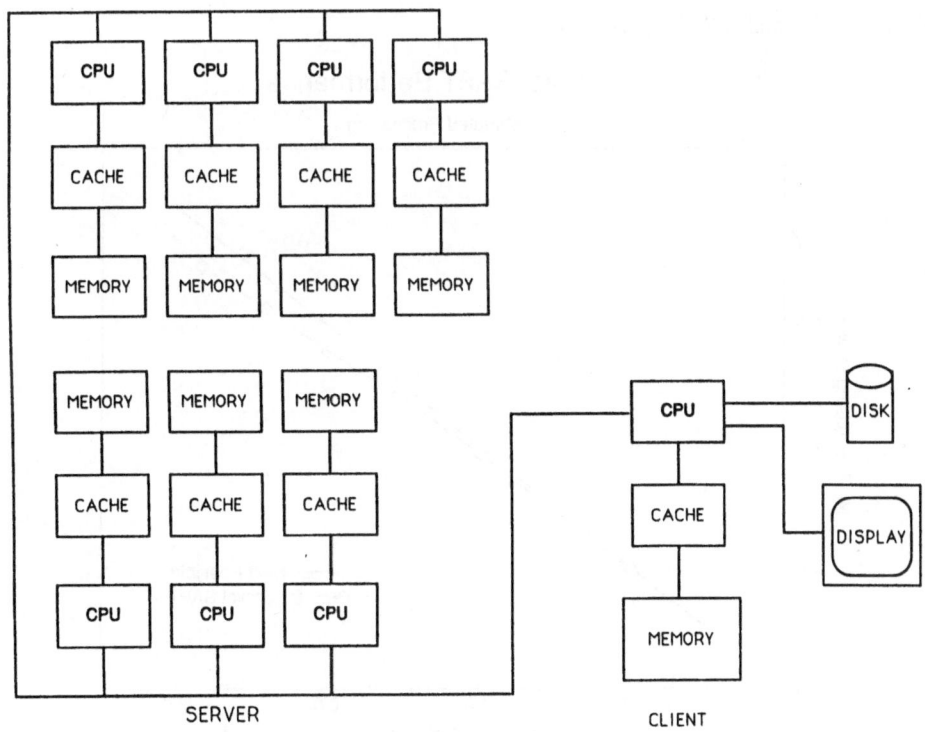

Figure 5. Distributed parallel SAFT configuration

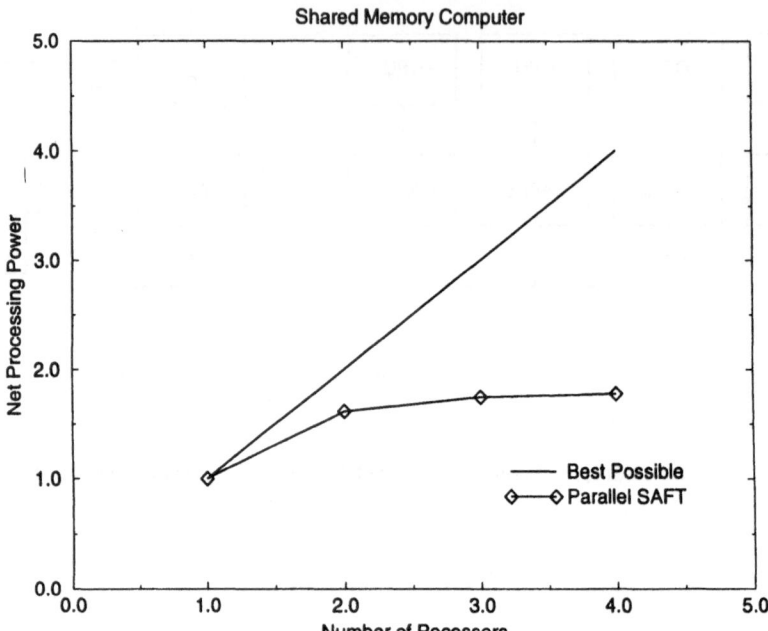

Figure 6. MP parallel SAFT performance

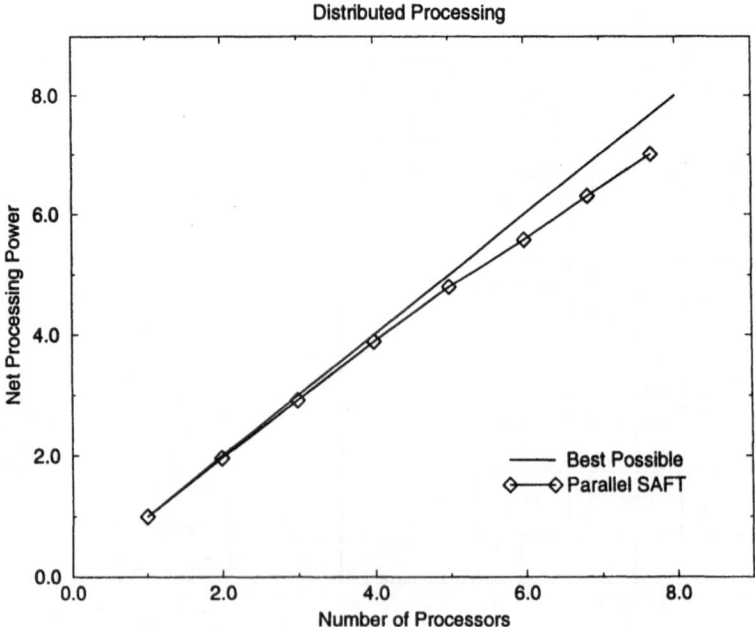

Figure 7. Distributed parallel SAFT performance

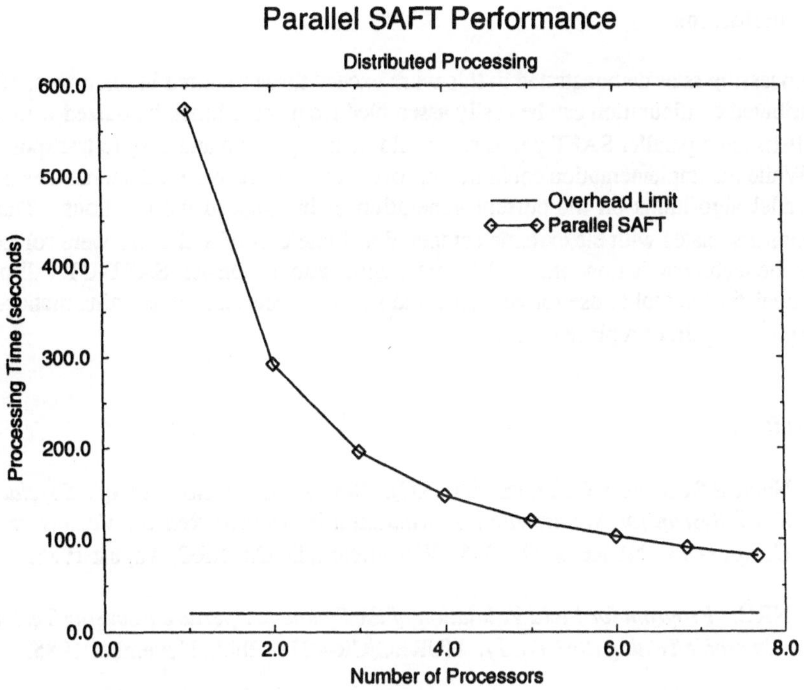

Figure 8. Time to process the test set, with the distributed parallel SAFT as a function of the number of processors together with the system overhead

5. Conclusions

The processing rates demonstrated in this work exceed those required in realtime applications. A distributed configuration can be easily assembled from small lunch-box-sized work stations. This distributed parallel SAFT processor would be inexpensive and easy to transport for field use. While the implementation could be improved, the current system demonstrates the power of parallel algorithms on the current generation of high-speed workstations. That is, the problems associated with the extreme computational intensity of SAFT-UT were solved, which means the technique is now practical for field applications. Onsite, SAFT-UT will provide a very sophisticated tool to use for verifying and sizing defects and for assisting management in making run, repair, or replace decisions.

6. References

1. Nuclear Regulatory Commission (NRC). *Design and Development of a Special-Purpose SAFT System for Nondestructive Evaluation of Nuclear Reactor Vessels and Piping Components.* NUREG/CR-4365. Washington, D. C.: NRC, August 1985.

2. NRC. *Program for Field Validation of the Synthetic Aperture Focusing Technique for Ultrasonic Testing (SAFT-UT).* NUREG/CR-4078. Ibid., November 1985.

3. NUREG/CR-1885. *Program for Field Validation of the Synthetic Aperture Focusing Technique.* Ibid., January 1981.

4. Hamlin, Dennis R. *Program for Performance Evaluation of Angle-Beam Synthetic Aperture Focusing Technique for Ultrasonic Testing (SAFT-UT).* Southwest Research Institute (SwRI), Final Report, SwRI Project 17-9330. San Antonio, Texas: SwRI, July 1984.

5. NRC. *Development and Validation of a Real-Time SAFT–UT System for the Inspection of Light-Water Reactor Components.* NUREG/CR-4583. Vol. 3. op. cit., July 1987.

6. Doctor, S. R., et al. "NDE Reliability and SAFT-UT Final Development." *Nuclear Engineering and Design* 118, 1990.

7. Müller, W., V. Schmitz, and G. Schafer. "Reconstruction by the Synthetic Aperture Focusing Technique (SAFT)." *Nuclear Engineering and Design* 94, 1986.

8. Gebhardt, W., and V. Schmitz. "Evaluation of Ultrasonic Volume and Underclad Indications at Long Distances." *Nuclear Engineering and Design* 112, 1989.

9. Yamamoto, S., M. Kishigami, and N. Ujuye (IHI Research Institute). "Measurement of IGSCC Depth by SAFT-UT." *Nuclear Engineering and Design* 94, 1986.

10. Yoshihiko, O., S. Hiroaki, T. Toshimasa, and T. Mutsuo. "A New System for Real-Time Synthetic Aperture Ultrasonic Imaging." *IEEE Transactions on Ultrasonics, Ferroelectrics, and Frequency Control* 35(6), November 1988.

11. Burch, S. F. "Comparison of SAFT and Two-Dimensional Deconvolution Methods for the Improvement of Resolution in Ultrasonic B-Scan Images." *Ultrasonics* 25, September 1987.

12. NRD. *Development and Validation of a Real-Time SAFT-UT System for the Inspection of Light-Water Reactor Components*. NUREG/CR-4583. Vol. 2. op. cit., May 1986.

13. Doctor, S. R., T. E. Hall, L. D. Reid, and G. A. Mart. "Development and Validation of a Real-Time SAFT-UT System for Inservice Inspection of LWRs." *Proceedings of the U.S. Nuclear Regulatory Commission Fifteenth Water Reactor Safety Information Meeting*. Washington, D. C: NRC, February 1988.

14. Benson, Amber, and Gary Aitken. *OI Programmer's Guide*, Englewood Cliffs, New Jersey: Prentice Hall, 1992.

15. Shames, P. B., and P. C. Dykstra. *PKG Interface Description*. Aberdeen, Maryland: U. S. Army Ballistic Research L   oratory and Space Telescope Science Institue, 1991.

APPLICATIONS OF HIGH RESOLUTION INVERSION TO ULTRASONIC IMAGING OF MULTI-LAYERED STRUCTURES

K.I. MCRAE
Defence Research Establishment Pacific
FMO, CFB Esquimalt
Victoria, B.C., Canada

ABSTRACT. A novel inversion algorithm has been developed which combines a "macro" layer-stripping approach with high resolution, L2-norm deconvolution. This algorithm has been implemented for real time data acquisition using a TMS 320C30 digital signal processor as well as for post-processing of the ultrasonic data. A further optimization-based procedure may also be applied, provided there exists adequate prior knowledge concerning the structure. Applications for these algorithms will be demonstrated for the detection of disbonds and delaminations in aircraft composite structures.

1. Introduction

The increased use of composite materials and adhesively bonded joints has resulted in the need for the development of inspection techniques appropriate for multi-layered structures. Ultrasonic pulse-echo imaging is a useful technique for this purpose. Resolution losses in the ultrasonic signal, however, result from the convolution of the transducer impulse response with the true acoustic impulse response of the layered material. The occurrence of multiple reflections in the image further complicates data interpretation. Although several deconvolution algorithms already exist for the purpose of removing the transducer impulse response from the ultrasonic data,[1-3] these techniques provide only a partial solution to the inverse problem for discretely layered media. A complete solution would estimate the acoustic impedance of each layer as a function of pulse travel time or layer thickness.

In order to be useful for NDE purposes, a practical inversion technique must satisfy the following criteria:

(1) it must be able to accommodate the mixed phase, band-limited wavelets that are generated by most ultrasonic transducers;

(2) it must be able to tolerate errors in the wavelet estimate and to withstand alterations in wavelet shape caused by propagation losses;

(3) the inversion result should be sufficiently accurate for imaging purposes;

X. P. V. Malague (ed.), Advances in Signal Processing for Nondestructive Evaluation of Materials, 435–447.
© 1994 *Kluwer Academic Publishers.*

(4) the algorithm should possess the ability to resolve closely spaced reflectors; and

(5) the algorithm should be capable of performing real time signal processing, i.e. to operate concurrently with the scanning operation.

Solutions to the inversion problem have already been applied to seismic signal processing and several possible inversion schemes have been proposed. These may be divided into two general categories; "exact" methods[4,5] which require the input data to be totally broadband and noise free and "estimation" methods which can be further classified into three categories:

(1) thresholding techniques;[6]

(2) maximum likelihood methods;[7] and

(3) optimization methods.[8-10]

The requirements for the exact techniques preclude their direct application to NDE data. Each of the estimation techniques has some practical value, however, none are capable of satisfying all of the criteria for successful NDE inversion applications. Some algorithms are unable to accommodate arbitrary wavelets, while others are extremely sensitive to the noise contained in most NDE ultrasonic data.

An estimation based high resolution inversion algorithm proposed by Zala[11] makes use of several features that have been used in other algorithms which are less successful for NDE applications, such as:

(1) layer stripping, used in exact surface calculation procedures;

(2) stable estimation of reflection coefficients by deconvolution; and

(3) thresholding techniques used in other estimation techniques.

This inversion algorithm and its implementation on a PC-based ultrasonic data acquisition system will be presented. It will be applied to impedance profile estimation on a graphite/epoxy - titanium adhesive bond.

A limitation of the high resolution algorithm mentioned above is the requirement to have complete information concerning all reflections from the material. For example, it is often not possible to include the reflection from the front surface of the material because this will severely limit the amount of information obtained from lower interfaces. Given sufficient *a priori* information concerning the impedances and approximate positions of all interfaces, an optimization based approach may be used.[10,12] This optimization approach and the application to imaging through reverberant interfaces will also be described.

2. High Resolution Inversion

2.1. FORWARD MODEL

Prior to any further consideration of the inverse problem, it is first necessary to develop an adequate forward model for the propagation of a normally incident plane wave through a layered medium. A particular solution to this problem has been previously developed by Bube and Burridge.[5] As shown in Figure 1, this solution assumes that the up-going and down-going wave components are represented by two characteristic variables u and d:

Time t

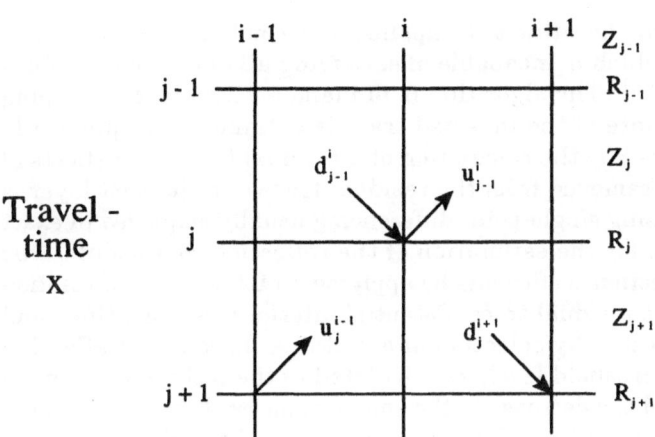

Figure 1

Schematic diagram showing up- and down-going wave components

$$d = 1/2\ (p + Zv)$$
$$u = 1/2\ (p - Zv)$$

where p is the pressure, v is the velocity and Z is the impedance of the medium. Assuming continuity of p and v continuous across an interface, it may be shown that the following set of relations are valid:[11]

$$d_1 + u_1 = d_2 + u_2$$

$$(1/Z_1)\,(d_1 - u_1) = (1/Z_2)\,(d_2 - u_2)$$

and
$$R_1 = (Z_2 - Z_1)\,/\,(Z_2 + Z_1)$$

where the subscripts 1 and 2 represent the two discrete layers and R_1 is defined as the downward reflection coefficient. By discretizing this continuous model using the convention of Goupillaud,[13] the up- and down-going wave components may then be given in terms of the reflection and transmission coefficients at each interface:

$$u_j^i = R_j\ d_j^i + T_j'\ u_{j+1}^{i-1}$$

$$d_{j+1}^{i+1} = T_j\ d_j^i + R_j'\ u_{j+1}^{i-1}$$

2.2 ALGORITHM

The high resolution inversion algorithm to be applied here combines components of three methods, any one of which is incapable of satisfying all of the criteria for a useful inversion technique.[10,11] The algorithm is fundamentally a layer stripping approach in which an estimate of the inverted trace is obtained by sequentially estimating each layer followed by the calculation of a residual trace. The effects of each layer are sequentially removed from the residual trace and the next layer is then estimated. Instead of using single-point differencing usually employed in exact layer stripping calculations for the estimation of the reflection coefficients, this algorithm estimates the reflection coefficients by applying an L2 norm deconvolution within a limited region of the residual trace. Potential interfaces are identified and then either accepted or rejected by selection of a suitable threshold. Reflection coefficients must exceed a threshold level, as calculated in the technique given by Koltracht and Lancaster[6] and the decrease in the sum of squares obtained by adding the new reflection coefficient to the model must exceed a second threshold. The first threshold rejects small reflections caused by noise or misfit, while the second threshold suppresses the estimation of layers which do not significantly increase the fit to the trace.

Figure 2 shows a flow diagram for the operation of this algorithm. The residual trace is first initialized to the input trace and a search region is located. If no further interfaces can be identified, the algorithm terminates at this point. An L2 norm deconvolution is then applied within this limited region and a potential reflection coefficient, R_i, is calculated. If this reflection satisfies the current Koltracht-Lancaster threshold, it is added to the model and the revised model is used to calculate a forward-modelled trace, which is then used to form a new residual trace, $\mathbf{r}$. If the decrease in the sum of squares exceeds a threshold value, R_i is retained in the current model. If the sum of squares of $\mathbf{r}$ is less than a specified tolerance, then the algorithm exits, otherwise, a new search region is defined and the algorithm iterates.

2.3 IMPLEMENTATION

A schematic diagram of the data acquisition and signal processing system is presented in Figure 3. In brief, all acquisition and processing functions, including the ultrasonic pulser/receiver and a 100 MHz A/D converter, are controlled by an IBM 80386 PC-compatible computer. The inversion signal processing may be performed using a TMS 320C30 digital signal processor which is capable of carrying out the complete inversion of ultrasonic traces (256 data elements per trace) at a rate of approximately 60 milliseconds per trace. It is possible, therefore, to perform the inversion processing during the scanning operation, however, it is usually preferable to transfer the data to a larger computer for subsequent processing and display.

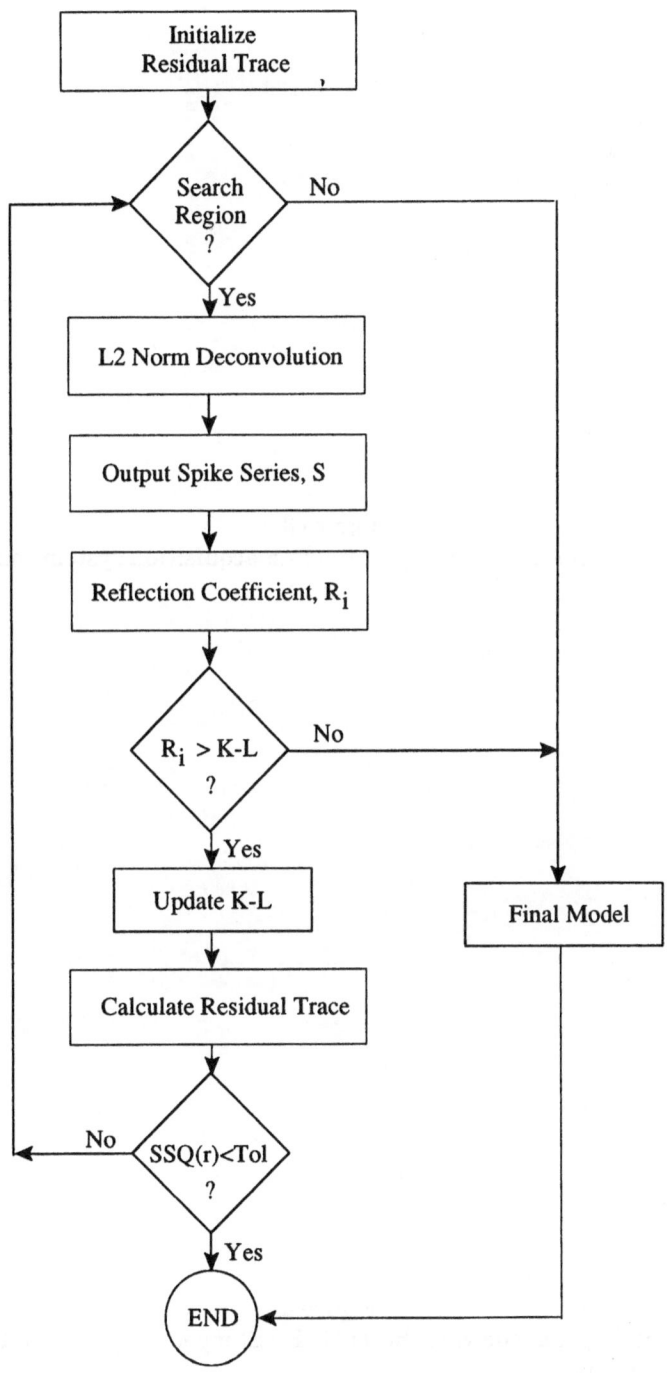

Figure 2
Flow diagram of the high resolution inversion algorithm.

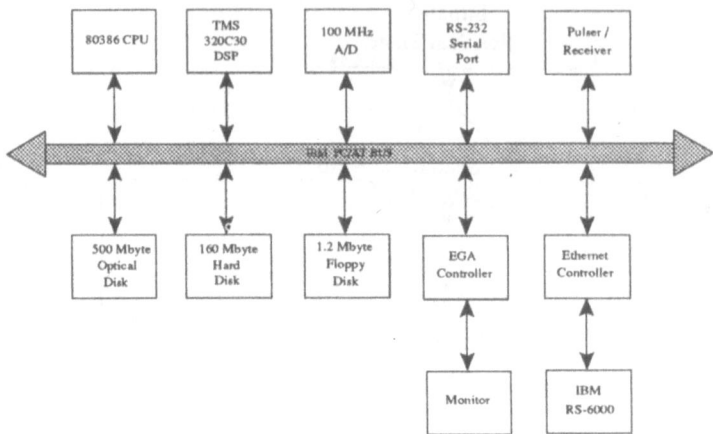

Figure 3
Schematic diagram showing the data data acquisition system configuration.

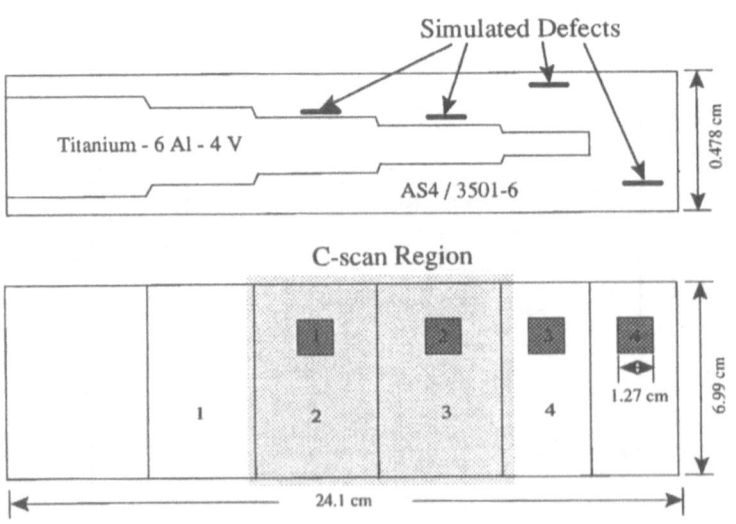

Figure 4
Schematic diagram showing the Ti-Gr/Ep Stepped lap joint and the region covered by the C-scan images.

2.4 APPLICATION TO ULTRASONIC IMAGING

In order to examine the application of this algorithm to realistic multi-layered bonded structures, pulse-echo ultrasonic data was first acquired from the Ti - Gr/Ep specimen shown in Figure 4. This specimen is a stepped lap joint which consists of a double-stepped wedge of Ti-6Al-4V bonded between two layers of AS-4/3501-6 graphite epoxy in a quasi-isotropic layup. The positions and depths of four Teflon inserts (1.27 x 1.27 cm) used to simulate disbonds and delaminations are also shown in Figure 4. In order to closely simulate actual flaws, the inserts include a thin layer of air (low acoustic impedance).

Figure 5 shows a single ultrasonic trace acquired from step level 2. The dotted line shows the calculated impedance profile. Starting at an initial value of relative impedance of 1.0 (relative to water, $Z = 1.484$ g/cm^2-sec), the detected reflection indicates a layer with a relative impedance value of 2.70, which corresponds well with the expected value of 2.87 for Gr/Ep. At the next interface, the calculated acoustic impedance rises to a value of 19.77 (actual value 18.78 for titanium) and then falls to a final value of 2.88, once again corresponding to Gr/Ep. Typically, the back surface of the specimen (Gr/Ep - water interface) could not be detected.

Reflection coefficient C-scan images may be formed by sequentially acquiring and processing individual traces in a raster pattern. Ultrasonic data representing a volume of the specimen was acquired from the region indicated by the shaded region of Figure 4. These data contain the individual A-scans (256 elements per trace sampled at a rate of 25 MHz) which were acquired at each point of a 256 x 256 element raster. The C-scan image obtained from these data in the absence of any

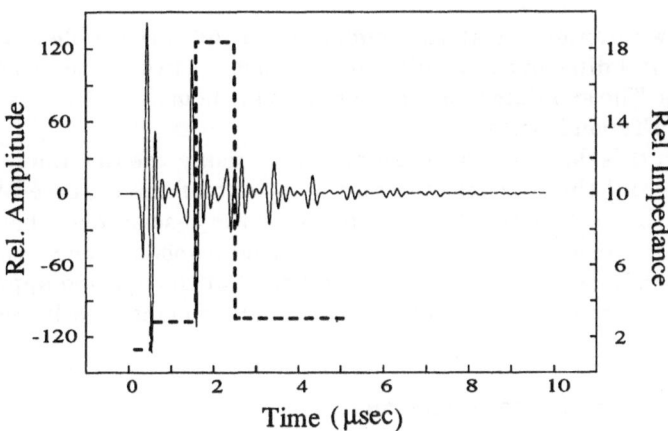

Figure 5

Diagram showing single A-scan and corresponding calculated impedance profile.

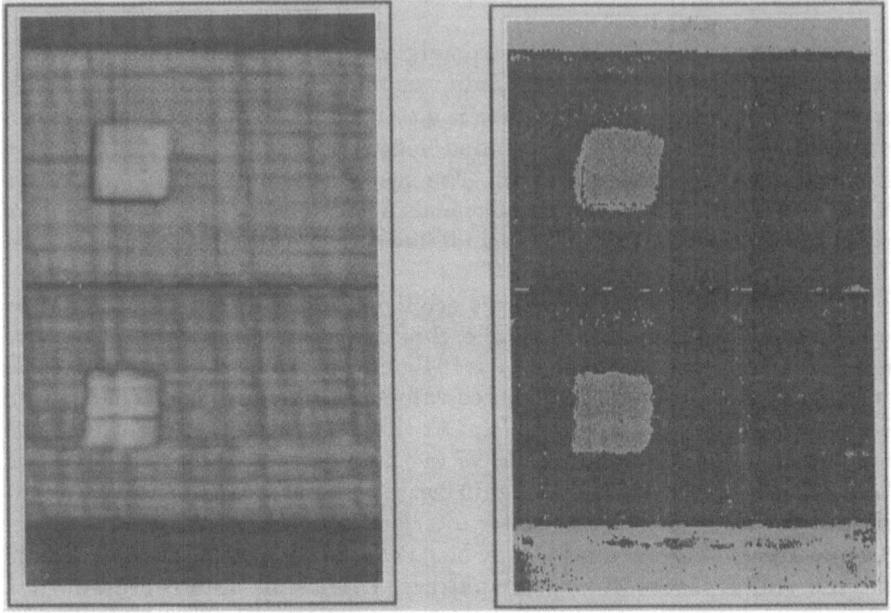

Figure 6
(a) Conventional amplitude C-scan image showing steps 2 and 3
(b) Reflection coefficient C-scan of the same data shown in (a)

processing is shown in Figure 6(a). Although the disbonds are visible in this image, there is only a small difference in reflection intensity between the disbonded and good bond regions. The simulated disbonds are also made more distinct by scattering at the edges of the Teflon inserts.

Each A-scan in this data set was then processed using the inversion algorithm described above and the magnitudes and positions of the detected reflection coefficients were then calculated. Figure 6(b) shows the C-scan image based on this inversion result. Because the reflection coefficient for a good adhesive bond is high and positive ($R = 0.73$) while the reflection coefficient at the disbond approaches -1, the effective dynamic range between good bond and bad bond is greatly increased.

3. Optimization Approach to Inversion

In the event that adequate *a priori* information is available or can be assumed, an improved inversion result may be obtained using the optimization algorithm originally proposed by Powell.[10,14] The model trace is parametrized in terms of (continuously variable) impedances z_j, j= 1,M and (discrete) thicknesses h_j, j = 1,M-1, where M is the number of layers. Given an initial estimate for $\mathbf{z}$ and $\mathbf{h}$, and fixed values for the attenuations in the first M-1 layers, the algorithm optimizes the fit to the measured trace, i.e. it optimizes:

$$\mathbf{f}(\mathbf{z},\mathbf{h}) \;=\; \sum_{n=N_1}^{N_2} [d_n - \mathbf{t}_n(\mathbf{z},\mathbf{h})]^2$$

where d is the measured trace, t is the model trace and N_1 and N_2 define the region of the trace to be used in the computation of the objective function. Upper and lower bounds on the impedances may also be specified for each layer.

Briefly, the algorithm proceeds in a series of stages, as follows. Given the initial estimate for the thicknesses, the impedances of all the layers are optimized subject to the bounds constraints. Then the thicknesses of the individual layers are sequentially incremented and decremented one unit at a time, with an optimization being performed at each set of thicknesses. The cycle of layer thickness adjustment continues until no adjustment can be made which results in a further decrease in the norm of the residuals. By this sequence of events, the optimal thicknesses and impedances of the layers are found.

3.1 APPLICATION OF INVERSION USING OPTIMIZATION

The Canadian Forces require a technique to detect surface defects, should they exist, on the surface of an aluminum alloy longeron. The inspection is complicated by the presence of a supporting strap (precipitation hardened stainless steel) which is fastened to the longeron. An intermediate adhesive sealant layer is used to inhibit corrosion and fretting damage. Each strap varies continuously in thickness from 0.038 cm at the outside edge to 0.160 cm at the centre. The adhesive sealant layer thickness is variable and unknown. The required inspection technique should be capable of detecting scratches 0.0127 cm - 0.0254 cm deep.

The difficulty of applying ultrasonic imaging to this inspection is illustrated schematically in Figure 7. The high reflectivity of both the stainless steel - sealant and stainless steel - water interfaces produces a sequence of multiple reflections within the upper steel layer. As a consequence, very little ultrasonic energy is transmitted through the sealant and subsequently reflected from the sealant - longeron interface. The sealant is itself highly attenuating, further reducing the total reflected energy signal from the longeron surface. The reflection from the steel - sealant interface and subsequent multiples totally dominates the data. In order to gain sufficient dynamic range for imaging purposes, it was necessary to commence acquisition after the front surface reflection. Reflections from the longeron surface are not readily detectable.

A specimen was manufactured in order to simulate the structure to be inspected. A series of grooves were machined into the surface of an aluminum plate at depths of 0.0127 cm, 0.0178 cm and 0.0254 cm, as illustrated in Figure 8. For each specific depth, a series of four grooves were machined with widths of 0.0228 cm, 0.051 cm, 0.160 cm and 0.239 cm. In order to simulate the presence of the steel strap, a small piece of 0.152 cm AISI 316 stainless steel was then bonded to the surface of the aluminum plate with adhesive sealant. No attempt was made to control the

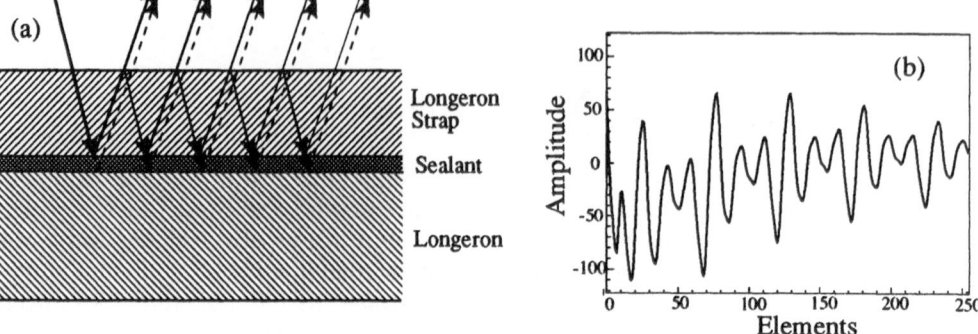

Figure 7
(a) Schematic diagram showing reflections from the strap, sealant and longeron.
(b) Typical A-scan from bonded structure showing multiple reflections.

thickness of the adhesive sealant bondline. The location and extent of the region for which ultrasonic data were obtained are also indicated in Figure 8.

Ultrasonic data were obtained by scanning the region noted above using a focused 5 MHz immersion transducer in pulse-echo mode. Individual A-scans were acquired at each point of a 256 x 256 raster. Each A-scan consisted of 256 elements, sampled at a frequency of 100 MHz and with 8-bit resolution. In this way, a three dimensional array of data which corresponds to the three dimensional volume of the specimen is obtained. An ultrasonic C-scan image of the adhesive bondline formed from these data is presented in Figure 9(a). None of the machined grooves are visible in this image. Clearly, it is necessary to eliminate the reflection response of the reverberant upper layer in order to obtain sufficient information to image the lower sealant - longeron interface.

One approach to the problem of elimination of multiple reflections from the ultrasonic image would be to perform a complete inversion of the ultrasonic data, i.e. to calculate either the impedance profile or, equivalently, the reflection coefficients of the individual interfaces. The high resolution algorithm described above could be applied to this problem, however, a limitation of this algorithm is that it requires all reflections to be included in the A-scan, including the large front surface reflection. Because the data acquisition system is restricted to 8-bit resolution, there is insufficient dynamic range to allow sufficient resolution for the reliable detection of the sealant - longeron interface. Although this algorithm has provided accurate and reliable inversion results for less reflective interfaces, no usable result was obtained from the application of this inversion algorithm.

In order to apply the optimization approach, the model was parametrized in terms of three layers: water, steel and sealant. The modelling procedure was applied to each trace in the data set, and the residual traces (containing information primarily from the surface of the aluminum longeron) were computed and displayed. The C-scan image formed from this residual data is shown in Figure 9(b). Although this technique cannot successfully image all of the simulated scratches on the longeron surface, it is now possible to at least detect most of the larger grooves.

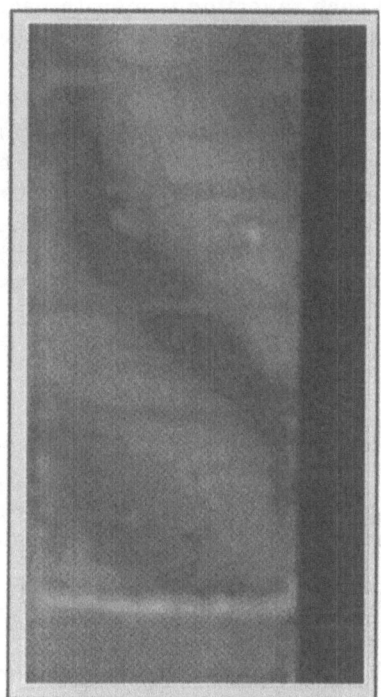

Figure 9
(a) C-scan obtained from region shown in Figure 8 from unprocessed data
(b) C-scan image formed from residuals following removal of the effects of the overlying steel and sealant layers.

4. Conclusions

The high resolution inversion algorithm meets all of the necessary criteria for successful application to nondestructive imaging. This inversion algorithm represents a stable method for the estimation of reflection coefficients at each interface of a planar, layered medium, such as bonded structures. Equivalently, the reflection coefficients may be converted and displayed as the impedance profile as a function of time.

The algorithm is still dependent, however, upon the correct selection of convergence tolerances and thresholds. The successful application of this inversion algorithm to NDE data, therefore, remains operator dependent. By the use of a floating point digital signal processor, ultrasonic traces may be inverted at a rate of approximately 60 milliseconds per trace.

An enhanced method for the identification of disbonds has been developed for the situation in which the bonded substrate has a high acoustic impedance. By imaging the estimated reflection coefficients, rather than the amplitude of the reflection, the polarity of each reflection is maintained. The reflection coefficient of disbonded regions approaches -1, whereas the reflection coefficient for the regions of "good" bond are large and positive.

An improvement in image quality was obtained if the ultrasonic C-scan image was formed from the residual data remaining after the elimination of the response of the reverberant upper layer. It is possible to estimate this residual with only limited *a priori* information using the optimization procedure described above. However, these residual based techniques did not provide truly satisfactory detection of the simulated "scratches" on the surface of the longeron test specimen because of the limited dynamic range of the A/D converter and the resulting low SNR of the residuals. An increase of the dynamic range of the data would be required to improve the success of these techniques.

5. References

1. M. Fatemi and A.C. Kak, "Ultrasonic B-Scan Imaging: Theory of Image Formation and a Technique for Restoration", Ultrasonic Imaging 2, 1980, pp. 1-47.

2. K.I. McRae and C.A. Zala, "Improved Axial Resolution of Ultrasonic B-Scans by L1-Norm Deconvolution", Review of Progress in Quantitative Nondestructive Evaluation, Vol. 7A, D.O. Thompson and D.E. Chimenti, eds., 1987, pp. 747-755.

3. K.I. McRae, T.L. Miller, C.A. Zala and I. Bailey, "Real-Time Super-Resolution Signal Processing Applied to the Ultrasonic Imaging of Adhesively Bonded Joints", Review of Progress in Quantitative Nondestructive Evaluation, Vol. 9, D.O. Thompson and D.E. Chimenti, eds., 1990, pp. 641-646.

4. B. Ursin and K.-A. Berteusen, "Comparison of Some Inverse Methods for Wave Propagation in Layered Media", Proc. IEEE, Vol. 74, 1986, pp. 389-400.

5. K.P. Bube and R. Burridge, "The One-Dimensional Inverse Problem of Reflection Seismology", SIAM Review, Vol. 25, 1983, pp. 497-559.

6. I. Kohltract and P. Lancaster, "Threshold Algorithms for the Prediction of Reflection Coefficients in a Layered Medium", Geophysics, Vol. 53, 1988, pp. 908-919.

7. J.M. Mendel and J. Goutsias, "One-Dimensional Normal Incidence Inversion: A Solution Procedure for Band-limited and Noisy Data", Proc. IEEE, Vol. 74. No. 3, 1986, pp. 401-414.

8. A. Bamberger, G. Chavent, Ch. Hemon and P. Lailly, "Inversion of Normal Incidence Seismograms", Geophysics, Vol.47, Number 5, 1982, pp. 757-770.

9. D.W. Oldenburg, S. Levy and K.J. Stinson, "Inversion of Band-Limited Reflection Seismograms: Theory and Practice", Proc. IEEE, Vol. 74, 1986, pp. 487-497.

10. C.A. Zala and K.I. McRae, "An Optimization Method for Acoustic Impedance Estimation of Layered Structures Using Prior Knowledge", Proceedings of the 18th International Symposium in Acoustical Imaging, Santa Barbara, 18 - 20 September 1989.

11. C.A. Zala, "High Resolution Inversion of Ultrasonic Traces", IEEE Trans. on Ultrasonics, Ferroelectrics and Frequency Controls, Vol. 39, No. 4, 1992, pp. 458-463.

12. K.I. McRae and C.A. Zala, "An Inversion Approach to Imaging Through Reflective Interfaces in Multi-Layered Structures", Review of Progress in Quantitative Nondestructive Evaluation, Vol. 12A, D.O. Thompson and D.E. Chimenti, eds., 1992, pp. 827-834.

13. P.L. Goupillaud, "An Approach to Inverse Filtering of Near-Surface Layer Effects from Seismic Records", Geophysics, Vol. 26, No. 6, 1961, pp. 754-760.

14. M.J.D. Powell, "TOLMIN: A Fortran Package for Linearly Constrained Optimization Calculations", DAMPT Report 1989/NA2, University of Cambridge, 1989.

THE EUROPEAN PROJECT TRAPPIST :
TRANSFER, PROCESSING AND INTERPRETATION OF 3D NDT DATA
IN A STANDARD ENVIRONMENT

B. GEORGEL
Électricité de France
Dir. des Études et Recherches,
dépt. Surveillance, Diagnostic, Maintenance
6 quai Watier
F-78400 CHATOU, FRANCE

C. NOCKEMANN
Bundesanstalt für Materialforschung
und -prüfung (BAM),

Unter den Eichen 87,
D-1000 BERLIN 45, DEUTSCHLAND

ABSTRACT. The European CEC-funded project TRAPPIST aims to provide the pre-requisites to combination of various NDT-methods. The key components to achieve this goal are a multi-method NDT standard data format and a platform-independent software environment. Another important feature is communication of both NDT data and expertise between remote workstations through state-of-the-art European ISDN broadband network. A full scale prototype is under development to demonstrate feasibility of this system. A survey on recent literature showing originality of the TRAPPIST features is included in the paper.

Introduction : the TRAPPIST consortium

The TRAPPIST consortium is born on the first of January, 1992 in the framework of the Commission of the European Communities (CEC) R&D program funding (RACE program for advanced communications in Europe). It aims at demonstrating the feasibility of transfering, processing, displaying and interpreting 3D NDT data through a standard environment.

It gathers the following companies and research institutions (see fig. 1) :

- Électricité de France, R&D Division (Paris, France),
- Bundesanstalt für Materialforschung und -prüfung (Berlin, Germany),
- Force Instituterne (Copenhagen, Denmark),
- the University of Strathclyde (Glasgow, United Kingdom),
- Deutsche Airbus (Bremen, Germany),
- Lufthansa and Philips (Hamburg, Germany),
- VTT (Helsinki, Finland).
- De Te Berkom (Berlin, Germany), project manager

X. P. V. Malague (ed.), Advances in Signal Processing for Nondestructive Evaluation of Materials, 449–455.
© 1994 Kluwer Academic Publishers.

450

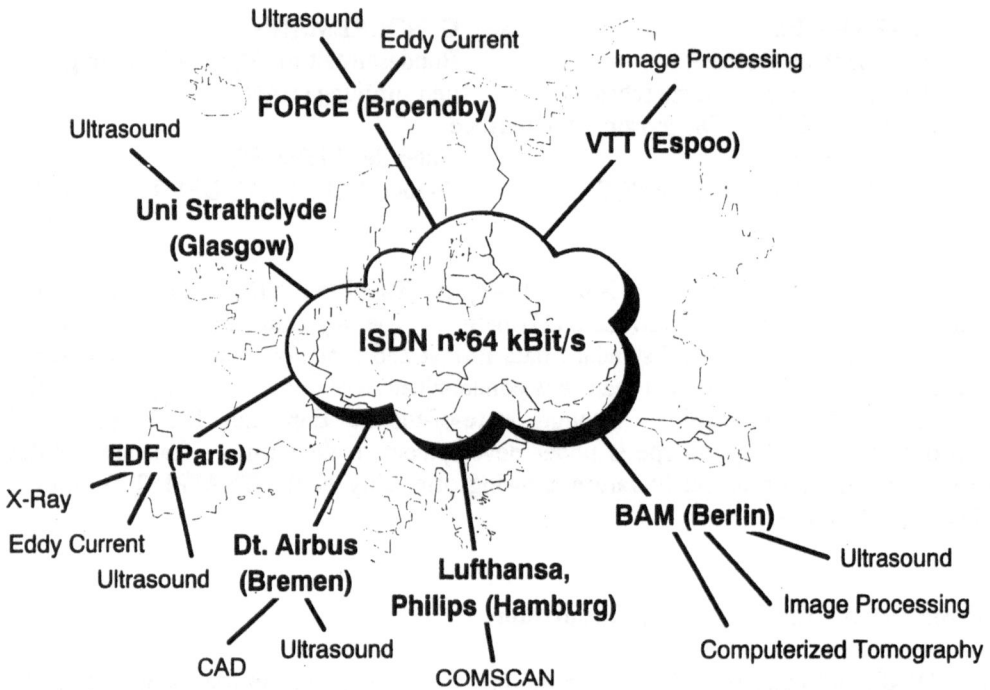

fig 1 : the partners of TRAPPIST

1. General description of the TRAPPIST prototype

The TRAPPIST prototype has been specified by the NDT experts of the different partners according to the SADT[1] methodology.

---

[1] SADT for Structured Analysis and Design Technique is a general method aimed to better communication between the designer and the user of a software. Originately recommended by the DOD of the USA, it is now widely used in all sectors of industry. It is mainly a graphical representation with boxes and links which are obtained by careful iterative review cycles between experts. It avoids misunderstandings and imprecision in designing a program.

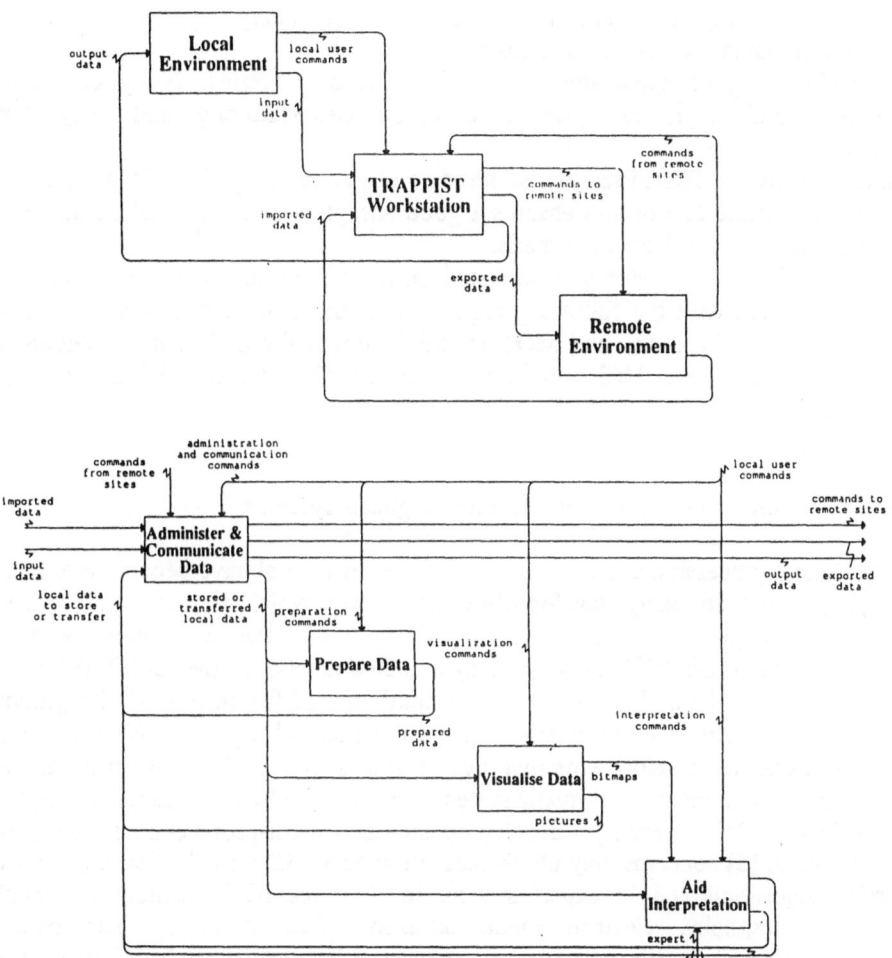

fig 2 : SADT block-diagram of the TRAPPIST prototype

It is decomposed into 4 main functional boxes (see fig.2) :
- administer and communicate data,
- prepare data,
- visualise data,
- aid interpretation.

The keywords of TRAPPIST are **communication, multi-method management, 3D, standard data format, inclusion of CAD data, data base management, image processing**.

The acquired data are geometrically transformed into a common spatial reference system and displayed using appropriate visualisation techniques. Dedicated 3D image processing and interpretation algorithms serve as an aid in restoring images and identifying component defects from the single and combined NDT information [6]. This method of combination can be applied with great benefit whenever a single NDT method is unable to detect a given defect with sufficient reliability or under very high safety requirements, such as in nuclear power plants and aerospace industry.

The TRAPPIST pilot application deals with ultrasonic testing, X-ray computerised tomography, multi-angle radiography, Compton backscattering and eddy current inspection.

Another challenge of TRAPPIST is also standardisation : participation of big NDT centers and customers in the consortium ensures a good acceptance of the standard data format and of many choices that have been made.

The TRAPPIST demonstrator will be tested on manufactured samples with flaws. It is under development with the following requirements and commercial products : UNIX[2] workstation with X-Window[3] and MOTIF[4], SET[5] format for CAD data, INGRES[6] data base management system, AVS[7] graphical system, SIMPA[8] signal and image processing software package.

2. State-of-the-art survey about automatic diagnosis systems

Although signal processing and computer techniques in general have become vital to NDT [1],[2],[3],[4], not so many developments are known which aim the same issues as TRAPPIST. A world wide literature survey, mainly based on the 13th Conference on NDT Proceedings (Brazil, Oct.,'92) show that only a limited number of the TRAPPIST features are addressed to by the authors (either 3D or multi-method fusion or CAD integration or data base management or aid to interpretation, etc but not all together). We will sum up a number of interesting developments that involve at least two of the above features (ie we won't comment on numerous automatic systems for radiography or ultrasonic testing).

The EXTRACSION software [5] is being updated in a multi-probe eddy current system. The testbench of [6] concerns only ultrasonics but includes 3D visualisation and many data processing algorithms and an expert system. In [7] three NDT methods are involved (ultrasonic holography, electromagnetic ultrasonic flaw detector, linac computer tomography scanner), in 3D, with use of a 'bibliographic' data base. In [8] two types of sensors are accepted as inputs : X-ray films and real time video. Automatic classification is implemented, together with knowledge based management. A 3D eddy current inspection system is described in [9]. The UltraSIM system of [10] combines CAD and 3D-simulation/display modules for ultrasonic inspection preparation. SUMIAD [11] covers all the range from acquisition to display of ultrasonic signals. In [12] not only 3D and CAD are included, but also data bases for US testing complex samples.

---

[2] UNIX is a trademark of ATT
[3] X-Window is a trademark of MIT
[4] MOTIF is a trademark of OSF
[5] SET is a CAD standard format
[6] INGRES is a trademark of INGRES Corp.
[7] AVS is a trademark of Stardent Computer Inc.
[8] SIMPA is propriety of CNRS Groupement de recherche TdSI, Paris, France

3. Standard data format and format conversions

A common format enabling to manage data coming from different 1, 2 or 3-D NDT techniques is the central part of TRAPPIST. This format include not only the measured values (for example A-scans) but also numerous parameters describing the context (apparatus, probe, experimental set up, scanning process, etc).
Different converters from and to the formats presently used by the partners have been realised.

4. Data Base Management System in TRAPPIST

Once the data have been acquired, a crucial issue is to ensure and to guarantee an efficient management and access to them in the TRAPPIST workstation. Moreover, TRAPPIST is dealing with heterogeneous data and broad variety of information such as material description, acquisition and sensor characterisation.
In this framework, it has been decided to use a relational DBMS in order to achieve a fast and reliable access to any useful piece of information, and to extract NDT-meaningful data value subsets.
On the one hand, INGRES and WINDOWS 4 GL (its man-machine interface generator) give you information answering to a broad number of more or less sophisticated queries. The queries are links with the various actions performed during an NDT diagnosis stage (to know the past acquisitions on an object, to compare the acquisition performed with the same probes...). The queries can be one of the following questions :

- what are the data corresponding a name of an inspected object ?
- what are the acquisitions using the computer tomography technique ?
- give me the ultrasonic data values acquired by Mr. Dupont in 1992 with the probe number STX23.

In fact the key point of the TRAPPIST DBMS is the inspected object which should be unique. CAD drawing, NDT data values and various acquisition parameters are attached to the object. The results of processing and even the reconstruction performed are plugged to the object. Any processed data is associated with a description of the processing history, allowing reiteration of the overall chained processes.
In the DBMS, only the information useful for the queries is stored in conjunction with links. These links enables us to find the rest of the information which is stored elsewhere using the TRAPPIST standard format. It avoids overloading the DBMS system with vast amount of data.
On the other hand, a second data base is in charge of managing the scripts available in the TRAPPIST environment. What is a script in TRAPPIST ? It is a chained list of actions in the TRAPPIST environment. Let us consider the simple script called "visualise 3D reconstructed data". This script will first call the DBMS of NDT data so that you will be able to send your queries. Then, it will call an adapted visualisation module.

The great advantage of having the script stored in a database is an easy calling mode. It is also easy to update the script, add new ones. Moreover, the scripts list can be adapted, changed according to the use of the TRAPPIST environment without any problem.

5. Image processing

Classical image processing (1 and 2-D) will be performed by the routines in SIMPA through a 'toolbox' mechanism. More sophisticated 3D routines are under development by the Consortium. They include averaging, median and low pass filtering, histogram stretching, edge detection, segmentation (split and merge, histogram relaxation, flood filling, texture analysis) and measuring algorithms.

An important feature of the architecture designed is its versatility : new routines can be incorporated easily at any time.

Some compression (LZW, JPEG) will be carried out before communication.

6. Visualisation

It is important to provide to the end user easy to interpret graphics. Several features are under development in the prototype : change of point of view, adjustment of light sources, slicer, excavation, wrapping and elevation, ray-tracing, isosurface, and also combined visualisation (ie visualisation of a combination of two or more images).

Conclusion

A demonstration software named TRAPPIST is under development in the framework of the CEC RACE program. It aims at efficient communication, processing and visualisation of NDT data stored in a standard format. 3D, CAD and combined display are the keywords of TRAPPIST. The partners involved in the project believe that wide acceptance of the format and of the choices made is possible and it will enhance reliability and productivity of in service inspection in a near future.

Aknowledgements

The authors wish to thank Mr. E. Fleuet and Mr. A. Schumm (from BAM) for contributing to this paper.
They also thank all the partners of the project and the CEC for funding it.

References

[1] B. Georgel, "Digital Signal Processing for NDT", 13th World Conference on NDT, São Paulo, Brazil, October 92, C.Hallai and P. Kulcsar eds., Elsevier.

[2] NATO ASI series F, vol.44, " Signal Processing and Pattern Recognition in Nondestructive Evaluation of Materials ", Québec, Canada, August 87, C.H. Chen ed., Springer Verlag.

[3] NDT International, " Special issue on Digital Signal Processing ", vol.19 n°3, June 86, Butterworths.

[4] B. Raj, " Reliable solutions to engineering problems in testing through acoustic signal analysis ", 13th World Conference on NDT, São Paulo, Brazil, October 92, C.Hallai and P. Kulcsar eds., Elsevier.

[5] B. Georgel, R. Zorgati, " EXTRACSION : a system for automatic Eddy Current diagnosis of steam generator tubes in nuclear power plants ", 13th World Conference on NDT, São Paulo, Brazil, October 92, C.Hallai and P. Kulcsar eds., Elsevier (see also Zorgati, Georgel, Duvaut, Doligez, Garreau in Actes du colloque GRETSI, Juan-les-Pins, septembre 93, —in French—).

[6] A. Mac Nab, I. Dunlop, " Advanced visualisation and interpretation techniques for the evaluation of ultrasonic data : the NDT workbench ", British Journal of NDT, vol.35 n°5, May 93.

[7] S. Miyoshi et alii, " Development of defect evaluation system in JAPEIC ", 1993.

[8] G. Montini et alii, " EXACT : an expert system for real time X-ray image analysis and classification of weldings ", 13th World Conference on NDT, São Paulo, Brazil, October 92, C.Hallai and P. Kulcsar eds., Elsevier.

[9] T. Shiraiwa et alii, " Three dimensional eddy current inspection system ", id p366.

[10] K. Jellensen, " UltraSIM ", id p901.

[11] R. Martínez-Oña, " SUMIAD : a workstation for ultrasonic inspection ", id p959.

[12] H. Matsumura, " Development of a 3D ultrasonic examination system for complex geometry ", id p969.

OPTICS AND INFORMATION PROCESSING AT NOI:
A GOOD MATCH FOR NDE[1]

DENIS GINGRAS
Information processing
National Optics Institute
369, Franquet, Sainte-Foy, Qc (Canada)
G1P 4N8

In this presentation, I will provide an illustrative overview of some research projects going on at NOI, which are relevant to NDE. Areas covered in particular will be in information processing and in optical metrology. Example of projects presented will include various optical machine vision systems applied to industrial inspection, a fringe pattern analysis software applied to Moiré and holographic interferometry and some of the signal processing methods such as classifiers in optical spectrometric instrumentation and in remote sensing multi-spectral images.

A. Bergeron, H. H. Arsenault, J. Gauvin, D.J. Gingras, "Computer-generated holograms improved by a global iterative coding," *Opt. Eng.* **32** [9]: 2216-2226, 1993.

Éric Harvey, Michel Bouchard, Pierre Langlois, "Holographic interferometry and Moiré deflectometry for visualization and analysis of low gravity experiments on laser materials processing," *Opt. Eng.* **32** [9]: 2143-2155, 1993.

[1]this contribution: abstract only

X. P. V. Malague (ed.), Advances in Signal Processing for Nondestructive Evaluation of Materials, 457.
© 1994 *Kluwer Academic Publishers.*

NON-INVASIVE MEASUREMENT OF TEMPERATURE CHANGES IN TETHERED FLYING BLOWFLIES BY THERMAL IMAGING

D.G. STAVENGA[‡], J. TINBERGEN[*], P.B.W. SCHWERING[†]
‡Department of Biophysics, University of Groningen,
Nijenborgh 4, NL-9747 AG Groningen,
*Audiological Institute, University of Groningen,
Oostersingel 59, NL-9713 EZ Groningen,
†TNO Physics and Electronics Laboratory,
P.O. Box 96864, NL-2509 JG the Hague,
the Netherlands

ABSTRACT. The changes in temperature occurring in the body of a flying blowfly are measured with the non-invasive technique of thermal imaging. It is found that during rest the temperatures of the three main body compartments, i.e. head, thorax and abdomen, approximately equal the ambient temperature. Upon flight onset the thorax temperature increases about exponentially, with a time constant $\approx$ 30 s. In steady flight, the thorax temperature is $\approx$ 5 °C higher than the ambient temperature ($\approx$ 25 °C). After flight, the temperature of the thorax decreases, again about exponentially, with a time constant of $\approx$ 50 s. A three compartment model of the insect body allows a quantitative description of these temperature changes, thus yielding values for the blowfly's thermal parameters.

1. Introduction

The temperature of an insect's body can strongly depend on its activity, notably in the larger insects, like the bumblebee or the locust (Heinrich 1979, 1993). Usually, the body temperature of an insect is measured with thermistors or thermocouples, but this invasive approach has some obvious drawbacks, especially in small insects. The 'grab and stab' technique is damaging to the animal, the spatial resolution is rather limited, and the readings have a systematic error (e.g. Stone and Willmer 1989). Furthermore, when this technique is applied after the insect is caught, individual variability complicates the analysis (e.g. Morgan and Heinrich 1987). An attractive alternative approach (though substantially more expensive) is the non-invasive technique of temperature measurements with a thermal imaging camera.

Thermal imaging of insects has been successfully applied to honey bees by Cena and Clark (1972), Schmaranzer (1983), Schmaranzer and Stabentheiner (1988) and Stabentheiner and Schmaranzer (1987, 1988), thus demonstrating that the infrared radiation emitted by the tiny body of an insect can be accurately monitored in time with considerable spatial resolution (Heinrich 1993).

Here, thermal imaging is applied to investigate the thermal effects of blowfly flight: the energetic performance of the flight muscles in the thorax is accompanied by heat production that can be visualized and quantitatively measured in a non-

459

X. P. V. Malague (ed.), Advances in Signal Processing for Nondestructive Evaluation of Materials, 459–467.
© 1994 *Kluwer Academic Publishers.*

invasive way. The experimental results are interpreted with a simple model that treats the insect body as consisting of three compartments, i.e., head, thorax and abdomen.

2. Materials and Methods

2.1. TETHERED FLYING BLOWFLIES

Blowflies (*Calliphora vicina*) were tethered by glueing a V-shaped aluminium wire to a small drop of wax put at the junction between head and thorax; the wire was clamped in a holder at a stand. A blowfly, thus tethered dorsally, remains at rest when it is offered a substratum, e.g. a small ball of paper tissue. Taking away the substratum releases flight (Fig. 1A), and this results in warming-up. The reverse process, end of flight and cooling down, was induced by reoffering the substratum.

2.2. THE THERMAL IMAGING SYSTEM

The temperature changes occurring during and after flight were monitored by adjusting stand plus fly so that the fly was in focus of a thermal imaging camera system. This was a Barr & Stroud IR-18 equipped with a telescope. The background, behind the fly, consisted of a temperature controlled plane, adjusted at 24.5 °C (approximately the ambient temperature). Adjacent to the fly, also in the focal plane, was another temperature controlled plane (Fig. 1B-F, left); this was set at 25.7 °C. These temperatures were chosen in the range of the fly body temperature of the experiment, and allow for a two-point calibration of temperature of the infrared images. Offset and gain of the IR-18 were chosen to avoid saturation in the 8 bit images. The measured apparent temperatures can be transformed into body temperatures when the emissivity of the fly body for infrared radiation is known. Because Stabentheiner and Schmaranzer (1987) found an emissivity of 0.994 ± 0.028 for the bee, we assume an emissivity of 100%.

The IR-18 images are CCIR compatible and represent intensities in the 7.3-10.7 μm band. This wavelength band contains the peak of the natural Planck function [$\lambda_{max}(\mu m)=2898/T(K)$]. The images were grabbed directly, sampling each 0.52 s, and analyzed with a Compaq 486/33L equipped with a DT2851 frame grabber (Data Translation). The data were stored at a Microtech OR650 optical disk system and processed further off-line. By using the characteristic spectral response curve of the IR-18 camera the radiance at the temperatures of the calibration planes is predicted. The absolute calibration is performed by comparing this with the actual measured bit levels on those planes. The final image of the camera (as those presented in Fig. 1) consists of 512x512 pixels, 8 bits deep; resolution 53 μm/pix horizontally and 36 μm/pix vertically with a total frame image of 27 mm x 18 mm. The temperature of the fly body compartments (Fig. 2) was calculated by taking the average of a selected area, as indicated by the boxes in Fig. 1E. R.A.W. Kemp wrote the software for grabbing and producing images.

2.3. A THREE COMPARTMENT MODEL FOR THE FLY BODY

We conceived a three compartment model for the fly body in order to quantitatively describe the measured temperature changes. The model is based on three assumptions: 1. heat production occurs in the thorax due to flight muscle activity; 2. heat loss occurs to the surroundings from all three compartments; 3. heat is transferred from the thorax to the head and to the abdomen. Formally:

$$\frac{dH_h}{dt} = -L_h + F_h \; ; \quad \frac{dH_{th}}{dt} = M - L_{th} - F_h - F_{ab} \; ; \quad \frac{dH_{ab}}{dt} = -L_{ab} + F_{ab} \quad (1)$$

or, equivalently:

$$W_h \frac{dT_h}{dt} = -C_h(T_h - T_a) + E_h(T_{th} - T_h) \quad (2a)$$

$$W_{th} \frac{dT_{th}}{dt} = M - C_{th}(T_{th} - T_a) - E_h(T_{th} - T_h) - E_{ab}(T_{th} - T_{ab}) \quad (2b)$$

$$W_{ab} \frac{dT_{ab}}{dt} = -C_{ab}(T_{ab} - T_a) + E_{ab}(T_{th} - T_{ab}) \quad (2c)$$

Here T_i, $i = h$, th, ab, a, is the temperature of head, thorax, abdomen and ambient; dH_i/dt is the rate of heat change of body compartment i; W_i is the heat capacity; $W_i = c_i m_i$, with c_i the specific heat and m_i the mass; M is the rate of heat produced by the thoracic muscles; C_i is the thermal conductance; E_h and E_{ab} is the thermal exchange coefficient of thorax to head and abdomen, respectively. The terms $L_i = C_i(T_i - T_a)$, $i = h$, th, ab, of Eqs 2a-c represent the rate of heat loss of the three compartments to the surroundings. The terms $F_i = E_i(T_{th} - T_i)$, $i = h$, ab, represent the rate of heat flux from thorax to head and abdomen, respectively.

Note that in the steady state the rate of heat production by the thoracic muscles is:

$$M = L_h + L_{th} + L_{ab} \quad (3)$$

3. Results

3.1 THERMAL IMAGING OF FLIES

Observation of a flying blowfly (Fig. 1A) with a thermal imaging camera immediately shows that the three main compartments of the body, i.e., head, thorax and abdomen, differ in temperature. Fig. 1B shows that at rest the heat radiation from the body compartments is about equal to that of the surroundings, indicating that the temperature of head, thorax and abdomen equals the ambient temperature. Figs 1C-F are images obtained at $t = 15$ s, 30 s, 60 s and 90 s after flight-onset. Indeed, the thermal imaging shows that upon initiation of flight, the temperature of the thorax rapidly increases. This is followed by a more moderate radiation increase from both the head and abdomen.

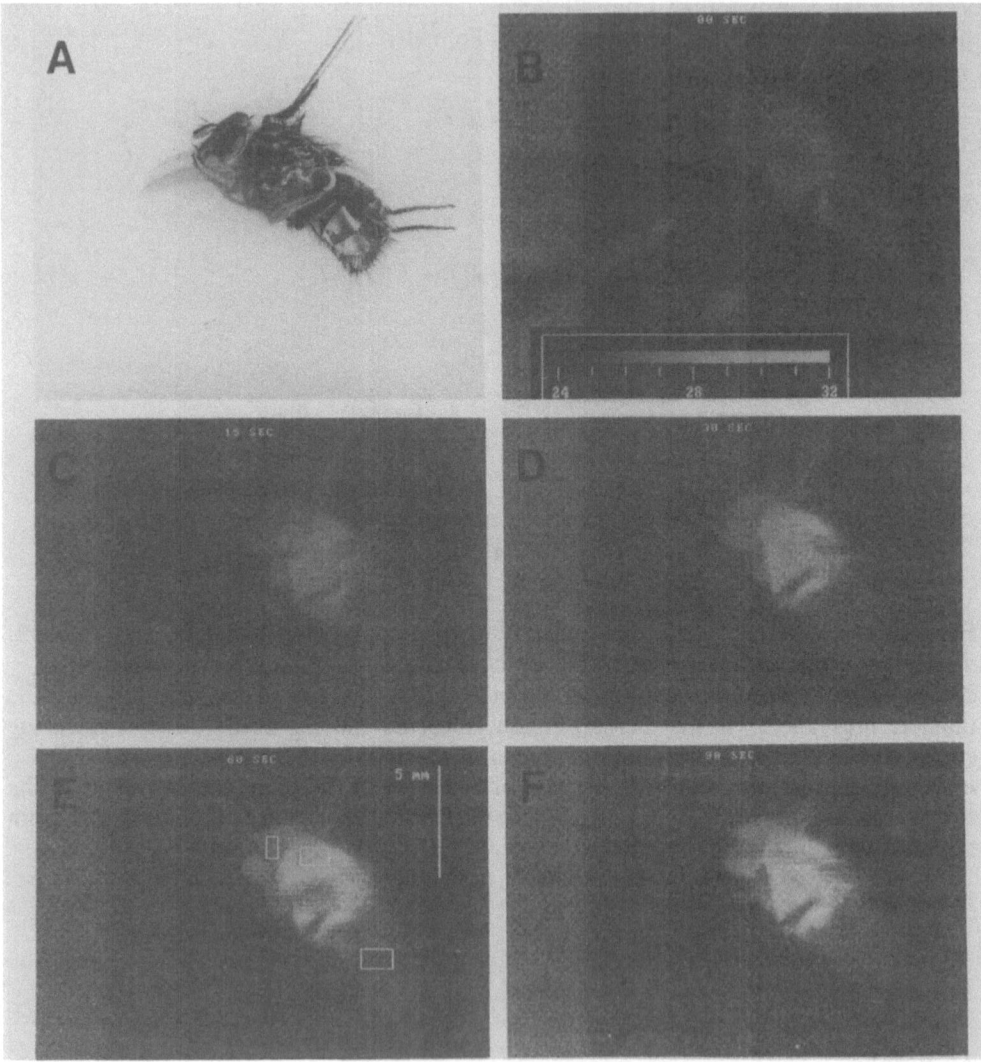

Fig. 1. *Photograph of a tethered-flying blowfly* (**A**) *and thermal images at flight onset, t = 0 s* (**B**), *and at t = 15* (**C**), *30* (**D**), *60* (**E**), *and 90 s* (**F**), *respectively. The tether, i.e. the aluminium wire, is seen upwards from the thorax. The background, behind the fly, consisted of a temperature controlled plane, adjusted at 24.5 °C. Adjacent to the fly was another temperature controlled plane, set at 25.7 °C. The grey scale in* **B** *indicates the temperature in °C. The rectangles in* **E** *indicate the areas sampled for obtaining the temperatures of the three body compartments. Note that the legs obstruct the radiation emitted by the thorax.*

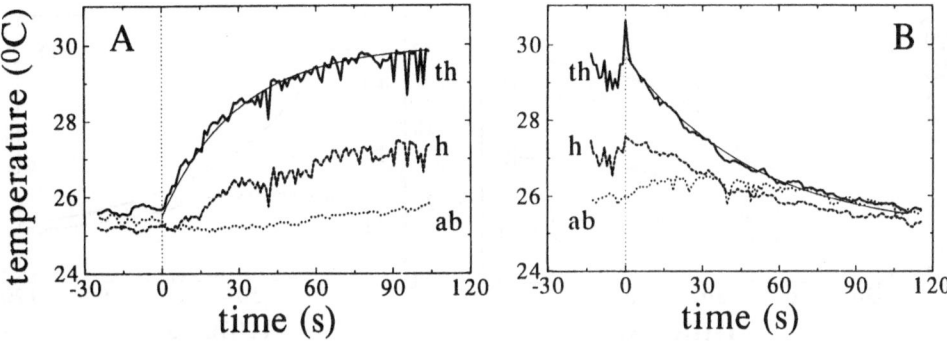

Fig. 2. Time course of temperature of thorax (th), head (h) and abdomen (ab) showing warming up during flight (A) and cooling down after flight (B). Thin lines are simple exponential functions with time constant 30 s (A) and 50 s (B).

Quantitative evaluation of the thermal images yields the time course of the temperature changes. The three body compartments were sampled in each frame as indicated by the rectangles in Fig. 1E. Fig. 2A demonstrates that the temperatures of the three body compartments in the resting state are close to the ambient temperature of 25.0 °C. Flying causes the thoracic temperature to rise to ≈ 30 °C in steady flight. The time course approximates a simple exponential function with a time constant of 30 s (thin line in Fig. 2A). An exponential is to be expected from a body that is suddenly heated by a constant source and that loses heat by thermal contact with a surrounding medium, i.e. the air.

Fig. 2A further shows that the temperature of head and abdomen also increase. The latter body parts do hardly produce heat themselves and thus must be heated indirectly; the thorax hence must transfer heat to the head and abdomen. For, the temperature time course of both head and abdomen more or less approximate an exponential with a similar time constant of about 30 s; a delay has to be included, however.

At the end of flight, the temperatures of the fly's body compartments fall back to the ambient temperature (Fig. 2B). The time course of the cooling phase of the thorax again has an approximately exponential shape, but the process is distinctly slowed down, as the time constant now is about 50 s (thin line Fig. 2B). The time courses of head and abdomen indicate that heat transfer from the thorax continues during the cooling phase.

3.2. MODELING THE TEMPERATURE DISTRIBUTION IN A BLOWFLY

The findings from the thermal imaging experiments have been fitted with the three compartment model described above. The heat capacities of head, thorax and abdomen can be calculated from their mass, measured as m_h = 5.0 mg and m_{th} = m_{ab} = 40 mg (average) and the specific heat (c_h = c_{th} = c_{ab} = 3.4 J g^{-1} K^{-1}, following Nachtigall et al. 1989 and Goller and Esch 1991), yielding W_h = 17 mJ K^{-1} and W_{th} = W_{ab} = 136 mJ K^{-1}.

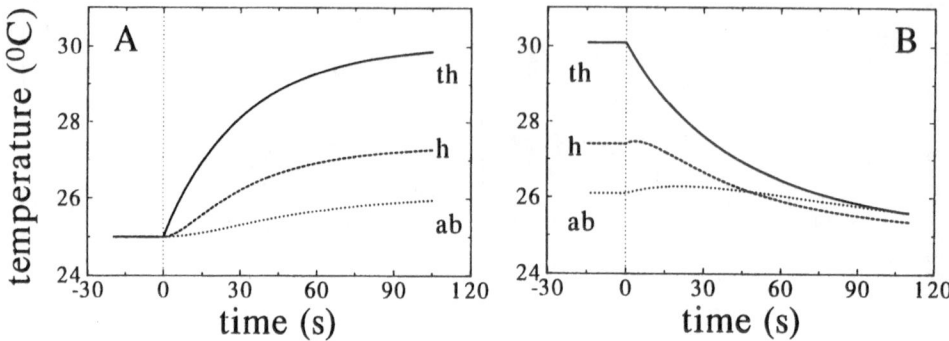

Fig. 3. Calculated time courses of the temperature of the three body compartments similar to Fig. 2 and using Eqs 1 and 2.

Assuming that the thermal exchange coefficients of thorax to head and abdomen are equal, i.e., $E_h = E_{ab}$, and that the heat power produced by the thoracic muscles, M, is a constant in the limited temperature range of our experiments, yields a quite satisfactory fit to the warming-up curves of Fig. 2A with a produced heat power $M = 1.38$ J min^{-1} = 23.0 mW, thermal conductances $C_h = 75$ mJ min^{-1} K^{-1} = 1.25 mW K^{-1}, $C_{th} = 183$ mJ min^{-1} K^{-1} = 3.05 mW K^{-1} and $C_{ab} = 244$ mJ min^{-1} K^{-1} = 4.07 mW K^{-1}, and thermal exchange coefficients $E_h = E_{ab} = 67$ mJ min^{-1} K^{-1} = 1.12 mW K^{-1}. Implementation of these values in Eq. 1 yields Fig. 3A.

The distinct difference of the cooling-down curves of Fig. 2B with respect to the curves of the warming-up phase can be readily interpreted by assuming slight changes in the thermal conductances. Whereas the thorax heats up with a time constant of ≈ 30 s, it cools with a time constant of ≈ 50 s. This indicates that the thermal conductance of the thorax has dropped by a factor of 0.6, or, after flight $C_{th} = 1.83$ mW K^{-1}. Fig. 2B then can be approximated by keeping the exchange coefficients constant and taking $C_h = 1.00$ mW K^{-1} and $C_{ab} = 1.63$ mW K^{-1}; or, the thermal conductances of head and abdomen after flight have decreased by a factor of 0.8 and 0.4, respectively, yielding Fig. 3B, closely corresponding to Fig. 2B.

Consequently, in the steady state after the warming-up phase, the heat loss terms are $L_{th} = 15.6$ mW, $L_h = 3.0$ mW and $L_{ab} = 4.4$ mW, so that the total heat loss equals the produced heat power, M (Eq. 3). The heat exchange from thorax to head and abdomen in the steady state, of course, is $F_h = L_h = 3.0$ mW and $F_{ab} = L_{ab} = 4.4$ mW, respectively.

4. Discussion

Non-invasive measurements of the body temperatures of tethered-flying blowflies can be successfully performed with a thermal imaging camera system. Similar experiments on tethered flying bees by Stabentheiner and Schmaranzer (1988) demonstrated that the temperature increase of the thorax during flight in still air is ≈ 8 °C above an ambient temperature of ≈ 27 °C; slightly higher than the 5 °C increase of the blowfly.

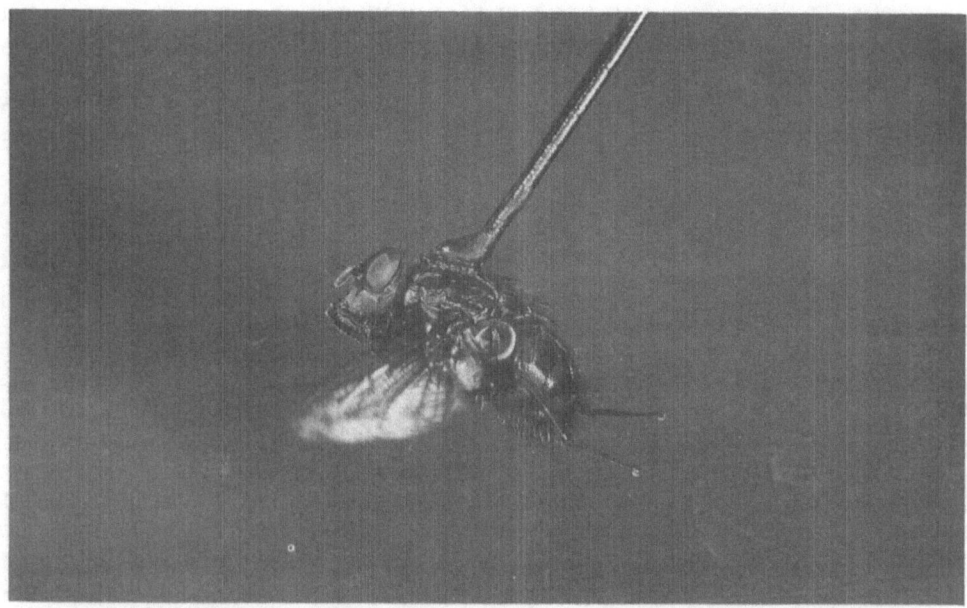

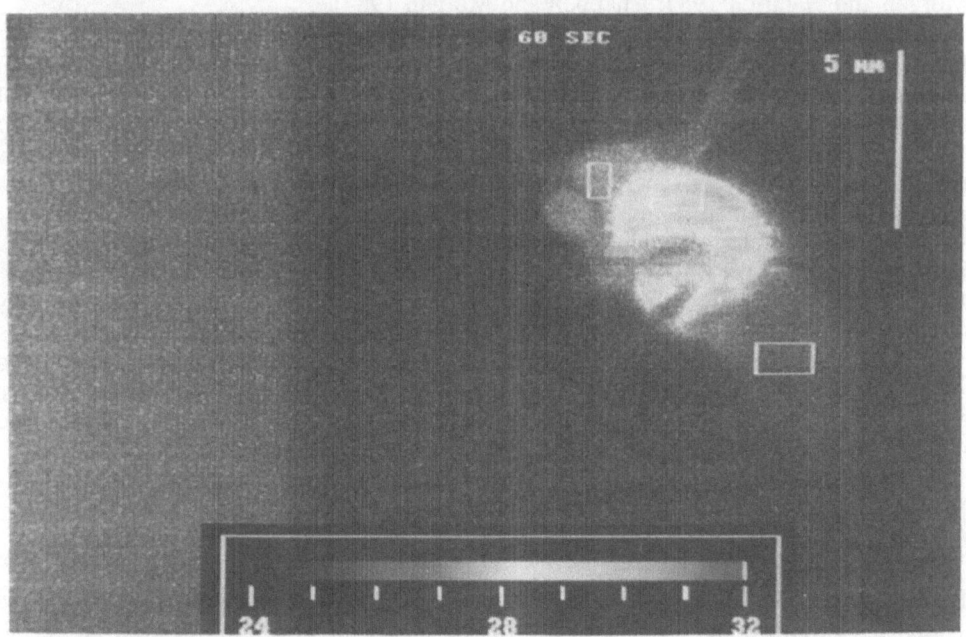

A tethered flying blowfly (top) is heated up due to the activity of the flight muscles. As shown by the thermogram (bottom), after 1 min flight the thorax (middle compartment) is heated up to about 5 °C above ambient. Due to the circulating blood flow also the head and (to a lesser extent) the abdomen have an increased temperature.

Strong air currents around the fly's body, similar as occur in normal flight, are produced by the wings during tethered flight. Actually, by assuming that the thermal conductances are suddenly reduced at the end of flight we could simply explain the slower time course of the cooling process. We have assumed that after flight the thermal conductance of head, thorax and abdomen decreased by a factor of 0.8, 0.6 and 0.4, respectively; in other words, the effect of air currents during flight is assumed to be progressively stronger from front to back.

The experimental traces of Fig. 2 show that the body temperatures are not necessarily equal to the ambient temperature, even when the fly is at rest. Indeed, the thoracic muscles of flies can be active without producing wing movements and flight, as occurs in other insects (Surholt et al. 1990; Goller et al. 1991). Inevatibly, or often purposefully (Heinrich 1974, 1981), this causes an increase in temperature. In the present modeling we have considered a case where the muscular activity is negligible at rest. Of course, it is a straightforward matter to incorporate pre-flight metabolic activity in our model. All the same, our preliminary results indicate that the derived values are representative and compatible with related data for other insects (e.g. Digby 1955; Morgan and Heinrich 1987; Nachtigall et al. 1989).

Bees can heat up their thorax endothermically, without flight, almost 20 °C above ambient (e.g. Heinrich 1980b; Stabentheiner and Schmaranzer 1988). One therefore can suspect that flies, when warming-up endothermically before flight, also will reach more elevated temperatures. Indeed, several examples of endothermic heating in flies are already known (e.g. Heinrich and Pantle 1975; Bartholomew and Lighton 1986; Morgan and Heinrich 1987; Chappell and Morgan 1987).

We have to note that heat production during tethered flight can be less than during free flight (e.g. Heinrich 1974). The measured temperatures therefore are perhaps not fully realistic for free flight. Furthermore, the air flow experienced during free flight substantially enhances the thermal conductance (e.g. Stabentheiner and Schmaranzer 1988; Nachtigall et al. 1989), and this will diminish the temperature excess. To assess the real temperature and energy relations it will be necessary to repeat the experiments with unrestrained flying animals.

The three compartment model of heat transfer in the fly body, presented here, heavily draws on previous work on other insects, especially moths and bees (e.g. Heinrich 1974, 1979, 1981; Casey 1992). So far, in formal treatments of body temperature and heat loss of insects, and in the experimental, quantitative estimations of the physical parameters involved in the thermal processes, the body is considered as a uniform physical body (e.g. Bartholomew 1981; Casey 1992), although the essential differences between the three body compartments have been recognized already for some time (e.g. Heinrich 1980a,b).

5. References

Bartholomew GA (1981) A matter of size: an examination of endothermy in insects and terrestrial vertebrates. In: *Insect Thermoregulation* (Heinrich B, ed.), pp 45-78. John Wiley & Sons: New York

Bartholomew GA, Lighton JRB (1986) Endothermy and energy metabolism of a giant tropical fly, *Pantophtalmus tabaninus* Thunberg. J Comp Physiol B 156:461-467

Casey TM (1992) Biophysical ecology and heat exchange in insects. Amer Zool 32: 225-237

Cena K, Clark JA (1972) Effect of solar regulation on temperatures of working honey bees. Nature New Biology 236:222-223

Chappell MA, Morgan KR (1987) Temperature regulation, endothermy, resting metabolism, and flight energetics of tachinid flies (*Nowickia* sp.). Physiol Zool 60:550-559

Digby PSB (1955) Factors affecting the temperature excess of insects in sunshine. J Exp Biol 32:279-298

Goller F, Esch HE (1991) Oxygen consumption and flight muscle acitivity during heating in workers and drones of *Apis mellifera*. J Comp Physiol B 161:61-67

Goller F, Esch HE, Heinrich B (1991) How do bees shiver? Naturwissenschaften 78:325-328

Heinrich B (1974) Thermoregulation in endothermic insects. Science 185:747-756

Heinrich B (1979) *Bumblebee Economics*. Harvard University Press: Cambridge Mass

Heinrich B (1980a) Mechanisms of body-temperature regulation in honeybees, *Apis mellifera*. I. Regulation of head temperature. J Exp Biol 85:61-72

Heinrich B (1980b) Mechanisms of body-temperature regulation in honeybees, *Apis mellifera*. II. Regulation of thoracic temperature at high air temperatures. J Exp Biol 85:73-87

Heinrich B (1981) Ecological and evolutionary perspectives. In: *Insect Thermoregulation* (Heinrich B, ed.), pp 235-302. John Wiley & Sons: New York

Heinrich B (1993) *The Hot-Blooded Insects. Strategies and Mechanisms of Thermoregulation*. Springer: Berlin

Heinrich B, Pantle C (1975) Thermoregulation in small flies (*Syrphus* sp.): basking and shivering. J Exp Biol 62:599-610

Morgan K, Heinrich B (1987) Temperature regulation in bee- and wasp-mimicking syrphid flies. J Exp Biol 133:59-71

Nachtigall W, Rothe U, Feller P, Jungmann R (1989) Flight of the honey bee. III. Flight metabolic power calculated from gas analysis, thermoregulation and fuel consumption. J Comp Physiol B 158:729-737

Schmaranzer S (1983) Thermovision bei trinkenden und tanzenden Honigbienen (*Apis mellifera carnica*). Verh Dtsch Zool Ges 76:319

Schmaranzer S, Stabentheiner A (1988) Variability of the thermal behavior of honeybees on a feeding place. J Comp Physiol B 158:135-141

Stabentheiner A, Schmaranzer S (1987) Thermographic determination of body temperatures in honey bees and hornets: calibration and applications. Thermology 2:563-572 (1987)

Stabentheiner A, Schamaranzer S (1988) Flight-related thermobiological investigations of honeybees (*Apis mellifera carnica*). In: *The Flying Honeybee* (Nachtigall W, ed), BIONA-report 6, pp 89-102. Gustav Fischer, Stuttgart

Stone GN, Willmer PG (1989) Endothermy and temperature regulation in bees: a critique of 'grab and stab' measurement of body temperature. J Exp Biol 143:211-223

Surholt B, Greive H, Baal T, Bertsch A (1990) Non-shivering thermogenesis in asynchronous flight muscles of bumblebees? Comparative studies on males of *Bombus terristris, Xylocopa sulcatipes* and *Acherontia atropos*. Comp Biochem Physiol 97A:493-499

SUBSTITUTING STATISTICAL FOR PHYSICAL DECOMPOSITION: ARE THERE APPLICATIONS FOR PARALLEL FACTOR ANALYSIS (PARAFAC) IN NON-DESTRUCTIVE EVALUATION?

RICHARD A. HARSHMAN
Department of Psychology
University of Western Ontario
London, Canada N6A 5C2

ABSTRACT. We describe the three-way factor analysis method PARAFAC (PARAllel FACtor analysis) and its possible application to Non-destructive Evaluation (NDE). Because standard analysis methods usually separate a signal into abstract, mathematically convenient parts such as frequency bands, orthogonal variance components, etc., irrelevant signals often remain mixed with important ones. In contrast, Parallel Factor Analysis separates mixtures into functionally distinct parts, i.e., parts showing distinct patterns of variation in magnitude across varying measurement conditions. It is used, for example, to decompose the mixtures of curves generated by fluorescence spectroscopy of complex samples into the individual spectra of the constituent chemical compounds in the mixture; this is feasible because each compound shows distinct patterns of variation in fluorescence intensity across varying stimulus frequencies. In NDE, PARAFAC could similarly separate the mixture of signals (or image patches) from an object under test into those arising from each causally/physically distinct component of interest in the object, provided that there are several parallel test conditions, or multiple time slices, where the signals from different components of interest show distinct patterns of variation in magnitude across the conditions. Various normal and anomalous signals can then be isolated onto distinct factors, allowing anomalous ones to be identified and linked to their physical sources.

1. Usage Characteristics of Parallel Factor Analysis

Parallel factor analysis can decompose a complex mixture of signals into components that correspond to *functionally distinct* parts of the system under study. It does so without benefit of an explicit model of the system, and so is complementary to (and possibly a useful adjunct to) the model- based methods of signal analysis discussed in many of the papers at this meeting.

1.1 AN EXAMPLE APPLICATION

To see how it works, we begin with an example—an implicit 'non-destructive testing' problem involving spectroscopic analysis of mixtures. Only recently have chemometricians realized that the trilinear model of parallel factor analysis provides a way to solve such mixture problems (e.g., Burdick et al., 1990; Sanchez & Kowalski, 1990). However, since this method of spectral analysis has many possible applications, it is now being presented in journals directed at much wider audiences (Leurgans & Ross, 1992).

1.1.1 *The data collection method.* Our example concerns a method called time-resolved fluorescence spectroscopy. The basic process is shown in Figure 1. (For a more adequate introductory discussion of the method and its relation to PARAFAC, see Leurgans & Ross, 1992.) Each data vector is obtained by directing a brief light flash of a particular wavelength at the specimen, then resolving the fluorescence this evokes into its spectrum and measuring the energy at a fixed set of emission

X. P. V. Malague (ed.), Advances in Signal Processing for Nondestructive Evaluation of Materials, 469–483.
© 1994 *Kluwer Academic Publishers.*

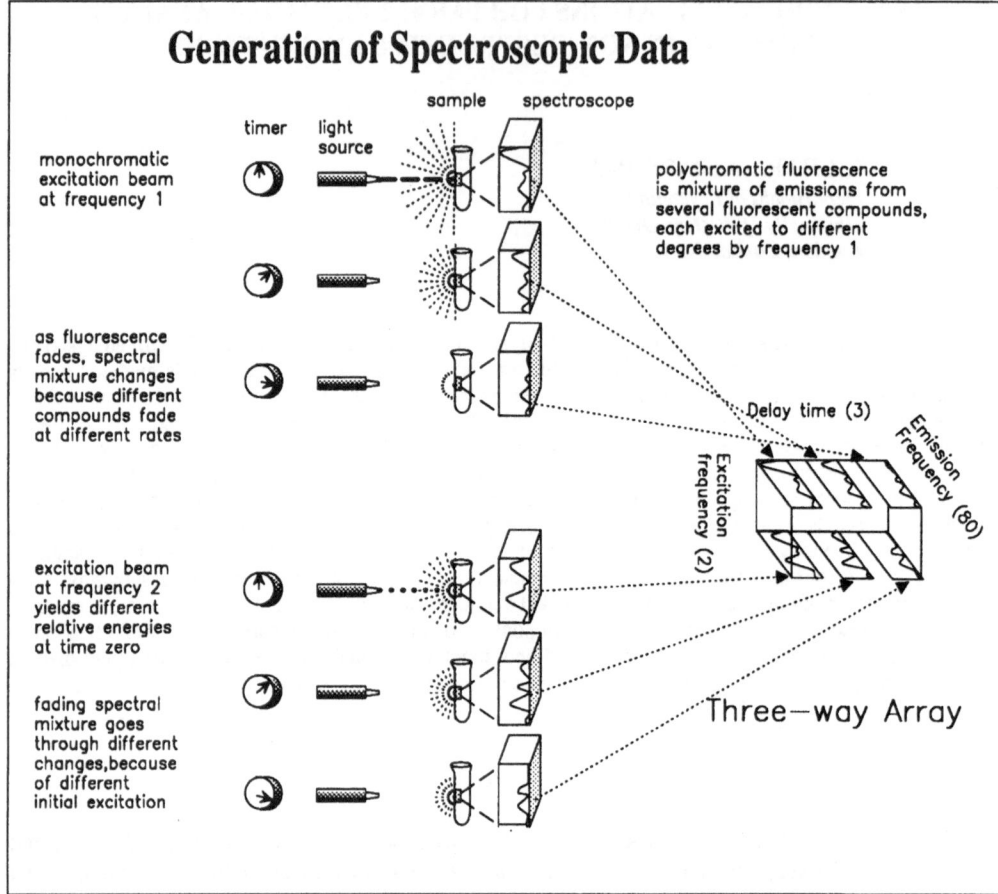

Figure 1. Three-mode data from fluorescence spectroscopy.
The fluorescence is excited by a light pulse; it is a mixtue of simpler spectra, one for each compound. The relative weighting of these spectra changes as a function of the excitation frequency and time after the pulse.

frequencies. Additional spectra are obtained from this *same* flash by remeasuring the emitted energies at successive time intervals after the flash. Finally, after the fluorescence has fully decayed, this entire measurement process is repeated, using a stimulus flash of a different wavelength, and repeated yet again at regular increments in excitation wavelength until curves can be constructed showing the fluorescent energy at each emission frequency as a function of excitation wavelength.

1.1.2 *The scientific question, and the analysis problem it creates.* Suppose we want to obtain the fluorescence spectra of certain transient compounds present only in a living plant. Our problem is that the compounds of interest co-occur with each other, and with an unknown number of less interesting compounds, in the cell organelles that we are studying. We cannot dissolve the organelles and physically isolate the desired chemical components, since killing the cell would destroy the biological processes that we are trying to study (cf. Stavenga, this volume).

1.1.3 *A solution.* To non-destructively recover the needed information, we must restrict ourselves to obtaining the fluorescence spectrum of the intact organelle in the living cell. This spectrum consists of some weighted combination of the spectra of all the fluorescing compounds in the organelle, so we will refer to it as a 'combination spectrum.' To recover the individual curves of interest, we obtain a combination spectrum in several different testing conditions, where the curves of individual compounds have different relative weights. By analyzing the several combinations simultaneously, using parallel factor analysis, we find a unique solution that decomposes the mixtures into the spectral curves of their constituents. We can then study the individual spectrum of each interesting compound.

1.2 THREE IMPORTANT DATA CHARACTERISTICS

As noted above, we must compare "several different testing conditions, where the curves of individual compounds have different relative weights." This deserves some further elaboration. In order for PARAFAC to correctly resolve mixtures into their underlying constituents, the data must meet the following three important requirements.

1.2.1 *Three-way structure.* The data must consist of repeated measures (except in 'indirect fitting,' see below); however, measurements should not be repeated under identical conditions, but rather under conditions that vary on three characteristics, so that the final data are a function of three measurement parameters. The values obtained should then be collected into an array where each observation has three subscripts. These three subscripts correspond to three 'modes' of classification of the data. It is conventional to call them Mode A, Mode B and Mode C, for short. Each mode has several 'levels,' corresponding to the different values the subscript for that mode can take. For example, see Figure 2.

1.2.2 *Proportional variation within each factor.* The pattern of values in a given mode corresponding to each single underlying component or factor—in our example, the values making up the spectrum of one single compound—should maintain its 'shape,' changing proportionally from one level to the next of the other two modes.

1.2.3 *Nonproportional variation between factors.* The two patterns of values characterizing two different factors in a given mode, must change by distinctively different proportions or ratios from one level to the next of the other two modes. That is, their overall changes should maintain a fixed ratio of values within each factor, but change the ratios of values between factors, as we compare one level to the next of the other two modes. This nonproportionality must hold across at least two

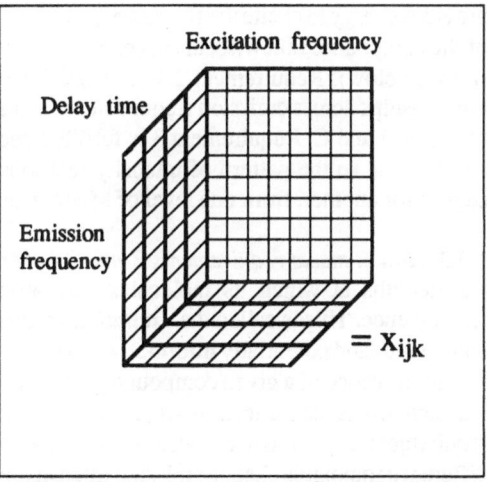

Figure 2. Three-way array of spectroscopic data.

levels (preferably more) of each mode. Thus, in our example, the spectra of two different compounds must have different relative weights in at least two combination spectra; in the excitation mode, the emission mode, and across at least two time intervals. The nonproportional levels that distinguish one pair of factors need not be the same two or more levels that distinguish a different pair of factors. Also, a factor may take on a zero weight at some levels of any mode.

1.3 DATA VARIATION IN THE FLUORESCENCE SPECTROSCOPY CASE

To make the three PARAFAC data requirements more concrete and understandable, it will be useful to see how they are fulfilled in the spectral analysis example. As frequently happens, all three requirements are met as a natural consequence of the behavior of the physical system under study. (Sometimes, however, ingenious manipulations of the physical system are needed to produce the necessary independent variation of the signals from separate factors of interest.)

1.3.1 How it meets requirement 1. In this example there are three measurement parameters: stimulus wavelength, emission wavelength, and time delay after excitation. The resulting three-way or three-mode data array is diagramed in Figure 2. In the diagram, the mode labeled 'Delay time' takes on 4 distinct values, so we say that it has 4 'levels'. Likewise, 'Excitation frequency' has 8 levels, and 'Emission frequency' has 10. Thus there are a total of 320 observed fluorescence intensity values, which can be considered as 8 combination emission spectra at 10 excitation frequencies and 4 delay times, or 10 combination excitation spectra, at 8 emission frequencies and 4 delay times, or 4 excitation-emission two-way arrays of spectral values, as observed at 4 different delay times after the flash.

1.3.2 How it meets requirements 2 and 3 in Mode A. Each compound in the sample has a distinctive set of frequencies at which it fluoresces in response to the light stimulation; this is its *emission spectrum*. The shape of this spectrum is independent of the stimulating wavelength, which merely provides energy to excite the fluorescence process. It is determined instead by the molecular structure of the compound. Likewise, at successive time delays the pattern grows dimmer, but proportionally so (see below). Requirement 2 is fulfilled because, for any given factor, the pattern of intensity relationships across emission frequencies maintains its shape, changing proportionally across levels of Mode B and C. Requirement 3 is fulfilled because different compounds have different emission spectra, and so the pattern of intensity relationships for one compound is not proportional to the pattern for another, from one level of Mode A to the next.

1.3.3 How it meets requirements 2 and 3 in Mode B. Each compound also has a set of excitation wavelengths at which its molecules best absorb energy from the light pulse and remit it as fluorescence. This is called the *excitation spectrum* of the compound; it, too, is characteristic of the compound (and potentially informative about it). As the wavelength of the excitation pulse is varied, the fluorescence of a given compound grows brighter or dimmer—but proportionally throughout all wavelengths as determined by its emission spectrum; and this remains true at each time delay. Thus requirement 2 is satisfied. Because different compounds have different excitation spectra, two different compounds change emission intensity in nonproportional ways as the excitation wavelength is varied. Thus requirement 3 is satisfied.

1.3.4 How it meets requirements 2 and 3 in Mode C. Finally, for the type of fluorescence spectroscopy discussed here, measurements are made at several time slices after each flash. As the surplus energy in a compound is lost, the fluorescence intensity decays —but for any single compound, all parts of the emission spectrum decay together, and so a given compound's contribution across time slices remain proportional. Thus the data satisfy requirement 2. Since different compounds lose energy, and hence brightness, at different rates, their contribution to the combination spectrum changes nonproportionally across time. This satisfies requirement 3.

2. The PARAFAC Model and Method, Compared to Two-Way Methods

2.1 THE MATHEMATICAL MODEL

2.1.1 *Scalar representation.* PARAFAC can be written in scalar form as follows:

$$(1) \quad x_{ijk} = \sum_{r=1}^{R} a_{ir} b_{jr} c_{kr} + e_{ijk}$$

where a_{ir} represents the relative degree of effect of the ith excitation frequency on the rth underlying spectral curve r, b_{jr} is a multiplier giving the relative size of the jth emission frequency (compared to other emission frequencies) in the rth latent spectral curve (this defines the shape of the rth latent spectrum), and c_{kr} is a multiplier giving the relative intensity of latent spectrum r in the kth time slice—this expresses the decay pattern for the rth latent spectrum.

As is apparent from (3), the data is modeled as the sum of R triple products. Hence the model is trilinear: the effect of changes across levels of any one mode—such as Mode A—are linear when the other two modes are held constant.

2.1.2 *Matrix representation.* Standard matrix notation is not well suited for three way arrays. In order to overcome this limitation, a convention is adopted in which the form of the three-way array is implied by giving the kth two-way slice:

$$(2) \quad \mathbf{X}_k = \mathbf{A}\mathbf{D}_k \mathbf{B}' + \mathbf{E}_k$$
$$\text{where } \mathbf{D}_k \text{ is diagonal, with } \{d_{ii,k}\} = [c_{k1} c_{k2} \dots c_{kR}]$$

One of the advantages of matrix notation is that it gives us familiar representation of the factors or dimensions. $\mathbf{A}$ and $\mathbf{B}$ are typical factor loading matrices, with rows representing the levels of Mode A and B respectively, and columns representing the factors that vary across those levels. In this case, the factors are the latent spectral curves of individual compounds.

2.1.3 *Tensor notation.* The most compact and elegant representation is in terms of tensor notation. For each of the R factors, a rank-1 tensor or three-way outer-product array is formed (see Figures 3).

$$(3) \quad \underline{\mathbf{X}} = \sum_{r=1}^{R} \mathbf{a}_r \otimes \mathbf{b}_r \otimes \mathbf{c}_r + \underline{\mathbf{E}} \qquad \text{where } \underline{\mathbf{X}} \text{ is the three-way array.}$$

The sum of these rank-1 arrays approximates the data array. We fit the model with a modified Alternating Least Squares procedure, so that the final converged solution minimizes the sum of squared residuals for any given R.

2.1.4 *Some background information.* The model described in this section is the standard or 'core' trilinear model for parallel factor analysis. It was independently formulated by Carroll and Chang (1970) as the basis for INDSCAL, and Harshman (1970); the key idea was proposed even earlier by Cattell (1944, 1955). Recently, the trilinear model was independently discovered once again by chemometricians (e.g., Sanchez & Kowalski, 1990).

Other models for three-way data have been proposed. The earliest and still most important is the three-mode factor analysis model of Tucker (1963,1966). It has the following form:

$$(4) \quad x_{ijk} = \sum_{r=1}^{R} \sum_{s=1}^{S} \sum_{t=1}^{T} a_{ir} b_{jr} c_{kr} g_{rst} + e_{ijk}$$

or, in tensor notation,

$$(5) \quad \underline{X} = \sum_{r=1}^{R} \sum_{s=1}^{S} \sum_{t=1}^{T} a_r \otimes b_s \otimes c_t \, g_{rst} + \underline{E} \quad ,$$

where g_{rst} is an element of $\underline{G}$, a small three-way array of interactions among the factors. This is probably the most general three-way model proposed to date, but, in part because of its generality, the parameters of the model are not identifiable by simply finding the best fit to a given set of data. Indeed, the Tucker model is rotationally indeterminate in parameters for all three modes.

Recent work in our laboratory has included the development of models with much of the generality of the Tucker model, but which retain the important identifiability or uniqueness property of the basic trilinear parallel factor analysis model (e.g., see Harshman & Lundy, in press b).

2.2 COMPARISON WITH TWO-WAY DECOMPOSITIONS.

The direct precursor of PARAFAC is conventional two-way factor analysis (e.g., Harman, 1976; McDonald, 1985). However, for those unfamiliar with factor analysis, PARAFAC can be considered a three-way generalization of the singular value decomposition (Figure 3) or principal components analysis (Figure 5), but without the orthogonality of factors used in the two-way methods to obtain identifiability. These are all mathematical/statistical methods of decomposing a data matrix into the additive sum of simpler parts—specifically, rank-1 parts given by the outer product of row and column weight vectors.

2.2.1 *Singular value decomposition.* The singular value decomposition of a matrix X is often written as $X = UDV'$ where U and V are the left and right singular vectors, respectively and D is diagonal and contains the singular values in descending order of size. To obtain a unique decomposition it is required that both sets of singular vectors be columnwise orthogonal. To show the parallel with our three-mode decomposition, we can rewrite this in terms of Mode A and B of a two way matrix. Furthermore, from factor analysis there is a convention to absorb the scale, or the singular vectors, into the set of weights corresponding to the variables (in this case let us assume this is Mode A), and set the other mode to have unit mean square. The decomposition could then be written in scalar, matrix, and tensor form as follows:

$$(6) \quad x_{ij} = \sum_{r=1}^{R} a_{ir} b_{jr} + e_{ij} \quad or \quad X = AB' + E \quad or \quad X = \sum_{r=1}^{R} a_r \otimes b_r + E$$

Two-way direct fitting: Singular value decomposition

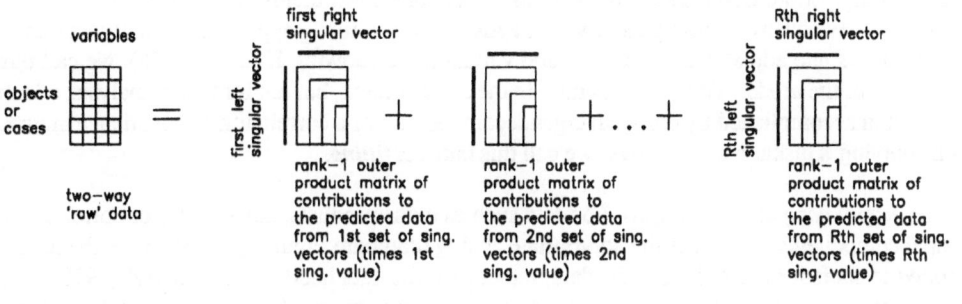

Three-way direct fitting: PARAFAC

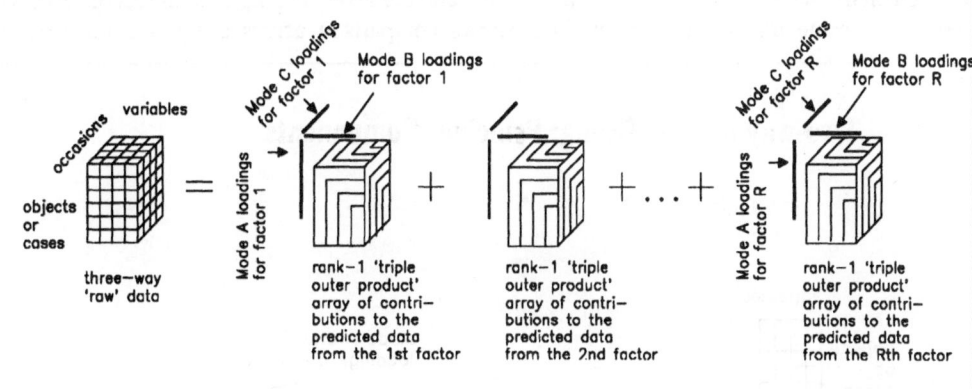

Figure 3. PARAFAC compared to SVD

Both PARAFAC and the SVD do direct fitting to data. Both extract rank-1 components which are outer products of vectors of dimension weights of factor loadings. Only for PARAFAC are these unique without further restrictions.

PARAFAC simply accomplishes the same sort of decomposition on a three-way array. It is very useful to think of PARAFAC as performing a series of SVD's in parallel, $X_k = AD_kB'$. For some purposes, however, a more fundamental perspective is that shown in Figure 3. Each component extracted by PARAFAC is trilinear rather than bilinear, so the three-way data is decomposed into a series of rank-1 three-way latent patterns (for a simple discussion of the rank of three-way arrays, see Kruskal, 1989). These three-way rank-1 components have a very important property that the two-way ones do not: under relatively modest independence conditions a sum of such patterns has a unique decomposition. Given a sum, the individual rank one constituents are identifiable.

So far, we have been discussing cases where PARAFAC is directly applied to the raw data matrix (perhaps after some adjustments of means and variances). Following Kruskal (1978), we call this *direct fitting* of the model. This seems natural in the SVD context, but not for factor analysis, which is more often accomplished by doing an eigendecomposition of a correlation or covariance matrix. Again applying Kruskal's terminology, we call this *indirect fitting*.

2.2.2 *Principal components analysis.* For data such as cross-products, covariances or correlations, principal components analysis is the appropriate analog (compare Figure 4 and 5). Here, the model is applied to statistics derived from the data, rather than the data itself, hence Kruskal (1978) calls this *indirect fitting*. In such an application, two of the three modes (i.e., rows and columns) are identical, and will give identical factor weight vectors. The third mode contains weights which are the squares of weights obtained in direct fitting. This method of analysis has several advantages and disadvantages. One advantage is that the covariances for one level of the three-way array need not be based on the same cases as the covariances from another level, so long as both sets of cases will involve the same underlying factors. This allows comparison across samples from different

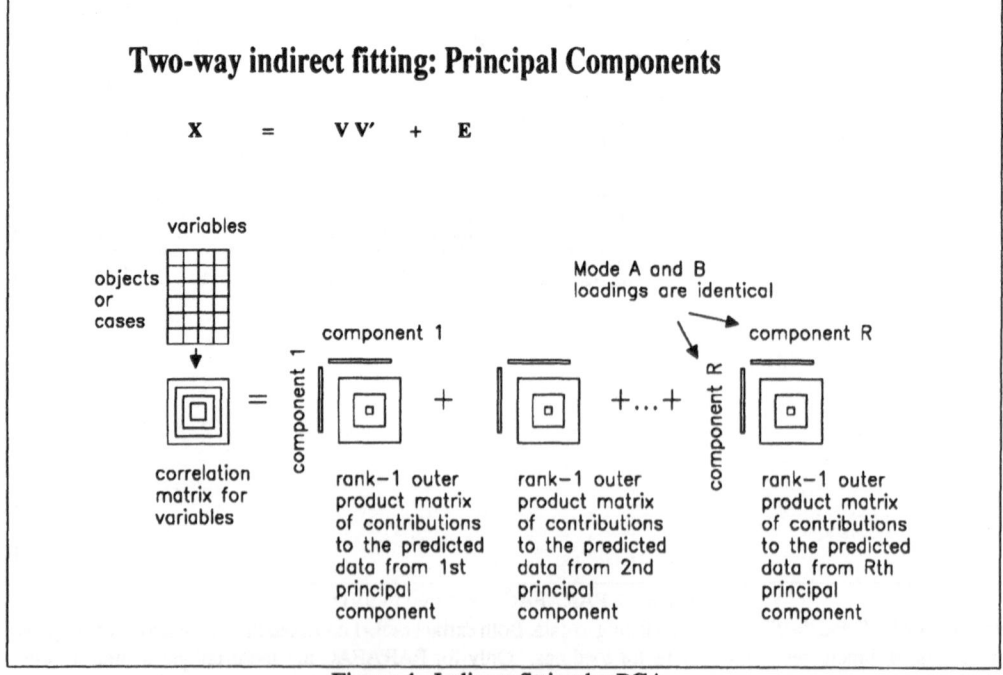

Figure 4. Indirect fitting by PCA.

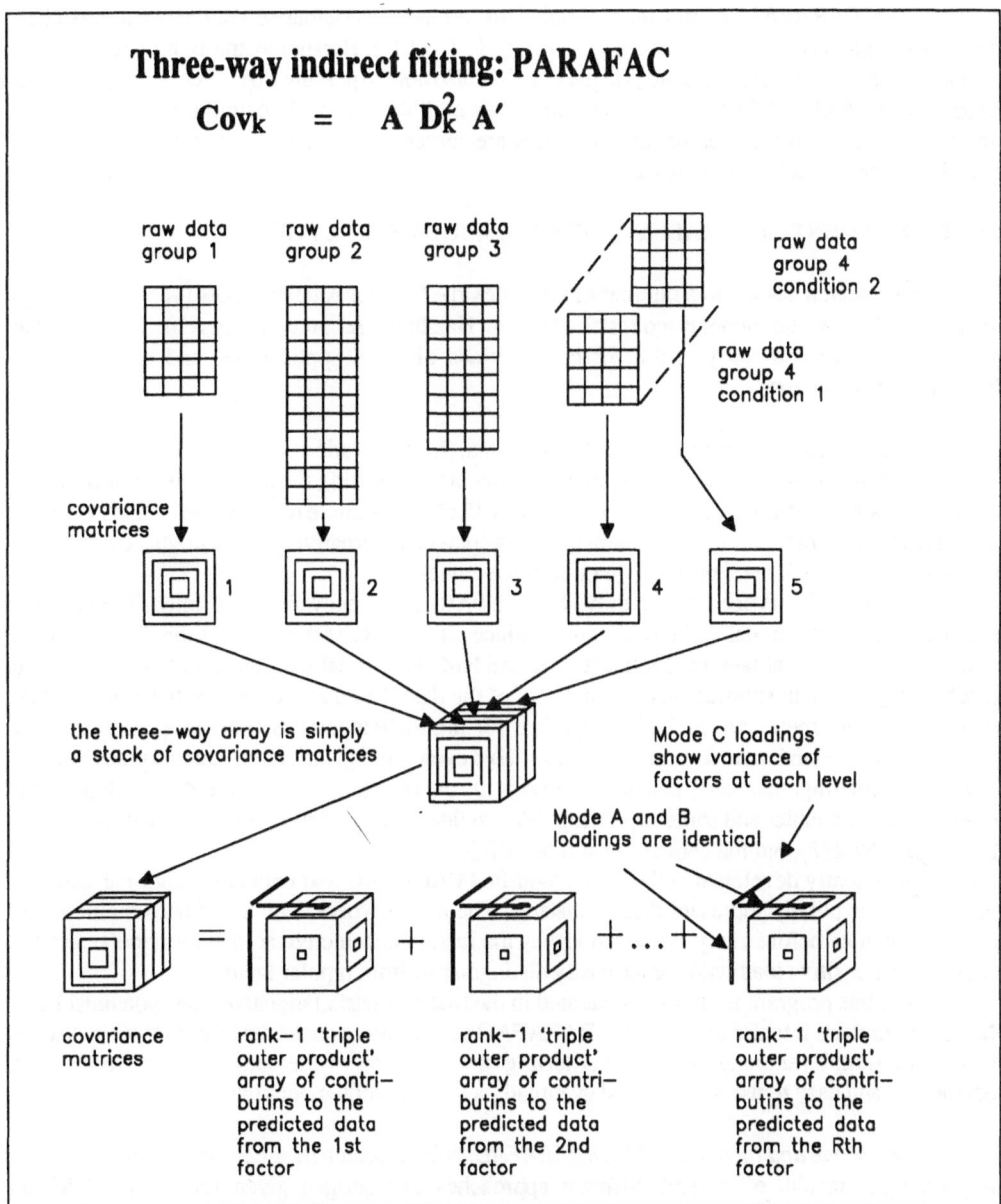

Three-way indirect fitting: PARAFAC
$$Cov_k = A\, D_k^2\, A'$$

raw data group 1   raw data group 2   raw data group 3

raw data group 4 condition 2

raw data group 4 condition 1

covariance matrices

1   2   3   4   5

the three–way array is simply a stack of covariance matrices

Mode C loadings show variance of factors at each level

Mode A and B loadings are identical

covariance matrices

rank–1 'triple outer product' array of contri– butins to the predicted data from the 1st factor

rank–1 'triple outer product' array of contri– butins to the predicted data from the 2nd factor

rank–1 'triple outer product' array of contri– butins to the predicted data from the Rth factor

Figure 5. Several varieties of three-way indirect fitting

The flexibility of indirect fitting is demonstrated in this figure. Three different samples of different sizes are included in the same analysis analysis. In adition, a single sample measured under two different conditions is also included. Converting these later two datasets to covariances for indirect fitting removes the requirement of proportional changes across cases.

subpopulations. It also allows one to deal with cases where the variation across the third mode is not going to be strictly proportional. On the other hand, there are some implicit constraints of factor independence that are not present in direct fitting (although these can be relaxed by using the more general model PARAFAC2). For a more detailed discussion and comparison of the two approaches, see, for example, Harshman and Lundy (1984a, pp. 137 to 139), Harshman and Berenbaum (1981), and particularly Harshman and Lundy (in press a). We also note in passing that the popular INDSCAL program of Carroll and Chang (1970) accomplishes multidimensional scaling by indirect fitting, where matrices of similarities or dissimilarities are turned into distances, then into covariances (pseudo-scalar products) for analysis.

2.3 SOME RECENT GENERALIZATIONS POSSIBLY USEFUL FOR NDE

Continuing research has led to a number of more flexible Parallel Analysis tools for classes of data where the strict trilinear model is too restrictive (e.g., Harshman &Lundy, in press b). Some of these generalizations seem directly useful for certain kinds of NDE applications, and so are briefly described in this section.

2.3.1 *PARAFAC with additive constants.* In some data, the observations are not strictly 'ratio scale.'
That is, there are artificial and/or extraneous baseline offsets shifting the data points up or down by a certain amount. All the points may have the same offset, for example due to unwanted background radiation or an arbitrary overall adjustment of the equipment. Alternatively, there might be a different constant offset on each level of one, two or all three modes.

These constant offsets need to be removed before PARAFAC analysis is performed. They disturb proportionality of the signal variations, and introduce artifactual constant factors that interfere with the extraction of the real factors of interest. The standard way to deal with this problem has been by specific kinds of mean subtraction or 'centering' of the data. But use caution, we have proven that not all such adjustments are in fact appropriate for multilinear models (see the section on data preprocessing in Harshman & Lundy, 1984b). But even among the mathematically appropriate methods of centering, there are three different ways the data can be singly-centered, plus all possible combinations of double- and triple- centering. We provide some guidelines in Harshman and Lundy (1984a, pp. 257-258), but the choice can still be difficult.

We have recently developed and tested a modified PARAFAC that contains additional terms in the model to explicitly represent these constants. The model works. This eliminates the need for centering the data before analysis. It also leaves the factors uncentered, so that monopolar factors (ones with all positive loadings) can be easily distinguished from bipolar factors.

Currently, this program is only implemented in the Matlab matrix language. Thus you must have Matlab or purchase it to use the program. PARAFAC with additive constants is also rather awkward to use at present. Eventually, after further testing and evolution of the code, we hope to create a documented and easy to use version, and eventually a Fortran implementation.

2.3.2 *Linked-Mode and Incomplete Mode PARAFAC.* NDE researchers are sometimes interested in
integrating the results of several different approaches to testing a given flaw. Linked Mode PARAFAC allows the analysis of several different three-way arrays of data that share only one mode in common. For the common mode, the analysis forces a single set of loadings. This method lets the joint information in the two different datasets be viewed in a common space, and uses each three-way array to strengthen the uniqueness and reduce the error of estimate of the other.

The Linked-Mode program also allows the user to relax the strong requirement of a fully crossed three-way array, by allowing major parts of the array to be empty. In this application we call it the Incomplete-Mode program/model. Instead of being considered missing data, the large empty spaces

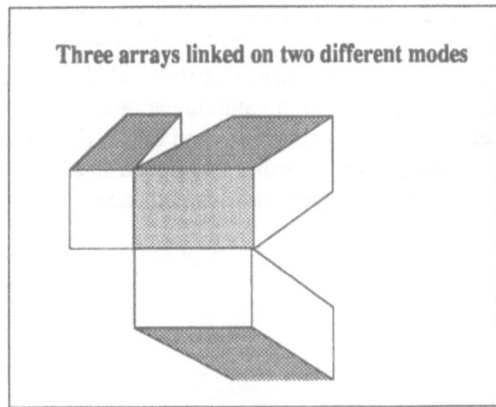

Three arrays linked on two different modes

Figure 6. Three Linked-mode datasets.
For Linked-Mode analysis, the several modes do not have to all share the same common mode.

in the array are considered undefined and not estimated. This program is currently being further developed and exhaustively tested in our laboratory, so documentation is still limited.

2.3.3 *Lagged PARAFAC*. When the NDE signal is a spectrum or image, it sometimes happens that the component of interest can shift 'horizontally' from one subset of levels to another set of levels of the sequentially organized mode or modes. For a spectrum, this would mean that some variation of the testing conditions caused the signal of interest to shift up or down in the frequency domain. For an image, this would mean that some lighter or differently colored region associated with a distinct physical aspect under study was displaced in position from one image to the next.

Changes in the set of variables or levels representing the process under study violate the proportionality required by the model. However, we are developing a modified version of PARAFAC to deal with this problem. It is called *Lagged PARAFAC*, and it contains special parameters that represent shifts of the factors across the set of sequentially organized levels in modes such as time, frequency, or position in an image. Tests for proportional change are performed only after shifts are done that attempt to align the factors across different levels of each mode. The program alternates between finding shifts that minimize the sum of squared errors for a given set of estimated factors, and improving the estimates of the factors to further minimize the sum of squared errors.

We anticipate that this model will have some very interesting properties. For example, in the analysis of images, the investigator could deliberately include images from slightly different horizontal and vertical angles. By recovering the different shifts that this introduces for components at different depths, the program would be able to distinguish parts of the image as if by a kind of stereoscopic depth perception (or quadroscopic if both horizontal and vertical shifts are used).

We also believe that if all the factors have distinct patterns of shifts, as well as distinct patterns of changes in magnitude, then shifts may provide the same information as a third mode, providing a fully identifiable factor model for two-way data.

2.4 FURTHER ISSUES AND INFORMATION

2.4.1 *Degenerate solutions*. When the data are not quite appropriate for the PARAFAC model, one will sometimes obtain degenerate solutions, in which factors become very highly negatively correlated and difficult to interpret. This phenomenon is discussed in more detail in Harshman and Lundy (1984b), Kruskal et al. (1989), and Lundy et al (1989). One method of dealing with this is to place zero-correlation or orthogonality constraints on one mode. To see an example of degenerate factors which emerge only after several nondegenerate ones have been extracted, and how these are made interpretable by constraints, see Harshman and DeSarbo (1984).

2.4.2 *Key references*. Additional information on PARAFAC should be consulted if you intend to try applications to your data. In particular, you will want to consult the standard reference articles

"The PARAFAC model for three- way factor analysis and multidimensional scaling" (Harshman & Lundy, 1984a) and "Data preprocessing and the extended PARAFAC model" (Harshman & Lundy, (1984b). You should also examine the sample application in the same volume (Harshman & DeSarbo, 1984), the overview chapter by the editors, and the article by Kruskal on multilinear models (Law et al., 1984). Another useful source is the more recent treatment by Harshman and Lundy (in press a) and the discussion by Leurgans and Ross (1992). You should pay particular attention in all of these sources to discussions of uniqueness, data preprocessing, and the degeneracy phenomenon.

3. Applying PARAFAC in NDE

3.1 TYPES OF APPLICATIONS

The application of PARAFAC and related models to NDE can have at least three different uses: exploration, sensitization/clarification, and simplification.

3.1.1 *Exploration.* It will sometimes be possible to perform NDE in a novel situation, where the signal/image characteristics of the flaw are not known. This should permit deviant signals related to physical defects or anomalies to be discovered even if previously unknown, and isolated for better interpretation. One must be able to construct a three-way array of measurements in which it is likely that the signals/images related to the flaw and other physical components will show the right kind of variation across test conditions (as described above).

3.1.2 *Sensitization/clarification.* When the nature of the signal or image produced by the flaw is fairly well understood, but of necessity is mixed in with other signal components that tend to obscure it, use of parallel factor analysis could sometimes increase the sensitivity of the test procedure by isolating the functionally distinct signal components onto different factors, and describing the variations in the intensity of the components across test conditions. This could be viewed as a special kind of 'functional filter' which increases sensitivity of the analysis by isolating the signal of interest, and clarifies details of the form of the signal/image component related to the flaw.

3.1.3 *Simplification.* If the signal or image identifying the flaw is adequately distinct for experts or highly trained technicians to make reliable assessments, it might nonetheless be desirable to use parallel factor analysis to isolate the flaw and make it easier to recognize by non-experts. The test procedure could then be less expensive for a company to apply. (This use of PARAFAC for simplification of use of thermogram and subsequently other methods of NDE is one objective of collaboration between the author and Dr. Chris Hobbs, of The National NDT Center, United Kingdom, initiated at this workshop.)

3.2 IMAGES

When the data are images, there is a temptation to consider the x and y axes of the image to be two different data modes. The problem with this is that the x and y axes do not multiply together to produce images; the light and dark shapes in the images are not usually representable as outer products of an x and y intensity vector. Some colleagues and I are in fact exploring the possible value of ignoring the complexity of the image and treating it as composed of outer product patterns, but we do not recommend this as a first and mathematically most appropriate approach. Instead, the image should be vectorized by taking successive columns of pixels and appending them one after the other; thus a 400 by 600 pixel image would convert to a 240,000 element vector. In this way, all the points on the image constitute levels of a single mode, allowing full representation of details. However, this also means that the researcher must come up with two other modes in which the meaningfully related sets of pixels are likely to show the desired patterns of variation. One must have multiple images, with two different kinds of changes in the conditions under which successive images are taken (one set for Mode B and an intersecting or fully crossed set of different conditions for Mode C).

3.3 HOW TO GET PARAFAC

The PARAFAC program in Fortran source code form is available free of charge by FTP (File Transfer Protocol — a way of getting files from distant locations over the worldwide internet of interconnected computers). There are at least three locations, the University of Western Ontario main Computing Center Unix server, the Social Science server, and the NDE Internet archive.. The Telnet absolute addresses for the first two computers are (i) 129.100.2.12 (central computer) and (ii) 129.100.38.15 (Social Science). You probably will not need these numeric addresses if your installation has updated tables of computer names. For further details of how to use FTP to obtain PARAFAC, see the Appendix to this article.

4. Conclusion

There may indeed be applications of Parallel Factor Analysis in Non-Destructive Evaluation. However, it will require that the investigator pay close attention to the special data requirements of the model. Given that appropriate three-way data can be collected, either for the standard or some generalized PARAFAC model, the unusual power of this analysis method to isolate functionally distinct components of the signals could be of considerable value for NDE. The field waits for some imaginative researcher(s) who will meet the challenge of generating appropriate data and give us the first examples of how to put the power of PARAFAC analysis to work in this area.

Acknowledgements

Fuad Gwadry went to extraordinary lengths in helping to get this article written and finished, including joining me in periods of 48 hours without sleep. For his hard work, and his many excellent ideas and suggestions, I am most grateful. A number of stimulating conversations with Chris Hobbs at the NATO workshop have no doubt contributed to a various ideas expressed in this article.

Most important of all, this article would never have appeared if it were not for the supreme patience and indefatigable good will of Xavier Maldague. As the preparation of this paper kept failing to meet one deadline after another, he never communicated unsupportive or harsh attitudes. Instead he was supportive and kindhearted throughout. My deepest thanks for the opportunity to finish this manuscript in the highest quality condition that I could provide.

References

Burdick, D.S., Tu, X.M., McGown, L.B., & Millican, D.W.(1990). Resolution of multicomponent fluorescent mixtures by analysis of the excitation-emission- frequency array. *Journal of Chemometrics, 4,* 15-28.

Carroll, J.D., & Chang, J.J. (1970). Analysis of individual differences in multidimensional scaling via N-way generalization of "Eckart-Young" decomposition. *Psychometrika, 35,* 282-319.

Cattell, R.B. (1944). "Parallel proportional profiles" and other principles of determining the choice of factors by rotation. *Psychometrika, 9,* 267-283.

Cattell, R.B., & Cattell, A.K.S. (1955). Factor rotation for proportional profiles: Analytic solution and an example. *British Journal of Statistical Psychology, 8,* 83-92.

Harman, H.H. (1976). *Modern factor analysis.* Chicago: University of Chicago Press.

Harshman, R.A., & Berenbaum, S.A., (1981). Basic concepts underlying the PARAFAC-CANDECOMP three-way factor analysis model and its applications to longitudinal data. In D.H. Eichorn, J.A. Clausen, N. Haan, M.P. Honzik, & P.H. Mussen (Eds.), *In present and past in middle life.* New York: Academic Press.

Harshman, R.A., & DeSarbo, W.S. (1984). An application of PARAFAC to a small sample problem, demonstrating preprocessing, orthogonality constraints, and split-half diagnostic techniques. In H.H. Law, C.W. Snyder, Jr., J.A. Hattie, & R.P. McDonald (Eds.), *Research methods for multimode data analysis* (pp. 602-642). New York: Praeger.

Harshman, R., Ladefoged, P., & Golstein, L. (1977). Factor analysis of tongue shapes. *Journal of the Acoustical Society of America, 62,* 693-707.

Harshman, R.A., & Lundy, M.E. (1984a). The PARAFAC model for three-way factor analysis and multidimensional scaling. In H.G. Law, C.W. Snyder, Jr., J.A. Hattie, & R.P. McDonald (Eds.), *Research methods for multimode data analysis* (pp. 122-215). New York: Praeger.

Harshman, R.A., & Lundy, M.E. (1984b). Data preprocessing and the extended PARAFAC model. In H.G. Law, C.W. Snyder, Jr., J.A. Hattie, & R.P. McDonald (Eds.), *Research methods for multimode data analysis* (pp. 216-284). New York: Praeger.

Harshman, R.A., & Lundy, M.E. (in press a). PARAFAC: Parallel factor analysis. *Computational Statistical Data Analysis.*

Harshman, R.A., & Lundy, M.E. (in press b). Uniqueness proof for a three-way factor analysis or MDS model with Tucker-like "interactions" and PARAFAC-like parallel profiles (2D case). *Psychometrika.*

Kruskal, J.B. (1978). Factor analysis and principal components: Bilinear methods. In W.H. Kruskal and J.M. Tanur (Eds.), *International encyclopedia of statistics.* New York: Free Press.

Kruskal, J.B. (1989). Rank, decomposition, and uniqueness for 3- way and N-way arrays. In R. Coppi & S. Bolasco (Eds.), *Multiway Data Analysis* (pp. 7-18). New York: North-Holland.

Kruskal, J.B., Harshman, R.A., & Lundy, M.E. (1989). How 3-MFA data can cause degenerate PARAFAC solutions, among other relationships. In R. Coppi & S. Bolasco (Eds.), *Multiway Data Analysis* (pp. 115-122). New York: North-Holland.

Law, H.G., Snyder, C.W., Jr, Hattie, J.A., & McDonald, R.P. (1984). *Research methods in multimode data analysis.* New York: Praeger.

Leurgans, S., & Ross R.T. (1992). Multilinear models: Applications in spectroscopy. *Statistical Science, 7,* 289-319.

Lundy, M. E., Harshman, R. A., & Kruskal, J. B. (1989). A two stage procedure incorporating good features of both trilinear and quadrilinear models. In R. Coppi & S. Bolasco (Eds.), *Multiway Data Analysis* (pp. 123-130). New York: North-Holland.

McDonald, R.P. (1985). *Factor analysis and related methods*. Hillsdale, New Jersey: Lawrence Erlbaum Associates.

Sanchez, E., & Kowalski, B.R. (1990). Tensorial resolution: A direct trilinear decomposition. *Journal of Chemometrics, 4*, 29-45.

Tucker, L.R. (1963). Implications of factor analysis of three-way matrices for measurement of change. In C.W. Harris (Eds.), *Problems in measuring change*. Madison, Wisc.: University of Wisconsin Press.

Tucker, L.R. (1966). Some mathematical notes on three-mode factor analysis. *Psychometrika*, 31, 279-311.

APPENDIX: Detailed instructions on use of FTP

The following explains in detail how to get PARAFAC by FTP. Note that unix machines are case sensitive, so be sure to type the directory and file names in appropriate case.

The retrieval procedure is as follows: (i) start your system's FTP software; (ii) ask for the ftp address of the host machine, e.g. for the uwo main computer ftp.area, ask for **ftp.uwo.ca**; (iii) ask to log on — usually type "log" unless the software has done this for you; (iv) you get the prompt "USERNAME:";(v) type : anonymous; (vi) you will receive the prompt "PASSWORD:"; (vii) protocol would be to give your e-mail address, eg., yourusername@yourplace.yournet; (viii) you will get a message about guest logon accepted, some restrictions will apply; (ix) type "dir" to get the directory. There should be a directory called: **pub**; (x) change to that directory, e.g., type "cd pub"; (xi) another "dir" should reveal a directory labeled **"misc"**; (xii) "cd misc" then "dir" should reveal a directory labeled **"parafac"**; (xiii) "cd parafac"; (xiv) type "binary"; (xv) use the "get" command, or whatever is appropriate at your installation, to retrieve the Parafac file; (xvi) type "ascii" and "get" the Readme file; (xvii) you may need to get a recent version of the unzipping software --as explained in the Readme file -- to unzip or uncompress the parafac.zip file into a set of fortran source and data files.

The same procedure is used to retrieve PARAFAC from the social science server (except that the PARAFAC subdirectory is only one level down instead of two) and the NDE Internet archive. The three ftp addresses are

for the uwo server:

host=ftp.uwo.ca directory=/pub/misc/parafac

for the Social Science server:

host=phobos.sscl.uwo.ca, directory= pub/parafac/

for the NDE Internet archive server:

host= swrinde.nde.swri.edu, directory =pub/nde/parafac

If you have any questions, Dr. Keith S. Pickens of the Southwest Research Institute (ksp@maxwell.nde.swri.edu) is the system operator on the NDE e-mail network. He archived PARAFAC, and in the process made it easier for unix users to unzip.

List of Participants and Contributors

Participants per country:

Canada:	14
United States:	10
Fance:	6
Germany:	4
Italy:	2
United Kingdom:	2
Belgium:	1
The Netherlands:	1
Russia:	1
Saudi Arabia:	1
Sweden:	1
Turkey:	1
total:	*44*

* Dr. Jean-Daniel Aussel
UltraOptec
27 de Lauzon
Boucherville, Qc **Canada**
J4B 1E7
ph: (514) 449-2096, fax: (514) 449-2463

Prof. Robert Bergevin
Electrical Engineering Dept.
Université Laval
Québec, Qc **Canada**
G1K 7P4
ph: (418) 656-5173, fax: (418) 656-3594
email: bergevin@gel.ulaval.ca

* Dr. Douglas D. Burleigh
General Dynamics
Space Systems Division
P.O. Box 85990 MZ 23-8560
San Diego, California **USA**
ph: (619) 573-8839, fax: (619) 974-4000

Prof. C.H. Chen
Electrical & Computer Eng. Dept.
University of Massachussetts Darthmouth
North Darthmouth, MA 02747 **USA**
ph: (508) 999-8475, fax: (508) 999-8485
email: CCHEN@UMASSD.EDU

* Dr. Marc Choquet
Institut des Matériaux Industriels
Conseil National de la Recherche
75, de Mortagne
Boucherville, Qc **Canada** J4B6Y4
ph: (514) 641-2280, fax: (514) 641-4627

Dr. Raphaël Danjoux
Mecica
3, rue du Val Clair
Z. I. La Pompelle
51100 Reims **France**
ph: (33) 26 82 04 00, fax: (33) 26 85 28 48

Prof. J.F. de Belleval
Université de Technologie Compiegne
UTC Département de génie mécanique
Laboratoire de Génie Mécanique pour les
Matériaux et les Structures
Centre de Recherches de Royallieu
B.P. 649
60206 Compiegne Cédex **France**
ph: (33) 44 23 44 23
fax: (33) 44 20 48 13

Prof. Ferial El-Hawary
Technical Univ. of Nova Scotia
P.O. Box 1000
Halifax, N.S. **Canada**
B3J 2X4
ph: (902) 420-7541, fax: (902) 420-7551
email: elhawary@tuns.ca

Dr. Lawrence D. Favro
Institute for Manufacturing Research
Wayne State University
Detroit, MI 48202 **USA**
ph: (313) 577-2970, fax: (313) 577-7743

Dr. E. A. Fuchs
Physical Properties of Materials
Division 8715
Sandia National Laboratory
Livermore, CA 94551-0969 **USA**
ph: (510) 294-2027, fax: (510) 294-3231

* **Dr. Yves Gagnon**
Centre de Recherches pour la défense
Val Cartier,
Division de l'Armement
C.P. 8800 Courcelette, Qc **Canada**
G0A 1R0
ph: (418) 844-4321

Dr. B. Georgel
Electricité de France
Direction des Études et Recherches
Dépt. SDM, 6 quai Watier
B.P. 49
F 78401 Chatou Cedex **France**
ph: (33) 1 30 87 78 74
fax: (33) 1 30 87 80 80

Dr. Denis Gingras
Traitement de l'Information
Institut National d'Optique
369, rue Franquet
Sainte-Foy, Qc **Canada**
G1P 4N8
ph: (418) 657-7006, fax: (418) 657-7009
email: gingras@ino.qc.ca

Dr. Ermanno Grinzato
Consiglio Nazionale delle Ricerche
Instituto per la Technica del Freddo
Corso Stati Uniti, 4
35100 Padova **Italy**
ph: 49 829-5722, fax: 49 829-5728

Mr. Emin Çagatay Güler
Biomedical Engineering Institute
Bogaziçi University
Bebek
80815 Istanbul **Turkey**
ph/fax: (90) 1 257 5030

Prof. Richard A. Harshman
Dept. of Psychology
University of Western Ontario
London, Ontario **Canada**
N6A 5C2
ph: (519) 679-2111 x 4691
fax: (519) 661-3213
email: harshman@uwovax.uwo.ca

Dr. S. Haykin
Communications Research Lab.
McMaster University
1280 Main Street West
Hamilton, Ontario **Canada**
L8S 4K1
ph: (416) 525-9140 x 4291
fax: (416) 521-2922

Prof. J. Henriette
Faculté Polytechnique de Mons
Service de thermique et d'Emploi des
Combustibles
rue de l'Epargne, 56
B-7000 Mons **Belgium**
ph: (32) 65 37 44 61
fax: (32) 65 37 44 00

Dr. B. V. Hess
Physical Properties of Materials
Division 8715
Sandia National Laboratory
Livermore, CA 94551-0969 **USA**
ph: (510) 294-2027, fax: (510) 294-3231

Dr. Chris Hobbs
The National NDT Centre
B 447
Harwell Laboratory
Didcot Oxfordshire **United Kingdom**
OX11 0RA
ph: (44) 235 43 52 53
fax: (44) 235 43 52 48

Prof. Dr.-Ing. W. Jüptner
BIAS
Klagenfurter Str. 2
W-28359 Bremen 33 **Germany**
ph: 0421 218-01, fax: 0421 218-5063

Dr. Doron Kishoni
College of William and Mary
NASA Langley Research Center
Mail Stop 231
Hampton, VA 23681-0001 **USA**
ph: (804) 864-4787, fax: (804) 864-4914
email: D.Kishoni@LaRC.NASA.GOV

* **Prof. Michel Lecours**
Electrical Engineering Dept.

Université Laval
Québec, Qc **Canada**
G1K 7P4
ph: (418) 656-2966, fax: (418) 656-3159
email: mlecours@gel.ulaval.ca

Dr. F. Lepoutre
ONERA
B.P. 72
Châtillon Cedex
F-92322 **France**
ph: (33) 1 46 73 48 73
fax: (33) 1 46 73 48 91

* Dr. Helmut Licht
Engineering Technology Div.
AECL - Research
Chalk River Labs.
Chalk River, Ontario **Canada**
K0J 1J0
ph: (613) 584-3311 , fax: (613) 584-4523

Prof. Xavier Maldague
Electrical Engineering Dept.
Université Laval
Québec, Qc **Canada**
G1K 7P4
ph: (418) 656-2962, fax: (418) 656-3594
email: maldagx@gel.ulaval.ca

Mr. S. Marinetti
Consiglio Nazionale delle Ricerche
Instituto per la Technica del Freddo
Corso Stati Uniti, 4
35100 Padova **Italy**
ph: 49 829-5722, fax: 49 829-5728

Dr. K. I. McRae
Defence Research Establishment Pacific
FMO, CFB Esquimalt
Victoria, BC **Canada**
V0S 1B0
ph/fax: (604) 363-2925

Dr. Rick Murphy
Engineering Technology Div.
AECL - Research
Chalk River Labs.
Chalk River, Ontario **Canada**
K0J 1J0

ph: (613) 584-3311 x 3754
fax: (613) 584-4523

Dr. C. Nockemann
Federal Institute for Materials Research
and Testing
Unter den Eichen 87
12205 Berlin 45 **Germany**
ph: (49) 30 8104 1624
fax: (49) 30 811 5089

Mr. Stephan Offermann
Faculté des Sciences de Reims
GRSM/Laboratoire d'Énergétique et
d'Optique
 BP 347
F-51062 Reims Cédex **France**
ph: (33) 26 05 33 55, fax: (33) 26 05 32 50

Dr. G. Pelx
Laboratoire de CND par Rayonnement
Ionisant
Bat. 303 - INSA de Lyon
20, av. Albert-Einstein
69621 Villeurbanne Cedex **France**
ph: (33) 72 43 82 62
fax: (33) 72 43 83 54

Prof. J. Pelzl
Arbeitsgruppe Festkörperspektroskopie
Fakultät für Physik und Astronomie
Institut für Experimentalphysik III
Ruhr-Universität
Universitätsstrabe 150
GebäudeNB 3/134
D-44780 Bochum **Germany**
ph: (02 34) 700 60 51
fax: (02 34) 70 94 172
email: PELZL@RUBA.RZ.
RUHR-UNI-BOCHUM.DBP.DE

Dr. Keith S. Pickens
Principal Scientist
Dept of NDE Research and Development
Southwest Research Institute
6220 Culebra Road
Post office Drawer 28510
San Antonio, TX 78228-0510 **USA**
ph: (210) 684-5111 x 3100

fax: (210) 684-4822

* **Prof. Jean-F. Rivest**
Département de génie électrique
Faculté de génie
Université d'Ottawa
550 Cumberland, Ottawa **Canada**
K1N 6N5
ph: (613) 564-3311, fax: (613) 564-5829
email: rivest@trix.genie.uottawa.ca

Prof. Jafar Sanlie
Ultrasonic Information Processing Lab
Dept. of Elect. & Computer Eng.
Armour College of Engineering
Illinois Institute of Technology
3301 South Dearborn
Chicago, Illinois 60616-3793 **USA**
ph: (312) 567-3400, fax: (312) 567-8976

Dr. L.W. Schmerr Jr.
Center for NDE
Iowa State University, ASCII
1915 Scholl Road
Ames, IA 50011 **USA**
ph: (515) 294-8152, fax: (515) 294-7771

Prof. D.G. Stavenga
Biophysics Department
University of Groningen
Nijenborgh 4
NL-947 AG Groningen **The Netherlands**
ph: (31) 50 63 47 85
fax: (31) 50 63 47 40
email: D.G.Stavenga@bcn.rug.nl

Prof. Tadeusz Stepinski
Uppsala University
School of Engineering
Dept. of Technology
Circuits and System Group
Box 354
S-751 21 Uppsala **Sweden**
ph: (46) 18 18 25 00
fax: (46) 18 55 50 96
email: Tadeusz.Stepinski@mercur.tekni-
kum.uu.se

Prof. Y. C. Trivedi
University of North Florida
11935 Nicobar Court
Jacksonville, Florida 32223 **USA**
ph: (904) 262-5913

Prof. Vladimir Vavilov
Tomsk Polytechnic Institute
Krylova 17-10
634029 Tomsk 29 **Russia**
ph: 7 382 2 23 02 41
fax (in Moscow): 7 095 292 65 11 (write
down INFRARED Box 9504 on cover
sheet)
email: denisk@infrared.tomsk.su

Prof. Kevin Warwick
Head of Department
Department of Cybernetics
School of Engineering and Information
Sciences
The University of Reading
P.O. Box 225
Whiteknights, Reading **United Kingdom**
RG6 2AY
ph: (44) 734) 31 8210
fax: (44) 734 31 82 20

Dipl. Ing. H. Wessel
Federal Institute for Materials Research
and Testing
Unter den Eichen 87
12205 Berlin 45 **Germany**
ph: 49 30 8104 3676
fax: 49 30 811 5089

Prof. Ahmed Yamani
King Fahd University of Petroleum and
Minerals
P.O. Box 1811
Dhahran 31261 **Saudi Arabia**
ph: (966) 3 860 2463
fax: (966) 3 860 3535
email: FACC034@SAUPM.BITNET

* participation only

Index